Lecture Notes in Physics

Volume 892

AF768146

The Lecture Notes in Physics

The series Lecture Notes in Physics (LNP), founded in 1969, reports new developments in physics research and teaching-quickly and informally, but with a high quality and the explicit aim to summarize and communicate current knowledge in an accessible way. Books published in this series are conceived as bridging material between advanced graduate textbooks and the forefront of research and to serve three purposes:

- to be a compact and modern up-to-date source of reference on a well-defined topic
- to serve as an accessible introduction to the field to postgraduate students and nonspecialist researchers from related areas
- to be a source of advanced teaching material for specialized seminars, courses and schools

Both monographs and multi-author volumes will be considered for publication. Edited volumes should, however, consist of a very limited number of contributions only. Proceedings will not be considered for LNP.

Volumes published in LNP are disseminated both in print and in electronic formats, the electronic archive being available at springerlink.com. The series content is indexed, abstracted and referenced by many abstracting and information services, bibliographic networks, subscription agencies, library networks, and consortia.

Proposals should be sent to a member of the Editorial Board, or directly to the managing editor at Springer:

Christian Caron
Springer Heidelberg
Physics Editorial Department I
Tiergartenstrasse 17
69121 Heidelberg/Germany
christian.caron@springer.com

More information about this series at
http://www.springer.com/series/5304

Eleftherios Papantonopoulos
Editor

Modifications of Einstein's Theory of Gravity at Large Distances

Springer

Editor
Eleftherios Papantonopoulos
Department of Physics
National Technical University of Athens
Athens, Greece

ISSN 0075-8450 ISSN 1616-6361 (electronic)
Lecture Notes in Physics
ISBN 978-3-319-10069-2 ISBN 978-3-319-10070-8 (eBook)
DOI 10.1007/978-3-319-10070-8
Springer Cham Heidelberg New York Dordrecht London

Library of Congress Control Number: 2014954249

Printed on acid-free paper

Springer is part of Springer Science+Business Media (www.springer.com)

Preface

This book is an edited version of the review talks given in the Seventh Aegean Summer School on *Beyond Einstein's Theory of Gravity*, held in Parikia on Paros Island, Greece, from 23 to 28 September 2013. The aim is to present an advanced multiauthored textbook meeting the needs of both postgraduate students and young researchers, in the fields of gravity, relativity, cosmology and quantum field theory.

In the past few years gravity theories were proposed which can be considered as extensions of Einstein's theory of gravity. Their main motivation was to explain the latest cosmological and astrophysical data on dark energy and dark matter. Advances in string theory also motivated the study of gravity theories in higher dimensions and higher curvature. These theories introduced large scale modifications of General Relativity giving a plethora of new gravity theories based mainly on various forms of couplings of matter to gravity and to the introduction of high curvature terms in the gravity action. Also they renewed the interest of the community to the long standing problem if the graviton has a mass leading to a fast growing field of massive gravity. Higher spin fields were also motivated leading to the study of higher spin gravity theories. Finally, motivated by string theory, holography was applied to modified gravity theories in a hope to understand perplexed strong coupled phenomena using the gauge/gravity duality.

The selected contributions to this volume discuss the main ideas and models of modified gravity. The long standing problem of massive graviton is discussed in detail and the fast growing field of massive gravity is explored. Higher spin theories and their connection to gravity are discussed and also Chern–Simons theories are presented and their holographic perspective is explored. Finally, dynamical processes like scattering amplitudes in gravity are discussed. The aim of this volume is to introduce postgraduate students and young researchers to these very challenging topics which constitute modifications of Einstein's theory of gravity and recently have attracted much interest.

In the first part of the book modifications of General Relativity at large distances are discussed mainly due to various forms of matter coupled to gravity and to the introduction of higher curvature terms. The first chapter by Thomas Sotiriou discusses gravity theories with non-minimally coupled scalar fields to demonstrate the challenges and future perspectives of considering alternatives to general relativity and reviews the generalized scalar-tensor theories. Next, the second chapter by Christos Charmousis reviews the recent progress in Lovelock and Horndeski theories, discusses how the Kaluza-Klein reduction of Lovelock theory can lead to scalar-tensor actions of the Horndeski type and presents black hole solutions of these theories. The third chapter by Christof Wetterich discusses the equivalence of models of modified gravity to couple quintessence and presents a modified gravity model by introducing a field dependent Planck mass, discussing also its cosmological implications. Finally the chapter by Shinji Tsujikawa introduces first an effective field theory of cosmological perturbations, applies it to Horndeski theories, and also it studies the equations of matter density perturbations based on Horndeski theory in connection to observations.

In the second part of the book the basic ideas and models of massive gravity are presented. In the first chapter by Claudia de Rham recent progress on massive gravity is reviewed. Special emphasis is paid to the ghost problem and its resolution and also drawbacks on superluminalities and strong coupling and their consequences are discussed. In the second chapter by Mikhail Volkov black hole solutions in ghost-free bigravity and massive gravity are presented. The next chapter by Eric Bergshoeff, Paul Townsend and collaborators introduces a wide class of three-dimensional gravity models which can be put into "Chern–Simons-like" form and then specializes these models to general massive gravity. Finally the last chapter in this part of the book is by Andrew Tolley in which an overview of cosmological solutions in extensions of massive gravity such as bi-gravity and quasi-dilaton massive gravity is presented.

The last part of the book deals with high spin theories, Chern–Simons theories and applications of holography to gravity theories. The first chapter is by Mikhail Vasiliev in which higher-spin gauge theory is introduced with the emphasis given on qualitative features of the higher-spin gauge theory and peculiarities of its space-time interpretation. The chapter by Ricardo Troncoso and collaborators reviews recent results in higher spin black holes in three-dimensional spacetimes, focusing for simplicity on the case of gravity nonminimally coupled to spin-3 fields, which nonperturbatively are described by a Chern–Simons theory. Next the chapter by Jorge Zanelli presents a review of the role of Chern–Simons forms in gravitation theories while the chapter by Daniel Grumiller and collaborators shows that Chern–Simons theories in three dimensions being topological field theories may have a holographic interpretation for suitable chosen gauge groups and boundary conditions on the fields. The last two chapters of the book deal with holographic aspects of gravity theories. The chapter by Marios Petropoulos discusses self-duality in Euclidean gravitational set ups which allows holographically to relate the boundary energy-momentum tensor and the boundary Cotton tensor and shows that this relation results from a topological mass term for gravity boundary dynamics. The chapter by Diana Vaman discusses stringy excitations of the graviton and using the AdS/CFT correspondence studies their dynamics.

The Seventh Aegean Summer School and the present book became possible with the kind support of many people and organizations. The Seventh Aegean Summer School was organized and supported by Paris-Sud (Orsay) University, University Francois Rabelais-Tours, Groningen University, and the National Technical University of Athens. It was sponsored by Paris-Sud (Orsay) University, University Francois Rabelais-Tours, Groningen University, National Technical University of Athens, Springer Lecture Notes in Physics, Municipality of Paros and Preservation Society of the Traditional Settlement of Parikia.

We specially thank the Municipality of Paros and the Preservation Society of the Traditional Settlement of Parikia for their kind hospitality in the island of Paros and their support. We also thank George Roussos for his valuable help in organizing the school in Paros. Without his endless help and support the organization of the Aegean School in Parikia would have been impossible. The administrative support of the Seventh Aegean Summer School was taken up with great care by Maria Kazadei and Katerina Papantonopoulou. We acknowledge the help of Vassilis Zamarias who designed and maintained the website of the School.

Last, but not least, we are grateful to the staff of Springer-Verlag, responsible for the Lecture Notes in Physics, whose abilities and help contributed greatly to the appearance of this book.

Athens, Greece
June 2014

Eleftherios Papantonopoulos

Contents

Part II Massive Gravity

Part I
Modification of General Relativity: General Scalar-Tensor Theories

Chapter 1
Gravity and Scalar Fields

Thomas P. Sotiriou

Abstract Gravity theories with non-minimally coupled scalar fields are used as characteristic examples in order to demonstrate the challenges, pitfalls and future perspectives of considering alternatives to general relativity. These lecture notes can be seen as an illustration of features, concepts and subtleties that are present in most types of alternative theories, but they also provide a brief review of generalised scalar-tensor theories.

1.1 Introduction

The predictions of general relativity are in impressive agreement with experiments whose characteristic length scale ranges from microns (μm) to about an astronomical unit (AU). On the other hand, the theory is expected to break down near the Planck length, $l_p \approx 1.6 \times 10^{-35}$ m, and a quantum theory of gravity is needed in order to adequately describe phenomena for which such small length scales are relevant. There are really no gravitational experiments that give us access to the region between the Plack length and the micron, so one has to admit that we have no direct evidence about how gravity behaves in that region.[1]

It was perhaps much more unexpected that experiments probing length scales much larger than the solar system held surprises related to gravity. General relativity can only fit combined cosmological and galactic and extragalactic data

[1]However, one can infer certain properties of gravity indirectly. Matter couples to gravity and we understand and probe the structure and behaviour of particles and fields at scales much smaller than the micron, so if one is given a model that describes how gravity interacts with matter then one could in principle gain insight into some aspects of gravity through the behaviour of matter. Applying this logic to the quantum aspects of gravity has given rise to what is called Quantum Gravity Phenomenology [1, 2]. The fact that the gravitational coupling is very weak poses a particular challenge in such an approach, but smoking gun signals can still exist in certain models.

T.P. Sotiriou (✉)
School of Mathematical Sciences & School of Physics and Astronomy, University of Nottingham, University Park, Nottingham NG7 2RD, UK
e-mail: Thomas.Sotiriou@nottingham.ac.uk

E. Papantonopoulos (ed.), *Modifications of Einstein's Theory of Gravity at Large Distances*, Lecture Notes in Physics 892, DOI 10.1007/978-3-319-10070-8_1

well if there is a non vanishing cosmological constant and about six times more Dark Matter—matter which we have so far detected only through its gravitational interaction—than visible matter (see, for instance, [3]). Moreover, the value of the cosmological constant has to be very small, in striking disagreement with any calculation of the vacuum energy of quantum fields, and mysteriously the associated energy density is of the same order of magnitude as that of matter currently [4, 5]. These puzzles have triggered the study of dynamical Dark Energy models, that come to replace the cosmological constant.

Since general relativity is not a renormalizable theory, it is expected that deviations from it will show up at some scale between the Planck scale and the lowest length scale we have currently accessed. It is tempting to consider a scenario where those deviation persist all the way to cosmological scales and account for Dark Matter and/or Dark Energy. After all, we do only detect these dark component through gravity. However, there is a major problem with this way of thinking. There is no sign of these modifications in the range of scales for which we have exhaustively tested gravity. So, they would have to be relevant at very small scales, then somehow switch off at intermediate scales, then switch on again at larger scales. It is hard to imagine what can lead to such behaviour, which actually contradicts our basic theoretical intuition about separation of scales and effective field theory. Nonetheless, intuition is probably not a good enough reason to not rigorously explore an idea that could solve two of the major problem of contemporary physics at once. This explain the considerable surge of interest in alternative theories of gravity in the last decade or so.

Considering alternatives to a theory as successful as general relativity can be seen as a very radical move. However, from a different perspective it can actually be though of as a very modest approach to the challenges gravity is facing today. Developing a fundamental theory of quantum gravity from first principle and reaching the stage where this theory can make testable predictions has proved to be a very lengthy process. At the same time, it is hard to imagine that we will gain access to experimental data at scales directly relevant to quantum gravity any time soon. Alternative theories of gravity, thought of as effective field theories, are the phenomenological tools that provide the much needs contact between quantum gravity candidates and observations at intermediate and large scales.

The scope of these notes is to briefly review the challenges one in bound to face when considering alternatives to general relativity and discuss various ways to overcome (some of) them. Instead of providing rigorous and general but lengthy arguments, I will mostly resort to the power of examples. The examples will be based on gravity theories with additional scalar degrees of freedom, so these notes will also act as a brief review of generalised scalar-tensor theories and their properties.

I have made extensive reference to various length scales in the arguments presented so far and one can rightfully feel uncomfortable talking about length scales when it comes to gravity. The strength of the gravitational interaction has to do with curvature and lengths are not even invariant under coordinate transformations. Indeed, the Planck length can only be understood as a fundamental invariant length as the inverse of the square root of a fundamental curvature scale (which has dimensions of 1 over a length square). In this spirit, it would be preferable to talk about the range of curvatures in which we have tested gravity. Actually, the experiments that span the range of lengths μm–AU all lie in a very narrow band of curvatures. This is not so surprising, as they are all weak-field experiments. This applies to binary pulsars as well as, even though the two companions that form the binary are compact enough to exhibit large curvatures in their vicinity, the gravitational interaction between them is still rather weak as the two stars are not close enough to be in the region of strong curvature. Hence, if we think in terms of curvatures, the range in which we have tested general relativity appears even more restricted. Neutron stars and stellar and intermediate mass black holes can exhibits curvatures which are many orders of magnitudes larger than the usual weak-field experiments. It is, therefore, particularly interesting to understand the structure of such objects and the phenomena that take place in their vicinity in alternative theories of gravity. They are most likely the new frontier in gravitational physics.

The rest of these notes is organised as follows: In Sect. 1.2 I lay out the basic assumption of general relativity and very briefly (and intuitively) discuss the consequences of relaxing these assumptions. The main scope of this section is to give an idea of what alternative theories of gravity are about and what kind of problems one usually faces when deviating from general relativity. In Sects. 1.3 and 1.4 I attempt to support the statements made in the previous section by considering characteristics examples from (generalised) scalar-tensor gravity theories. Section 1.5 focuses on black hole physics in scalar-tensor gravity. The final section contains conclusions.

1.2 General Relativity and Beyond

1.2.1 General Relativity: Basic Assumptions and Uniqueness

The action of general relativity is

$$S = \frac{1}{16\pi G} \int d^4x \sqrt{-g}(R - 2\Lambda) + S_m(g_{\mu\nu}, \psi), \tag{1.1}$$

where G is Newton's constant, g is the determinant of the spacetime metric $g_{\mu\nu}$, R is the Ricci scalar of the metric, Λ is the cosmological constant, and S_m is the matter action. ψ collectively denotes the matter fields, which are understood to couple minimally to the metric.

Coupling the matter fields ψ only to the metric and with the standard prescription of minimal coupling guaranties that the Einstein Equivalence Principle is satisfied. That is, test particles follow geodesics of the metric and non-gravitational physics is locally Lorentz invariant and position invariant [6]. The reason why the last two requirements are satisfied once matter is minimally coupled is that in the local frame the metric is flat to second order in a suitably large neighborhood of a space-time point and S_m reduces to the action of the Standard Model. It is worth elaborating a bit more on universality of free fall and how this is related to the form of the matter action.

Consider the stress-energy tensor $T_{\mu\nu}$ of a pressure-less fluid, usually referred to as dust. An infinitesimal volume element of such a fluid is as close as one can get to a test particle. A rather straightforward calculation reveals that the conservation of the stress-energy tensor, $\nabla^\mu T_{\mu\nu} = 0$, implies that the 4-velocity of the fluid satisfies the geodesic equation. That is, $\nabla^\mu T_{\mu\nu} = 0$ implies that test particles follow geodesics. On the other hand, the conservation of the stress-energy tensor can be shown to follow from diffeomorphism invariance of the matter action S_m, *provided* that the matter fields are on shell (they satisfy their field equations).

Let ξ^μ be the generator of a diffeomorphism and $\mathcal{L}_\xi$ denote the associated Lie derivative. Diffeomorphism invariance of the matter action implies

$$\mathcal{L}_\xi S_m = 0\,. \tag{1.2}$$

One can express the action of the Lie derivative in terms of functional derivatives of S_m with respect to the fields, i.e.

$$\frac{\delta S_m}{\delta g^{\mu\nu}}\mathcal{L}_\xi g^{\mu\nu} + \frac{\delta S_m}{\delta \psi}\mathcal{L}_\xi \psi = 0\,. \tag{1.3}$$

However, $\delta S_m/\delta\psi = 0$ are actually the field equations for ψ. So, on shell we have

$$\frac{\delta S_m}{\delta g^{\mu\nu}}\mathcal{L}_\xi g^{\mu\nu} = 0\,. \tag{1.4}$$

With the usual definitions for the stress-energy tensor and for the action of a Lie derivative on the metric and after some manipulations, the above equation can take the form

$$\int d^4x\sqrt{-g}T_{\mu\nu}\nabla^\mu\xi^\nu = 0\,. \tag{1.5}$$

Finally, integrating by parts and taking into account that ξ^μ vanishes at the boundary yields

$$\int d^4x\sqrt{-g}(\nabla^\mu T_{\mu\nu})\xi^\nu = 0\,. \tag{1.6}$$

Since, ξ^μ is a generic diffeomorphism, Eq. (1.6) implies that $\nabla^\mu T_{\mu\nu} = 0$.

In conclusion, diffeomorphism invariance of the matter action allows one to link geodesic motion with the requirement that the matter fields are on shell. An important assumption here is that there is no field other than the metric that couples to the matter fields ψ and at the same time enters the gravitational action as well. This assumption is reflected in the condition that $\delta S_m/\delta\psi = 0$, i.e. *all* fields other than the metric are on shell. If there were a field, say ϕ entering both S_m and the gravitational action, then $\delta S_m/\delta\phi = 0$ would not actually be its field equation and it would not be sensible to impose it as a condition by assuming that this field is on shell.

One more point that is worth stressing is that in the arguments and calculations shown above one only makes reference to the matter action. This implies that they are not specific to general relativity. Instead, they will apply to any theory in which the matter couples only to the metric through minimal coupling.

In conclusion, the requirement to satisfy the Einstein Equivalence Principle, which has been experimentally tested to very high accuracy, pins down the matter action and the coupling between matter and gravity. What is left is to argue why the dynamics of $g_{\mu\nu}$ should be governed by the first integral in Eq. (1.1), known as the Einstein–Hilbert action. Luckily, this requires less work as Lovelock has provided us with a theorem [7, 8] stating that this is indeed the *unique* choice, provided that the following assumptions hold true:

1. The action is diffeomorphism-invariant;
2. it leads to second-order field equations for the metric;
3. we are restricting our attention to four dimensions;
4. no fields other than the metric enter the gravitational action.

1.2.2 Less Assumptions Means More Degrees of Freedom!

We now consider what would be the implications of giving up one of the assumptions listed above. Let us start by relaxing the assumption that the gravitational action depends only on the metric, and allow a dependence on a new field ϕ. Obviously, we would need to dictate how the gravitational action depends on ϕ in order to pin down the theory we are considering. However, as it should be clear from the analysis in the previous section, if we were to allow this new field to enter the matter action and couple to the matter fields then we would have violations of the Einstein Equivalence Principle and signatures of this coupling would appear in non-gravitational experiments. Constraints on universality of free fall, local Lorentz symmetry in the matter sector, and deviations from the standard model in general are orders of magnitude more stringent than constraints coming from gravitational experiments. This explains why in the literature the common approach is to assume that any new fields do not enter the matter action, or at least that the coupling between these field and matter is weak enough to be irrelevant at low energies. We will follow the same line of thought in what comes next. It should, however, be clear that if there are new fields in the gravity sector at the classical level, then one would expect that quantum corrections will force them to couple to the matter fields. Hence, a consistent theory should actually include a mechanism that naturally suppressed the coupling between these new fields and matter. This is required in order to theoretical justify what phenomenologically seems to be the only option.

A thorny issue is that of field redefinitions. Note that all of the assumptions, conditions, and requirements discussed above, should in principle be posed as "there exists a choice of fields where...". This becomes particularly relevant when one has extra fields mediating gravity. Suppose, for instance, that ϕ does couple to matter but in such a way that one can introduce a new metric, $\tilde{g}_{\mu\nu}$, which can be given in closed form in terms of $g_{\mu\nu}$ and ϕ (and potentially its derivatives), so that matter actually couples minimally to $\tilde{g}_{\mu\nu}$. Then, the whole theory can be re-written in terms of $\tilde{g}_{\mu\nu}$ and ϕ and the matter action will be the conventional one with matter coupling only to a metric with minimal coupling.

What would happen if we kept the field content unchanged and we instead relaxed any of the other three assumptions of Lovelock's theorem?

We could consider more than four-dimensions. However, so far we experimentally detect only 4. Moreover, as long as we are interested in low energies and a phenomenological description, one is justified to expect that for any higher-dimensional theory there exist a four-dimensional effective theory. If this theory is not general relativity, then it will have to contradict one of the other three assumptions. Going beyond the four-dimensional effective description will be necessary in order to explain various characteristics of the theory which might seems ad hoc or unnatural when one is judging naively based on the four-dimensional picture (e.g. why the action has a certain form or why some couplings have specific values). But the four-dimensional effective description should usually be adequate to discuss low-energy phenomenology and viability.

If we were to allow the equation of motion to be higher than second order partial differential equations (PDEs), then we would be generically introducing more degrees of freedom. This can be intuitively understood by considering the initial data one would have to provide when setting up an initial value problem in this theory (assuming that an initial value problem would be well posed). For instance, consider for simplicity a fourth order ordinary differential equation: to uniquely determine the evolution one would need to provide the first 3 time derivatives as initial data. So, a theory with higher order equation will generically have more propagating modes. Increasing the differential order is actually quite unappealing, as it leads to serious mathematical complications—higher-order PDEs are not easy to deal with—and serious stability issue. These will be discussed shortly.

Finally, one could give up diffeomorphism invariance. However, it has been long known that symmetries can be restored by introducing extra fields. This procedure is known as the Stueckelberg mechanism, see [9] for a review. In Stueckelberg's work the new field was a scalar field introduced to restore gauge invariance in a massive Abelian gauge theory. By choosing the appropriate gauge one does away with the Stueckelberg field (it becomes trivial) but the theory is no longer manifestly gauge invariant. The Stueckelberg mechanism can be generalised to other symmetries, and specifically to diffeomorphism invariance.[2] Hence, one can choose to think of theories that are not invariant under diffeomorphisms as diffeomorphism-invariant theories with extra Stueckelberg fields.

In the previous section we demonstrated that diffeomorphism invariance has a central role in relating energy conservation and geodesic motion to the requirement that matter fields are on shell. From this discussion it also follows that if Stueckelberg fields are required in order to write a theory in a manifestly diffeomorphism invariant formulation, then these fields should not appear in the matter action, as is the case for any field that coupled non-minimally to gravity.

To summarised, we have argued that irrespectively of which of the 4 assumption of Lovelock's theorem one chooses to relax, the outcome is always the same: one ends up with more degrees of freedom. The name of the game in alternative theories of gravity is, therefore, to tame the behaviour of these degrees of freedom.

Clearly, many of the statements made in this section where rather heuristics and we relied heavily on the reader's intuition. In Sect. 1.3 convincing examples from scalar-tensor gravity that demonstrate all of the above will be presented.

1.2.3 *Taming the Extra Degrees of Freedom*

Consider a simple system of two harmonic oscillators, describe by the lagrangian

$$L = \frac{1}{2}\dot{q}_1^2 - \frac{1}{2}q_1^2 + \frac{1}{2}\ddot{q}_2^2 - \frac{1}{2}q_2^2\,. \tag{1.7}$$

If we were to flip the sign of q_1^2 in the lagrangian q_1 would have to exhibit exponential growth. If instead, we were to flip the sign of $\dot{q}_1^2$, the corresponding hamiltonian would not be bound from below. Having the wrong sign in front of certain terms renders the system unstable, but luckily in simple systems such as harmonic oscillators it is easy to know which sign to choose. In fact, coupling the two oscillators minimally would not affect this choice. Things become significantly more complicated though when one has degrees of freedom that couple non-minimally. Imagine adding a term such as $q_1^2 q_2^2$, $f(q_1)\dot{q}_2^2$ or $\dot{q}_1\dot{q}_2$. It is no longer obvious whether you system is stable or not.

The situation is no different in a field theory. Fields whose hamiltonian is not bound from below are called ghosts and sensible theories are expected to be free of them. At the perturbative level this means that excitation around a certain configuration should have the right sign in front of the kinetic term. One also expects that physical configurations are classically stable, i.e. all excitation around them

[2] Erich Kretschmann argued in 1917 that any theory can be put in a generally covariant form, which led to a famous debate with Einstein. A covariant version of Newtonian gravity can be found in [10].

have real propagation speeds. A complication that is always present in alternative theories of gravity is that the extra degrees of freedom are always non-minimally coupled to gravity (else there would be matter fields by definition). So, when constructing an action for a theory with a given field content it is nontrivial to judge whether it will satisfy the stability criteria mentioned above. As a result, one of the first calculations one does in every alternative theory of gravity is to check if all excitations satisfy these criteria around flat space (or some maximally symmetric space—the vacuum solution of the theory).

In Sect. 1.2.2 we mentioned that theories that lead to higher-order equations are generically plagued by instabilities. These instabilities are essentially due to the presence of ghosts. It has been shown by Ostrogradski in 1850 that non-degenerate Lagrangians with higher-order derivatives generically lead to Hamiltonians that are linear in at least one of the momenta [11]. Such Hamiltonians are not bound from below. A detailed discussion can be found in [12]. Obviously, Ostragradski's instabilities make higher-order theories particularly unappealing. However, higher-order theories which can be explicitly re-written as second-order theories with more fields evade such instabilities. We will see an example of such a theory below.

Once stability issues have been addressed, and the behaviour of the new degrees of freedom has been tamed, the next step is to find a mechanism that hides them in regimes where general relativity is well tested and no extra degrees of freedom have been seen, but still allows them to be present and lead to different phenomenology in other regimes. How challenging a task this is and how inventive we have been in order to circumvent the difficulties will be demonstrated by the examples from scalar-tensor gravity presented in the next section.

It should be mentioned that a road less taken is to consider alternative theories with non-dynamical extra degrees of freedom. In fact, one could circumvent Lovelock's theorem by considering a gravity theory where fields other than the metric are present, but they are auxiliary fields, so that they do not satisfy dynamical equation but can be instead algebraically eliminated. This way ones has the same degrees of freedom as in general relativity and does not have to worry about instabilities associated with new dynamical fields. However, such an approach is not without serious shortcomings, see [13] for a discussion and references therein. For the rest of these notes we will focus one theories with dynamical new degrees of freedom, as most popular alternative theories of gravity fall under this category.

1.3 Scalar-Tensor Gravity

1.3.1 The Prototype: Brans–Dicke Theory

The action for Brans–Dicke theory is

$$S_{\rm BD} = \frac{1}{16\pi G}\int d^4x\sqrt{-g}\left(\varphi R - \frac{\omega_0}{\varphi}\nabla^\mu\varphi\nabla_\mu\varphi - V(\varphi)\right) + S_m(g_{\mu\nu},\psi), \quad (1.8)$$

where φ is a scalar field and ω_0 is known as the Brans–Dicke parameter. After some manipulations, the corresponding field equations can take the form

$$R_{\mu\nu} - \frac{1}{2} R g_{\mu\nu} = \frac{8\pi G}{\varphi} T_{\mu\nu} + \frac{\omega_0}{\varphi^2} \left(\nabla_\mu \varphi \nabla_\nu \varphi - \frac{1}{2} g_{\mu\nu} \nabla^\lambda \varphi \nabla_\lambda \varphi \right)$$

$$+ \frac{1}{\varphi} \left(\nabla_\mu \nabla_\nu \varphi - g_{\mu\nu} \Box \varphi \right) - \frac{V(\varphi)}{2\varphi} g_{\mu\nu} , \tag{1.9}$$

$$(2\omega_0 + 3) \Box \varphi = \varphi V' - 2V + 8\pi G T , \tag{1.10}$$

where $\Box = \nabla^\lambda \nabla_\lambda$ and a prime denotes differentiation with respect to the argument. In its original formulation Brans–Dicke theory did not have a potential.

It is straightforward to see that in vacuo, where $T_{\mu\nu} = 0$, the theory admits solutions where with $\varphi = \varphi_0 =$ constant, provided that $\varphi_0 V'(\varphi_0) - 2V(\varphi_0) = 0$. For such solutions the metric actually satisfies Einstein's equations with an effective cosmological constant $V(\phi_0)$. So, one could be misled to think that, as long as $V(\varphi_0)$ has the right value, the predictions of the theory could be the same as those of general relativity. For instant, the space-time around the Sun could be described by such a solution, and then solar system constraints would be automatically satisfied. What invalidates this logic is that the $\varphi = \varphi_0$ solutions are not unique. φ could actually have a nontrivial configuration, which would also force the metric to deviate for the corresponding solution of general relativity.

This is indeed the case for spherically symmetric solution that describe the exterior of stars, and in particular the Sun. Consider for concreteness the case where $V = m^2(\varphi - \varphi_0)^2$. Performing a newtonian expansion one can calculate the newtonian limit of the metric. The perturbations of the metric are

$$h_{00} = \frac{GM_s}{\varphi_0 r} \left(1 - \frac{1}{2\omega_0 + 3} exp \left[-\sqrt{\frac{2\varphi_0}{2\omega_0 + 3}} mr \right] \right) , \tag{1.11}$$

$$h_{ij} = \frac{GM_s}{\varphi_0 r} \delta_{ij} \left(1 + \frac{1}{2\omega_0 + 3} exp \left[-\sqrt{\frac{2\varphi_0}{2\omega_0 + 3}} mr \right] \right) , \tag{1.12}$$

where M_s is the mass of the Sun. There is a Yukawa-like correction to the standard $1/r$ potential, with effective mass $m_{\text{eff}} = \sqrt{\frac{2\varphi_0}{2\omega_0+3}} m$ and range m_{eff}^{-1}. The ratio of the perturbations of the time-time component h_{00} over any space-space diagonal component $h_{ij}|_{i=j}$, which is also known as the γ (Eddington) parameter is then given by [14]

$$\gamma \equiv \frac{h_{ij}|_{i=j}}{h_{00}} = \frac{2\omega_0 + 3 - exp \left[-\sqrt{\frac{2\varphi_0}{2\omega_0+3}} mr \right]}{2\omega_0 + 3 + exp \left[-\sqrt{\frac{2\varphi_0}{2\omega_0+3}} mr \right]} . \tag{1.13}$$

It is clear that in order for γ to be close to 1, which is the value it has in general relativity, either ω_0 or $m_{\rm eff}$ should be very large. Indeed, in the limit where $\omega_0 \to \infty$ or $m \to \infty$ the equation imply that $\varphi \to \varphi_0$ and the constant φ solutions with $g_{\mu\nu}$ satisfying Einstein's equation become unique. Current constraints on γ require that $\gamma - 1 = (2.1 \pm 2.3) \times 10^{-5}$ [15]. For $m = 0$, this constraint would require ω_0 to be larger than 40,000, which would make the theory indistinguishable from general relativity at all scales. For $\omega_0 = O(1)$, the range of the Yukawa correction would have to be below the smaller scale we have currently tested the inverse square law, i.e. a few microns. But if this is indeed the case, then this correction will never play a role at large scales.

The main message here is that weak gravity constraints are very powerful. It seems very hard to satisfy them and still have a theory whose phenomenology differs from that of general relativity at scales where we currently test gravity. One would have to circumvent this problem in order to construct a theory which is phenomenologically interesting.

1.3.2 Scalar-Tensor Theories

Scalar-tensor theories are straightforward generalisations of Brans–Dicke theory in which ω_0 is promoted to a general function of φ. Their action is

$$S_{\rm st} = \frac{1}{16\pi G}\int d^4x\sqrt{-g}\Big(\varphi R - \frac{\omega(\varphi)}{\varphi}\nabla^\mu\varphi\nabla_\mu\varphi - V(\varphi)\Big) + S_m(g_{\mu\nu},\psi)\,. \tag{1.14}$$

This is the most general action one can write for a scalar field non-minimally coupled to gravity which is second order in derivatives of the scalar. It can, therefore, be thought of as an effective field theory which captures, at some appropriate limit, the phenomenology of a more fundamental theory that contains a scalar field. The corresponding field equations are, after some manipulations

$$R_{\mu\nu} - \frac{1}{2}Rg_{\mu\nu} = \frac{8\pi G}{\varphi}T_{\mu\nu} + \frac{\omega(\varphi)}{\varphi^2}\left(\nabla_\mu\varphi\nabla_\nu\varphi - \frac{1}{2}g_{\mu\nu}\,\nabla^\lambda\varphi\nabla_\lambda\varphi\right)$$

$$+\frac{1}{\varphi}\left(\nabla_\mu\nabla_\nu\varphi - g_{\mu\nu}\Box\varphi\right) - \frac{V(\varphi)}{2\varphi}g_{\mu\nu}\,, \tag{1.15}$$

$$[2\omega(\varphi)+3]\,\Box\varphi = -\omega'(\varphi)\,\nabla^\lambda\varphi\nabla_\lambda\varphi + \varphi\,V' - 2V + 8\pi G\,T\,. \tag{1.16}$$

Scalar-tensor theories have been extensively studied and we will not review them here. See [16, 17] for detailed reviews. The behaviour of the theories in the weak field limit will be no different than that of Brans–Dicke theory, though allowing ω

to be a function of φ will lead to a novel way of getting exciting phenomenology in the strong gravity regime, as we will see shortly.

We have given the action and field equations of scalar-tensor theory in terms of the metric that minimally couples to matter, $g_{\mu\nu}$. This is referred to as the Jordan frame. It is fairly common to re-write them in a different conformal frame, know as the Einstein frame, in which the (redefined) scalar couples minimally to gravity but it also couples to the matter.

The conformal transformation $\hat{g}_{\mu\nu} = \varphi\, g_{\mu\nu}$, together with the scalar field redefinition $4\sqrt{\pi}\varphi d\phi = \sqrt{2\omega(\varphi)+3}\, d\varphi$, brings the action (1.14) to the form

$$S_{\rm st} = \int d^4x\sqrt{-\hat{g}}\Big(\frac{\hat{R}}{16\pi} - \frac{1}{2}\hat{g}^{\nu\mu}\partial_\nu\phi\partial_\mu\phi - U(\phi)\Big) + S_m(g_{\mu\nu},\psi)\,, \tag{1.17}$$

where $U(\phi) = V(\varphi)/\varphi^2$, $\hat{g}_{\mu\nu}$ is Einstein frame metric and all quantities with a hat are defined with this metric. The field equations in the Einstein frame take the form

$$\hat{R}_{\mu\nu} - \frac{1}{2}\hat{R}\hat{g}_{\mu\nu} = 8\pi G\, T^{\phi}_{\mu\nu} + \frac{8\pi G}{\varphi(\phi)}T_{\mu\nu}\,, \tag{1.18}$$

$$\hat{\Box}\phi - U'(\phi) = \sqrt{\frac{4\pi G}{(2\omega+3)}}T\,, \tag{1.19}$$

where

$$T^{\phi}_{\mu\nu} = \nabla_\mu\phi\nabla_\nu\phi - \frac{1}{2}g_{\mu\nu}\nabla_\lambda\phi\nabla^\lambda\phi - U(\phi)g_{\mu\nu}\,, \tag{1.20}$$

whereas $T_{\mu\nu}$ and T are the Jordan frame stress-energy tensor and its trace respectively.

The fact that ϕ couples minimally to $\hat{g}_{\mu\nu}$ in the Einstein frame makes calculations much simpler in many cases, especially in vacuo, where the theory becomes general relativity with a minimally coupled scalar field. One can use any of the two frames to perform calculations but some care is needed when interpreting results that do not involve conformally invariant quantities. The physical significance of the two metrics, $g_{\mu\nu}$ and $\hat{g}_{\mu\nu}$, should be clear: the former is the metric whose geodesics will coincide with test particle trajectories, as it couples minimally to matter. The latter is just a special choice which brings the action in a convenient form. See [18] and references therein for more detailed discussions.

1.3.3 Hiding the Scalar Field, Part I

We will now briefly discuss some mechanisms that can hide the scalar field in the weak field regime near matter but still allow the theory to deviate significantly from general relativity in cosmology or in the strong gravity regime.

The first and oldest of these mechanisms is present in theories were $\omega(\varphi)$ diverges for some constant value of φ [19, 20]. Consider theories without a potential. In configurations where $\omega \to \infty$ one essentially ends up with a constant scalar and metrics that satisfy Einstein's equations. This follows intuitively by the analysis of the newtonian limit of Brans–Dicke theory when $\omega_0 \to \infty$, or more rigorously by inspecting the field equations or the action. It is more convenient and straightforward to consider the Einstein frame. In the absence of a potential, Eq. (1.19) admits $\phi = \phi_0$ =constant solutions with $\omega(\phi_0) \to \infty$ even inside matter.[3] For such solutions Eq. (1.18) reduce to Einstein's equations (with a rescaled coupling inside matter). Going back to the Jordan frame, such solutions correspond to φ =constant with $g_{\mu\nu}$ satisfying Einstein's equations.

A key difference with Brans–Dicke theory with very large ω_0 is that here ω diverges only in the specific configuration for the scalar, so one needs to check under which circumstances such configurations are solution of the physical system of interest. In other words, one has to check that φ will be dynamically driven into this configuration in situations where one would like to recover general relativity.

It has been indeed shown in [19, 20] that there exist theories where in principle both $\phi = \phi_0$ and non-trivial ϕ solutions exist for stars. Which of the two configurations will be realised after gravitational collapse depends (roughly speaking) on the compactness of the star. For ordinary stars, such as the Sun, the constant scalar solution is the one realised. The metric describing their exterior is then the same as in general relativity and this makes the theories indistinguishable from the latter in the Solar system. For compact stars instead, such as neutron stars, the non-trivial scalar configuration becomes energetically favourable and the metric significantly deviates from the one general relativity would yield. Hence, the strong-field phenomenology will be distinct from that of general relativity. The importance of this result lies on the fact that it was the first demonstration that one can construct a theory which agrees with general relativity in the weak field limit but still gives distinct and testable predictions in the strong field regime. There is a very sharp transition from the ϕ =constant to the non-trivial ϕ configurations as one increases the compactness of the star, so the mechanism that causes this transition has been dubbed "spontaneous scalarization" [19, 20].

This mechanism relies entirely on the functional form of ω, which turned out to be intimately related to how the scalar field is sourced by matter. There is a different type of mechanism to hide the scalar field that relies on the potential V, or U, and is called the chameleon mechanism [21]. In terms of the newtonian limit of Brans–Dicke theory that was given in Sect. 1.3.1 the chameleon mechanism can be thought of as a dependence of the effective mass, and the corresponding range of the Yukawa-like correction, on the characteristics of a given matter configuration. As discussed earlier, when the effective mass gets large enough, the range of the Yukawa-like correction becomes short enough to be negligible in any known experiment. But if one wants the scalar field to have any effect in cosmology, for

[3] If there is a potential $\phi = \phi_0$ solutions are only admissible if $U'(\phi_0) = 0$ as well.

example to account for dark energy, then the range of the correction should actually be long. The dependence of the mass on the nearby matter configuration makes it possible to have it both ways.

For a scalar field that experiences only self interactions one defines as the mass the value of the second derivative of its potential at the minimum of the potential. However, things are slightly more complicated for non-minimally coupled scalar fields. It is easier to resort to the Einstein frame and consider Eq. (1.19). Then ϕ's dynamics are governed by en effective potential $U_{\rm eft} = U(\phi) + (\ln\varphi)T/2$ [as $U'_{\rm eft} = U'(\phi) + \sqrt{4\pi G/(2\omega+3)}T$]. By choosing U appropriately (the behaviour of ω is much less relevant) one can arrange that ϕ have a very small mass when T is small and a very large mass when T is large, as the term $(\ln\varphi)T/2$ clearly deforms the potential. The most characteristic example is when choosing $U \sim e^{-\phi}$ and ω is a constant, so that the T-dependent deformation is linear in ϕ. Without this deformation the range of the force would be infinite. But the deformation introduces a minimum that leads to a short range force.

There are two subtleties in the line of reasoning we just laid out, which are sometimes not given enough attention in the literature. Firstly, we used the Einstein frame, but the mass that determines the range of the Yukawa-like correction is not actually the one associated with the effective potential of ϕ in this frame (neither the one defined as $V''(\varphi_0)$ in the Jordan frame actually, hence the use of $m_{\rm eff}$ in Sect. 1.3.1). However, one can show that the various masses are intimately related [22]. Secondly, Solar system test are not really performed in a high density environment but in vacuo, outside a high density matter configuration. On the other hand, continuity of the scalar field profile implies that, even outside the star, there will be a region for which the configuration will be influenced more by the interior configuration through boundary conditions that by the asymptotic configuration. We refer the reader to a recent review on the chameleon mechanism for a thorough discussion [23].

A third mechanism for hiding the scalar field in the Solar system is the symmetron mechanism [24]. Here both the form of ω and the form of the potential are important. In the Einstein frame the potential U is assumed to have the form

$$U(\phi) = -\frac{1}{2}\mu^2\phi^2 + \frac{1}{4}\lambda\phi^4 \,. \tag{1.21}$$

In the absence of matter $U(\phi)$ would then have a minimum at $\phi_0 = \mu/\sqrt{\lambda}$. The value of the potential at the minimum is related to an effective cosmological constant, which one can tune to the desired value by appropriately choosing μ and λ. Assume now that ω has such a functional dependence on φ (and implicitly on ϕ) that in the presence of matter the effective potential would be

$$\begin{aligned} U_{\rm eff}(\phi) &= -\frac{1}{2}\mu^2\phi^2 + \frac{1}{4}\lambda\phi^4 + (1+\frac{\phi^2}{M^2})\frac{T}{2} \\ &= \frac{1}{2}\left(\frac{T}{M^2} - \mu^2\right)\phi^2 + \frac{1}{4}\lambda\phi^4 + \frac{T}{2}\,, \end{aligned} \tag{1.22}$$

where M is a characteristic mass scale, and

$$U'_{\text{eff}}(\phi) = -\mu^2\phi + \lambda\phi^3 + \frac{\phi}{M^2}T\,, \tag{1.23}$$

For such a choice, $\omega(\phi = 0) \to \infty$. Provided that $T/M^2 > \mu^2$, $\phi = 0$ becomes the minimum of the effective potential and Eq. (1.19) admits $\phi = 0$ solution in the presence of matter.

In a certain sense, there is some similarity between the symmetron mechanism and the models that exhibit spontaneous scalarization in compact stars discussed earlier. In fact, one could see the symmetron mechanism as a cosmological scalarization. The way the symmetron mechanism works in a realistic matter configuration is actually more complicated than the simplistic description given above. For example, in a realistic matter configuration, the scalar has to smoothly change from being zero inside the matter to obtaining its non-zero asymptotic value outside the matter. We refer the reader to [24] for more details.

1.3.4 The Horndeski Action

The action of scalar-tensor theory in Eq. (1.14) is the most general action that is quadratic in derivatives of the scalar, up to boundary terms. It is not, however, the most general action that can lead to second order field equations for the metric and the scalar. Horndeski has shown that the most general action with this property is [25]

$$S_H = \int d^4x\sqrt{-g}\,(L_2 + L_3 + L_4 + L_5)\,, \tag{1.24}$$

where

$$L_2 = K(\phi, X), \tag{1.25}$$

$$L_3 = -G_3(\phi, X)\Box\phi, \tag{1.26}$$

$$L_4 = G_4(\phi, X)R + G_{4X}\left[(\Box\phi)^2 - (\nabla_\mu\nabla_\nu\phi)^2\right], \tag{1.27}$$

$$L_5 = G_5(\phi, X)G_{\mu\nu}\nabla^\mu\nabla^\nu\phi - \frac{G_{5X}}{6}\left[(\Box\phi)^3 - 3\Box\phi(\nabla_\mu\nabla_\nu\phi)^2 + 2(\nabla_\mu\nabla_\nu\phi)^3\right], \tag{1.28}$$

the G_i are unspecified functions of ϕ and $X \equiv -\frac{1}{2}\nabla^\mu\phi\nabla_\mu\phi$ and $G_{iX} \equiv \partial G_i/\partial X$. Scalar fields described by this action are also known as Generalised Galileons [26]. The name comes from a particular class of scalar theories in flat space which enjoy Galilean symmetry, i.e. symmetry under $\phi \to \phi + c_\mu x^\mu + c$, where c_μ is a constant one-form and c is a constant [27]. These fields are known as Galileons.

A certain subclass of Generalised Galileons reduce to Galileons in flat space. But galilean symmetry itself does not survive the passage to curved space [28] (it is local symmetry) and the full Horndeski action does not reduce to the Galileon action in flat space.[4]

Horndeski's theory is intrinsically interesting as a field theory, as it contains more than two derivatives in the action but still leads to second order equations. That comes at the price of having highly nonlinear derivative (self-)interactions. It is worth noting that, even though Horndeski's actions includes second derivatives of the fields, it avoids Ostrogradski's instability because it does not satisfy the non-degeneracy assumption.[5]

A more detailed discussion about the characteristics of the theory goes beyond the scope of these lecture notes, so we refers the reader to [29] for a recent review.

1.3.5 Hiding the Scalar Field, Part II

The high degree of non-linearity in the scalar field equations of Hordenski's theory certainly makes them mathematically complicated. However, it does not come without advantages. In regimes where these highly non-linear terms will dominate over the standard Brans–Dicke-like terms the behaviour of the scalar field will be significantly different from that of the Brans–Dicke scalar discussed above. In fact, such theories can exhibit the "Vainshtein effect": solutions of the linearised version of the theory—in which the higher derivative terms would give no significant contribution—can be very different from solutions of general relativity, but fully non-linear solutions might be indistinguishable from those of the latter. The term "Vainshtein effect" originates from massive gravity theory where the mechanism was first demonstrated by Vainshtein in [30]. A detailed introduction to the Vainshtein mechanism can be found in [31].

1.4 Scalar-Tensor Gravity in Disguise

In Sect. 1.2.2 it was argued that allowing for higher-order field equations or giving up diffeomorphism invariance leads to more degrees of freedom. In this section we provide two examples that support this claim. In both cases the new degree of

[4]The numbering of the terms in the Lagrangian, L_2 to L_5, is also a remnant of the original flat space Galileons [27]. The index indicates there the number of copies of the field in each term. In the Generalised Galileons the L_i term contains $i-2$ second derivatives of the scalar.

[5]The Einstein–Hilbert action also contains second derivatives of the metric and is degenerate, thus avoiding Ostrogradski's instability.

freedom is a scalar field and this can be made explicit, either by field redefinitions, or via the Stueckelberg mechanism.

1.4.1 $f(R)$ Gravity

The action of $f(R)$ gravity is

$$S = \frac{1}{16\pi G}\int d^4x\sqrt{-g}f(R) + S_m(g_{\mu\nu}, \psi)\,, \tag{1.29}$$

where f is some function of the Ricci scalar of $g_{\mu\nu}$. Variation with respect to the metric $g^{\mu\nu}$ yields

$$f'(R)R_{\mu\nu} - \frac{1}{2}f(R)g_{\mu\nu} - [\nabla_\mu\nabla\nu - g_{\mu\nu}\Box]f'(R) = 8\pi G T_{\mu\nu}\,. \tag{1.30}$$

Provide that $f''(R) \neq 0$, in which case the theory would be general relativity, these are clearly fourth-order equations in $g_{\mu\nu}$. One would then expect the theory to suffer from the Ostrogradski instability mentioned earlier.

Consider now the action

$$S = \frac{1}{16\pi G}\int d^4x\sqrt{-g}\,[f(\phi) + \varphi(R-\phi)] + S_m(g_{\mu\nu}, \psi)\,. \tag{1.31}$$

Variation with respect to φ yields $\phi = R$. Replacing this algebraic constraint back into the action yields the action of $f(R)$ gravity. Hence, the two actions are (classically) dynamically equivalent. If instead one varies with respect to ϕ one gets $\varphi = f'(\phi)$. Replacing this algebraic relation back in the action one gets another dynamically equivalent action

$$S = \frac{1}{16\pi G}\int d^4x\sqrt{-g}\,[\varphi R - V(\varphi)] + S_m(g_{\mu\nu}, \psi)\,, \tag{1.32}$$

where $V(\varphi) \equiv f(\phi) - \phi f'(\phi)$ (V is essentially the Legendre transform of f). This theory is actually a Brans–Dicke theory with vanishing ω_0, also known as the O'Hanlon action [32].

This simple exercise establishes that $f(R)$ gravity can be recast into the form of a special Brans–Dicke theory, something that has been known for quite a while, see e.g. [33]. It demonstrates both how higher-order theories propagate more degrees freedom—in this case a scalar—and how such theories avoid Ostrogradski's instability when they can be recast into second-order theories with more degrees of freedom.

1.4.2 Hořava Gravity

Hořava gravity [34] is a theory with a preferred spacetime foliation. The action of the theory is [35]

$$S_H = \frac{1}{16\pi G_H}\int dTd^3x\, N\sqrt{h}\left(L_2 + \frac{1}{M_\star^2}L_4 + \frac{1}{M_\star^4}L_6\right), \tag{1.33}$$

where

$$L_2 = K_{ij}K^{ij} - \lambda K^2 + \xi\,{}^{(3)}R + \eta a_i a^i, \tag{1.34}$$

where T is the preferred time, K_{ij} is the extrinsic curvature of the surfaces of the foliation and K its trace, ${}^{(3)}R$ is the intrinsic curvature of these surfaces, N is the lapse function, h_{ij} is the induced metric and h is the determinant of the induced metric, $a_i \equiv \partial_i \ln N$, G_H is a coupling constant with dimensions of length squared and λ, ξ, and η are dimensionless couplings. Since the action is written in a preferred foliation the theory does not enjoy invariance under diffeomorphisms. It is still invariant under the subset of diffeomorphisms that respect the foliation, $T \to T' = f(T)$ and $x^i \to x'^i = x'^i(T, x^i)$. L_4 and L_6 include all possible terms that respect this symmetry and contain up to four and six spatial derivatives respectively. $M_\star$ is a characteristic mass scale suppressing these higher order terms.

Hořava gravity has been proposed as a power-counting renormalizable gravity theory and the presence of the higher-order terms in L_4 and L_6 is crucial in order to have the right UV behaviour [34]. However, these terms will not concern us here, as we intend to consider the low energy part of the theory, L_2, as an example of a gravity theory that does not respect diffeomorphism invariance. For a brief review on the basic features of Hořava gravity see [36].

Consider now the action

$$S' = \frac{1}{16\pi G'}\int \sqrt{-g}(-R - M^{\alpha\beta}{}_{\mu\nu}\nabla_\alpha u^\mu \nabla_\beta u^\nu)d^4x \tag{1.35}$$

where

$$M^{\alpha\beta}{}_{\mu\nu} = c_1 g^{\alpha\beta}g_{\mu\nu} + c_2\delta^\alpha_\mu\delta^\beta_\nu + c_3\delta^\alpha_\nu\delta^\beta_\mu + c_4 u^\alpha u^\beta g_{\mu\nu}, \tag{1.36}$$

c_i are dimensionless coupling constants and u_μ is given by

$$u_\mu = \frac{\partial_\mu T}{\sqrt{g^{\lambda\nu}\partial_\lambda T\partial_\nu T}}. \tag{1.37}$$

This is a scalar-tensor theory where the scalar field T only appears in the action in the specific combination of Eq. (1.37). Therefore, u^μ can be thought of as a hypersurface orthogonal, unit, timelike vector (as $u^\mu u_\mu = 1$). The theory can be thought of as a restricted version of Einstein-aether theory [37, 38] where the aether is forced to be hyper surface orthogonal before the variation.

Now, following the lines of [39], one can observe that T always has a timelike gradient, so it can be a good time coordinate for any solution. Then one can give up some of the gauge freedom in order to re-write the theory in terms of this time coordinate. This involves introducing a foliation of T =constant hyper surfaces, to which u^μ will be normal, and re-writing the action in this foliation. Then $u_\mu = N\delta^0_\mu$, where N is the lapse of this foliation, and action (1.35) takes the form

$$S' = \frac{1}{16\pi G_H}\int dTd^3x\, N\sqrt{h}\left(K_{ij}K^{ij} - \lambda K^2 + \xi^{(3)}R + \eta a_i a^i,\right), \tag{1.38}$$

where

$$\frac{G_H}{G'} = \xi = \frac{1}{1-(c_1+c_3)}, \quad \lambda = \frac{1+c_2}{1-(c_1+c_3)}, \quad \eta = \frac{c_1+c_4}{1-(c_1+c_3)}. \tag{1.39}$$

Action (1.38) is clearly the infrared (L_2) part of action (1.33), which means that the initial action (1.35) is just the diffeomorphism invariant version of the infrared limit of Hořava gravity. T can then be thought of as the Stueckelberg field one needs to introduce in order to restore full diffeomorphism invariance in Hořava gravity. It is clearly a dynamical field and in the covariant picture one can think of it as having a nontrivial configuration which defines the preferred foliation in every solution. When the theory is written in the preferred foliation, as in Eq. (1.33), then the scalar degree of freedom is no longer explicit, but one can expect its existence because the action has less symmetry.

1.5 Scalar Fields Around Black Holes

As already mentioned in the introduction, black holes and compact stars are of particular interest in alternative theories of gravity as potential probes of the strong gravity regime. Black holes in particular have the advantage of being vacuum solutions, so one need not worry about matter, and of containing horizons, hence they have a very interesting causal structure.

One could argue that the existence of extra degrees of freedom—in this case a scalar field—in a gravity theory will generically lead to black hole solutions that differ from their general relativity counterpart. They could then be used as probes for deviation from Einstein's theory, or even for the very existence of scalar fields. However, there are "no-hair" theorems is scalar-tensor gravity that suggest otherwise [40, 41]. In particular, according to these theorems stationary, asymptotically flat black holes in the theories described by the action of Eq. (1.14) are identical to black holes in general relativity. This is because the scalar field is forced to have a ϕ = constant configuration in stationary, asymptotically flat space times with a horizon. Quiescent astrophysical black holes that are the endpoints of gravitational collapse are stationary. They are also asymptotically flat to a very good approximation. Hence, one is tempted to believe that black holes in scalar-tensor theories will be indistinguishable from black holes in general relativity.

Such an interpretation of the no-hair theorems would be misleading for several reasons. First of all, a perturbed Kerr spacetime in a scalar-tensor theory would differ from a perturbed Kerr spacetime in general relativity, a characteristic example being the existence of a scalar mode in the gravitational wave spectrum [42]. Secondly, cosmological asymptotics do induce scalar hair in principle [43], though the deviation from the Kerr geometry is unlikely to be detectable [44]. Finally, astrophysical black holes tend to be surrounded by matter in various forms—companion stars, accretion disks, or the galaxy as a whole. Equation (1.16) or (1.19) imply that, in the presence of matter, constant scalar solutions are only allowed in theories for which ω diverges at the minimum of the potential. This has been already discussed in Sect. 1.3.3 (theories that exhibit "spontaneous scalarization" [19, 20]). Hence, generically the presence of matter around the black hole will tend to induce scalar hair and the pending question is to determine how important this effect might be.

So, when put in astrophysical context, the no-hair theorems tell us that black holes that are endpoints of collapse will be rather close to the Kerr solution and that we can use perturbative techniques in order to study phenomena around them (which provides an important simplification). They do not, however, imply that astrophysical black holes in scalar-tensor gravity are indistinguishable from astrophysical black holes in general relativity. In fact, it has been suggested that there might be smoking gun effects associated with the scalar field in scalar-tensor theories. For example, in [45] it has been shown that there exist floating orbits around Kerr black holes in these theories, i.e. particles can orbit the black holes without "sinking" into it even though gravitational radiation is emitted. The loss of energy of the emission is balanced by loss of angular momentum of the black hole. In [46] instead, it was shown that, in theories that admit a constant scalar configuration in the presence of matter, black holes can undergo spontaneous scalarization or exhibit instabilities related to superradiance and very large amplification factors for superradiant scattering.

We now move on to black holes in generalised scalar-tensor gravity, i.e. theories described by the Horndeski action in Eq. (1.24). There are no no-hair theorems covering the complete class of theories. On the contrary, there are already known black hole solutions that have non-trivial scalar field configurations in theories that belong to this class, see e.g. [47]. It has been claimed in [48] that in the subclass of theories in which the scalar enjoys shift symmetry, i.e. symmetry under $\phi \to \phi+$ constant, only trivial scalar configuration are admissible for static, spherically symmetric and asymptotically flat black holes, and, hence, these black holes are described by the Schwarzschild solutions. It has been argued in [49] that, when valid, the no-hair theorem of [48] can straightforwardly be generalised to slowly rotating black holes. However, it has been also been shown there that the theorem holds in the first place only if one forbids a linear coupling between the scalar field and the Gauss–Bonnet invariant. Such a coupling is allowed by shift symmetry, since the Gauss–Bonnet term is a total diverge. A term that contains this coupling is implicitly part of the Horndeski action, even though the representation of Eq. (1.24) does not make that manifest. One can impose symmetry under $\phi \rightarrow -\phi$ in order to do away with this term (together with various others in the action). However, the conclusion is that the subclass of theories for which one can have a no-hair theorem is more limited than originally claimed.

We close this section with a few remarks on black holes in Lorentz-violating theories, since, as we argued above Hořava gravity can be re-written as a scalar-tensor theory. One could question whether black holes can actually exist in this theory, as well as in other Lorentz-violating theories, as one can have perturbations that travel with arbitrarily high speed and could, therefore, penetrate conventional horizons.[6] However, it has been shown that a new type of horizon that shields its exterior from any signal that comes from its interior, irrespectively of how fast it propagates, can exist in theories with a preferred foliation, called the universal horizon [50–53]. The existence of such a horizon implies that the notion of a black hole can exist in Lorentz-violating theories. For a thorough discussion on this topic see [54].

Conclusions

In these lecture notes I have attempted to highlight some interesting concepts, pitfalls and subtleties that appear when one goes beyond general relativity. Perhaps it is helpful to list the most important ones:

- Any attempt to modify the action of general relativity will generically lead to extra degrees of freedom (carefully engineered exceptions can exist);
- These degrees of freedom may be manifest as extra dynamical fields or may be implicit because of higher order equations or less symmetry;
- The actual number of degrees of freedom might be quite obscure in some specific field representation;
- Taming the behaviour of these extra degrees of freedom is what constructing viable and successful (in terms of some desirable phenomenological signature) models is about;
- One should constantly be seeking for new constraints on deviations from general relativity, and the strong gravity regime is particularly promising in this respect.

(continued)

[6] Hořava gravity exhibits instantaneous propagation even at low energies [50], and on general grounds one would expect the UV completion of any Lorentz violating theory to generically introduce higher order dispersion relations.

A brief review of gravity theories with an extra scalar degree of freedom has been given and some of their basic features have been discussed. Even though I touched upon virtually all such theories, these lecture notes do not constitute a thorough review of the theories and their phenomenology. I have simply selectively discussed specific aspects of each theory in an attempt to provide useful examples for the points listed above.

Acknowledgements The research leading to these results has received funding from the European Research Council under the European Union's Seventh Framework Programme (FP7/2007–2013)/ERC Grant Agreement n. 306425 "Challenging General Relativity".

References

1. G. Amelino-Camelia, Lect. Notes Phys. **669**, 59 (2005) [gr-qc/0412136]
2. S. Liberati, Class. Quantum Grav. **30**, 133001 (2013) [arXiv:1304.5795 [gr-qc]]
3. P.A.R. Ade et al. [Planck Collaboration], Astron. Astrophys. (2014)
4. S. Weinberg, Rev. Mod. Phys. **61**, 1 (1989)
5. S.M. Carroll, Living Rev. Relat. **4**, 1 (2001)
6. C. Will, *Theory and Experiment in Gravitational Physics* (Cambridge University Press, Cambridge, 1993)
7. D. Lovelock, J. Math. Phys. **12**, 498 (1971)
8. D. Lovelock, J. Math. Phys. **13**, 874 (1972)
9. H. Ruegg, M. Ruiz-Altaba, Int. J. Mod. Phys. A **19**, 3265 (2004)
10. C.W. Misner, K.S. Thorne, J.A. Wheeler, *Gravitation* (W.H. Freeman, San Francisco, 1973)
11. M. Ostrogradski, Mem. Ac. St. Peterbourg **VI 4**, 385 (1850)
12. R.P. Woodard, Lect. Notes Phys. **720**, 403 (2007)
13. P. Pani, T.P. Sotiriou, D. Vernieri, Phys. Rev. D **88**, 121502 (2013)
14. L. Perivolaropoulos, Phys. Rev. D **81**, 047501 (2010)
15. B. Bertotti, L. Iess, P. Tortora, Nature **425**, 374 (2003)
16. V. Faraoni, *Cosmology in Scalar-Tensor Gravity* (Springer, Berlin, 2004)
17. Y. Fujii, K. Maeda, *The Scalar-Tensor Theory of Gravitation* (Cambridge University Press, Cambridge, 2003)
18. T.P. Sotiriou, V. Faraoni, S. Liberati, Int. J. Mod. Phys. D **17**, 399 (2008)
19. T. Damour, G. Esposito-Farese, Phys. Rev. Lett. **70**, 2220 (1993)
20. T. Damour, G. Esposito-Farese, Phys. Rev. D **54**, 1474 (1996)
21. J. Khoury, A. Weltman, Phys. Rev. Lett. **93**, 171104 (2004)
22. V. Faraoni, Class. Quantum Grav. **26**, 145014 (2009)
23. J. Khoury, Class. Quantum Grav. **30**, 214004 (2013)
24. K. Hinterbichler, J. Khoury, Phys. Rev. Lett. **104**, 231301 (2010)
25. G.W. Horndeski, Int. J. Theor. Phys. **10**, 363 (1974)
26. C. Deffayet, S. Deser, G. Esposito-Farese, Phys. Rev. D **80**, 064015 (2009)
27. A. Nicolis, R. Rattazzi, E. Trincherini, Phys. Rev. D **79**, 064036 (2009)
28. C. Deffayet, G. Esposito-Farese, A. Vikman, Phys. Rev. D **79**, 084003 (2009)
29. C. Deffayet, D.A. Steer, Class. Quantum Grav. **30**, 214006 (2013)
30. A.I. Vainshtein, Phys. Lett. B **39**, 393 (1972)
31. E. Babichev, C. Deffayet, Class. Quantum Grav. **30**, 184001 (2013)
32. J. O' Hanlon, Phys. Rev. Lett. **29**, 137 (1972)

33. P. Teyssandier, P. Tourrenc, J. Math. Phys. **24**, 2793 (1983)
34. P. Hořava, Phys. Rev. D **79**, 084008 (2009)
35. D. Blas, O. Pujolas, S. Sibiryakov, Phys. Rev. Lett. **104**, 181302 (2010)
36. T.P. Sotiriou, J. Phys. Conf. Ser. **283**, 012034 (2011)
37. T. Jacobson, D. Mattingly, Phys. Rev. D **64**, 024028 (2001)
38. T. Jacobson, PoS QG **-PH**, 020 (2007)
39. T. Jacobson, Phys. Rev. D **81**, 101502 (2010) [Erratum-Phys. Rev. D **82**, 129901 (2010)]
40. S.W. Hawking, Commun. Math. Phys. **25**, 167 (1972)
41. T.P. Sotiriou, V. Faraoni, Phys. Rev. Lett. **108**, 081103 (2012)
42. E. Barausse, T.P. Sotiriou, Phys. Rev. Lett. **101**, 099001 (2008)
43. T. Jacobson, Phys. Rev. Lett. **83**, 2699 (1999)
44. M.W. Horbatsch, C.P. Burgess, J. Cosmol. Astropart. Phys. **1205**, 010 (2012)
45. V. Cardoso, S. Chakrabarti, P. Pani, E. Berti, L. Gualtieri, Phys. Rev. Lett. **107**, 241101 (2011)
46. V. Cardoso, I.P. Carucci, P. Pani, T.P. Sotiriou, Phys. Rev. Lett. **111**, 111101 (2013) [arXiv:1308.6587 [gr-qc]]
47. P. Kanti, N.E. Mavromatos, J. Rizos, K. Tamvakis, E. Winstanley, Phys. Rev. D **54**, 5049 (1996) [hep-th/9511071]
48. L. Hui, A. Nicolis, Phys. Rev. Lett. **110**, 241104 (2013)
49. T.P. Sotiriou, S.-Y. Zhou, Phys. Rev. Lett. **112**, 251102 (2014)
50. D. Blas, S. Sibiryakov, Phys. Rev. D **84**, 124043 (2011)
51. E. Barausse, T. Jacobson, T.P. Sotiriou, Phys. Rev. D **83**, 124043 (2011)
52. E. Barausse, T.P. Sotiriou, Phys. Rev. Lett. **109**, 181101 (2012) [Erratum-Phys. Rev. Lett. **110**, 039902 (2013)]
53. E. Barausse, T.P. Sotiriou, Phys. Rev. D **87**, 087504 (2013)
54. E. Barausse, T.P. Sotiriou, Class. Quantum Grav. **30**, 244010 (2013) [arXiv:1307.3359 [gr-qc]]

Chapter 2
From Lovelock to Horndeski's Generalized Scalar Tensor Theory

Christos Charmousis

Abstract We review and discuss some recent progress in Lovelock and Horndeski theories modifying Einstein's General Relativity. Using as our guide the uniqueness properties of these modified gravity theories we then discuss how Kaluza-Klein reduction of Lovelock theory can lead to effective scalar-tensor actions including several important terms of Horndeski theory. We show how this can be put to practical use by mapping analytic black hole solutions of one theory to the other. We then elaborate on the subset of Horndeski theory that has self-tuning properties and review a generic method giving scalar-tensor black hole solutions.

2.1 Introduction

General Relativity (GR) is a classical or effective theory of gravity which is based on very solid mathematical and physical foundations. It agrees with overwhelming accuracy local[1] observational tests both for weak and strong gravity [1] including laboratory tests of Newton's force law. GR, is not only a very successful physical theory. It is theoretically very robust and as it turns out mathematically a unique metric theory. Indeed if one considers a theory depending on a massless metric and up to its second derivatives endowed with a Levi-Civita connection then,

$$S^{(4)} = \frac{{m_{Pl}}^4}{2} \int d^4x \sqrt{-g^{(4)}} \left[R - 2\Lambda \right], \tag{2.1}$$

is the unique action giving equations of motion of second order in the metric field variable. This theorem, as we will see, is a consequence of Lovelock's theorem [2] (see also [3, 4]). In other words, GR plus a cosmological constant is the

[1]Local distance scales range up to 30 odd astronomical units, size of the solar system, but also size of typical binary pulsar systems. The astronomical unit is a rough earth to sun distance.

C. Charmousis (✉)
Laboratoire de Physique Théorique (LPT), Univ. Paris-Sud, CNRS UMR 8627, 91405 Orsay, France
e-mail: christos.charmousis@th.u-psud.fr

E. Papantonopoulos (ed.), *Modifications of Einstein's Theory of Gravity at Large Distances*, Lecture Notes in Physics 892, DOI 10.1007/978-3-319-10070-8_2

unique gravity theory constructed out of a single massless metric with a Levi-Civita connexion (which is also defined uniquely). This means that any other curvature scalar would necessarily yield either trivial or higher order than second derivatives in the field equations. Higher order than second, ∇^2-derivatives, lead directly to a theory with ghost vacua, clearly an important setback for any classical physical theory [5]. The only other term evading this problem is the one associated to the cosmological constant. We will encounter consequences of this very shortly.

As we emphasized GR is an *effective* metric field theory. As such we expect Einstein's theory to break down at very high energies (strong curvatures) close to the Planck scale, $m_{Pl}^2 = \frac{1}{16\pi G}$, where higher order curvature terms can no longer be neglected and are even dominant compared to the leading Einstein-Hilbert term. What is maybe more surprising is that recent cosmological observations, point towards the tantalizing possibility that GR may also be modified at very low energy scales deep in the infra-red [6]. A tiny positive cosmological constant generates an inversely proportional enormous cosmological horizon and can account very simply for such a dark energy component. After all, as we saw in (2.1), it is a mathematically allowed term in the metric action. However, the difference in between cosmological and local scales corresponds to an enormous number, of magnitude of 10^{15}, in other words we are very deep in the infra red and physics may well differ from scales where we control gravity observationally. Furthermore, although a cosmological constant provides a phenomenologically correct and economic way to put away the dark energy problem it suffers from a theoretical short-coming, the cosmological constant problem [7]. Indeed, from very simple field theory considerations, GR, from its founding strong equivalence principle perceives all forms of matter in time and space including vacuum energy. Vacuum energy gravitates just as does radiation or matter. The cosmological constant, for example, receives zero point energy contributions from each particle species up to the UV cut-off of the relevant QFT. These contribute to the total value of the cosmological constant which has to be fine-tuned to almost zero by the arbitrary bare contribution we saw above (2.1). This fact only gets worse once we realize that phase transitions in the early universe will actually shift this value around, and again each time some miraculous fine-tuning will be required to tune the overall cosmological constant to its tiny but non-zero value we observe today. The "big" cosmological constant problem is precisely how all these vacuum energies associated to the GUT, SUSY, the standard model etc. are fined-tuned each time to zero by an exactly opposite in value bare cosmological constant Λ_{bare} appearing in (2.1) and being the net result of the universe acceleration today. The unexplained small value of the cosmological constant Λ_{now} is then an additional two problems to add to the usual "big" cosmological constant problem [7], namely, why the cosmological constant is not cancelled exactly to zero and why do we observe it now. In a later section we will see of such an attempt to classically[2] evade this problem [9].

[2]For a interesting proposal tackling the cosmological constant problem including the crucial radiative corrections see, [8].

Although this is not really a scientific argument, one can make reference to a historically parallel situation. At the advent of General Relativity, observational evidence pointed towards shortcomings of Newton's gravitational theory in the 'strong gravity' regime. Amongst these was the advance of the perihelion of Mercury, which deviated from Kepler's laws describing planetary motion. As such, the existence of a small planet in an even closer orbit to the sun, Vulcan, was hypothesized. Alternatively, the presence of an unknown substance, aether, was put forward, mediating and slightly modifying the prediction of Kepler's laws to account for observational data. Indeed a simple and slight correction to the established laws of the time could account correctly for the advance of the perihelion. The solution to the puzzle was, however, not as simple or economic as initially considered. In fact, it was only after the theory of GR was put forward that this slight difference was accounted for as, rather, a fundamental modification of gravity theory. As often in physics, a modified physical theory is attained upon reaching a critical energy scale; here, the critical scale in question is the strong gravity field of the sun applied to its closest planet. There is in recent times an observational parallel to the above, in the context of type Ia supernovae explosions, pointing towards an accelerating universe [6]. Friedmann's laws, in order to remain valid, require the addition of an as yet unknown dark energy component, which is the dominant component in the Universe. The addition of a small cosmological constant gives very good agreement with observational data and is the most economic (in terms of additional degrees of freedom) phenomenological explanation of the acceleration phenomenon. Given, however, the above example, it seems to us important to entertain the following question: could it be that recent observations are pointing towards a fundamental modification of gravity rather than a modification in the unknown matter sector? Are novel observations indications of a new gravity theory beyond GR? This question is even more compelling since we know that dark matter is so far unaccounted for and in the ultraviolet GR needs to be modified anyway. A second important point concerns the predictions and motivation of a modified gravity theory. Indeed, as we argued above, the initial conditions calling for a modified gravity theory are in order to account on the one hand for the late-time acceleration of the universe and to provide on the other hand a well-defined limit at local scales where the theory at hand should be indistinguishable from GR. This is of course an important and difficult initial step that provides a filter for possible theories under consideration, but this is not all. Since observations can be accounted for by a small cosmological constant put in by hand, one needs to go further in order to make new accurate predictions theoretically. These novel predictions are the real motivation in a modified theory of gravity. Indeed, General Relativity's great successes are not the explanation of the advance in the perihelion of Mercury or its classical limit to Newtonian theory, but rather, completely novel ideas and solutions stemming from the theory itself, such as black holes, Big Bang inflationary theory and so forth.

So how do we go about modifying such a robust theory such as GR (see the review [10])? Not surprisingly it is extremely hard both observationally but also theoretically, the windows of modification are rather narrow. This is at the same time fortunate because at the end not there are not too many possibilities left over. In rather loose terms following (2.1) and not breaking some fundamental symmetry like Lorentz invariance (see for example [11]), there are four at least routes emanating from a Lagrangian formulation. First, suppose we keep the single massless metric character of the theory. Then inevitably we have to consider higher dimensions. We will show that the relevant theory is then Lovelock theory (see for example [12]). Secondly suppose we stick to 4 space-time dimensions. Then inevitably we consider the existence of additional fields, in other words we add novel gravitational degrees of freedom in four dimensional space-time. Here the prototype is scalar-tensor theory and we know its most general form, Horndeski theory [13]. We will study basics of this theory here. All the terms present in Horndeski theory have been shown to be originating from Galileons i.e. scalar tensor terms having Gallilean symmetry in flat space-time [14] and the latter equivalent theory to Horndeski has been elegantly given for curved space-time in [15]. Thirdly we can consider that the elementary particle mediating spin 2 gravity, the graviton, has a finite range of application. In other words it is not a massless field but has some (small) mass. This is the theme of massive gravity [16] which will also be covered in later lectures. Lastly we can consider the possibility of allowing for other geometric constructions such as a differing connexion than that of Levi-Civita. This allows for torsion i.e. non zero parallel transport of scalars (rather than vectors) or first order formalism, Palatini formalism (see for example [17]). These four directions are not independent of each other in fact often they are related and it is useful to use information from one to the other. We will give such relations during these lectures. We will discuss in fact Lovelock and Horndeski theory and relate the two via the Kaluza-Klein formalism.

Using as our guide uniqueness theorems we will discuss certain elements of Lovelock and Horndeski theory. We will see in what sense these theories are unique. We will focus throughout on recent elements of Lovelock theory that we will be using in relation to Horndeski theory. We will omit some basic properties diverting the interested reader to [12]. We will then go on to discuss Horndeski theory which is the most general scalar-tensor theory in four dimensions. We will then move on to review some black hole solutions of Lovelock theory and see how, by toroidal Kaluza-Klein reduction we can construct four dimensional scalar-tensor black holes. In this way we will establish a clear and practical connection in between Lovelock and Horndeski theory. In the fifth section we will discuss the cosmological constant problem and define a theory which is a subset of Horndeski theory and has interesting self-tuning properties. This theory dubbed fab four [9], will at least classically provide a partial solution to the big cosmological constant problem. We will then sketch a recent and relatively simple way to obtain black hole solutions in such scalar-tensor theories [18].

2.2 The Lovelock and Horndeski Uniqueness Theorems

2.2.1 Lovelock Theory

Our purpose in this lecture is to present Lovelock theory in relation to Horndeski theory. To this end it is mostly sufficient to truncate Lovelock theory to what is usually called Einstein-Gauss-Bonnet (EGB) theory in the literature. Unlike the name suggests, this is the five or six dimensional version of Lovelock theory originally discussed by Lanczos [3]. Let us start with the uniqueness theorem defining Lovelock theory and stick to six dimensions in order to fix notation. The five dimensional theory is identical. Consider $\mathscr{L} = \mathscr{L}(\mathscr{M}, g, \nabla, \nabla^2)$ where $(\mathscr{M}, g)$ is a six dimensional locally differentiable Lorentzian manifold without boundary[3] and ∇ is the Levi-Civita metric connexion over $\mathscr{M}$. The field equations obtained upon metric variation of the action,

$$S^{(6)} = \frac{{M_{(6)}}^4}{2} \int \sqrt{-g^{(6)}} \left[R - 2\Lambda + \alpha \hat{G} \right], \tag{2.2}$$

are unique and admit the following properties:

- they depend on a symmetric two-tensor $\mathscr{E}_{AB}$
- the equations of motion are second-order PDE's with respect to the metric field variables
- satisfying Bianchi identities.

Here, $M_{(6)}$ is the fundamental mass scale in six-dimensional spacetime, $\hat{G}$ is the Gauss-Bonnet density reading,

$$\hat{G} = R_{ABCD} R^{ADCB} - 4 R_{AB} R^{AB} + R^2 , \tag{2.3}$$

and Λ is the cosmological constant. The field equations in vacuum are

$$\mathscr{E}_{AB} = G_{AB} + \Lambda g_{AB} + \alpha H_{AB} = 0 , \tag{2.4}$$

where G_{AB} stands for the standard Einstein tensor. Uppercase Latin indices will refer to six-dimensional coordinates whereas Greek indices will always refer to four dimensional space-time. We have also introduced the Lanczos or second order Lovelock tensor,

$$H_{AB} = \frac{g_{AB}}{2} \hat{G} - 2 R R_{AB} + 4 R_{AC} R^{C}{}_{B} + 4 R_{CD} R^{C}{}_{A}{}^{D}{}_{B} - 2 R_{ACDE} R_{B}{}^{CDE} . \tag{2.5}$$

[3] The result depicted here is easily extended to manifolds with boundaries [12].

Naturally, the Lanczos tensor is also divergence free, $\nabla^A H_{AB} = 0$. It is important to note that, just like GR in four dimensions, i.e. under the same set of hypotheses, EGB theory is the unique and most general metric theory with second order PDE's in five or six space-time dimensions. This is a non-trivial statement since the terms appearing in the action already contain second order derivatives. In a moment we will see that in higher than six dimensions this property is generalized by adding the relevant higher order Lovelock terms. Furthermore in four dimensions the tensor (2.5) is identically zero. Therefore we can note as a prelude that Lovelock theory is the unique massless metric theory in arbitrary dimensions identical to GR with a cosmological constant in four dimensional spacetime.

Before moving on it is useful to discuss some tensorial properties. The Lanczos tensor (2.5) can be elegantly written (in arbitrary dimension) using the following rank four tensor that will be useful to us later on,

$$P_{ABCD} = R_{ABCD} + R_{BC}\, g_{AD} - R_{BD}\, g_{AC} - R_{AC}\, g_{BD} + R_{AD}\, g_{BC} + \frac{1}{2} R\, g_{AC}\, g_{BD} - \frac{1}{2} R\, g_{BC}\, g_{AD}, \tag{2.6}$$

as

$$H_{AB} = -2 P_{ACDE} R_B{}^{CDE} + \frac{g_{AB}}{2} \hat{G}\,. \tag{2.7}$$

The 4 index tensor P_{ABCD} has several interesting tensorial properties. For a start it is divergence free (in all indices) since Bianchi identities of the curvature tensor are simply written as $\nabla^D P_{ABCD} = 0$. It has the same index symmetries as the Riemann curvature tensor. Its bi-tensor obtained by tracing two of its non-consecutive indices yields

$$P^B{}_{ACB} = (D-3) G_{AC}, \tag{2.8}$$

the Einstein tensor. In fact divergence freedom of the Einstein tensor can be seen to originate from this relation. In a nutshell, one can say that P_{ABCD} is the curvature tensor whose bi-tensor is the Einstein tensor, just as the Ricci tensor is the bi-tensor of the Riemann tensor. A last interesting property is that a metric is an Einstein space, $R_{AB} = \frac{g_{AB}}{D} R$, if and only if $P_{ABCD} = R_{ABCD}$.

In four dimensions, the $P_{\mu\nu\rho\sigma}$ tensor is even more very special. Indeed it has all the above properties but, additionally it can be pictured in a very similar way to the Faraday tensor in electromagnetism,

$$\star\, F_{\mu\nu} = \tfrac{1}{2} \epsilon_{\mu\nu\kappa\rho}\, F^{\kappa\rho} \tag{2.9}$$

In analogy here, $P_{\mu\nu\rho\sigma}$ is a 4 tensor, and coincides with the double dual (i.e. for each pair of indices) of the Riemann tensor defined as,

$$P^{\mu\nu}{}_{\rho\sigma} = ({}^\star R^\star)^{\mu\nu}{}_{\rho\sigma} \doteq -\frac{1}{2} \epsilon^{\rho\sigma\lambda\kappa}\, R_{\lambda\kappa}{}^{\xi\tau}\, \frac{1}{2} \epsilon_{\xi\tau\mu\nu}, \tag{2.10}$$

where $\epsilon_{\mu\nu\rho\sigma}$ is the rank 4 Levi-Civita tensor. Finally since in four dimensions we have that $H_{\mu\nu} = 0$ we obtain the Lovelock identity,

$$P_{\alpha\nu\rho\sigma} R_{\beta}{}^{\nu\rho\sigma} = \frac{g_{\alpha\beta}}{4}\hat{G} \tag{2.11}$$

which will be useful to us later on (see [19] for extensions).

In order to define the generic Lovelock densities one can use the elegant language of differential forms [12]. Alternatively we take the route taken by Lovelock using the generalized Kronecker delta symbols; the same route taken later on by Horndeski,

$$\delta^{A_1 \dots A_h}_{B_1 \dots B_h} = \begin{vmatrix} \delta^{A_1}_{B_1} & \dots & \delta^{A_1}_{B_h} \\ \vdots & & \vdots \\ \delta^{A_h}_{B_1} & \dots & \delta^{A_h}_{B_h} \end{vmatrix} \tag{2.12}$$

$$= h!\delta^{A_1}_{[B_1} \dots \delta^{A_h}_{B_h]} \tag{2.13}$$

which is antisymmetric in any pair of upper or lower indices. In fact we have $\delta^{A_1 \dots A_h}_{B_1 \dots B_h} = \epsilon_{B_1 \dots B_h}\epsilon^{A_1 \dots A_h}$ with respect to the Levi-Civita symbols. Once this has been digested the Lovelock densities are the complete contraction of the above with the Riemann curvature tensor,

$$L_{(h/2)} = \frac{1}{2^h}\delta^{A_1 A_2 \dots A_h}_{B_1 B_2 \dots B_h} R_{A_1 A_2}{}^{B_1 B_2} \dots R_{A_{(h-1)} A_h}{}^{B_{(h-1)} B_h} \tag{2.14}$$

As such we can check that $L_{(1)}$ is the Einstein-Hilbert term whereas $L_{(2)}$ is the Gauss-Bonnet combination. This immediately means that for $h > D$ all Lovelock densities vanish. Therefore the Lovelock Langrangian is given by,

$$L = \sum_{h=0}^{k} c_h L_h \tag{2.15}$$

where $k = [(D-1)/2]$. The case $h = D$ is quite special because then the Lovelock density is a topological one. Indeed we can query what is special about Lovelock densities. The answer lies in differential geometry (see for example [20]). One can trace the origin of such terms in the early works of Gauss who measuring geodesic distances noted that scalar (Gauss) curvature of two dimensional surfaces depended only on the first fundamental form, in other words the intrinsic metric of the surface and its derivatives. This was the basis of what he called the Egregium theorem; scalar curvature (unlike other extrinsic curvature components) does not depend on the variation of the normal vector field on the surface i.e. on how the surface is embedded in three dimensional space. Then later on Euler in his work on surface triangulations noted that two dimensional surfaces can be topologically classified by their "Euler" number, χ: $\chi[\mathscr{M}] = 2 - 2h$ where h is the number of topological handles. So one can take an arbitrary surface with no boundary and continuously deform it to a sphere, a torus a double torus and so on.[4] This completely classifies topologically two dimensional surfaces. In other words all topological properties of two dimensional surfaces can be understood or characterized by their Euler number. Gauss and Bonnet essentially related this topological number to a differentiable geometric quantity, the scalar curvature, resulting in the celebrated relation,

$$\chi[\mathscr{M}_2] = \frac{1}{4\pi} \int_{\mathscr{M}} R. \tag{2.16}$$

The Gauss-Bonnet theorem on surfaces has nothing to do with the Gauss-Bonnet term given above (2.3). For our purposes the above Gauss-Bonnet relation means that the Einstein-Hilbert term is in two dimensions is a topological invariant i.e. the Einstein tensor in two dimensional space-time is identically zero. This analogy goes through for all Lovelock terms as a corollary to the works of Chern [21] who generalized the theorem of Gauss and Bonnet to higher dimensions finding the relevant higher order curvature scalars. For example we have,

$$\chi[\mathscr{M}_4] = \frac{1}{32\pi^2} \int_{\mathscr{M}} \hat{G} \tag{2.17}$$

and thus the Lanczos or Gauss-Bonnet density is a topological invariant in four dimensions whose integral is the generalised Euler or Chern topological number. Beware this does not mean that the Gauss-Bonnet scalar is zero or constant in four dimensions. It means that the Lanczos density is identically zero $H_{\mu\nu} = 0$ as we admitted earlier.

Dimensionally extending the Chern scalar densities we obtain the Lovelock densities (2.14) i.e. just those densities whose variation leads to second order field equations. Any higher order derivatives present in the variation of Lovelock densities conveniently end up as total divergent terms and thus do not contribute to the field equations. In a similar way for example, in seven or eight dimensions, the six-dimensional Euler density will be promoted to a Lovelock density of third power in the curvature tensor and so forth. This explains the nice and unique properties of the Lovelock densities and Lovelock theory in general. For more details the reader can consult [12].

2.2.2 Horndeski Theory

So much for the moment concerning higher dimensional metric theories. In four space-time dimensions we know that the unique classical metric theory is GR with

[4] When a surface has a boundary an analogous result holds.

a cosmological constant. Hence any four dimensional modification of gravity will have to involve some other non-trivial field. The simplest of cases is when this extra field is a scalar. The prototype of scalar tensor gravity is Brans-Dicke theory [22] which has been studied extensively throughout the years (see [23] and references within). We should note that in the class of scalar-tensor theories fall also other modified gravity theories like $f(R)$ or $f(\hat{G})$ which [24] are just particular scalar-tensor theories in disguise. Furthermore other interesting GR modifications such as bigravity or massive gravity theories [16] admit scalar tensor theories as particular limits, for example the decoupling limit for massive gravity [25]. Hence scalar tensor theories are a consistent prototype of GR modification and their important properties are expected in some form, in other consistent gravity theories. Hence the particular recent interest in scalar-tensor theories concerning modification of gravity. So in this section we reiterate the question: what is the most general scalar tensor theory in four dimensional space-time yielding second order field equations? The answer has been given by Horndeski a long-time ago [13] but has remained unnoticed since only recently [9], and states a similar theorem to that of Lovelock for four dimensional scalar-tensor theories. Consider a single scalar field ϕ and a metric $g_{\mu\nu}$ as the gravitational degrees of freedom of some Lorentzian manifold endowed with a Levi-Civita connection. Consider a theory that depends on these degrees of freedom and an arbitrary number of their derivatives,

$$\mathscr{L} = \mathscr{L}(g_{\mu\nu}, g_{\mu\nu,i_1}, \ldots, g_{\mu\nu,i_1\ldots i_p}, \phi, \phi_{,i_1}, \ldots, \phi_{,i_1\ldots i_q}) \tag{2.18}$$

with $p, q \geq 2$. The finite number of derivatives signifies that we have again an effective theory since we have a finite number of degrees of freedom. Here just like in usual Brans Dicke theory we consider that matter couples only to the metric and not to the scalar field thus fixing the metric and the frame as the physical one. In this frame the metric will continue to verify the weak equivalence principle. In a nutshell the metric in question can always be put locally in a normal frame where by definition the Christophel symbols are identically zero. This frame is locally equivalent to an inertial frame. The Hornedski action can be written in such a way to involve only second derivatives and reads,

$$\begin{aligned}
\mathscr{L}_H &= \kappa_1(\phi,\rho)\delta^{\alpha\beta\gamma}_{\mu\nu\sigma}\nabla^\mu\nabla_\alpha\phi R_{\beta\gamma}{}^{\nu\sigma} - \frac{4}{3}\kappa_{1,\rho}(\phi,\rho)\delta^{\alpha\beta\gamma}_{\mu\nu\sigma}\nabla^\mu\nabla_\alpha\phi\nabla^\nu\nabla_\beta\phi\nabla^\sigma\nabla_\gamma\phi \\
&\quad +\kappa_3(\phi,\rho)\delta^{\alpha\beta\gamma}_{\mu\nu\sigma}\nabla_\alpha\phi\nabla^\mu\phi R_{\beta\gamma}{}^{\nu\sigma} - 4\kappa_{3,\rho}(\phi,\rho)\delta^{\alpha\beta\gamma}_{\mu\nu\sigma}\nabla_\alpha\phi\nabla^\mu\phi\nabla^\nu\nabla_\beta\phi\nabla^\sigma\nabla_\gamma\phi \\
&\quad +[F(\phi,\rho)+2W(\phi)]\delta^{\alpha\beta}_{\mu\nu}R_{\alpha\beta}{}^{\mu\nu} - 4F(\phi,\rho)_{,\rho}\delta^{\alpha\beta}_{\mu\nu}\nabla_\alpha\phi\nabla^\mu\phi\nabla^\nu\nabla_\beta\phi \\
&\quad -3[2F(\phi,\rho)_{,\phi}+4W(\phi)_{,\phi}+\rho\kappa_8(\phi,\rho)]\nabla_\mu\nabla^\mu\phi + 2\kappa_8\delta^{\alpha\beta}_{\mu\nu}\nabla_\alpha\phi\nabla^\mu\phi\nabla^\nu\nabla_\beta\phi \\
&\quad +\kappa_9(\phi,\rho), \\
\rho &= \nabla_\mu\phi\nabla^\mu\phi,
\end{aligned} \tag{2.19}$$

The action (2.19) is rather general and depends on four arbitrary functions $\kappa_i(\phi, \rho)$, $i = 1, 3, 8, 9$ of the scalar field ϕ and its kinetic term denoted as ρ. Furthermore,

$$F_{,\rho} = \kappa_{1,\phi} - \kappa_3 - 2\rho\kappa_{3,\rho} \tag{2.20}$$

with $W(\phi)$ an arbitrary function of ϕ, which means we can set it to zero without loss of generality by absorbing it into a redefinition of $F(\phi, \rho)$. According to Horndeski's theorem, [13], the action (2.19) is the unique[5] action whose variation with respect to the scalar and metric yields second order field equations and Bianchi identities. In his original work, Horndeski makes just like Lovelock, systematic use of the anti-symmetric Kronecker deltas (2.12). The equations of motion are obtained by variation of the metric and scalar field, are parametrized by the arbitrary functions $\kappa_i(\phi, \rho)$ and read respectively,

$$\mathscr{E}^{\mu\nu} = \tfrac{1}{2}T^{\mu\nu}, \; \mathscr{E}_\phi = 0 \tag{2.21}$$

where $T^{\mu\nu} = \frac{2}{\sqrt{-g}}\frac{\delta S_m}{\delta g_{\mu\nu}}$ is the matter energy-momentum tensor. The tensor $\mathscr{E}^{\mu\nu}$ is divergent free. A rather more intuitive and economic way of obtaining the Horndeski action is given in terms of the general Galileon covariant action [15] and reads,

$$\begin{aligned}\mathscr{L}_{DGSZ} = {} & K(\phi,\rho) - G_3(\phi,\rho)\nabla^2\phi + G_4(\phi,\rho)R + G_{4,\rho}\left[(\nabla^2\phi)^2 - (\nabla_\mu\nabla_\nu\phi)^2\right] \\ & + G_5(\phi,\rho)G_{\mu\nu}\nabla^\mu\nabla^\nu\phi - \frac{G_{5,\rho}}{6}\left[(\nabla^2\phi)^3 - 3\nabla^2\phi(\nabla_\mu\nabla_\nu\phi)^2\right. \\ & \left. + 2(\nabla_\mu\nabla_\nu\phi)^3\right]\end{aligned} \tag{2.22}$$

In this version it is far easier to recognize subsets of this theory, GR, Brans-Dicke, K-essence etc. and to figure out the most common Galileon terms. Again the theory depends on four free potentials. It was shown in [26] that in four dimensions Horndeski's theory is equivalent to the generalised galileon theory with the potentials given by,

$$K = \kappa_9 + \rho\int^\rho d\rho'\left(\kappa_{8,\phi} - 2\kappa_{3,\phi\phi}\right) \tag{2.23}$$

$$G_3 = 6(F + 2W)_{,\phi} + \rho\kappa_8 + 4\rho\kappa_{3,\phi} - \int^\rho d\rho'\left(\kappa_8 - 2\kappa_{3,\phi}\right) \tag{2.24}$$

$$G_4 = 2(F + 2W) + 2\rho\kappa_3 \tag{2.25}$$

$$G_5 = -4\kappa_1 \tag{2.26}$$

[5] In the action one can always add terms that can be written as a total divergence. Therefore the term "unique action" refers to the unique class of equivalence which is in turn defined modulo total divergence terms. In other words two actions are equal if and only if they are in the same class of equivalence or they differ only by a totally divergent term.

The uniqueness proof by Horndeski is quite technical and can be found in his original paper. Here we simply sketch its important steps. The proof is based on the property relating the metric and scalar field equations,

$$\nabla^{\mu}\mathscr{E}_{\mu\nu} = \tfrac{1}{2}\mathscr{E}_{\phi}\nabla_{\nu}\phi \tag{2.27}$$

This is of course an identity and shows explicitly that the scalar field equation results from the metric equations of motion as a Bianchi identity. Now starting from (2.18) and requiring that $\mathscr{E}_{\mu\nu}$ and $\mathscr{E}_{\phi}$ have second at most derivatives automatically means that this will also have to hold for $\nabla^{\mu}\mathscr{E}_{\mu\nu}$. In general if $\mathscr{E}_{\mu\nu}$ is of second order this is not true for $\nabla^{\mu}\mathscr{E}_{\mu\nu}$ but here it is required from (2.27). So Horndeski starts by finding the most general symmetric, second order 2-tensor $A_{\mu\nu}$ whose divergence $\nabla^{\mu}A_{\mu\nu}$ is also of second order. This places constraints on the form of $A_{\mu\nu}$ leaving a solution parametrized with ten free functions. These tensors include of course $\mathscr{E}_{\mu\nu}$ but not all of them verify (2.27). Finally then Horndeski imposes (2.27) on the former family. This leaves him with four free functions at the end giving his final result (2.19).

Now we have at hand the general scalar-tensor and higher dimensional metric gravity framework we will move on to see some of their solutions and how the theories are in fact related in practical terms. We will in particular use known solutions from Lovelock theory in order to construct Horndeski solutions.

2.3 Seeking Exact Solutions in Lovelock Theory

One of the nice characteristics of Lovelock theory is that despite its additional technical difficulties related to the higher order nature of the theory, certain uniqueness black hole theorems of GR remain valid; at least under some weaker hypotheses. In particular, a generalization of Birkhoff's theorem remains true apart from a case of fine tuning of coupling parameters[6] [28]. Let us review the higher dimensional version of this result, for this will lead us to some relatively simple yet interesting solutions where Lovelock theory even circumvents problems of higher dimensional general relativity. The solutions we will consider will also have a nice application to Galileon/Horndeski theories leading us to black hole solutions for four dimensional scalar tensor theories.

The Birkhoff theorem states that, in four dimensions, any spherically symmetric solution to Einstein's equations in the vacuum is necessarily locally static. In other words there exists a local time like Killing vector. This leads to the celebrated Scharszchild metric as the unique GR solution of spherical symmetry in vacuum.

[6]The special relation between the coupling parameters corresponds to the strong coupling limit of EGB-literally the case where the Gauss-Bonnet term is of maximal relative strength to the Einstein-Hilbert term and gives a very special theory with enhanced symmetries, usually referred to as Chern–Simons theory (see the nice review [27]).

The theorem is not modified when one includes a negative or positive cosmological constant but the solution itself is slightly more general. Indeed a negative cosmological constant allows also for exotic horizon topologies of flat or hyperbolic geometry. The general solution of the Einstein field equations with a cosmological constant in $D = 4$ dimensions assuming a constant curvature 2-space (rather than a 2-sphere) reads,

$$ds^2 = -V(r)dt^2 + \frac{dr^2}{V(r)} + r^2\left(\frac{d\chi^2}{1-\kappa\chi^2} + \chi^2 d\phi^2\right). \tag{2.28}$$

where the constant (t, r) sections are two-dimensional constant curvature spaces parametrized by normalized curvature $\kappa = 0, \pm 1$. For linguistic simplicity we will call the surfaces of constant (t, r), horizon sections, preluding the presence of a black hole. The lapse function in (2.28) reads, $V(r) = \kappa - \frac{\Lambda}{3}r^2 - \frac{2M}{r}$. Note then that since the metric is static, zeros of $V = V(r)$ correspond to Killing horizons and exist for $\Lambda < 0$ even when $\kappa = 0$ or $\kappa = -1$. This can be explicitly checked by going to an Eddington-Finkelstein chart,

$$v = t \pm \int \frac{dr}{V(r)} \tag{2.29}$$

These black holes are often called topological due to the fact that special identifications have to be made in order for the horizon to be compact [29]. For $\Lambda \geq 0$ only the spherical topologies give regular solutions with the presence of an extra cosmological horizon. So much for four dimensional GR with a cosmological constant.

In higher, D dimensional, GR Birkhoff's theorem remains valid not only for constant curvature sections, but also for horizon sections which are Einstein spaces [30]. Substituting the constant curvature surface of the horizon sections with a $(D-2)$-dimensional Einstein manifold will not alter locally the black hole lapse function and the general solution is static. The structure of space-time locally[7] transverse to the horizon sections is in this way not affected by the details of the internal geometry, as long as the latter continues to be an Einstein space. In particular the horizon structure is the same. To picture this let us take a particular example: consider the, for example, six dimensional solution,

$$ds^2 = -V(r)dt^2 + \frac{dr^2}{V(r)} + r^2\left(f(\rho)d\tau^2 + \frac{d\rho^2}{f(\rho)} + \rho^2 d\Omega_{II}^2\right)$$

where $f(\rho) = 1 - \frac{\mu}{\rho}$. Hence the horizon sections in four dimensions are given by a Euclidean Schwarzschild black hole obtained by Wick rotating the time coordinate of the original four dimensional black hole. The metric (2.30) is a valid

[7]We will see that when the horizon sections carry non-zero curvature there is a global change in the topology of the solution related to the presence of a solid angle deficit. This will end up having important consequences that we will discuss in detail later with the solution at hand.

six dimensional solution since the horizon sections are Ricci flat. On the hand we can consider a second solution of the form,

$$ds^2 = -V(r)dt^2 + \frac{dr^2}{V(r)} + r^2 dT_{IV}^2$$

with now toroidal horizon sections, i.e. a locally flat four dimensional metric. In both cases we have the same lapse function $V(r) = k^2 r^2 - \frac{m}{r^3}$ independently whether our horizon is of flat or Euclidean Schwarzschild geometry (which is of course Ricci flat but has non zero Weyl curvature)-Ricci flatness of both the horizon sections means that $\kappa = 0$ for the lapse function $V(r)$. The former exotic black holes often have classical instabilities [31] in a similar fashion to those of the black string [32]. In fact black string metrics can be Wick rotated to a subclass of metrics with exotic horizons. The exotic horizon section in this case is nothing but the Euclidean version of four dimensional Schwarzschild (as in the example above). We see therefore that in higher dimensional GR a certain kind of degeneracy appears in the possible solutions which are not completely fixed by the symmetries and the field equations. Therefore one could entertain the possibility that the additional unphysical exotic black holes are just an artifact of not considering the full classical gravity theory in higher dimensions. Indeed we will provide clear indications that this is the case at least for six dimensional Lovelock theory in the sense that the possible horizon geometries will be seen to be far more constrained [33] and asymptotically non-trivial.

So how are these results translated in Lovelock theory? In order to answer this question [34], we start by considering an appropriate anzatz for the metric and stick to $D = 6$ dimensions and EGB theory. We have a transverse 2-space, which carries the timelike coordinate t and the radial coordinate r, and an internal 4-space, which is going to represent the horizon sections of the possible six-dimensional black holes. The metric of the internal four dimensional space we note $h_{\mu\nu}$ and we take to be an arbitrary metric of the internal coordinates $x^\mu, \mu = 0, 1, 2, 3$ only. We furthermore impose that the internal and transverse spaces are orthogonal to each other. This is immediately true for GR as a result of the theorem of Frobenius but not true for Lovelock theory. It is an additional assumption we have make in order to make the problem tractable [34]. The quite general metric anzatz for which we want to solve the EGB field equations boils down to,

$$ds^2 = e^{2\nu(t,z)} B\,(t,z)^{-3/4} \left(-dt^2 + dz^2\right) + B\,(t,z)^{1/2}\, h_{\mu\nu}^{(4)}\,(x)\, dx^\mu dx^\nu \,. \tag{2.30}$$

Using light-cone coordinates,

$$u = \frac{t-z}{\sqrt{2}}, \quad v = \frac{t+z}{\sqrt{2}}. \tag{2.31}$$

the metric reads

$$ds^2 = -2e^{2\nu(u,v)} B\,(u,v)^{-3/4}\, dudv + B\,(u,v)^{1/2}\, h_{\mu\nu}^{(4)}\,(x)\, dx^\mu dx^\nu \,. \tag{2.32}$$

We want to solve Lovelock's equations (2.4) for metric (2.32). The key to doing so boils down to the following two equations: the (uu) and (vv) equations that literally play the role of integrability conditions for the full system of equations of motion [35],

$$\mathscr{E}_{uu} = \frac{2\nu_{,u}B_{,u} - B_{,uu}}{B}\left[1+\alpha\left(B^{-1/2}R^{(4)} + \frac{3}{2}e^{-2\nu}B^{-5/4}B_{,u}B_{,v}\right)\right], \tag{2.33}$$

$$\mathscr{E}_{vv} = \frac{2\nu_{,v}B_{,v} - B_{,vv}}{B}\left[1+\alpha\left(B^{-1/2}R^{(4)} + \frac{3}{2}e^{-2\nu}B^{-5/4}B_{,u}B_{,v}\right)\right]. \tag{2.34}$$

The above permit to classify and eventually completely solve the full system of field equations [34]. We have three classes of solutions depending on wether the second, the first factor is zero, or again a third class for constant B. Here we concentrate on the class of most interest, class II, corresponding to,

$$\frac{2\nu_{,v}B_{,v} - B_{,vv}}{B} = 0, \qquad \frac{2\nu_{,u}B_{,u} - B_{,uu}}{B} = 0 \tag{2.35}$$

The other classes are degenerate and occur for special relations of couplings only. Most importantly class II solutions are directly connected to GR since (2.35) is independent of the coupling constant α. Solving (2.35) immediately shows that we have a locally static space-time [36] and thus a somehow weaker version of Birkhoff's theorem still holds.

Solving the remaining field equations leads eventually to the metric solution [33, 34],

$$ds^2 = -V(r)\,dt^2 + \frac{dr^2}{V(r)} + r^2 h^{(4)}_{\mu\nu}(x)\,dx^\mu dx^\nu, \tag{2.36}$$

with lapse function,

$$V(r) = \frac{R^{(4)}}{12} + \frac{r^2}{12\alpha}\left[1 \pm \sqrt{1 + \frac{12\alpha\Lambda}{5} + \frac{\alpha^2\left(R^{(4)2} - 6\hat{G}^{(4)}\right)}{r^4} + 24\frac{\alpha M}{r^5}}\right], \tag{2.37}$$

Note first that there are two branches of solutions. This is true generically in EGB theory and results from the higher order nature of the theory [12]. In EGB there are generically two vacua for a given theory.[8] The upper $'+'$ branch does not have a well-defined GR limit ($\alpha \to 0$) and turns out to be unstable (for a full recent

[8]In higher order Lovelock theory there are more according to the order of the highest order Lovelock term [12].

discussion on stability of EGB vacua see [37]). The lower branch is ghost free [38] and is the branch that we will consider from now on omitting the '+' branch. The horizon sections, parametrized by the four dimensional metric $h^{(4)}_{\mu\nu}$ are constrained by the field equations to be Einstein spaces,

$$R^{(4)}_{\mu\nu} = \frac{R^{(4)}}{D-2} h^{(4)}_{\mu\nu} \tag{2.38}$$

But as we can see from the lapse function which has to be a function of the radial variable r, a new geometric condition now appears whereupon the horizon quantity ${R^{(4)}}^2 - 6\hat{G}^{(4)}$ has to be constant. Combining the two conditions gives,

$$C^{\alpha\beta\gamma\mu} C_{\alpha\beta\gamma\nu} = \Theta \delta^{\mu}_{\nu} \tag{2.39}$$

where Θ is a given constant and $C_{\alpha\beta\gamma\nu}$ is the four dimensional Weyl tensor associated to $h^{(4)}_{\mu\nu}$. This is a supplementary condition for EGB theories, the Dotti-Gleiser condition, (2.39) imposed in addition to the usual Einstein space condition (2.38) for higher dimensional general relativity. Clearly then, for EGB theory, the lapse function for the black hole carries a supplementary information particular to the type of horizon section for the black hole solution. For example, the Euclidean Schwarszchild metric is not a legitimate internal metric anymore since it does not verify (2.39). Both of the conditions (2.38) and (2.39) present a geometric similarity in that we ask for (part of) *the curvature tensor to be analogous to the spacetime metric*. The main difference being that the curvature tensor in (2.39) is the Weyl tensor and, given its symmetries, it is actually its square which has to be analogous to the spacetime metric. Clearly horizons with $\Theta \neq 0$ will not be homogeneous spaces and not even asymptotically so. Another interesting point is that the Gauss-Bonnet scalar, whose spacetime integral is the Euler characteristic of the horizon, has to be constant for these solutions to be valid. The Gauss-Bonnet scalar of the internal space then reads $\hat{G}^{(4)} = 4\Theta + 24\kappa^2$ and the potential [33, 34],

$$V(r) = \kappa + \frac{r^2}{12\alpha}\left(1 - \sqrt{1 + \frac{12}{5}\alpha\Lambda - 24\frac{\alpha^2\Theta}{r^4} + 24\frac{\alpha M}{r^5}}\right) \tag{2.40}$$

since $R^{(4)} = 12\kappa$. For $\Theta = 0$, we obtain the black holes first discussed by Boulware and Deser (see [39]). In this case since the Weyl curvature is zero the horizon sections are geometries of constant curvature. Taking $\Lambda = 0$ we note that these black holes are asymptotically flat and are an extension of the higher dimensional version of the Schwarzschild solution. In fact taking the limit of α small and large r one obtains precisely the latter GR solution. Recently these black holes have been reported to have a spin 2 instability for small enough mass parameter [40]. This result has been extended to Lovelock black holes [41]. It is not yet understood what is the physical nature of this "short distance scale instability" and if it is somehow related to thermodynamic instability and quantum Hawking radiation.

Putting this aside we now want to examine cases of Einstein metrics whose squared Weyl curvature is not zero but constant. This is a special case of Einstein metric and the Dotti-Gleiser condition is much like a supplementary requirement. What is already clear is that any such solution will not be asymptotically "usual" as the fall off the relevant term is 5 rather than six dimensional. Indeed notice that the Θ-term in (2.40) has a fall off rate of a five-dimensional Boulware-Deser black hole [39] and is therefore dominant over the "usual" mass term contribution, [42]. We now investigate a simple example which will have interesting four dimensional consequences.

Consider a four-dimensional space which is a product of two 2-spheres,

$$ds^2 = \rho_1^2 \left(d\theta_1^2 + \sin^2\theta_1 d\phi_1^2\right) + \rho_2^2 \left(d\theta_2^2 + \sin^2\theta_2 d\phi_2^2\right) , \tag{2.41}$$

where the (dimensionless) radii ρ_1 and ρ_2 of the spheres are constant. The entire six-dimensional metric reads,

$$ds_{(4)}^2 = -V(r)\,dt^2 + \frac{dr^2}{V(r)} + r^2\rho_1^2 \left(d\theta_1^2 + \sin^2\theta_1 d\phi_1^2\right) + r^2\rho_2^2 \left(d\theta_2^2 + \sin^2\theta_2 d\phi_2^2\right) , \tag{2.42}$$

with lapse function

$$V(r) = \frac{R^{(4)}}{12} + \frac{r^2}{12\alpha}\left(1 - \sqrt{1 - 24k^2\alpha - 24\Theta\frac{\alpha^2}{r^4} + 24\alpha\frac{M}{r^5}}\right) . \tag{2.43}$$

In order for (2.42) to be a solution to the equations of motion the spheres have to be of equal radius, $\rho_1 = \rho_2$. This ensures that (2.42) is an Einstein space. The second condition is then immediately verified for a product of 2-spheres. We have $\kappa = \frac{1}{3\rho_1^2} > 0$ and $\Theta = \frac{4}{3\rho_1^4}$. Note that even when the 2-sphere curvature is normalized to $\rho_1 = 1$ then $\kappa \neq 1$. A linear redefinition of the r coordinate then shows that the area of the four dimensional space is reduced compared to the homogeneous 4-sphere. In other words space-time is asymptotically altered by an overall solid angle deficit. This results in a genuine curvature singularity at $r = 0$. Of course when we have $M \neq 0$ there is central curvature singularity at $r = 0$ anyway. But, for (2.42) the $r = 0$ singularity is present *even for zero mass whenever* $\Theta \neq 0$! This is not an artefact of EGB theory. In fact it is easy to see, taking the combined limit of $\alpha \to 0$ and large r, that the resulting GR black hole with $V(r) = \frac{R^{(4)}}{12} + r^2k^2 - \frac{M}{r^3}$ has exactly the same problem at the origin independently of the value of M. For $M = 0$ the GR solution has a naked singularity at the origin. Note again that the lapse function for higher dimensional GR is the same with the higher dimensional Schwarzschild black hole modulo the horizon curvature term. The zero mass solution is singular at the origin wether we are in GR or Lovelock theory. But for Lovelock theory an interesting effect occurs due to the presence of the Θ term in the lapse function.

To see this consider for the moment $M = 0$ in the lapse function (2.42). Then the Θ term in (2.42) is identical to the mass term in the Boulware Deser black hole in five dimensions. Therefore, as we know from the Boulware-Deser solution [39] this extra Θ term generically generates an event horizon cloaking the central $r = 0$ singularity as long as $\alpha \neq 0$! In fact the length scale of this event horizon is given by the coupling constant $\alpha \sim length^2$ which we know from string theory effective actions [43] is related to the fundamental string tension α'. One then can interpret this horizon as a higher order 'quantum' cloak of an otherwise naked singularity present in GR. Details for the horizon structure can be found in [34]. We will come back to this solution in order to construct a Galileon black hole.

Let us, before moving on, make some final remarks regarding these solutions. First we should note that most probably these multiple sphere solutions can be unstable to linear perturbations. It has been shown in GR [31] that there is a "balloon instability" whereupon one of the spheres wants to deflate with respect to the other. This geometric effect may remain true in the above EGB version [33] although the perturbation equations for EGB in lesser symmetry change completely compared to GR. It is also probable that this instability may be stabilized by the inclusion of a magnetic field in the relevant solution [44]. Secondly we should note that the above construction involving multiples of equal radius spheres, can be undertaken in arbitrary even dimensional spacetime as long as we truncate Lovelock theory to EGB. If one considers higher order Lovelock terms it is not known under what geometric conditions the horizon sections will be admissible. One may expect a higher order condition of the type (2.39) in third and higher curvature order... We expect the horizon sections to be more and more constrained as higher order Lovelock terms come into play. At the same time since horizon sections will be of higher dimension this will allow for a richer geometry. This is also an open question. Finally, putting it all together we have arrived to the following result concerning EGB theory: *given the anzatz (2.30), the only asymptotically flat solution of six dimensional EGB theory with zero cosmological constant, is the Boulware Deser solution [39].* This is because whenever $\Theta \neq 0$ the solution is not asymptotically flat for six dimensional space-time. Therefore we can deduce that EGB theory is very similar in this aspect to four dimensional GR lifting the degeneracy present in higher dimensional GR due to the additional elegant geometric condition (2.39).

2.4 From Lovelock to Horndeski Theory: Kaluza-Klein Reduction

In order to apply higher dimensional Lovelock theory to cosmology or gravity in four dimensional space-time one needs some means of approach to four dimensional gravity. There are at least two routes, braneworlds and Kaluza Klein reduction. In the recent past Lovelock theory had an important implication in the braneworld paradigm [45]. Braneworlds consist of higher dimensional spacetimes endowed with a distributional brane where standard matter is localized. The idea "inspired" in rather loose terms from string theory, is that gravity perceives all the space-time dimensions while matter is localized on a four dimensional braneworld. Since the set-up involves junction or matching conditions an essential feature is the number of extra dimensions, namely codimension, yielding for example a wall or string type of defect. There is a long literature of articles on the subject treating codimension one [12] and codimension two braneworlds (see [46] and references within) involving respectively five and six dimensional EGB theory. In particular Lovelock theory permits, due to the generalized junction conditions [47], well defined codimension two braneworld cosmology [48]. This leads to important consequences since in GR one cannot consider distributional sources for cosmological symmetry and codimension 2. Again, the richer structure of Lovelock theory permits solutions with distributional sources not available in higher dimensional GR. For more details on these aspects see [12] and [46] and references within. Here, we will focus on the more classical Kaluza-Klein compactification since it will give us a direct connection to higher order scalar tensor terms, found in Galileon/Horndeski theory. It will also provide a way to obtain exact black hole solutions [49].

It has been known since a long time [50] that a consistent Kaluza Klein reduction of Lovelock theory gives a scalar-tensor theory with higher order derivatives, but crucially, with second order equations of motion. In this sense many of the Galileon terms discussed later on were known from previous work on Kaluza Klein compactifications and braneworlds [51]. This is the direction we will take here. The most generic of Kaluza-Klein reduction to four dimensions has recently been given in the nice paper of [52]. There it has been shown that only up to the third order Lovelock terms contribute to the Kaluza-Klein compactification in four dimensions. Here we will concentrate on EGB theory i.e. up to second order Lovelock theory. We will consider the simplest consistent toroidal compactification giving rise to one extra scalar degree of freedom.

Start by taking D-dimensional Einstein Gauss-Bonnet theory which is the five or six-dimensional Lovelock theory truncated to arbitrary dimension. The arbitrary dimension D will be important when we end up promoting dimension from a positive integer to a real parameter once we have undertaken a consistent Kaluza-Klein reduction. We have the EGB action with a cosmological constant,

$$S = \frac{1}{16\pi G_N} \int d^D x \sqrt{-g}\left[-2\Lambda + R + \hat{G}\right] \tag{2.44}$$

Consider now the simplest but consistent diagonal reduction along some arbitrary n-dimensional internal curved space $\tilde{\mathbf{K}}$. We aim to reduce this theory down to four space-time dimensions with $D = 4 + n$:

$$ds^2_{(4+n)} = d\bar{s}^2_{(4)} + e^{\phi} d\tilde{K}^2_{(n)} \,. \tag{2.45}$$

This particular frame is chosen in such a way as so there is no conformal factor of ϕ in front of the four-dimensional metric. As such the asymptotic character (i.e. radial fall off) of a Lovelock D dimensional solution will be similar to the four dimensional one. All terms with a tilde refer to the curved n-dimensional internal space, while terms with a bar refer to the (4)-dimensional space-time. One can show for the given metric Anzatz that the KK reduction for n arbitrary is consistent, i.e. that the reduced equations of motion are derived from the reduced action [53]. This reduction is therefore generalised in the manner defined in [54, 55]. The important result of this is that the integer n corresponding to the compact Kaluza-Klein space can be analytically continued to a real parameter of the reduced action. Naturally n corresponds to a dimension only for n integer. The solutions from the four dimensional point of view are still solutions of the resulting effective action for arbitrary n. The four dimensional effective action reads after integrating out the internal space,

$$\begin{aligned}\bar{S}_{(4)} = \int d^4x\,\sqrt{-\bar{g}}\,e^{\frac{n}{2}\phi}\Big\{&\bar{R} - 2\Lambda + \alpha\bar{G} + \frac{n}{4}(n-1)\partial\phi^2 - \alpha n(n-1)\bar{G}^{\mu\nu}\partial_\mu\phi\partial_\nu\phi \\ &-\frac{\alpha}{4}n(n-1)(n-2)\partial\phi^2\nabla^2\phi + \frac{\alpha}{16}n(n-1)^2(n-2)\left(\partial\phi^2\right)^2 \\ &+ e^{-\phi}\tilde{R}\left[1 + \alpha\bar{R} + \alpha 4(n-2)(n-3)\partial\phi^2\right] + \alpha\tilde{G}e^{-2\phi}\Big\}\,,\end{aligned} \tag{2.46}$$

For $\alpha = 0$ this effective action is just the usual toroidal KK effective action. The higher order Gauss-Bonnet term gives rise to several higher order scalar-tensor Galileon (or equivalently Horndeski) terms, [14, 15, 56, 57] with very particular potentials. The Galileon field ϕ can then simply be understood to be the scalar field parametrising the volume of the internal space.

Indeed, apart from the usual lower order terms appearing in standard Kaluza Klein compactification of Einstein dilaton theories, we see the emergence of several higher order terms. For a start we have the four dimensional Gauss Bonnet term $\bar{G}$ which will contribute to the scalar field variation although it is a topological term for four dimensional GR. Secondly we have $\bar{G}^{\mu\nu}\partial_\mu\phi\partial_\nu\phi$ involving the coupling of the Einstein tensor with the kinetic term. Here, rather than metric-scalar interaction, as for the standard kinetic term of ϕ we have a curvature-scalar interaction which we will see has very interesting consequences in the forthcoming section. This term has equations of motion of second order essentially due to the divergence free property of the Einstein tensor $G_{\mu\nu}$. For example if one considers $\bar{R}^{\mu\nu}\partial_\mu\phi\partial_\nu\phi$ this is not true. It has also shift symmetry in the scalar field typical of certain Galileon terms. Furthermore we have, what is often called the DGP term, $\partial\phi^2\nabla^2\phi$ appearing in the decoupling limit of the DGP model [25] and then the standard Galileon term $\left(\partial\phi^2\right)^2$ which are also shift symmetric in ϕ. The last line in the effective action takes part only for a curved internal space in the face of Ricci and Gauss Bonnet curvature. Reducing from the EGB action yields terms up to quartic order in derivatives (either of the metric or the scalar, or a mixed combination of the two). Reducing higher order Lovelock densities yields terms with a higher number of derivatives. A typical example is the higher order permissible curvature-scalar interaction,

$$P^{\mu\nu\alpha\beta}\nabla_\mu\phi\nabla_\alpha\phi\nabla_\nu\nabla_\beta\phi \tag{2.47}$$

which involves six derivatives and one can show [9] originates from the Kaluza-Klein reduction of the third order Lovelock density [52]. Again the reader will recollect the divergence freedom of the double dual tensor giving second order field equations. Taking for example $R^{\mu\nu\alpha\beta}\nabla_\mu\phi\nabla_\alpha\phi\nabla_\nu\nabla_\beta\phi$ would fail this Galileon property.

Although this effective action is very complex, and its field equations even more so, it is "simple" to generate solutions for the above (2.46) in four dimensions [53]. One starts from a convenient Lovelock solution in D dimensions. Since we want the four dimensional solution to have, at least locally, spherical horizon sections we have to consider a solution where the $(D-2)$ dimensional horizon sections are $(D-2)/2$-products of two spheres. This is precisely the extension of the six dimensional solution we discussed in the previous section (2.42) generalized to arbitrary dimensions [53]. The solution reads,

$$d\bar{s}^2_{(4)} = -V(R)dt^2 + \frac{dR^2}{V(R)} + \frac{R^2}{n+1}d\bar{K}^2_{(2)}, \tag{2.48}$$

$$V(R) = \kappa + \frac{R^2}{\tilde{\alpha}_r}\left[1 \mp \sqrt{1 - \frac{2\tilde{\alpha}_r}{l^2} - \frac{2\tilde{\alpha}_r^2\kappa^2}{(n-1)R^4} + \frac{4\tilde{\alpha}_r m}{R^{3+n}}}\right], \tag{2.49}$$

$$\tilde{\alpha}_r = 2\alpha n(n+1), \qquad \frac{1}{\ell^2} = \frac{-2\Lambda}{(n+2)(n+3)} \tag{2.50}$$

$$e^\phi = \frac{R^2}{n+1}, \tag{2.51}$$

Here, n is the dimension of the internal space minus one 2-sphere in other words, $n = D-4$. This is the higher dimensional interpretation of n but once the solution is written out we simply take n an arbitrary real number and (2.48) is still an exact solution and n parametrizes the theory. In our notation here $\kappa = 0, 1, -1$ is the normalised horizon curvature and we have redefined for this section the constants $\tilde{\alpha}_r$ and ℓ. Taking carefully the $\tilde{\alpha}_r \to 0$ limit, gives a standard Einstein dilaton solution with a Liouville potential [49]. Set $\Lambda = 0$, $\kappa = 1$ and let us start by making some qualitative remarks describing properties of the solution without entering into technical details. Note that, taking carefully the $n = 0$ limit switches off the scalar field and the higher-derivative corrections, and we obtain pure GR in (2.46) and a Schwarzschild black hole (2.48). This is particularly interesting since the scalar-tensor solution given above for arbitrary n is a continuous deformation of the Schwarzschild solution. When n is in the neighborhood of zero we are closest to the GR black hole. As we hinted in the previous section the topology of the solution is not that of GR. Indeed the warp factor of the 2-sphere, in (2.48), is recovered only at $n = 0$, i.e. the GR limit. Otherwise the area of the reduced spherical horizon is given by $\frac{4\pi R^2}{n+1}$ rather than the 2-sphere area, $4\pi R^2$. This is again a solid deficit angle (and not a conical deficit angle) the same one we encountered for the Lovelock solution in the previous section. As stressed in the previous sections this will give, at $R = 0$, a true curvature singularity even if $m = 0$. For large R, we have a spacetime metric very similar to that of a gravitational monopole [58]. Expanding 2.49 for small $\tilde{\alpha}_r$ and large R gives,

$$V(R) = 1 + \frac{\tilde{\alpha}_r}{(n-1)R^2} - \frac{2m}{R^{n+1}} + \ldots \tag{2.52}$$

This solution is reminiscent of a RN black hole solution where the role of the electric charge is undertaken by the leading horizon curvature correction in $\tilde{\alpha}_r$. This is the particular Θ term we discussed in the previous section. This term dominates the mass term close to the horizon and for $n < 1$. Note that it can be of negative sign depending on the value of n and $\tilde{\alpha}_r$. The further we are from $n = 0$, the GR limit, the further we deviate from a standard four-dimensional radial fall-off. The first important question we want to deal with is the central curvature singularity at $R = 0$, which is due to the solid deficit angle and is present even if $m = 0$. Also note that whenever the square root in the lapse function (2.49) is zero we also have a branch singularity which is also a dangerous curvature singularity. Setting $m = 0$, we find that for $-1 < n < 1$ and $\tilde{\alpha}_r > 0$ the singularity at $R = 0$ is covered by an event horizon created by the higher-order curvature correction. In its absence ($\tilde{\alpha}_r = 0$), this solution would have been singular…The UV (small R) behaviour of the solution is therefore regularised by the presence of the higher-order terms. If $n > 1$ or $n < -1$, then $\tilde{\alpha}_r < 0$ is needed in order to preserve the event horizon. The remaining cases are singular.

Now let us switch on the mass, $m \neq 0$. Whenever $\tilde{\alpha}_r > 0$, we have a single event horizon. When $-1 < n < 1$, there is no branch singularity however small m is. On the contrary, when $n > 1$, the mass is bounded from below in order to avoid a branch singularity:

$$m > \left(\frac{2}{n+3}\right)^{\frac{n+3}{4}} \frac{\tilde{\alpha}_r^{\frac{n+1}{2}}}{n-1}. \tag{2.53}$$

When $n < -1$, the solution is also a black hole but the mass term is not falling off at infinity. The region of most immediate interest is whenever n is small but not zero.

The black hole properties are rather different for $\tilde{\alpha}_r < 0$. When $-1 < n < 1$, there is an inner and an outer event horizon as long as the following condition is fulfilled:

$$\left(\frac{1}{2}\right)^{\frac{n+1}{2}} < \frac{m(1-n)}{|\tilde{\alpha}_r|^{\frac{n+1}{2}}} < \left(\frac{2}{n+3}\right)^{\frac{n+3}{4}}. \tag{2.54}$$

When $n > 1$, a single event horizon exists, covering a single branch singularity with $R_s < R_h$.

Overall we can say that the KK solution given here has an interesting horizon structure and presents again a quantum cloaking of an otherwise Einstein-Dilaton singular solution. It is however not of ordinary asymptotics bifurcating in this way the no hair paradigm for Galileons [59]. In the last section we will see a way to construct asymptotically ordinary solutions with fake black hole hair.

2.5 Self-tuning and the Fab 4

As we saw in the previous sections Horndeski or Galileon theory encompasses all the possible (single) scalar tensor terms one can consider in order for the equations of motion to be of second order. This is an essential requirement for a well-defined classical modification of gravity [5]. In this section we will question which of the terms in scalar-tensor theory have "self-tuning" properties. Self-tuning is a rather simple and quite old idea with application to the cosmological constant problem. The basic principle consists of finding solutions for flat (or possibly maximally symmetric vacua) of some gravity action endowed with a bulk cosmological constant, independently of the value of the cosmological constant in the action. In order for self-tuning (and not fine tuning) to be effective the cosmological constant should not be fixed with respect to any of the coupling constants in the gravitational action. The idea then is that the cosmological constant is absorbed by a dynamical solution involving the non-trivial scalar field without affecting the gravitational background. This can only be a "partial" solution to the cosmological constant problem since radiative corrections will destabilize this vacuum solution beyond a certain energy scale, the cutoff of the effective gravity theory. It is however an interesting first step especially since no theories were known, before [9], to have such a property without some hidden effective fine tuning of the action coupling constants as for example in codimension one braneworld models (see for example [60]). We should note that recently there has been considerable progress on protecting the cosmological constant from standard model radiative corrections [61] and we refer the interested reader to this article for the model in question which interestingly is a rather minimal extension of GR. Rather, for our purposes, having at hand the general scalar tensor theory we will formulate the following question: is there a subset of Horndeski theory with self-tuning properties? The answer is affirmative as shown in [9], yielding a rather simple and neat geometrical theory which was dubbed by the authors as Fab 4 theory. We start by presenting the theory and then give a specific self tuning solution which elegantly and non-technically gives the general idea. We close the section by showing a simple method to obtain regular black hole solutions in fab 4 and Horndeski theory, independently of self-tuning.

The Fab 4 potentials make up the most general scalar-tensor theory capable of self-tuning. They are given by the following geometric terms,

$$\mathscr{L}_{john} = \sqrt{-g}V_{john}(\phi)G^{\mu\nu}\nabla_\mu\phi\nabla_\nu\phi \tag{2.55}$$

$$\mathscr{L}_{paul} = \sqrt{-g}V_{paul}(\phi)P^{\mu\nu\alpha\beta}\nabla_\mu\phi\nabla_\alpha\phi\nabla_\nu\nabla_\beta\phi \tag{2.56}$$

$$\mathscr{L}_{george} = \sqrt{-g}V_{george}(\phi)R \tag{2.57}$$

$$\mathscr{L}_{ringo} = \sqrt{-g}V_{ringo}(\phi)\hat{G} \tag{2.58}$$

where R is the Ricci scalar, $G_{\mu\nu}$ is the Einstein tensor, $P_{\mu\nu\alpha\beta}$ is the double dual of the Riemann tensor (2.7), $\hat{G} = R^{\mu\nu\alpha\beta}R_{\mu\nu\alpha\beta} - 4R^{\mu\nu}R_{\mu\nu} + R^2$ is the Gauss-Bonnet combination. As we saw in the previous section all of these terms with particular potentials appear in Kaluza-Klein reduction of higher order Lovelock terms. Self tuning solutions exist for any of these potentials as long as either $\{V_{john}\} \neq 0$ or , $\{V_{paul}\} \neq 0$ or $\{V_{george}\}$ are not constant. Note that this constraint means that GR in accordance to Weinberg's no-go theorem [7] does not have self-tuning solutions. Also V_{ringo} cannot self-tune but does not spoil self-tuning, i.e. it cannot self-tune without (a little) help from his friends-hence the unfortunate name. Also note that taking $\{V_{george}\} = constant$ as for GR with $\{V_{john}\} \neq 0$ suffices for example to have a self-tuning theory. In fact pure GR does not exclude self-tuning of the theory as long as another non-trivial fab 4 term is present. This is also very interesting from a phenomenological point of view. We will see in what follows how all these facts come about.

The fab 4 terms are related to particular functionals of the Horndeski potentials,

$$\kappa_1 = 2V'_{ringo}(\phi)\left[1 + \frac{1}{2}\ln(|\rho|)\right] - \frac{3}{8}V_{paul}(\phi)\rho \tag{2.59}$$

$$\kappa_3 = V''_{ringo}(\phi)\ln(|\rho|) - \frac{1}{8}V'_{paul}(\phi)\rho - \frac{1}{4}V_{john}(\phi)\left[1 - \ln(|\rho|)\right] \tag{2.60}$$

$$\kappa_8 = \frac{1}{2}V'_{john}(\phi)\ln(|\rho|), \tag{2.61}$$

$$\kappa_9 = -\rho_\Lambda^{bare} - 3V''_{george}(\phi)\rho \tag{2.62}$$

$$F + 2W = \frac{1}{2}V_{george}(\phi) - \frac{1}{4}V_{john}(\phi)\rho\ln(|\rho|) \tag{2.63}$$

Notice that the self-tuning constraints fix completely the dependence on the kinetic term ρ. Notice also that the Fab 4 terms are scalar interactions with space-time curvature. No pure potential or kinetic terms are allowed for self-tuning. Again, we will see why their form has to be so special.

Weinberg's no-go theorem tells us that our vacuum solution must not be Poincaré invariant [7]. Hence if we consider cosmological symmetry with a time dependent background, the scalar field has to depend non-trivially on the time coordinate breaking Poincaré invariance for flat space-time. The self-tuning filter defining the self-tuning property and thus the form of Fab 4 terms is as follows:

- Fab 4 terms admit locally a Minkowski vacuum for *any* value of the net bulk cosmological constant
- this remains true before and after any phase transition in time where the cosmological constant jumps instantaneously by a finite amount. The scalar field

will have to be able to change accordingly in order to accommodate the novel value without affecting the flat space-time background.

- Fab 4 terms permit non-trivial cosmologies i.e. does not self-tune for any other matter backgrounds other than vacuum energy.

The last condition ensures that Minkowski space is not the only cosmological solution available, something that is certainly required by observation. The idea is that the cosmological field equations should be dynamical, with the Minkowski solution corresponding to some sort of fixed point. In other words, once we are on a Minkowski solution, we stay there—otherwise we evolve to it dynamically [62]. This last statement would indicate that the self-tuning vacuum is an attractive fixed point. Mathematically self-tuning under these conditions, and especially the second, translates to a junction condition problem where the metric is regular and C^2 whereas the second derivative of the scalar field contains Dirac distribution terms.

The full equations of motion are given by,

$$\mathcal{E}^{\mu\nu}_{john} + \mathcal{E}^{\mu\nu}_{paul} + \mathcal{E}^{\mu\nu}_{george} + \mathcal{E}^{\mu\nu}_{ringo} = \tfrac{1}{2}T^{\mu\nu} \tag{2.64}$$

$$\mathcal{E}^{\phi}_{john} + \mathcal{E}^{\phi}_{paul} + \mathcal{E}^{\phi}_{george} + \mathcal{E}^{\phi}_{ringo} = 0 \tag{2.65}$$

We have included the cosmological constant in the energy momentum tensor $T^{\mu\nu}$. The contribution of each term from variation of the metric is given by

$$\begin{aligned}\mathcal{E}^{\eta\epsilon}_{john} &= \frac{1}{2}V_{john}(\rho G^{\eta\epsilon} - 2P^{\eta\mu\epsilon\nu}\nabla_\mu\phi\nabla_\nu\phi) + \\ &\quad + \tfrac{1}{2}g^{\epsilon\theta}\delta^{\eta\alpha\beta}_{\theta\mu\nu}\nabla^\mu(\sqrt{V_{john}}\nabla_\alpha\phi)\nabla^\nu(\sqrt{V_{john}}\nabla_\beta\phi)\end{aligned} \tag{2.66}$$

$$\begin{aligned}\mathcal{E}^{\eta\epsilon}_{paul} &= \frac{3}{2}P^{\eta\mu\epsilon\nu}\rho V^{2/3}_{paul}\nabla_\mu\left(V^{1/3}_{paul}\nabla_\nu\phi\right) \\ &\quad + \tfrac{1}{2}g^{\epsilon\theta}\delta^{\eta\alpha\beta\gamma}_{\theta\mu\nu\sigma}\nabla^\mu\left(V^{1/3}_{paul}\nabla_\alpha\phi\right)\nabla^\nu\left(V^{1/3}_{paul}\nabla_\beta\phi\right)\nabla^\sigma\left(V^{1/3}_{paul}\nabla_\gamma\phi\right)\end{aligned} \tag{2.67}$$

$$\mathcal{E}^{\eta\epsilon}_{george} = V_{george}G^{\eta\epsilon} - (\nabla^\eta\nabla^\epsilon - g^{\eta\epsilon}\nabla^\rho\nabla_\rho)V_{george} \tag{2.68}$$

$$\mathcal{E}^{\eta\epsilon}_{ringo} = -4P^{\eta\mu\epsilon\nu}\nabla_\mu\nabla_\nu V_{ringo} \tag{2.69}$$

and from variation of the scalar by

$$\mathcal{E}^{\phi}_{john} = 2\sqrt{V_{john}}\nabla_\mu(\sqrt{V_{john}}\nabla_\nu\phi)G^{\mu\nu} \tag{2.70}$$

$$\mathcal{E}^{\phi}_{paul} = 3V^{1/3}_{paul}\nabla_\mu\left(V^{1/3}_{paul}\nabla_\alpha\phi\right)\nabla_\nu\left(V^{1/3}_{paul}\nabla_\beta\phi\right)P^{\mu\nu\alpha\beta} - \frac{3}{8}V_{paul}\rho\hat{G} \tag{2.71}$$

$$\mathcal{E}^{\phi}_{george} = -V'_{george}R \tag{2.72}$$

$$\mathcal{E}^{\phi}_{ringo} = -V'_{ringo}\hat{G} \tag{2.73}$$

Notice that the scalar equation of motion vanishes identically for flat space-time. This necessary condition can be traced back to the distributional origin of the scalar field and strongly characterizes these terms. Indeed note that a canonical kinetic term for the scalar is disqualified from self-tuning because there is no matter source to account for the distributional part of the scalar field. This is why fab 4 terms represent curvature-scalar interactions: so that their scalar field equations are redundant for the self-tuning background in question.

Instead of going through the detailed derivation of the self-tuning terms in Horndeski theory we will rather look at a simple cosmological example in order to see how self-tuning works in practice. For the details we refer the interested reader to the original papers, [9].

In order to evade Weinberg's no-go argument concerning the cosmological constant we have to break Poincaré invariance for the scalar field. As such we consider a time-dependent scalar field and the FRW family of cosmological metrics of the form,

$$ds^2 = -dT^2 + a^2(t)\gamma_{ij}dx^i dx^j \tag{2.74}$$

where γ_{ij} is the metric on the unit plane ($\kappa = 0$), sphere ($\kappa = 1$) or hyperboloid ($\kappa = -1$). The Friedmann equation reads $\mathscr{H} = -\rho_\Lambda^{bare}$ as we are supposing only vacuum energy to be present,

$$\mathscr{H} = \mathscr{H}_{john} + \mathscr{H}_{paul} + \mathscr{H}_{george} + \mathscr{H}_{ringo} + \rho_\Lambda^{bare} \tag{2.75}$$

and

$$\begin{aligned}
\mathscr{H}_{john} &= 3V_{john}(\phi)\dot{\phi}^2\left(3H^2 + \frac{\kappa}{a^2}\right)\\
\mathscr{H}_{paul} &= -3V_{paul}(\phi)\dot{\phi}^3 H\left(5H^2 + 3\frac{\kappa}{a^2}\right)\\
\mathscr{H}_{george} &= -6V_{george}(\phi)\left[\left(H^2 + \frac{\kappa}{a^2}\right) + H\dot{\phi}\frac{V'_{george}}{V_{george}}\right]\\
\mathscr{H}_{ringo} &= -24V'_{ringo}(\phi)\dot{\phi}H\left(H^2 + \frac{\kappa}{a^2}\right)
\end{aligned}$$

Self-tuning requires a flat space-time solution and a time dependent non-trivial scalar field whenever $\rho_m = \rho_\Lambda$ and for all Λ. Flat space in cosmological coordinates is given for a hyperbolic slicing $\kappa = -1$ with $a(T) = T$ and $H = 1/T$. This is Milne space-time, the cosmological slicing of flat Minkowski space-time. Therefore, plugging $H^2 = -\kappa/a^2$ into (2.75), we immediately see that

$$V_{john}(\phi)(\dot{\phi}H)^2 + V_{Paul}(\phi)(\dot{\phi}H)^3 - V'_{george}(\phi)(\dot{\phi}H) + \rho_\Lambda = 0 \tag{2.76}$$

Here, we immediately see that *ringo* or a constant *george* do not spoil self-tuning but require necessarily another non-trivial fab 4 term. Indeed we see that the scalar field ϕ is given locally (in space and time) with respect to the arbitrary bulk value of the cosmological constant. This is again an essential condition. For since the scalar field equation is redundant and the space-time metric given, the Friedmann equation has to fix the scalar field dynamically i.e. with respect to its derivative. Hence the first condition means that the Friedmann equation is not trivial; it depends on $\dot{\phi}$. Furthermore, the scalar equation of motion is actually redundant for flat space-time. This is important for otherwise under an abrupt change of the cosmological constant the scalar derivative could not be discontinuous disallowing self-tuning. This is the implementation of the second condition. Indeed the scalar equation $E_\phi = 0$, where

$$E_\phi = E_{john} + E_{paul} + E_{george} + E_{ringo} \tag{2.77}$$

and

$$\begin{aligned}
E_{john} &= 6\frac{d}{dt}\left[a^3 V_{john}(\phi)\dot{\phi}\Delta_2\right] - 3a^3 V'_{john}(\phi)\dot{\phi}^2\Delta_2 \\
E_{paul} &= -9\frac{d}{dt}\left[a^3 V_{paul}(\phi)\dot{\phi}^2 H\Delta_2\right] + 3a^3 V'_{paul}(\phi)\dot{\phi}^3 H\Delta_2 \\
E_{george} &= -6\frac{d}{dt}\left[a^3 V'_{george}(\phi)\Delta_1\right] + 6a^3 V''_{george}(\phi)\dot{\phi}\Delta_1 + 6a^3 V'_{george}(\phi)\Delta_1^2 \\
E_{ringo} &= -24V'_{ringo}(\phi)\frac{d}{dt}\left[a^3\left(\frac{\kappa}{a^2}\Delta_1 + \frac{1}{3}\Delta_3\right)\right]
\end{aligned}$$

with operator

$$\Delta_n = H^n - \left(\frac{\sqrt{-\kappa}}{a}\right)^n \tag{2.78}$$

vanishes on shell for $n > 0$. However, we should note that the third condition is implemented by the fact that the full scalar equation of motion should *not be independent* of $\ddot{a}$. This ensures that the self-tuning solution can be evolved to dynamically, and allows for a non-trivial cosmology. The second Friedmann equation results from the scalar and 1st Friedmann equation as a Bianchi identity.

In order to explicitly show a self-tuning solution consider some particularly simple potentials that can be obtained by Taylor expansion on ϕ.

$$V_{john} = C_j, \qquad V_{paul} = C_p, \tag{2.79}$$

$$V_{george} = C_g + C_g^1\,\phi, \qquad V_{ringo} = C_r + C_r^1\,\phi + C_r^2\,\phi^2, \tag{2.80}$$

This Taylor expansion corresponds to a slow varying late time scalar filed. Since (2.76) is homogeneous in $\dot{\phi}H$ it is quite easy to see that

$$\phi = \phi_0 + \phi_1 T^2, \tag{2.81}$$

is a solution where ϕ_0 and ϕ_1 are constants, with

$$-C_g^1\,\phi_1 + 2\,C_j\phi_1^2 - 4\,C_p\phi_1^3 + \frac{\rho_\Lambda}{6} = 0\,, \tag{2.82}$$

Therefore for arbitrary Λ there exists ϕ_1 satisfying locally (2.82) without fine tuning of the potentials, here C_{fab4}. If Λ jumps to a different value then so can do ϕ_1 and this corresponds to a discontinuous scalar field $\dot{\phi}$. The same mechanism occurs for arbitrary potentials, of course there the solution is more complex. An interesting question now arises: is it possible that self-tuning solutions exist for other vacuum metrics of the theory. Could we for example have Fab 4 with a cosmological constant and find a self-tuning vacuum black hole, in other words a black hole solution than rather than de-Sitter have flat space-time asymptotics. This is still an open problem for the theory, although a self-tuning solution has been recently found in the literature with a remnant effective cosmological constant [18].

Let us now move on into the direction of exact solutions, describing a method which will give black hole solutions in this theory [18]. Let us for simplicity consider two of the Fab 4 terms namely John and George and let us also consider their potentials to be constants. We have therefore the action,

$$S = \int d^4x\sqrt{-g}\left[\zeta R + \beta G^{\mu\nu}\partial_\mu\phi\partial_\nu\phi\right], \tag{2.83}$$

and here notice we have not included a cosmological constant. The relevant coupling constants are now ζ and β and as a result the above action is shift-symmetric for the scalar field ϕ. According to what we described above, this theory is a self-tuning theory for flat space-time as long as $\beta \neq 0$ had we had a cosmological constant in the action. The metric field equations read,

$$\zeta G_{\mu\nu} + \frac{\beta}{2}\Big[(\partial\phi)^2 G_{\mu\nu} + 2P_{\mu\alpha\nu\beta}\nabla^\alpha\phi\nabla^\beta\phi + g_{\mu\alpha}\delta^{\alpha\rho\sigma}_{\nu\gamma\delta}\nabla^\gamma\nabla_\rho\phi\nabla^\delta\nabla_\sigma\phi\Big] = 0, \tag{2.84}$$

where $P_{\alpha\beta\mu\nu}$ is the double dual of the Riemann tensor (2.7). The ϕ equation of motion can be rewritten in the form of a current conservation, as a consequence of the shift symmetry of the action,

$$\nabla_\mu J^\mu = 0, \quad J^\mu = \beta G^{\mu\nu}\partial_\nu\phi. \tag{2.85}$$

Note that (2.85) contains a part of the metric field equations, namely that originating from the Einstein-Hilbert term. We now consider a spherically symmetric Anzatz

$$ds^2 = -h(r)dt^2 + \frac{dr^2}{f(r)} + r^2 d\Omega^2. \tag{2.86}$$

where $f(r)$, $h(r)$ are to be determined from the field equations.

Let us make a slight pause in order to make connection with the flat self-tuning solution we exposed previously. In the (2.86) system of coordinates, Milne space-time is given as,

$$T = \sqrt{t^2 - r^2}, \qquad \coth X = \frac{t}{r} \tag{2.87}$$

and thus we note from (2.81) that the self tuning solution we depicted previously (2.81) is given by $\phi(t, r) = \phi_0 + \phi_1(t^2 - r^2)$. The scalar field therefore in this coordinate chart is a radial, time dependent function. Therefore any self-tuning black hole solution will have to have a time and radially dependent scalar.

Although we do not find a self-tuning solution for the forthcoming example (we have taken $\Lambda = 0$) we consider the Anzatz,

$$\beta G^{rr} = 0, \qquad \phi(t, r) = qt + \psi(r). \tag{2.88}$$

involving a linear time dependence in the scalar field.[9] Notice from the field equations (2.84) that due to shift symmetry no time derivatives are present, the equations of motion are ODE's. This condition (2.88) solves not only the scalar but also the (tr)-metric equation which is not trivial due to time dependence of the scalar field ϕ. Therefore (2.88) is a valid anzatz rendering the whole system integrable. Indeed the remaining equations are solved for, with $f = h = 1 - \frac{\mu}{r}$, whereas the scalar field is not trivial and reads,

$$\phi_\pm = qt \pm q\mu \left[2\sqrt{\frac{r}{\mu}} + \log \frac{\sqrt{r} - \sqrt{\mu}}{\sqrt{r} + \sqrt{\mu}} \right] + \phi_0 \tag{2.89}$$

The regularity of the metric and the scalar field at the horizon can be conveniently checked by use of the generalized Eddington-Finkelstein coordinates, with the advanced time coordinate, v,

$$v = t + \int (fh)^{-1/2} dr. \tag{2.90}$$

One finds from (2.86) and (2.90),

$$ds^2 = -hdv^2 + 2\sqrt{h/f}\, dvdr + r^2 d\Omega^2. \tag{2.91}$$

[9]Clearly, had we been seeking a self-tuning solution in the presence of an arbitrary cosmological constant this linear anzatz would not do. We know rather that there must be at large distance a t^2 dependence on the scalar field. This unfortunately renders the field equations t-dependent and the system cannot admit a non zero mass solution. In other words a self-tuning black hole would have to be part of a radiating space-time. Again this is an open problem.

One can explicitly check that the solution (2.89) with the plus sign does not diverge on the future horizon (whereas the solution with the minus sign is regular on the past horizon). Indeed the transformation (2.90) reads, $v = t + r + \mu \log(r/\mu - 1)$, and using (2.89) one finds,

$$\phi_+ = q\left[v - r + 2\sqrt{\mu r} - 2\mu \log\left(\sqrt{\frac{r}{\mu}} + 1\right)\right] + const, \tag{2.92}$$

which is manifestly regular at the horizon, $r = \mu$. This is therefore a regular GR black hole with a non-trivial scalar field which is also regular at the horizon. This method can be applied in differing Gallileon contexts yielding relatively simple and well-defined black hole solutions [63]. It seems that the linear time dependence of the scalar field, its shift symmetry and the presence of higher order terms is capital to the presence of regular black hole solutions. Indeed if there is no linear time-dependence then the scalar field can present singular behavior at the horizon and solutions are not asymptotically flat [64]. We can re-iterate the Anzatz (2.88) roughly as long as the Galileon scalar equation of motion gives the metric field equation of the lower order term. In other words gravitational terms go in pairs, as here in our example, the Einstein-Hilbert and the John term. One can show that a similar property holds for Ringo and Paul terms of the Fab 4. Indeed one can show that the scalar equation associated to Paul, $P^{\mu\nu\alpha\beta}\nabla_\mu\phi\nabla_\alpha\phi\nabla_\nu\nabla_\beta\phi$ with $V_{paul} = constant$ gives the metric field equations of $\phi\hat{G}$. Note also that the latter is also invariant under shift symmetry. This method bifurcates the no-hair arguments in [59] (see [18] and [65]).

Conclusions

In this lecture we have studied certain aspects of Lovelock and Horndeski theory that have been discussed very recently in the literature of modified gravity theories. The former theory, as we saw is the general metric theory of massless gravity in arbitrary dimensions and with a Levi-Civita connexion, whereas the latter is the general scalar-tensor theory in four dimensional space-time, again using a Levi-Civita connexion. Lovelock theory, is GR with a cosmological constant in four dimensions whereas Horndeski theory is GR once the scalar field is frozen. In this sense and given their unique properties the two theories are essential and very general examples of modified gravity theories. General because, for example, Horndeski theory includes known and widely studied $F(R)$ or $F(\hat{G})$ theories. General also since part of Horndeski theory is a limit of other fundamental modified gravity theories such as massive gravity [16] in its decoupling limit [25]. We saw that Lovelock and Horndeski theories are explicitly related via Kaluza-Klein reduction and one can map solutions from one theory to the other. This permitted to find analytic

(continued)

black hole solutions in Horndeski theory for the first time [53]. We then moved on to discuss a subset of Horndeski theory which has self-tuning properties. This particular theory consisting of four scalar-curvature interaction terms has been dubbed Fab 4 [9]. Although Fab 4 does not present a full solution of the cosmological problem since it does not account for radiative corrections [8], the theory itself has some very interesting integrability properties giving for the first time scalar-tensor black holes with regular scalar field on the black hole horizon. The method described briefly here is quite powerful since it can be applied in differing gravitational theories of the Galileon type or even with bi-scalar tensor theories [63]. We have depicted very recent ongoing research directions in these fields which have numerous open problems. We hope that these notes will help in tackling some of those in the recent future.

Acknowledgements I am very happy to acknowledge numerous interesting discussions with my colleagues during the course of the Aegean summer school. I am also indebted to E Papantonopoulos for the very effective and smooth organisation of the school. I am very happy to thank Eugeny Babichev for numerous comments as well as Stanley Deser for remarks on the first version of this paper. I also thank the CERN theory group for hosting me during the final stages of this work.

References

1. C.M. Will, *The Confrontation Between General Relativity and Experiment,* gr-qc/0510072; C.M. Will, *Theory and Experiment in Gravitational Physics* (Cambridge University Press, Cambridge) (1981)
2. D. Lovelock, J. Math. Phys. **12**, 498 (1971)
3. C. Lanczos, Ann. Math. **39**, 842 (1938)
4. B. Zumino, Phys. Rep. **137**, 109 (1986)
5. M. Ostrogradski, Mem. Ac. St. Petersbourg **4**, 385 (1850)
6. M. Betoule et al. [SDSS Collaboration], [arXiv:1401.4064 [astro-ph.CO]]; G. Hinshaw et al. [WMAP Collaboration], Astrophys. J. Suppl. **208**, 19 (2013) [arXiv:1212.5226 [astro-ph.CO]]
7. S. Weinberg, Rev. Mod. Phys. **61**, 1 (1989)
8. N. Kaloper, A. Padilla, Phys. Rev. Lett. **112**, 091304 (2014) [arXiv:1309.6562 [hep-th]]
9. C. Charmousis, E.J. Copeland, A. Padilla, P.M. Saffin, Phys. Rev. Lett. **108**, 051101 (2012) [arXiv:1106.2000 [hep-th]]; C. Charmousis, E.J. Copeland, A. Padilla, P.M. Saffin, Phys. Rev. D **85**, 104040 (2012) [arXiv:1112.4866 [hep-th]]
10. T. Clifton, P.G. Ferreira, A. Padilla, C. Skordis, Phys. Rep. **513**, 1 (2012) [arXiv:1106.2476 [astro-ph.CO]]
11. P. Horava, Phys. Rev. D **79**, 084008 (2009) [arXiv:0901.3775 [hep-th]]; D. Blas, O. Pujolas, S. Sibiryakov, Phys. Rev. Lett. **104**, 181302 (2010) [arXiv:0909.3525 [hep-th]]; B. Audren, D. Blas, J. Lesgourgues, S. Sibiryakov, JCAP **1308**, 039 (2013) [arXiv:1305.0009 [astro-ph.CO], arXiv:1305.0009]

12. N. Deruelle, J. Madore, gr-qc/0305004; C. Charmousis, Lect. Notes Phys. **769**, 299 (2009) [arXiv:0805.0568 [gr-qc]]; C. Garraffo, G. Giribet, Mod. Phys. Lett. A **23**, 1801 (2008) [arXiv:0805.3575 [gr-qc]]
13. G.W. Horndeski, Int. J. Theor. Phys. **10**, 363 (1974)
14. A. Nicolis, R. Rattazzi, E. Trincherini, Phys. Rev. D **79**, 064036 (2009) [arXiv:0811.2197 [hep-th]]
15. C. Deffayet, S. Deser, G. Esposito-Farese, Phys. Rev. D **80**, 064015 (2009) [arXiv:0906.1967 [gr-qc]]
16. C. de Rham, arXiv:1401.4173 [hep-th]; M.S. Volkov, arXiv:1405.1742 [hep-th]
17. G.J. Olmo, in *Open Questions in Cosmology*, ed. by G.J. Olmo (InTech Publishing, Rijeka, Croatia, 2012). ISBN 978-953-51-0880-1 (Downloadable from http://www.intechopen.com/books/open-questions-in-cosmology) [arXiv:1212.6393]
18. E. Babichev, C. Charmousis, arXiv:1312.3204 [gr-qc]
19. S.B. Edgar, A. Hoglund, J. Math. Phys. **43**, 659 (2002) [gr-qc/0105066]
20. M. Spivak, *A Comprehensive Introduction to Differential Geometry* (Publish or Perish, Houston, 1999)
21. S.-S. Chern, Ann. Math. **45**, 747–752 (1944); S.-S. Chern, Ann. Math. **46**, 674–684 (1945)
22. C. Brans, R.H. Dicke, Phys. Rev. **124**, 925 (1961)
23. T. Damour, G. Esposito-Farese, Class. Quantum Grav. **9**, 2093 (1992); G. Esposito-Farese, AIP Conf. Proc. **736**, 35 (2004) [gr-qc/0409081]; T.P. Sotiriou, arXiv:1404.2955 [gr-qc]
24. T.P. Sotiriou, V. Faraoni, Rev. Mod. Phys. **82**, 451 (2010) [arXiv:0805.1726 [gr-qc]]
25. N. Arkani-Hamed, H. Georgi, M.D. Schwartz, Ann. Phys. **305**, 96 (2003) [hep-th/0210184]
26. T. Kobayashi, M. Yamaguchi, J.'i. Yokoyama, Prog. Theor. Phys. **126**, 511 (2011) [arXiv:1105.5723 [hep-th]]
27. M. Banados, R. Troncoso, J. Zanelli, Phys. Rev. D **54**, 2605 (1996) [gr-qc/9601003]
28. R. Zegers, J. Math. Phys. **46**, 072502 (2005) [gr-qc/0505016]
29. D. Birmingham, Class. Quantum Grav. **16**, 1197 (1999) [hep-th/9808032]; S. Aminneborg, I. Bengtsson, S. Holst, P. Peldán, Class. Quantum Grav. **13**, 2707 (1996) gr-qc/9604005; J.P.S. Lemos, Phys. Lett. B **353**, 46 (1995) [arXiv:gr-qc/9404041]; J.P.S. Lemos, Class. Quantum. Grav. **12**, 1081 (1995) [arXiv:gr-qc/9407024]; D. Brill, J. Louko, P. Peldán, Phys. Rev. D **56**, 3600 (1997) gr-qc/9705012
30. G.W. Gibbons, S.A. Hartnoll, C.N. Pope, Phys. Rev. D **67**, 084024 (2003) [hep-th/0208031]
31. G. Gibbons, S.A. Hartnoll, Phys. Rev. D **66**, 064024 (2002) [hep-th/0206202]
32. R. Gregory, R. Laflamme, Phys. Rev. Lett. **70**, 2837 (1993) [hep-th/9301052]
33. G. Dotti, R.J. Gleiser, Phys. Lett. B **627**, 174 (2005) [hep-th/0508118]
34. C. Bogdanos, C. Charmousis, B. Gouteraux, R. Zegers, JHEP **0910**, 037 (2009) [arXiv:0906.4953 [hep-th]]
35. P. Bowcock, C. Charmousis, R. Gregory, Class. Quantum Grav. **17**, 4745 (2000) [hep-th/0007177]
36. C. Charmousis, J.-F. Dufaux, Class. Quantum Grav. **19**, 4671 (2002) [hep-th/0202107]
37. C. Charmousis, A. Padilla, JHEP **0812**, 038 (2008) [arXiv:0807.2864 [hep-th]]
38. B. Zwiebach, Phys. Lett. B **156**, 315 (1985); S. Deser, A.N. Redlich, Phys. Lett. B **176**, 350 (1986) [Erratum-Phys. Lett. B **186B**, 461 (1987)]
39. D.G. Boulware, S. Deser, Phys. Rev. Lett. **55**, 2656 (1985); R.-G. Cai, Phys. Rev. D **65**, 084014 (2002) [hep-th/0109133]
40. R.J. Gleiser, G. Dotti, Phys. Rev. D **72**, 124002 (2005) [gr-qc/0510069]
41. T. Takahashi, J. Soda, Prog. Theor. Phys. **124**, 711 (2010) [arXiv:1008.1618 [gr-qc]]
42. H. Maeda, Phys. Rev. D **81**, 124007 (2010) [arXiv:1004.0917 [gr-qc]]
43. D.J. Gross, J.H. Sloan, Nucl. Phys. B **291**, 41 (1987); R.R. Metsaev, A.A. Tseytlin, Nucl. Phys. B **293**, 385 (1987)
44. Y. Bardoux, C. Charmousis, T. Kolyvaris, Phys. Rev. D **83**, 104020 (2011) [arXiv:1012.4390 [gr-qc]]; Y. Bardoux, M.M. Caldarelli, C. Charmousis, JHEP **1205**, 054 (2012) [arXiv:1202.4458 [hep-th]]

45. N. Arkani-Hamed, S. Dimopoulos, G.R. Dvali, Phys. Lett. B **429**, 263 (1998) [hep-ph/9803315]; L. Randall, R. Sundrum, Phys. Rev. Lett. **83**, 4690 (1999) [hep-th/9906064]; L. Randall, R. Sundrum, Phys. Rev. Lett. **83**, 3370 (1999) [hep-ph/9905221]; R. Gregory, V.A. Rubakov, S.M. Sibiryakov, Phys. Rev. Lett. **84**, 5928 (2000) [hep-th/0002072]; G.R. Dvali, G. Gabadadze, M. Porrati, Phys. Lett. B **485**, 208 (2000) [hep-th/0005016]
46. S.M. Carroll, M.M. Guica, hep-th/0302067; P. Bostock, R. Gregory, I. Navarro, J. Santiago, Phys. Rev. Lett. **92**, 221601 (2004) [hep-th/0311074]; C. Charmousis, A. Papazoglou, JHEP **0807**, 062 (2008) [arXiv:0804.2121 [hep-th]]
47. C. Charmousis, R. Zegers, Phys. Rev. D **72**, 064005 (2005) [hep-th/0502171]; C. Charmousis, R. Zegers, JHEP **0508**, 075 (2005) [hep-th/0502170]
48. C. Charmousis, G. Kofinas, A. Papazoglou, JCAP **1001**, 022 (2010) [arXiv:0907.1640 [hep-th]]
49. C. Charmousis, B. Gouteraux, J. Soda, Phys. Rev. D **80**, 024028 (2009) [arXiv:0905.3337 [gr-qc]]
50. F. Mueller-Hoissen, Class. Quantum Grav. **3**, 665 (1986); C. Wetterich, Nucl. Phys. B **252**, 309 (1985); N. Deruelle, J. Madore, Mod. Phys. Lett. A **1**, 237 (1986); F. Mueller-Hoissen, Nucl. Phys. B **337**, 709 (1990); F. Mueller-Hoissen, Ann. Phys. **48**, 543 (1991)
51. A. Jakobek, K.A. Meissner, M. Olechowski, Nucl. Phys. B **645**, 217 (2002) [hep-th/0206254]; C. Charmousis, S.C. Davis, J.-F. Dufaux, JHEP **0312**, 029 (2003) [hep-th/0309083]; L. Amendola, C. Charmousis, S.C. Davis, JCAP **0612**, 020 (2006) [hep-th/0506137]
52. K. Van Acoleyen, J. Van Doorsselaere, Phys. Rev. D **83**, 084025 (2011) [arXiv:1102.0487 [gr-qc]]
53. C. Charmousis, B. Gouteraux, E. Kiritsis, JHEP **1209**, 011 (2012) [arXiv:1206.1499 [hep-th]]
54. I. Kanitscheider, K. Skenderis, JHEP **0904**, 062 (2009) [arXiv:0901.1487 [hep-th]]
55. B. Gouteraux, J. Smolic, M. Smolic, K. Skenderis, M. Taylor, JHEP **1201**, 089 (2012) [arXiv:1110.2320 [hep-th]]
56. C. Deffayet, G. Esposito-Farese, A. Vikman, Phys. Rev. D **79**, 084003 (2009) [arXiv:0901.1314 [hep-th]]
57. C. Deffayet, X. Gao, D.A. Steer, G. Zahariade, Phys. Rev. D **84**, 064039 (2011) [arXiv:1103.3260 [hep-th]]
58. M. Barriola, A. Vilenkin, Phys. Rev. Lett. **63**, 341 (1989)
59. L. Hui, A. Nicolis, Phys. Rev. Lett. **110**(24), 241104 (2013) [arXiv:1202.1296 [hep-th]]
60. N. Arkani-Hamed, S. Dimopoulos, N. Kaloper, R. Sundrum, Phys. Lett. B **480**, 193 (2000) [hep-th/0001197]
61. N. Kaloper, A. Padilla, Phys. Rev. Lett. **112**, 091304 (2014) [arXiv:1309.6562 [hep-th]]
62. E.J. Copeland, A. Padilla, P.M. Saffin, JCAP **1212**, 026 (2012) [arXiv:1208.3373 [hep-th]]
63. T. Kobayashi, N. Tanahashi, arXiv:1403.4364 [gr-qc]; M. Bravo-Gaete, M. Hassaine, arXiv:1312.7736 [hep-th]; C. Charmousis, T. Kolyvaris, E. Papantonopoulos, M. Tsoukalas, arXiv:1404.1024 [gr-qc]
64. M. Rinaldi, Phys. Rev. D **86**, 084048 (2012) [arXiv:1208.0103 [gr-qc]]; M. Minamitsuji, Phys. Rev. D **89**, 064017 (2014) [arXiv:1312.3759 [gr-qc]]; A. Anabalon, A. Cisterna, J. Oliva, Phys. Rev. D **89**, 084050 (2014) [arXiv:1312.3597 [gr-qc]]; A. Cisterna, C. Erices, Phys. Rev. D **89**, 084038 (2014) [arXiv:1401.4479 [gr-qc]]
65. T.P. Sotiriou, S.-Y. Zhou, arXiv:1312.3622 [gr-qc]

Chapter 3
Modified Gravity and Coupled Quintessence

Christof Wetterich

Abstract The distinction between modified gravity and quintessence or dynamical dark energy is difficult. Many models of modified gravity are equivalent to models of coupled quintessence by virtue of variable transformations. This makes an observational differentiation between modified gravity and dark energy very hard. For example, the additional scalar degree of freedom in $f(R)$-gravity or non-local gravity can be interpreted as the cosmon of quintessence. Nevertheless, modified gravity can shed light on questions of interpretation, naturalness and simplicity. We present a simple model where gravity is modified by a field dependent Planck mass. It leads to a universe with a cold and slow beginning. This cosmology can be continued to the infinite past such that no big bang singularity occurs. All observables can be described equivalently in a hot big bang picture with inflation and early dark energy.

3.1 Introduction

Einstein's equation

$$M^2(R_{\mu\nu} - \frac{1}{2}Rg_{\mu\nu}) = T_{\mu\nu} \tag{3.1}$$

expresses geometrical quantities on the left hand side in terms of matter and radiation on the right hand side. The basic geometrical quantity is the metric $g_{\mu\nu}$, with $R_{\mu\nu}$ and R the Ricci tensor and curvature scalar formed from the metric and its derivatives. The energy momentum tensor $T_{\mu\nu}$ contains contributions from the particles of the standard model ("baryons", neutrinos, radiation) and from dark matter.

The observation of the present accelerated expansion [1,2] as well as indications for an inflationary epoch in very early cosmology tell us that Eq. (3.1) cannot be complete despite the numerous successful predictions of general relativity. One may

C. Wetterich (✉)
Institut für Theoretische Physik, Philosophenweg 16, 69120 Heidelberg, Germany
e-mail: C.Wetterich@ThPhys.Uni-Heidelberg.DE

E. Papantonopoulos (ed.), *Modifications of Einstein's Theory of Gravity at Large Distances*, Lecture Notes in Physics 892, DOI 10.1007/978-3-319-10070-8_3

supplement terms on the left or right side, as indicated by the dots

$$R_{\mu\nu} - \frac{1}{2} R g_{\mu\nu} + \ldots = \frac{1}{M^2} (T_{\mu\nu} + \ldots). \quad (3.2)$$

Additional contributions to the energy momentum tensor are usually called dark energy, whereas a change on the left hand side is associated with a modification of gravity or general relativity.

It is obvious that this distinction cannot be a particularly strict one since the validity of an equation does not depend on where one writes terms. The most prominent candidate for the explanation of an accelerated expansion, the cosmological constant, can be interpreted as an additional contribution to the energy momentum tensor $\Delta T_{\mu\nu} = \lambda g_{\mu\nu}$. This interpretation is suggested by the contribution of the effective potential of the Higgs scalar to λ, or similar for other scalar fields. We could write the cosmological constant term also on the left hand side and consider it as a modification of gravity—after all it influences the gravitational equations in "empty space".

One may try a more concise definition of the meaning of modified gravity by requiring that the change of the Einstein tensor on the l.h.s. of Eq. (3.1) involves derivatives of the metric, while terms with additional fields and no derivatives of $g_{\mu\nu}$ would contribute to $T_{\mu\nu}$. We will see, however, that modified gravity models defined in this way can often be rewritten in terms of different fields, frequently additional scalar fields. What appears in one field basis as a modification of gravity with terms involving derivatives of the metric shows up as dark energy with new fields and without metric derivatives in an other field basis. In particular, modified gravity theories that are consistent with the observed evolution of the universe are often equivalent to dynamical dark energy or quintessence. The borderline between modified gravity and dark energy becomes rather fuzzy. In fact, the first model of quintessence has originally been formulated as a modification of gravity [3].

The reason for this ambiguity between modified gravity and dark energy is connected to a basic property: observables depend on the dynamical degrees of freedom, but not on the choice of fields used to describe them ("field relativity"). For example, the metric may contain a scalar degree of freedom besides the graviton. This scalar is not distinguished from a "fundamental scalar field" (cosmon) which is the basic ingredient of quintessence.

These lecture notes will present several examples for the equivalence of modified gravity and quintessence. In particular, $f(R)$ gravity or a large class of non-local gravity models are equivalent to coupled quintessence [4, 5]. We do not aim, however, to cover all possible modifications of gravity. More general modified gravity models may contain further non-scalar degrees of freedom (vectors of tensors), involve an infinite number of degrees of freedom, or give up the basic diffeomorphism symmetry underlying general relativity.

Recent reviews of modified gravity can be found in [6–11]. We concentrate here on the deep connection between modified gravity and coupled quintessence. This helps to understand many of the rich features of modified gravity in a simple and unified way. It also shows that many claims for observational distinguishability between modified gravity and quintessence are actually not justified.

In Sects. 3.2 and 3.3 we display our basic setting and discuss the field transformations that relate different versions of a given physical model. In Sect. 3.4 we describe the cosmology of Brans-Dicke theory in the language of coupled quintessence. This points to strong observational bounds on the effective coupling β between the cosmon and matter that will play an important role later. Section 3.5 discusses general scalar-tensor models with actions containing up to two derivatives. We highlight the importance of field-dependent particle masses in order to find models obeying the bounds on β. Section 3.6 discusses a simple three-parameter cosmological model along these lines which is compatible with all present observations from inflation to late dark energy domination. Formulated as a scalar-tensor theory (Jordan frame) it exhibits an unusual cosmic history. The universe shrinks during the radiation- and matter-dominated epochs and the evolution is always very slow. Cosmological solutions remain regular in the infinite past and there is no big bang singularity. On the other hand, the same model is characterized in the Einstein frame by a more usual big bang picture. This underlines that the field transformations that a crucial for these notes also incorporate important conceptual aspects.

In Sect. 3.7 we describe the equivalence of $f(R)$-modified gravity with coupled quintessence [12–14]. For constant particle masses the equivalent coupled quintessence models exhibit a large universal cosmon-matter coupling $\beta = 1/\sqrt{6}$. This issue is a major problem for the construction of realistic $f(R)$ models. We sketch in Sect. 3.8 how a vanishing coupling $\beta = 0$ can be obtained for $f(R)$-models with field-dependent particle masses. In Sect. 3.9 we turn to simple models of non-local gravity. Again, such models are equivalent to coupled quintessence. In Sect. 3.10 we ask the general question to what extent modified gravity models which lead to second order field equations, as Horndeski's models [15], can find an equivalent description as coupled quintessence models. We find a huge class of such modified gravity models for which the scalar-gravity part is given by the action for quintessence, while additional information is contained in the details of the effective cosmon-matter coupling. Our conclusions are drawn in the final section. Parts of Sects. 3.5 and 3.6 have overlap with work reported in [16, 17].

3.2 Basic Setting

We will assume that the theory which describes the late universe (say from radiation domination onwards) can be formulated as a quantum field theory. (This quantum field theory may be an effective theory embedded in a different framework as string theory.) We also restrict the discussion to the case where diffeomorphism symmetry (invariance under general coordinate transformations) is maintained. The most convenient way of specifying models is then the quantum effective action Γ from which the field equations can be derived by variation. It is supposed to include all effects from quantum fluctuations. We can perform arbitrary changes of variables in Γ. They correspond to changes of variables in the differential field equations. All predictions of the model are contained in the field equations. A change of variables can therefore not affect any observable quantities. We will in the following heavily rely on this property of "field relativity" in order to demonstrate the equivalence of many modified gravity theories with coupled quintessence. (Note that on the level of the functional integral for a quantum theory a change of variables has two effects. It transforms the classical action and induces a Jacobian for the functional measure. The effective action is already the result of functional integration such that no Jacobian plays a role in the variable transformation.)

We postulate that Γ is invariant under general coordinate transformations and write it in the form

$$\Gamma = \int d^4x \sqrt{g}(\mathscr{L}_g + \mathscr{L}_m). \tag{3.3}$$

Here $\mathscr{L}_g$ is the gravitational part, while the variation of $\sqrt{g}\mathscr{L}_m$ with respect to $g_{\mu\nu}$ yields the energy momentum tensor $T^{\mu\nu}$. Einstein's equation follows for

$$\mathscr{L}_g = -\frac{M^2}{2}R, \tag{3.4}$$

while $\mathscr{L}_m$ involves matter and radiation

$$\mathscr{L}_m = \mathscr{L}_{\text{standard model}} + \mathscr{L}_{\text{dark matter}}. \tag{3.5}$$

Modified gravity corresponds to a more general form of $\mathscr{L}_g$. The simplest form of quintessence adds to $\mathscr{L}_m$ the contribution from a scalar field $\varphi(x)$, consisting of a potential $V(\varphi)$ and a kinetic term,

$$\Delta\mathscr{L}_m = \frac{1}{2}\partial^\mu\varphi\partial_\mu\varphi + V(\varphi). \tag{3.6}$$

This scalar field is called the "cosmon".

Simple modifications of gravity add to $\mathscr{L}_g$ terms involving higher powers of the curvature scalar as R^2. They can play an important role for inflation as in Starobinski's model [18]. Within higher dimensional theories the higher order curvature invariants have been employed for a mechanism of spontaneous compactification [19] and for a description of inflation as an effective transition from higher dimensions to four "large" dimensions [20,21]. The field equations for actions where R is replaced by an arbitrary function $f(R)$ have been investigated long ago [22]. Modifications of gravity also arise if our four-dimensional world is a "brane" embedded in some higher-dimensional space [23]. Higher-dimensional scenarios can be described in an equivalent four-dimensional setting, involving in principle infinitely many fields and in some cases non-local interactions. In the four-dimensional language typically both $calL_g$ and $\mathscr{L}_m$ are modified simultaneously. We will concentrate in this lecture on simple four-dimensional models with only a few effective degrees of freedom. Many important aspects of modified gravity can be understood in this simple setting. We are mainly interested in the role of modified gravity for the present cosmological epoch and leave aside its potential relevance for the early inflationary epoch.

Modified gravity models have a long history. One of the most prominent historical models is Brans-Dicke theory [24], where the reduced Planck mass M in $\mathscr{L}_g$ is replaced by a scalar field $\chi(x)$. In this case both $\mathscr{L}_g$ and $\mathscr{L}_m$ get modified,

$$\mathscr{L}_g = -\frac{\chi^2}{2}R, \tag{3.7}$$

$$\Delta\mathscr{L}_m = \frac{1}{2}K\partial^\mu\chi\partial_\mu\chi. \tag{3.8}$$

(Our choice of a scalar field χ differs from the original formulation in [24]. The constant K is related to the ω-parameter in Brans-Dicke theory by $K = 4\omega$.) Many aspects that are crucial for these notes can already be seen in Brans-Dicke theory, and we will discuss them in the next two sections.

3.3 Weyl Scaling

It is possible to express Brans-Dicke theory as a type of coupled quintessence model. For this purpose we perform a Weyl scaling [25,26] by using a different metric field $g'_{\mu\nu}$, related to $g_{\mu\nu}$ by

$$g_{\mu\nu} = w^2 g'_{\mu\nu}. \tag{3.9}$$

Here the factor w^2 can be a function of other fields. Let us consider a scaling involving the scalar field χ without derivatives, $w = w(\chi)$. The new curvature scalar R' formed from $g'_{\mu\nu}$ and its derivatives is related to R by

$$R = w^{-2}\{R' - 6(\ln w)_{;}{}^{\mu}(\ln w)_{;\mu} - 6(\ln w)_{;}{}^{\mu}{}_{\mu}\}. \tag{3.10}$$

Here we denote by semicolons covariant derivatives, in particular

$$(\ln w)_{;\mu} = \partial_\mu \ln w\ ,\ \ (\ln w)_{;}^{\mu} = g'^{\mu\nu}\partial_\nu \ln w. \tag{3.11}$$

The square root of the determinant of the metric, $g = -\det(g_{\mu\nu})$, transforms as

$$\sqrt{g} = w^4\sqrt{g'}. \tag{3.12}$$

We next make the specific choice

$$w^2 = \frac{M^2}{\chi^2}, \tag{3.13}$$

resulting in

$$\sqrt{g}\chi^2 R \to \sqrt{g'} M^2 R' + \text{ derivatives of } \chi. \tag{3.14}$$

The term $\mathscr{L}_g$ takes now the standard form (3.4) and the "modification of gravity" has been transformed away. As a counterpart, the kinetic term for χ is modified by replacing $\sqrt{g}\Delta\mathscr{L}_m \to \sqrt{g'}\Delta\mathscr{L}'_m$,

$$\Delta\mathscr{L}'_m = \frac{M^2}{2}(K+6)\partial^\mu \ln\chi \, \partial_\mu \ln\chi. \tag{3.15}$$

For $K > -6$ the model describes gravity coupled to a scalar field. A canonical form of the scalar kinetic term $\Delta\mathscr{L}'_m = \partial^\mu\varphi\partial_\mu\varphi/2$ obtains for

$$\varphi = \sqrt{K+6}\, M \ln\left(\frac{\chi}{M}\right). \tag{3.16}$$

The choice of the metric $g'_{\mu\nu}$ is called the Einstein frame. In the Einstein frame the Planck mass M is a fixed constant that does not depend on any other fields. Cosmologies of two effective actions related by Weyl scaling are strictly equivalent, with all observables taking identical values [27]. For a quantum field theory the concept of the quantum effective action Γ is crucial for this statement. Its first functional derivatives, the field equations, describe exact relations between expectation values of quantum fields. Variable transformations as the Weyl scaling are transformations among these field values—they may be associated with "field coordinate transformations". Observables that can be expressed in terms of field values have to be transformed according to these variable transformations. For cosmology it is crucial that all quantities, including temperature T, particle masses m, or the coupling of particles to fields β, are transformed properly under Weyl scaling. It can then be established that suitable dimensionless ratios, as T/m, remain invariant under Weyl scaling [27]. Dimensionless quantities are the only ones accessible to measurement and observation. One is therefore free to use the Einstein frame with metric $g'_{\mu\nu}$ or the "Jordan frame" (3.7) with metric $g_{\mu\nu}$ - both are equivalent, yielding the same results for dimensionless observable quantities. This has been verified by detailed studies of many observables [27–32]. We may summarize that physical observables cannot depend on the choice of fields used to describe them, a principle called "field relativity" [32]. This principle extends to observables involving correlations, which can be found from higher functional derivatives of Γ.

It is crucial that also the matter and radiation part $\mathscr{L}_m$ is transformed under Weyl scaling, due to the presence of the factor $\sqrt{g}$, or $g^{\mu\nu}$ in derivative terms. In general, not only the metric but also other fields appearing in $\mathscr{L}_m$ need to be transformed under Weyl scaling. The electromagnetic gauge field A_μ needs no rescaling. Indeed the Maxwell kinetic term remains invariant since a factor w^4 from $\sqrt{g}$ cancels two factors w^{-2} from the inverse metric $g^{\mu\nu}$ appearing in

$$\mathscr{L}_F = \frac{1}{4} F_{\mu\nu} F_{\rho\sigma} g^{\mu\rho} g^{\nu\sigma}. \tag{3.17}$$

For fermions, the factors of w drop out of the kinetic term provided we combine the Weyl scaling (3.9) with a transformation of the fermion field

$$\psi = w^{-\frac{3}{2}} \psi'. \tag{3.18}$$

This yields

$$\sqrt{g}\bar{\psi}\gamma^\mu \partial_\mu \psi \rightarrow \sqrt{g'}\bar{\psi}'\gamma^\mu \partial_\mu \psi' + \dots, \tag{3.19}$$

where the dots denote a term containing a derivative of χ, i.e. $\sqrt{g'}\bar{\psi}'\gamma^\mu\psi'\partial_\mu\chi$. For a model containing only massless gauge bosons and fermions the Weyl scaled version of Brans-Dicke theory describes standard gravity and a massless scalar field that has only derivative couplings. In this case φ can be associated with the Goldstone boson of spontaneously broken dilatation or scale symmetry.

For massive fermions the situation changes drastically. A mass term $m_F\sqrt{g}\bar{\psi}\psi$ transforms according to

$$m_F\sqrt{g}\bar{\psi}\psi \rightarrow m'_F\sqrt{g'}\bar{\psi}'\psi' = m_F\frac{M}{\chi}\sqrt{g'}\bar{\psi}'\psi'. \tag{3.20}$$

We end with a non-derivative coupling of φ to the fermion mass

$$\mathscr{L}_{F,m} = m_F \exp\left(-\frac{\beta\varphi}{M}\right)\bar{\psi}'\psi', \tag{3.21}$$

with cosmon-matter coupling[4, 27]

$$\beta = \frac{1}{\sqrt{K+6}} = \frac{1}{\sqrt{4\omega+6}}. \tag{3.22}$$

3.4 Brans-Dicke Cosmology

For understanding the cosmological role of the coupling β it is instructive to study the cosmology of the Brans-Dicke theory in the Einstein frame. We assume a homogeneous and isotropic Schwarzschild metric with scale factor $a(t)$, $H = \partial \ln a/\partial t$, and vanishing spatial curvature, coupled to a homogeneous scalar field $\varphi(t)$. The field equations for a fluid of massive particles read [4, 5]

$$H^2 = \frac{1}{3M^2}(\rho + \frac{1}{2}\dot{\varphi}^2), \tag{3.23}$$

$$\dot{\rho} + 3H(\rho + p) + \frac{\beta}{M}(\rho - 3p)\dot{\varphi} = 0, \tag{3.24}$$

$$\ddot{\varphi} + 3H\dot{\varphi} = \frac{\beta}{M}(\rho - 3p). \tag{3.25}$$

For the radiation dominated epoch with $p = \rho/3$ the coupling β plays no role. The field φ settles rapidly to an arbitrary constant value and one finds standard cosmology. Additional massless fields for which β vanishes do not change this situation.

Once particles become non-relativistic, however, and matter starts to dominate over radiation, the coupling β leads to a modified cosmology. The field φ evolves and particle masses change. After a transition period cosmology reaches a scaling solution which reads ($p = 0$)

$$H = \frac{\eta}{t}, \quad \dot{\varphi} = \frac{cM}{t}, \quad \rho = \frac{fM^2}{t^2}. \tag{3.26}$$

Equations (3.23)–(3.25) become algebraic equations for η, c and f, with solution

$$\eta = \frac{2}{3 + 2\beta^2}, \quad f = \frac{12 - 8\beta^2}{(3 + 2\beta^2)^2}, \quad c = \frac{4\beta}{3 + 2\beta^2}. \tag{3.27}$$

This asymptotic solution exists for

$$\beta < \sqrt{\frac{3}{2}}, \quad \omega > -\frac{4}{3}. \tag{3.28}$$

For β of the order one one finds a scalar field dominated cosmology that is not compatible with observation. This becomes even more drastic for $\beta > \sqrt{3/2}$ where matter can be neglected as compared to the scalar kinetic energy. In contrast, for small β the modification of the expansion remains small, with η close to the standard value $2/3$. The most prominent cosmological effect concerns the time variation of the ratio of nucleon mass over Planck mass. Indeed, the field φ has changed between matter-radiation equality and today by $\Delta\varphi = \varphi(t_0) - \varphi(t_{eq})$,

$$\Delta\varphi \approx 4\beta M \ln\left(\frac{t_0}{t_{eq}}\right), \tag{3.29}$$

with a corresponding change of the nucleon mass

$$\frac{m_n(t_{eq})}{m_n(t_0)} \approx \exp\left(\frac{\beta}{M}\Delta\varphi\right) = \left(\frac{t_0}{t_{eq}}\right)^{4\beta^2} = z_{eq}{}^{6\beta^2} \approx (1100)^{\frac{3}{2\omega}}. \tag{3.30}$$

The relative change of the nucleon mass $R_n \approx (3/2\omega)\ln(1100)$ bounds ω as a function of the observational bound $R_n < \bar{R}_n$ on the relative variation of the nucleon mass,

$$\omega > \frac{10}{\bar{R}_n} \gtrsim 100. \tag{3.31}$$

The upper bound on the relative variation of the nucleon mass $\bar{R}_n$ can be estimated from nucleosynthesis. (For Brans-Dicke theory no substantial change of the nucleon mass occurs between nucleosynthesis and matter radiation equality.) We evaluate

$$R_n = \frac{\Delta m_n}{m_n} = -\frac{1}{2}\frac{\Delta G_N}{G_N}, \tag{3.32}$$

with $\Delta m_n = m_n(t_n) - m_n$, $m_n = m_n(t_0)$ and t_n the time of nucleosynthesis. The second equation involves Newton's constant G_N. It reflects the fact that all particle masses vary $\sim m_n$ and only dimensionless ratios as $m_n^2 G_N$ can influence the element abundancies produced during nucleosynthesis [33]. We may use the bound from [33]

$$-0.19 \leq \frac{\Delta G_N}{G_N} \leq 0.1 \tag{3.33}$$

for a constraint $\bar{R}_n = 0.1, \omega > 100$. This cosmological bound is weaker than the bound from solar system gravity experiments $\omega > 4 \cdot 10^4$ [34]. On the other hand, this bound restricts the overall cosmological evolution. More precisely, the cosmological bound constrains a combination of β and the change in the normalized cosmon field since nucleosynthesis,

$$-0.05 \leq \frac{\beta}{M}\big(\varphi(t_n) - \varphi(t_0)\big) \leq 0.1. \tag{3.34}$$

3.5 Scalar Tensor Models

The problem with φ-dependent particle masses in the Einstein frame persists for many scalar tensor models. There are two types of general solutions for this issue:

(i) Particle masses in the Jordan frame are dependent on χ and scale $\sim \chi$. In the Einstein frame the particle masses are then independent of χ and β vanishes [3,27].
(ii) The scalar field φ changes very little, both in cosmology and locally.

The simplest way to realize the second alternative is to add a potential $V(\chi)$ in the Jordan frame. After Weyl scaling one finds in the Einstein frame

$$\sqrt{g}V(\chi) = \sqrt{g'}V'(\chi), V' = w^4 V = \frac{M^4}{\chi^4} V = \exp\left\{-\frac{4\varphi}{\sqrt{K+6}\,M}\right\} V. \quad (3.35)$$

If $V'(\varphi)$ has a minimum at φ_0 the cosmological solution will typically settle at this minimum at early time, such that there is no residual cosmic time variation of the ratio m_n/M. On the other hand, if φ settles to φ_0 only after nucleosynthesis or continues evolving, the cosmological bound (3.34) has to be respected.

A local mass distribution acts as a source for the scalar field with strength β/M. This induces an additional scalar-mediated attraction. For a massless scalar field the relative strength of this interaction as compared to Newtonian gravity is $2\beta^2$. If the scalar mass

$$m_\varphi = \sqrt{\frac{\partial^2 V}{\partial \varphi^2}(\varphi_0)} \quad (3.36)$$

is smaller than the inverse size of the solar system the presence of this scalar interaction would be visible in post-Newtonian gravity experiments, limiting $1.2 \cdot 10^{-5}$, cf. Eq. (3.22). For larger m_φ the additional exponential suppression of a Yukawa interaction allows for larger β. If m_φ exceeds the inverse size of a massive object the scalar field φ tends to settle inside the object at a value different from φ_0. Then the nucleon mass becomes density dependent, implying again upper bounds on β [35]. For models predicting large β and a small cosmological mass m_φ there remains still the possibility that the local mass inside an object is substantially higher than the cosmological mass outside the object, due to non-linear effects. This is called chameleon effect [36]. We will see that many popular $f(R)$-theories lead to large β and small m_φ.

In the remainder of this section we will concentrate on the alternative (i) with χ-dependent particle masses. We will investigate a general class of scalar tensor theories with an effective action

$$\Gamma = \int_x g^{\frac{1}{2}} \left\{-\frac{1}{2}F(\chi)R + \frac{1}{2}K(\chi)\partial^\mu\chi\partial_\mu\chi + V(\chi)\right\}. \quad (3.37)$$

This is the most general form for a scalar coupled to gravity which preserves diffeomorphism symmetry, provided that terms with four or more derivatives can be neglected. For a homogenous and isotropic Universe (and for vanishing spatial curvature) the field equations take the form [16,27]

$$K(\ddot{\chi} + 3H\dot{\chi}) + \frac{1}{2}\frac{\partial K}{\partial \chi}\dot{\chi}^2 = -\frac{\partial V}{\partial \chi} + \frac{1}{2}\frac{\partial F}{\partial \chi}R + q_\chi, \tag{3.38}$$

$$FR = F(12H^2 + 6\dot{H}) = 4V - \left(K + 6\frac{\partial F}{\partial \chi^2}\right)\dot{\chi}^2 \tag{3.39}$$

$$-6\frac{\partial F}{\partial \chi^2}(\ddot{\chi} + 3H\dot{\chi})\chi - 12\frac{\partial^2 F}{(\partial \chi^2)^2}\chi^2\dot{\chi}^2 - T^\mu_\mu,$$

$$F(R_{00} - \frac{1}{2}Rg_{00}) = 3FH^2 = V + \frac{1}{2}K\dot{\chi}^2 - 6\frac{\partial F}{\partial \chi^2}H\chi\dot{\chi} + T_{00}. \tag{3.40}$$

The r.h.s. of the field equations involves the energy-momentum tensor $T_{\mu\nu}$ and the incoherent contribution to the scalar field equation q_χ. The general consistency relation between q_χ, $T_{00} = \rho$ and $T_{ij} = p\delta_{ij}$ reads

$$\dot{\rho} + 3H(\rho + p) + q_\chi\dot{\chi} = 0. \tag{3.41}$$

For an ideal fluid of particles with a χ-dependent mass $m_p(\chi)$ the explicit form of q_χ is given by

$$q_\chi = -\frac{\partial \ln m_p}{\partial \chi}(\rho - 3p). \tag{3.42}$$

In particular, for $m_p(\chi) \sim \chi$ and $\rho - 3p = m_p n_p$, with n_p the number density of particles, Eq. (3.42) reads

$$q_\chi = -\frac{\rho - 3p}{\chi} = -\frac{m_p}{\chi}n_p. \tag{3.43}$$

Let us consider the case where particle masses scale $m_p \sim \chi$ and concentrate on

$$F(\chi) = \chi^2\ ,\ K(\chi) = K. \tag{3.44}$$

A particular case is $V = \lambda\chi^4$. In this case the effective action (3.37) contains no parameter with dimension of mass or length. If, furthermore, all particle masses in $\mathcal{L}_m$ scale precisely $\sim\chi$ no mass scale appears in $\mathcal{L}_m$ either. Such models are scale invariant or dilatation invariant [3, 37]. Scale symmetry can be realized by a fixed point in the "running" of dimensionless couplings and mass ratios as a function of χ. If the strong gauge coupling, normalized a momentum scale $q^2 = \chi^2$, is independent of χ, the "confinement" scale Λ_{QCD} scales $\sim \chi$. For a scale invariant potential for the Higgs doublet

$$\mathcal{L}_h = \frac{\lambda_h}{2}(h^\dagger h - \epsilon_h\chi^2)^2 \tag{3.45}$$

the minimum occurs for

$$h_0 \sim \chi, \tag{3.46}$$

such that for constant Yukawa couplings one has

$$m_e \sim \chi, \tag{3.47}$$

and similar for quark and other charged lepton masses.

The cosmology of a model with exact scale symmetry is simple. After Weyl scaling the potential becomes $V' = \lambda M^4$ and particle masses are constant. The model describes a standard cosmology with cosmological constant λM^4, coupled to an exactly massless Goldstone boson with derivative couplings, the dilaton. The dilaton settles to an arbitrary constant value in early cosmology and is not relevant for late cosmology [3]. In particular, this type of model cannot account for dynamical dark energy.

The situation changes profoundly if we allow for violations of scale symmetry (dilatation anomaly) [3]. For example, we may consider a cosmological constant in the Jordan frame, $V = V_0$, or a quadratic potential $V = \mu^2\chi^2$. In both cases the potential in the Einstein frame decays exponentially,

$$V = M^4 \exp\left(-\frac{\alpha\varphi}{M}\right), \tag{3.48}$$

with $\alpha = 4/\sqrt{K+6}$ for $V = V_0$ and $\alpha = 2/\sqrt{K+6}$ for $V = \mu^2\chi^2$. (We absorb a multiplicative constant by a shift in φ.) The scalar "cosmon" field will roll down the potential, $\varphi(t \to \infty) \to \infty,\ V(t \to \infty) \to 0$. Models of this type with constant particle masses in the Jordan frame lead to non-trivial cosmologies [38,39]. They are excluded, however, by the bounds on the time variation of m_n/M since the coupling β is large.

At this point a simple setting for a realistic dynamical dark energy becomes visible. One may combine a dilatation anomaly in the potential, say $V = V_0$ or $V = \mu^2\chi^2$, with a scale invariant standard model of particle physics. If the charged lepton masses and quark masses as well as Λ_{QCD} all scale proportional to χ, the nucleon and charged lepton masses as well as binding energies and cross sections become independent of φ in the Einstein frame. All observational bounds on time varying fundamental couplings and apparent violations of the equivalence principle are obeyed. The first realistic model of dynamical dark energy or quintessence was actually a "modified gravity" of this type [3]. Models of this type can also explain the recent increase in the fraction of dark energy Ω_h [40,41]. Scale symmetry violation in the neutrino sector induced by a dilatation anomaly in the sector of heavy singlet fields entering by the seesaw mechanism can account for an increasing neutrino mass in the Einstein frame, $\beta < 0$. This stops the evolution of φ as soon as neutrinos become non-relativistic, typically around $z = 5$. From this time on the cosmology looks very similar to a cosmological constant.

Scalar tensor models that lead to dynamical dark energy for the present cosmological epoch [3,27,42–47] are sometimes called "extended quintessence". By virtue of Weyl scaling they are equivalent to a subclass of "coupled quintessence" [4,5,48–56]. Constant particle masses in the Jordan frame imply in the Einstein frame a universal coupling β for all massive particles, while χ-dependent masses offer more realistic perspectives. As compared to constant particle masses in extended quintessence, coupled quintessence is a more general concept where the cosmon-matter coupling can vary from one species to another. While the effective coupling β_n to nucleons has to be very small, more sizeable couplings to dark matter are allowed ($\beta_{dm} \lesssim 0.1$), and the cosmon-neutrino coupling can be large, say $\beta_\nu \approx 100$. Present data slightly favor a non-zero coupling, $\beta \approx 0.07$ [57].

3.6 Slow Freeze Universe

In this section we briefly describe a simple scalar-tensor model with only three cosmologically relevant dimensionless parameters [17]. It is based on the effective action

$$\Gamma = \int d^4x\sqrt{g}\left\{-\frac{\chi^2}{2}R + \left(\frac{2}{\alpha^2} - 3\right)\partial^\mu\chi\partial_\mu\chi + V(\chi)\right\}. \tag{3.49}$$

The potential

$$V = \frac{\mu^2\chi^4}{m^2 + \chi^2}\,,\ \lambda = \frac{\mu^2}{m^2}. \tag{3.50}$$

shows a crossover between two scale invariant limits, one for $\chi \to 0$ with $V \approx \lambda\chi^4$, and the other for $\chi \to \infty$ with $V/\chi^4 \to \mu^2/\chi^2 \to 0$. The mass scales μ and m violate scale symmetry. We take

$$\mu = 2 \cdot 10^{-33}\,\text{eV} \tag{3.51}$$

and $m \approx 10^6\mu$. The Planck mass χ being dynamical, no tiny dimensionless parameter for the cosmological constant appears in this model.

For "late cosmology" after inflation we can approximate

$$V = \mu^2\chi^2. \tag{3.52}$$

During radiation domination the universe shrinks [32] according to a de Sitter solution with negative constant Hubble parameter

$$H = -\frac{\alpha}{2}\mu. \tag{3.53}$$

In this period the value of the cosmon field χ increases exponentially according to

$$\dot{s} = \frac{\dot{\chi}}{\chi} = \alpha\mu \ , \ \chi \sim \exp(\alpha\mu t). \tag{3.54}$$

Due to the shrinking of the universe with scale factor $a \sim 1/\sqrt{\chi}$ the energy density in radiation increases $\sim \chi^2$,

$$\rho_r = 3\left(\frac{\alpha^2}{4} - 1\right)\mu^2\chi^2, \tag{3.55}$$

similar to the potential and kinetic energy in the homogeneous scalar field which obey

$$\rho_h = V + \frac{2}{\alpha^2}\dot{\chi}^2 = 3\mu^2\chi^2. \tag{3.56}$$

This results in a constant fraction of early dark energy [58, 59]

$$\frac{\rho_h}{\rho_r + \rho_h} = \Omega_e = \frac{4}{\alpha^2}. \tag{3.57}$$

While the temperature increases during radiation domination, $T \sim (\rho_r)^{\frac{1}{4}} \sim \sqrt{\chi}$, the particle masses increase even faster $\sim \chi$. The equilibrium number density of a given species gets strongly Boltzmann-suppressed once a particle mass exceeds T. With Fermi scale $\langle h \rangle \sim \chi$ and $\Lambda_{QCD} \sim \chi$, as well as constant dimensionless couplings, the decay rates scale $\sim \chi$, and all cross sections and interaction rates scale with the power of χ corresponding to their dimension. As a consequence, nucleosynthesis proceeds as in usual cosmology, now triggered by nuclear binding energies and the neutron-proton mass difference exceeding the temperature as χ increases. The evolution of all dimensionless quantities is the same as in standard cosmology, once we measure time in units of the (decreasing) inverse nucleon mass. The resulting element abundancies are essentially the same as in standard cosmology. The only difference arises from the presence of a fraction of early dark energy (3.57). This acts similarly to the presence of an additional radiation component, resulting in a lower bound on α from nucleosynthesis [3, 4, 60, 61]. Later on, protons and electrons combine to hydrogen once the atomic binding energy (increasing $\sim \chi$) exceeds the temperature $T \sim \sqrt{\chi}$. Up to small effects of early dark energy the quantitative properties of the CMB-emission are the same as in standard cosmology. The effect of early dark energy on the detailed distribution of CMB-anisotropies gives so far the strongest bound on α, $\alpha \gtrsim 10$ [62–67].

The ratio of matter to radiation energy density increases as $\rho_m/\rho_r \sim \chi a$, with $a \sim \chi^{-\frac{1}{2}}$ during radiation domination ($Ta = const.$). This triggers the transition to a matter dominated scaling solution once ρ_m exceeds ρ_r, given again by a shrinking de Sitter universe

$$H = -\frac{\alpha\mu}{3\sqrt{2}}\,,\ \dot{s} = \frac{\alpha\mu}{\sqrt{2}}\,,\ \rho_m = \frac{2}{3}(\alpha^2 - 3)\mu^2\chi^2, \tag{3.58}$$

with a constant fraction of early dark energy $\Omega_e = 3/\alpha^2$. Observations of redshifts of distant galaxies are explained by the size of atoms shrinking faster than the distance between galaxies [32, 68–70], resulting in an increase of the relevant ratio $\sim a\chi$.

The transition to the present dark energy dominated epoch can be triggered by neutrinos. Assume that the heavy singlet scale entering the neutrino masses by the seesaw mechanism decreases with increasing χ. Neutrino masses will then grow faster than χ, with positive

$$\tilde{\gamma}(\chi) = \frac{1}{2}\frac{\partial \ln \left(m_\nu(\chi)/\chi\right)}{\partial \ln \chi}. \tag{3.59}$$

The value of $\tilde{\gamma}$ in the present epoch will be the third dimensionless cosmological parameter of our model besides α and μ/m. Together with the present neutrino mass it determines the present dark energy density.

In a rather recent cosmological epoch ($z \approx 5$) the neutrinos become non-relativistic. For $\tilde{\gamma} \gg 1$ the increase of their mass faster than χ stops effectively the time evolution of the cosmon field. The dark energy density ρ_h remains frozen at the value it had at this moment, relating it to the average neutrino mass. More precisely, the cosmological solution oscillates around a very slowly evolving "average solution" for which the r.h.s. of Eq. (3.38) vanishes to a good approximation, $V = \tilde{\gamma}\rho_\nu$. This yields for the homogeneous dark energy density ρ_h the interesting quantitative relation [40]

$$\rho_h^{\frac{1}{4}} = 1.27\left(\frac{\tilde{\gamma} m_\nu}{\mathrm{eV}}\right)^{\frac{1}{4}} 10^{-3}\mathrm{eV}. \tag{3.60}$$

(Present neutrino masses on earth may deviate from the value of m_ν according to the cosmological average solution, due to oscillations and a reduction factor for neutrinos inside large neutrino lumps [71, 72]. Cosmological bounds on m_ν are modified due to the mass variation.)

For low redshift $z \lesssim 5$ cosmology is very similar to the ΛCDM-model with an effective equation of state for dark energy (more precisely the coupled cosmon-neutrino fluid) very close to -1,

$$w = -1 + \frac{\Omega_\nu}{\Omega_h} = -1 + \frac{m_\nu(t_0)}{12\mathrm{eV}}. \tag{3.61}$$

An important observational distinction to the ΛCDM-model is the clumping of the neutrino background on very large scales which may render it observable [71, 73–75]. The parameter μ in Eq. (3.51) obtains from the observed value of the present dark energy density $\sqrt{\rho_h} = (2 \cdot 10^{-3}\,\mathrm{eV})^2 \approx \sqrt{V} = \mu M$. This also fixes $\tilde{\gamma} m_\nu = 6.15\,\mathrm{eV}$.

Primordial cosmology corresponds to an inflationary epoch. Matter and radiation play no role and we solve the field equations (3.38)–(3.40) with $T_{\mu\nu} = 0, q_\chi = 0$. One finds a scaling solution without a big bang singularity that can be continued to $t \to -\infty$,

$$\chi = \left(\frac{-2m^3}{\sqrt{3}\alpha^2 \mu t} \right)^{\frac{1}{3}}, \quad H = \left(\frac{-2\mu^2}{9\alpha^2 t} \right)^{\frac{1}{3}}, \quad \frac{\dot{\chi}}{\chi} = -\frac{1}{3t}. \tag{3.62}$$

The spectrum of primordial density fluctuations generated during inflation will be discussed below.

A universe shrinking during radiation and matter domination was much colder in the past than the present background radiation. Its shrinking was very slow, with $|H| \approx \alpha\mu$ only slightly faster than the present expansion rate. During inflation the expansion was even slower, cf. Eq. (3.62). The typical time scale of the universe was never much shorter than 10^{10} yr. Despite the unusual aspects of such a "slow freeze" picture of the evolution of the universe no present observation is in contradiction to it.

For a quantitative discussion of observables it is useful to perform a Weyl scaling in order to bring this model to the form (3.3)–(3.6). In the Einstein frame the potential decays exponentially for large φ

$$V' = \lambda M^4 \left[1 + \exp\left(\frac{\alpha\varphi}{M} \right) \right]^{-1}. \tag{3.63}$$

Particle masses except for the neutrinos do not depend on φ, while the cosmon-neutrino coupling

$$\beta = -M \frac{\partial \ln m_\nu}{\partial \varphi} = -\frac{\tilde{\gamma}}{\alpha} \tag{3.64}$$

realizes growing neutrino quintessence.

A quantitative discussion of the spectrum of density fluctuations is straightforward in the Einstein frame. For the inflationary epoch, our model can be treated in the slow roll approximation. For fluctuations corresponding to the present scale of galaxies or clusters, which have crossed the horizon N e-foldings before the end of inflation, one finds for the spectral index n

$$n = \frac{1}{2N} = 0.96 - 0.967, \tag{3.65}$$

while the tensor amplitude r is very small

$$r = \frac{8}{N^2\alpha^2} < 3 \cdot 10^{-5}. \tag{3.66}$$

A realistic amplitude for the primordial density fluctuations is found for

$$\frac{\mu}{m} = \frac{5}{N\alpha} \cdot 10^{-4}. \tag{3.67}$$

The spectrum of primordial density fluctuations of our model is compatible with Planck-results [67].

Our model has no more free parameters than the ΛCDM-model and is therefore subject to many observational tests. Its compatibility with all present observations demonstrates how a simple modification of gravity can lead to a rather natural setting with a unified description of inflation and present dark energy. The naturalness of the simple quadratic potential for large χ, $V = \mu^2\chi^2$, may look less obvious if the model would be originally formulated in the Einstein frame with a potential (3.63). While we could add a cosmological constant to $V(\chi)$ without affecting the late time behavior for large χ, an addition of a constant to Eq. (3.63) would drastically change the late time cosmology. Thus the issue of naturalness of an asymptotically vanishing cosmological constant looks very different in modified gravity (Jordan frame) or the associated standard gravity (Einstein frame).

3.7 Modified Gravity with $f(R)$

Let us next discuss $f(R)$-theories, where $\mathscr{L}_g$ takes the form

$$\mathscr{L}_g = -\frac{M^4}{2} f(y) \, , \; y = \frac{R}{M^2}. \tag{3.68}$$

We will see that they are equivalent to models of coupled quintessence with a coupling $\beta = 1/\sqrt{6}$. Due to their rather simple structure they are among the most popular models of modified gravity [22, 76–84].

We start with a simple example where f contains terms linear and quadratic in R,

$$\Gamma[g_{\mu\nu}] = \int_x \sqrt{g}\left\{-\frac{cM^2}{2}R - \frac{\alpha}{2}R^2\right\} \, , \; f(y) = cy + \alpha y^2. \tag{3.69}$$

This includes the model used by A. Starobinski [18] in his early discussion of the inflationary universe. It is straightforward to see that this model is equivalent to a scalar model with

$$\Gamma[\phi, g_{\mu\nu}] = \int_x \sqrt{g}\left\{-\frac{cM^2}{2}R - \frac{\alpha}{2}R^2 + \frac{\alpha}{2}\left(\frac{\phi}{\alpha} - R\right)^2\right\}. \tag{3.70}$$

Indeed, the scalar field equation,

$$\frac{\delta \Gamma}{\delta \phi} = 0, \tag{3.71}$$

has a general solution

$$\phi = \alpha R. \tag{3.72}$$

Reinsertion into the effective action yields Eq. (3.69). Expanding the last term in Eq. (3.70) yields the equivalent scalar-gravity model

$$\Gamma[\phi, g_{\mu\nu}] = \int_x \sqrt{g} \left\{ V(\phi) - \frac{M^2}{2} \left(c + \frac{2\phi}{M^2} \right) R \right\}, \tag{3.73}$$

with potential

$$V(\phi) = \frac{1}{2\alpha} \phi^2. \tag{3.74}$$

At this stage the modified gravity model (3.69) has been transformed into a scalar-tensor model (3.73).

We next perform a Weyl scaling with

$$w^2 = \frac{1}{c + \frac{2\phi}{M^2}}, \tag{3.75}$$

resulting in

$$\Gamma[\phi', g'_{\mu\nu}] = \int_x \sqrt{g'} \left\{ V' - \frac{M^2}{2} \left(R' - \frac{3}{2} (\ln w^2)_{;}{}^{\mu} (\ln w^2)_{;\mu} \right) \right\}, \tag{3.76}$$

with

$$V' = w^4 V = \frac{\phi^2}{2\alpha \left(c + \frac{2\phi}{M^2} \right)^2}. \tag{3.77}$$

The canonical normalization of the scalar kinetic term obtains for

$$\varphi = \sqrt{\frac{3}{2}} M \ln \left(c + \frac{2\phi}{M^2} \right), \tag{3.78}$$

corresponding to

$$w^2 = exp\left\{-\sqrt{\frac{2}{3}}\frac{\varphi}{M}\right\}. \tag{3.79}$$

The modified gravity model appears now as a model of quintessence without any modification of gravity,

$$\Gamma[\varphi, g'_{\mu\nu}] = \int_x \sqrt{g'}\left\{V' + \frac{1}{2}\partial^\mu\varphi\partial_\mu\varphi - \frac{M^2}{2}R'\right\}. \tag{3.80}$$

The potential decays exponentially for large φ

$$V'(\varphi) = \frac{M^4}{8\alpha}\left(1 - c\exp\left(-\sqrt{\frac{2}{3}}\frac{\varphi}{M}\right)\right)^2. \tag{3.81}$$

We take $\alpha > 0$ such that the potential is bounded from below.

It is instructive to expand the potential for small φ

$$V'(\varphi) = \frac{M^4}{8\alpha}\left\{(1-c)^2 + \sqrt{\frac{8}{3}}c(1-c)\frac{\varphi}{M} + \frac{2}{3}c(2c-1)\frac{\varphi^2}{M^2} + \dots\right\}. \tag{3.82}$$

For $c = 1$ the leading term is the quadratic

$$V'(\varphi) = \frac{M^4}{12\alpha}\varphi^2 + \dots \tag{3.83}$$

with scalar mass given by

$$m_\varphi = \frac{M}{\sqrt{6\alpha}}. \tag{3.84}$$

For α of the order one this mass turns out to be of the order of the Planck mass. In this case the scalar field settles very early in cosmology to the minimum of the potential at $\varphi = 0$. Subsequently, the potential V' plays no role for late cosmology. Cosmology is described by standard gravity coupled to a massive scalar field. The situation is similar for the corresponding modification of gravity. The term $\sim \alpha R^2$ in Eq. (3.69) can play a role during inflation [18], but is irrelevant for late cosmology. If one wants to have the term $\sim \alpha R^2$ to play a role in the present cosmological epoch one needs a huge value of α such that αR becomes comparable to M^2,

$$\alpha \approx 10^{60}. \tag{3.85}$$

This points to a very general issue for $f(R)$-theories: The deviations from Einstein's equation play a role in present cosmology only if the expansion in derivatives involves huge coefficients or diverges. In other words, any function $f(y)$

which admits a Taylor expansion around $f(y)$ with coefficients that are substantially smaller than 10^{60} leads to modifications of gravity that are not observable in the present cosmological evolution. This remark extends to more general effective actions, involving, for example, $R_{\mu\nu}R^{\mu\nu}$.

For $c > 0$ the potential has a minimum for a finite value of φ

$$\varphi_{\min} = \sqrt{\frac{3}{2}} M \ln c. \tag{3.86}$$

We observe that at the minimum the effective cosmological constant vanishes

$$V'(\varphi_{\min}) = 0. \tag{3.87}$$

The scalar mass (3.84) is independent of c. For $c < 0$ the minimum of V' occurs for $\varphi \to \infty$, with

$$V(\varphi \to \infty) = \frac{M^4}{8\alpha}. \tag{3.88}$$

In this case the scalar mass vanishes in the asymptotic limit. A realistic effective cosmological constant would require

$$\alpha \approx 10^{120}. \tag{3.89}$$

A major problem for $f(R)$-models is the universal large coupling $\beta = 1/\sqrt{6}$ of the cosmon to all massive particles in the Einstein frame. Indeed, the Weyl scaling will take for all $f(R)$-models the form (3.79). This implies for the nucleon mass in the Einstein frame

$$m'_n = w m_n = \exp\left\{-\frac{1}{\sqrt{6}}\frac{\varphi}{M}\right\} m_n, \tag{3.90}$$

resulting in a cosmon-nucleon coupling

$$\beta_n = -M\frac{\partial}{\partial\varphi}\ln m'_n = \frac{1}{\sqrt{6}}. \tag{3.91}$$

Thus $f(R)$-theories are equivalent to coupled quintessence. In order to obey the observational bound (3.34) on m_n/M the cosmon is allowed to vary only by a tiny amount since nucleosynthesis. Furthermore, unless the cosmon mass is large enough, the large value $\beta_n = 1/\sqrt{6}$ contradicts post-Newtonian gravity measurements in the solar system. The cosmological scalar mass is typically very small, however, if the modifications of gravity are important in present cosmology (e.g. Eq. (3.84) with huge α). Due to this clash,realistic models need to invoke the chameleon mechanism [36]. The combination of the absence of a Taylor expansion (with moderate coefficients) and the need for the chameleon mechanism limits severely the choice of realistic functions $f(y)$. At the end, realistic functions are very close to $f(y) = c_0 + y$, with $c_0 = \lambda/M^4$ related to the cosmological constant λ. In the next section we will sketch how part of these problems can be avoided for $f(R)$ theories with field dependent particle masses.

We end this section by a short discussion of the general map from an $f(R)$-theory to coupled quintessence. Consider a scalar-tensor theory with

$$\Gamma = \int_x \sqrt{g}\{-\phi R + V(\phi)\}. \tag{3.92}$$

The solution of the field equation for the scalar field expresses $\phi(R)$ as a function of R, given implicitly by

$$\frac{\partial V}{\partial \phi} = R. \tag{3.93}$$

For $\partial^2 V/\partial\phi^2 \neq 0$ this solution is unique. Insertion of $\phi(R)$ into the action (3.92) yields an equivalent $f(R)$-theory (3.68) with

$$f\left(\frac{R}{M^2}\right) = \frac{2}{M^4}\{R\phi(R) - V(\phi(R))\}. \tag{3.94}$$

By virtue of Eq. (3.93) the function $f(y) = f(R/M^2)$ obeys the relation

$$\frac{\partial f(y)}{\partial y} = \frac{2\phi(R)}{M^2}. \tag{3.95}$$

The construction above associates to a given potential $V(\phi)$ an equivalent $f(R)$-model. Inversely, for a given $f(y)$ Eqs. (3.94), (3.95) yield the potential $V(\phi)$ as a Legendre transform

$$V(\phi) = \frac{M^4}{2}\left(y\frac{\partial f(y)}{\partial y} - f(y)\right), \tag{3.96}$$

with $y(\phi)$ given by Eq. (3.95). This holds provided Eq. (3.95) has a unique solution, i.e. for $\partial^2 f/\partial y^2 \neq 0$.

A Weyl scaling brings finally the action (3.92) to the standard form (3.80). Due to the absence of a kinetic term in Eq. (3.92) the dependence of the conformal factor w on the normalized scalar field φ is universal,

$$w^2 = \frac{M^2}{2\phi} = \exp\left\{-\sqrt{\frac{2}{3}}\frac{\varphi}{M}\right\}. \tag{3.97}$$

As a consequence, $f(R)$-theories with constant particle masses are found to be equivalent to coupled quintessence, with a universal coupling $\beta = 1\sqrt{6}$ given by Eq. (3.91). For the normalized scalar field in the Einstein frame the potential is related to $f(y)$ by

$$V'(\varphi) = \frac{M^2}{2}\,\frac{Rf' - f}{(f')^2}. \tag{3.98}$$

As an example, we may consider

$$f(y) = f_0 y^\gamma. \tag{3.99}$$

Equation (3.95) implies

$$\phi = \frac{\gamma f_0 M^2}{2}\left(\frac{R}{M^2}\right)^{\gamma-1}, \quad R = M^2\left(\frac{2\phi}{\gamma f_0 M^2}\right)^{\frac{1}{\gamma-1}}, \tag{3.100}$$

and the potential in the scalar-tensor model reads

$$V(\phi) = \frac{M^4(\gamma-1)}{2} f(y) = \frac{M^4(\gamma-1) f_0}{2}\left(\frac{2\phi}{\gamma f_0 M^2}\right)^{\frac{\gamma}{\gamma-1}}. \tag{3.101}$$

Weyl scaling leads in the Einstein frame to an additional factor $(M^2/2\phi)^2$ for V', such that

$$V' = \frac{M^4(\gamma-1)}{2\gamma}(\gamma f_0)^{-\frac{1}{\gamma-1}}\left(\frac{M^2}{2\phi}\right)^{1-\frac{1}{\gamma-1}}. \tag{3.102}$$

For the particular "critical" value $\gamma = 2$ the potential V' is constant. For $1 < \gamma < 2$ the minimum of V' occurs for $\phi = 0$, $V'(\phi = 0) = 0$. On the other hand, for $\gamma > 2$ the potential takes its minimal value for $\phi \to \infty$, with

$$V'(\phi \to \infty) = 0. \tag{3.103}$$

With

$$\phi = \frac{M^2}{2}\exp\left\{\sqrt{\frac{2}{3}}\frac{\varphi}{M}\right\} \tag{3.104}$$

the limit $\phi \to \infty$ corresponds to $\varphi \to \infty$ and we observe that the potential $V'(\varphi)$ decays to zero exponentially. These models are of the same type as the one discussed in Sect. 3.6, using in (3.49) the identifications $\phi = 2\chi^2$, $\alpha^2 = 2/3$, and $V(\chi) = V'(\phi = 2\chi^2)$.

We observe that the addition of a cosmological constant $\bar{\lambda}_c$ in the effective action for modified gravity results in

$$f(y) = f_0 y^\alpha - e_0, \quad \bar{\lambda}_c = \frac{e_0 M^4}{2}. \tag{3.105}$$

After Weyl scaling this adds to V' a part

$$\Delta V' = \frac{e_0 M^8}{8\phi^2}. \tag{3.106}$$

This becomes irrelevant for large ϕ. Modified gravity theories with $\gamma > 2$ are an example for a self-tuning of the cosmological constant to zero as a consequence of the asymptotic cosmological solution for large time.

For $\gamma = 1$ one has Einstein gravity without an additional scalar degree of freedom. For $0 < \gamma < 1$ and $f_0 > 0$ the potential V' is negative, diverging for $\phi \to 0$. For negative f_0 one finds negative ϕ such that the gravitational constant would have a wrong sign, leading to instability. The range $0 < \gamma < 1$ does not seem to lead to a reasonable cosmology. We may, however, consider the values $\gamma < 0$, $f_0 < 0$, with positive γf_0 and ϕ. The potential V' is now again positive, decaying to zero for $\phi \to \infty$. The behaviour is similar as for $\gamma > 2$ and $f_0 > 0$. We conclude that $f(R)$-models could lead to interesting cosmologies with a dynamical self-tuning of the cosmological constant to zero if all particles are massless. For massive particles one has to find a way to avoid the universal large cosmon-matter coupling $\beta = 1/\sqrt{6}$, as we will discuss in the next section.

3.8 $f(R)$-Gravity with Varying Particle Masses

Having established the equivalence between $f(R)$-models and scalar-tensor theories a simple solution of the problem of a too large cosmon-matter coupling becomes visible. One may follow the strategy (i) in Sect. 3.5: If particle masses scale $\sqrt{\phi}$ in the Jordan frame, their mass will be constant in the Einstein frame, implying $\beta = 0$. Realistic models may therefore be found if the particle masses show an appropriate effective field dependence in the Jordan frame.

Let us consider the quarks and charged leptons. In the standard model of particle physics their masses are proportional to the expectation value h_0 of the Higgs doublet h. For cosmology, h_0 is replaced by the value of h according to the cosmological solution. If this solution implies that h_0 scales proportional to $\sqrt{\phi}$ we will find a vanishing cosmon-matter coupling $\beta = 0$ in the Einstein frame.

To be specific, we consider a first model where the effective action for gravity and the Higgs doublet is given by

$$\Gamma = \int_x \sqrt{g}\left\{-\frac{a}{2}\left(\frac{R-2\mu^2}{2\epsilon}\right)^2 - h^\dagger h\left(\frac{R-2\mu^2}{2\epsilon}\right) + \frac{Z_h}{2}\partial^\mu h^\dagger \partial_\mu h\right\}. \quad (3.107)$$

The parameters a and ϵ are dimensionless, such that scale symmetry is violated only by the parameter μ with dimension of mass. The function $f(y)$ is quadratic in y, with field dependent coefficient of the linear term,

$$f = a\left(\frac{y-2\mu^2/M^2}{2\epsilon}\right)^2 + \frac{2h^\dagger h}{M^2}\left(\frac{y-2\mu^2/M^2}{2\epsilon}\right). \quad (3.108)$$

We emphasize that the Planck mass M is not a parameter of the model (3.107). In Eq. (3.108) it is merely introduced by the conventions for y and f.

According to Eq. (3.95) the relation between ϕ and R reads

$$\phi = \frac{a}{4\epsilon^2}(R-2\mu^2) + \frac{h^\dagger h}{2\epsilon}, \quad (3.109)$$

and the corresponding potential of the equivalent scalar-tensor model becomes

$$V = \frac{1}{2a}(h^\dagger h - 2\epsilon\phi)^2 + 2\mu^2\phi. \quad (3.110)$$

Identifying $2\phi = \chi^2$ we can associate the first term in Eq. (3.110) with Eq. (3.45), for $\epsilon_h \sim \epsilon$ and $\lambda_h \sim 1/a$. For $h = h_0$ the potential becomes $V = 2\mu^2\phi = \mu^2\chi^2$, which coincides for large χ with the potential (3.50).

In the Einstein frame the Higgs doublet is rescaled according to

$$h' = wh\,,\ w^2 = \frac{M^2}{2\phi}. \quad (3.111)$$

This yields for the potential

$$V' = \frac{1}{2a}(h'^\dagger h' - \epsilon M^2)^2 + \frac{\mu^2 M^4}{2\phi}. \quad (3.112)$$

It is obvious that h' settles to a constant value at the minimum of V', implying constant particle masses if the dimensionless Yukawa couplings are constant, $\beta = 0$.

The kinetic terms for h' and ϕ in the Einstein frame read

$$\begin{aligned}\mathscr{L}_{\text{kin}} &= \frac{Z_h}{2}\{\partial^\mu h'^\dagger \partial_\mu h' + \frac{1}{2}\partial^\mu \ln\phi\ \partial_\mu(h'^\dagger h')\} \\ &\quad + \frac{1}{8}(6M^2 + Z_h h'^\dagger h')\partial^\mu \ln\phi\ \partial_\mu \ln\phi. \end{aligned} \quad (3.113)$$

For constant $h'^{\dagger}h' = \epsilon M^2$ the remaining kinetic term for ϕ becomes

$$\mathscr{L}_{\text{kin}} = \frac{M^2}{8}(6 + Z_h\epsilon)\partial^\mu \ln\phi\, \partial_\mu \ln\phi. \tag{3.114}$$

Neglecting the contribution $\sim Z_h\epsilon$ (see below) the normalized scalar field is related to ϕ by Eq. (3.104) and Eq. (3.79) remains valid. For $h' = h'_0$ the potential decays exponentially

$$V' = \mu^2 M^2 \exp\left(-\frac{\alpha\varphi}{M}\right)\ ,\ \alpha = \sqrt{\frac{2}{3}}. \tag{3.115}$$

The value of α is too small for allowing for the scaling solutions with constant early dark energy fraction $\Omega_e < 1$. This issue is related to the absence of a kinetic term for ϕ in Eq. (3.73) or (3.92). For initial values of ϕ_{in} much smaller than M^2 the universe becomes scalar dominated long before the present epoch, leading to unrealistic cosmology. For $\phi_{\text{in}} \gg M^2$ the scalar potential will play a role only in the far future and the model cannot account for dark energy. Realistic cosmology requires a particular initial value with ϕ_{in} close to $M^2/2$. Cosmology is then of the type of "thawing quintessence". The need for a particular choice of initial conditions makes the model perhaps less attractive than the scaling solution found in the model of Sect. 3.6.

Despite this shortcoming, the simple model (3.107) offers an interesting perspective on a dynamical fine tuning of the cosmological constant. Indeed, the effective cosmological constant vanishes asymptotically in the Einstein frame, even if we add an additional constant to the modified gravitational action (3.107). In the Einstein frame the resulting contribution to $V'(\varphi)$ decays exponentially for large φ. Scale symmetry becomes exact for $\varphi \to \infty$ and the cosmon corresponds in this limit to the dilaton, the Goldstone boson associated to the spontaneous breaking of scale symmetry.

It is also interesting to discuss the issue of dilatation symmetry in the framework of $f(R)$-models. For $\mu = 0$ the effective action (3.107) is scale invariant. The potential in the Einstein frame (3.110) has then one exactly massless direction, realizing the Goldstone boson. This demonstrates how the expected Goldstone boson arises in a model (3.107) that does not contain an explicit scalar singlet degree of freedom.

The model (3.107) contains large dimensionless parameters. The Fermi scale is given by the canonically normalized doublet in the Einstein frame, $h_R = Z_h^{1/2}h'_0 = 175\,\text{GeV}$. This implies $\epsilon_h = Z_h\epsilon = (h_R/M)^2 \approx 5 \cdot 10^{-33}$. The renormalized quartic Higgs coupling is $\lambda_h = 1/(aZ_h^2)$, such that the prefactor of R^2 in Eq. (3.107) becomes $a/(8\epsilon^2) = 1/(8\lambda_h\epsilon_h^2) \approx 10^{64}/(2\lambda_h)$, similar in size to Eq. (3.85).

More reasonable couplings arise if one associates h with a scalar field in some grand unified theory instead of the Higgs doublet. In this event ϵ_h could be roughly of the order one. The effective quark and lepton masses are then suppressed by the gauge hierarchy, i.e. the ratio between the Fermi scale and the scale h_0 which is now characteristic for grand unification. If gauge couplings take a fixed value for momenta given by h also the QCD-confinement scale and therefore the nucleon masses are proportional to h, completing our mechanism for vanishing β. If h is associated with a field characteristic for grand unification the parameter a/ϵ^2 can be taken to be of the order one, such that the prefactor of the term $\sim R^2$ in Eq. (3.107) is of the order one. In this case, however, ϕ is given essentially by $h^\dagger h/(2\epsilon)$ and the term $\sim R^2$ in Eq. (3.107) plays only a negligible role. (The limit $a \to 0$ has no qualitative influence on the late cosmology of this model.)

As a second example we consider a family of models

$$\Gamma = \int_x \sqrt{g}\left\{\sigma(R^2+\rho)^{\frac{\gamma}{2}} + \bar{\lambda}_c - \frac{1}{2\epsilon}h^\dagger h R + \frac{Z_h}{2}\partial^\mu h^\dagger \partial_\mu h\right\}. \tag{3.116}$$

The relation between ϕ, h and R reads

$$x = -\gamma\tilde{\sigma} y(y^2+\tilde{\rho})^{\frac{\gamma}{2}-1}, \tag{3.117}$$

with

$$x = \frac{2\epsilon\phi - h^\dagger h}{2M^2\epsilon}\ ,\ y = \frac{R}{M^2}\ ,\ \tilde{\sigma} = \sigma M^{2\gamma-4}\ ,\ \tilde{\rho} = \frac{\rho}{M^4}. \tag{3.118}$$

In terms of ϕ the effective action becomes

$$\Gamma = \int_x \sqrt{g}\left\{-\phi R + V(\phi,h) + \frac{Z_h}{2}\partial^\mu h^\dagger \partial_\mu h\right\}, \tag{3.119}$$

where

$$V = M^4\tilde{\sigma}(y^2+\tilde{\rho})^{\frac{\gamma}{2}-1}\left\{\tilde{\rho} + (1-\gamma)y^2\right\} + \bar{\lambda}_c, \tag{3.120}$$

and y is related to ϕ and h by Eq. (3.117). After Weyl scaling the effective action for the metric and the scalars ϕ and h' takes a standard form

$$\Gamma = \int_x \sqrt{g'}\left\{-\frac{M^2}{2}R' + V'(\phi,h') + \mathcal{L}_{\text{kin}}\right\}, \tag{3.121}$$

with

$$V'(\phi,h') = \frac{M^4 V}{4\phi^2}, \tag{3.122}$$

and $\mathcal{L}_{\text{kin}}$ given by Eq. (3.113). Again, y is related to x by Eq. (3.117) with

$$x = \frac{\phi(\epsilon M^2 - h'^{\dagger}h')}{\epsilon M^4}. \tag{3.123}$$

We may next investigate the field equation for h'. One finds a static solution with $h_0'^{\dagger} h_0' = \epsilon M^2$ provided $V(x)$ has its minimum for $x = 0$. Particle masses are then constant in the Einstein frame, $\beta = 0$. Inserting $h'^{\dagger}h' = \epsilon M^2$ and assuming $y(x = 0) = 0$ the potential gets a simple form

$$V' = \frac{M^4}{4\phi^2}\left(\bar{\lambda}_c + \sigma\rho^{\frac{\gamma}{2}}\right). \tag{3.124}$$

For positive $V_0 = \bar{\lambda}_c + \sigma\rho^{\gamma/2}$ it decays to zero for $\phi \to \infty$. For a canonical scalar field (neglecting the term $Z_h\epsilon$ in Eq. (3.114)) the potential decays exponentially

$$V' = V_0 \exp\left(-\frac{\alpha\varphi}{M}\right) \;, \; \alpha = \sqrt{\frac{8}{3}}. \tag{3.125}$$

Again, this value of α is too small in order to realize the scaling solution with $\Omega_e < 1$. Cosmology is similar to our first example, with realistic thawing quintessence realized for initial values ϕ_{in} close to $10^{60}\sqrt{V_0}$.

We notice that cosmology is the same for all ranges of γ, σ and ρ for which V has its minimum for $x = 0$. For $\rho > 0$ the effective action (3.116) and the potential V are analytic. A special case occurs for $\rho = 0$ which is similar to the model (3.99) except for the additional coupling to h. The potential is no longer analytic

$$V = M^4\tilde{\sigma}(1-\gamma)|y|^{\gamma} + \bar{\lambda}_c = M^4\tilde{\sigma}(1-\gamma)\left|\frac{x}{\gamma\tilde{\sigma}}\right|^{\frac{\gamma}{\gamma-1}} + \bar{\lambda}_c. \tag{3.126}$$

For $\rho > 0$ the potential (3.126) describes the behavior for large $y^2 \gg \tilde{\rho}$. We observe that for $\gamma < 1$ the limit $x \to 0$ can be reached for $|y| \to 0$ or $|y| \to \infty$. If the potential minimum corresponds to the second case the value $V_0 = \bar{\lambda}_c$ may only be reached for asymptotic time $t \to \infty$.

We conclude that the problematic universal cosmon-matter coupling β in the Einstein frame can be avoided if $f(R)$-theories allow for a suitable field dependence of particle masses. The other generic problem of $f(R)$-models, namely the need of large couplings multiplying the terms in a Taylor expansion of $f(y)$, will need a particular physics explanation which produces and stabilizes such large couplings appearing in the effective action. (In the generic case quantum fluctuations lead to a very fast running of very large dimensionless couplings, typically bringing them to values of the order one or making them divergent.) At present, we are still far from constructing an $f(R)$-model which would show a similar simplicity as the scalar-tensor theory discussed in Sect. 3.6. The benefit would be, of course, that no explicit scalar field χ is needed in modified gravity.

3.9 Non-local Gravity

For non-local gravity (see [85] for a recent review and references) the action involves the inverse of the covariant Laplacian $\mathscr{D}$, or similar operators that grow strongly for small covariant momenta. As a consequence, such modifications of gravity can play a role at long distances, without invoking very large dimensionless parameters as α in the preceding section. Already the first non-local gravity model in this spirit [86] has noted the equivalence to a model of a scalar field coupled to gravity.

Let us consider the effective action [86]

$$\mathscr{L}_g = \frac{M^2}{2}\left\{-R + \frac{\tau^2}{2} R\mathscr{D}^{-1}R\right\}, \tag{3.127}$$

with covariant derivative D_μ and covariant Laplacian

$$\mathscr{D} = -D^\mu D_\mu. \tag{3.128}$$

(In order to make Eq. (3.127) well defined one has to regularize the operator $\mathscr{D}^{-1}$ [86].) The model (3.127) admits an equivalent formulation as a scalar-tensor model with effective action

$$\Gamma = \int_x \sqrt{g}\left\{-\frac{M^2}{2}(1+\tau\phi)R - \frac{M^2}{4}\partial^\mu\phi\partial_\mu\phi\right\}. \tag{3.129}$$

Indeed, the field equation for ϕ,

$$\mathscr{D}\phi = -D^\mu D_\mu \phi = -\tau R, \tag{3.130}$$

expresses ϕ as a functional of the metric,

$$\phi = -\tau\mathscr{D}^{-1}R. \tag{3.131}$$

Inserting the formal solution (3.131) into the action (3.129) yields the equivalent effective action (3.127) of non-local gravity.

The scalar-tensor theory (3.129) can be brought to the standard form of a coupled quintessence model by use of a Weyl scaling with

$$w = (1+\tau\phi)^{-\frac{1}{2}}. \tag{3.132}$$

The resulting kinetic term,

$$\mathscr{L}_{\text{kin}} = \frac{M^2}{4}\left(\frac{3\tau^2}{(1+\tau\phi)^2} - \frac{1}{1+\tau\phi}\right)\partial^\mu\phi\partial_\mu\phi, \tag{3.133}$$

can be cast into a standard normalization (3.6) by defining φ with

$$\frac{\partial\varphi}{\partial\phi} = \frac{M}{\sqrt{2}(1+\tau\phi)}\sqrt{3\tau^2 - (1+\tau\phi)}. \tag{3.134}$$

The potential vanishes for this model, similar to Brans-Dicke theory.

The Weyl scaling typically leads to coupled quintessence. Consider non-local modified gravity (3.127) and a particle with constant mass m. One obtains in the Einstein frame a φ-dependent mass, $m' = w(\varphi)m$. Defining the φ-dependent coupling $\beta(\varphi)$ by

$$\beta(\varphi) = -M\frac{\partial \ln m'}{\partial\varphi} \tag{3.135}$$

one obtains

$$\beta = \left[6 - \frac{2}{\tau}\left(\phi + \frac{1}{\tau}\right)\right]^{-\frac{1}{2}}, \tag{3.136}$$

where ϕ can be expressed in terms of φ using Eq. (3.134).

We observe that stability requires a positive effective Planck mass and a positive kinetic term (3.133), which is realized for the range

$$0 \le 1 + \tau\phi \le 3\tau^2. \tag{3.137}$$

In this range β is well defined. The minimum value for β is

$$\beta_{\min} = \frac{1}{\sqrt{6}}, \tag{3.138}$$

resembling a Brans-Dicke theory with $\omega = 0$. Such a large coupling is not compatible with observation, such that the model (3.127) is not phenomenologically viable [86].

In summary, the gravitational part of non-local gravity models has no problem of consistency. It is equivalent to standard gravity coupled to a massless scalar, similar to Brans-Dicke theory. Adding relativistic particles as photons remains unproblematic. Issues of compatibility with observation arise, however, if massive particles are considered within non-local gravity. The coupling between the scalar field and massive particles typically turns out to be unacceptably large.

One may construct large classes of consistent non-local gravity models by starting from a local scalar-tensor model that only contains terms linear and quadratic in ϕ. Such generalizations of Eq. (3.129) can contain higher derivatives of ϕ, a coupling of ϕ to higher order curvature invariants, terms $\sim R\partial^\mu\phi\partial_\mu\phi$ or $\sim R^{\mu\nu}\partial_\mu\phi\partial_\nu\phi$ etc. The field equations for ϕ involve terms linear in ϕ as well as a ϕ-independent "source term". The general solutions are functionals of the metric. Inserting these solutions into the action yields consistent models of non-local gravity. Consistency does not imply compatibility with observation, however. It seems not easy to avoid a too large coupling between the scalar field and massive particles in the Einstein frame.

While non-local modifications of gravity are consistent, it is not easy to motivate why the quantum effective action for gravity should have this form. Unless one can identify some quantum effect producing such non-localities they may not look very natural, however. For the moment, the only physically well motivated origin of non-localities of the type discussed in this section that is known to us arises from the exchange of an effective massless degree of freedom, similar to the Coulomb interaction between electrons or the Newtonian interaction between massive particles. In this event it seems much simpler to use directly a field for the exchanged particle.

3.10 Higher Derivative Modified Gravity with Second Order Field Equations

We have seen that $f(R)$-theories and a large class of non-local gravity theories can be mapped to a quintessence model,

$$\Gamma = \int_x \sqrt{g'} \left\{ -\frac{1}{2} M^2 R' + \frac{1}{2} \partial^\mu \varphi \partial_\mu \varphi + V(\varphi) \right\}, \tag{3.139}$$

by an appropriate Weyl scaling. One may ask how large is the class of modified gravity theories that can be mapped to the simple action (3.139) by suitable field transformations. (See [87,88] for earlier work on this issue.) A large class of actions involving higher derivatives, that nevertheless lead to second order field equations, has been found by Horndeski [15]. One would like to know if they are equivalent to the action (3.139).

Part of the answer can be given by considering general field transformations

$$\begin{aligned} \varphi &= v(\chi, R, \partial^\mu \chi \partial_\mu \chi, \dots), \\ g'_{\mu\nu} &= w^{-2}(\chi, R, \partial^\mu \chi \partial_\mu \chi, \dots) g_{\mu\nu} + s_1(\chi, R, \partial^\mu \chi \partial_\mu \chi, \dots) \partial_\mu \chi \partial_\nu \chi \\ &\quad + s_2(\chi, R, \partial^\mu \chi \partial_\mu \chi, \dots) R_{\mu\nu} + \dots \end{aligned} \tag{3.140}$$

Here v, w, s_1, s_2 are functions of various possible scalars that can be formed from χ and $g_{\mu\nu}'$, with dots standing for additional scalars as $R_{\mu\nu} R^{\mu\nu}$, $\partial_\mu \chi \partial_\nu \chi R^{\mu\nu}$ etc. We only require that the objects on the r.h.s. of Eq. (3.140) have the correct tensor transformation properties.

It is obvious that a very large class of effective actions for modified gravity can be constructed by inserting Eq. (3.140) into Eq. (3.139).

$$\Gamma[\chi, g_{\mu\nu}] = \Gamma\big[\varphi[\chi, g_{\mu\nu}]\,,\ g'_{\rho\sigma}[\varphi, g_{\mu\nu}]\big]. \tag{3.141}$$

All these models have as physical degrees of freedom a scalar coupled to the graviton. Even though these actions can contain an arbitrary number of derivatives, the field equations will finally be second order field equations, equivalent to those derived from the action (3.139). The requirement of equivalence imposes, however, some mild conditions on the functions appearing in Eq. (3.140). What is needed is the invertibility of the variable transformation (3.140).

We may demonstrate this explicitly for transformations with $s_1 = s_2 = 0$. The field equations for the transformed action,

$$\frac{\partial \Gamma}{\partial \chi(x)} = 0\,,\ \frac{\partial \Gamma}{\partial g_{\mu\nu}(x)} = 0, \tag{3.142}$$

can be expressed as (∂ stands here for functional derivatives)

$$\int_y \left\{ \frac{\partial \Gamma}{\partial g'_{\mu\nu}(y)} \frac{\partial w^{-2}(y)}{\partial \chi(x)} g_{\mu\nu}(y) + \frac{\partial \Gamma}{\partial \varphi(y)} \frac{\partial v(y)}{\partial \chi(x)} \right\} = 0, \tag{3.143}$$

and

$$\int_y \left\{ \frac{\partial \Gamma}{\partial g'_{\rho\sigma}(y)} \frac{\partial w^{-2}(y)}{\partial g_{\mu\nu}(x)} g_{\rho\sigma}(y) + w^{-2}(y) \frac{\partial \Gamma}{\partial g'_{\mu\nu}(x)} \delta(y-x) + \frac{\partial \Gamma}{\partial \varphi(y)} \frac{\partial v(y)}{\partial g_{\mu\nu}(x)} \right\} = 0. \tag{3.144}$$

Obviously, the solutions of the field equations of the action (3.139),

$$\frac{\partial \Gamma}{\partial \varphi(y)} = 0\,,\ \frac{\partial \Gamma}{\partial g'_{\mu\nu}(y)} = 0, \tag{3.145}$$

are also solutions of the field equations (3.142). The conditions on the functions w and v have to ensure that no additional "spurious" solutions are generated by the transformation (3.140).

Consider, for example, the case $w = 1$. Then the matrix $\partial v(y)/\partial \chi(x)$ should be invertible, such that Eq. (3.143) implies $\partial \Gamma/\partial \varphi(y) = 0$. Invertibility means that a function $H(x,z)$ exists such that

$$\int_x \frac{\partial v(y)}{\partial \chi(x)} H(x,z) = \delta(y-z). \tag{3.146}$$

For $w = 1$ the gravitational field equation (3.144) reads

$$\frac{\partial \Gamma}{\partial g'_{\mu\nu}(x)} + \int_y \left\{ \frac{\partial \Gamma}{\partial \varphi(y)} \frac{\partial v(y)}{\partial g_{\mu\nu}(x)} \right\} = 0. \tag{3.147}$$

The second term vanishes for invertible $\partial v(y)/\partial \chi(x)$ since $\delta \Gamma/\delta \varphi(y) = 0$, such that both field equations (3.145) must be obeyed necessarily. Similarly, we may consider $v = \chi$ and an invertible matrix $\partial g'_{\rho\sigma}(y)/\partial g_{\mu\nu}(x)$. The field equation (3.144) implies then $\partial \Gamma/\partial g'_{\mu\nu}(x) = 0$, such that Eq. (3.143) guarantees $\partial \Gamma/\partial \varphi(x) = 0$. Again, the field equations (3.145) must be necessarily obeyed. This generalizes to arbitrary transformations $g_{\rho\sigma}(y)\big[g_{\mu\nu}(x)\big]$, as in Eq. (3.140). Invertible transformations with $v = \chi$ or $w = 1$ can be combined to yield more general invertible transformations. We conclude that invertibility of the transformation (3.140) guarantees the absence of spurious solutions, such that the effective action $\Gamma[g_{\mu\nu}, \chi]$ is fully equivalent to $\Gamma[g'_{\mu\nu}, \varphi]$ given by Eq. (3.139).

It may be instructive to discuss two simple examples of field transformations with $w = 1$. For the first we take $\varphi = v(\chi, R)$, such that

$$\frac{\partial v(y)}{\partial \chi(x)} = \frac{\partial v}{\partial \chi}\big(\chi(x), R(x)\big)\delta(y - x). \tag{3.148}$$

If $\partial v/\partial \chi$ is non-vanishing for all χ and R the transformation is invertible. On the other hand, if $\partial v/\partial \chi = 0$ has a solution $\chi_0(R)$, the configuration $\chi = \chi_0(R)$ solves the field equation $\partial \Gamma/\partial \chi(x) = 0$ without being a solution of Eq. (3.145). This is an example of a spurious solution. A second example with a spurious solution is

$$\varphi(x) = m^{-3}\big(\chi;^{\mu}{}_{\mu}(x) + m^2 \chi(x)\big)\chi(x). \tag{3.149}$$

While the solutions (3.145) remain solutions of the field equations (3.142), additional solutions of Eq. (3.142) are provided by $\chi;^{\mu}_{\mu} + m^2\chi = 0$. This model can still be cast into the form of an action with at most two derivatives, involving two scalar fields. Besides the solutions (3.145) one has new solutions for non-zero values of a free massive scalar field with mass m. (The last term in Eq. (3.144) ensures that the energy momentum tensor of the second scalar field is induced in the gravitational field equation.) Many transformations with higher derivatives are invertible and do not lead to spurious solutions, however.

It remains an interesting question if invertible transformations of the type (3.140) are sufficient in order to show the equivalence of a large class (or all) of Horndeski's models with the effective action (3.139). This seems very likely to us for models that contain no further physical degrees of freedom besides a scalar and the graviton. The effective action (3.141) obtained by inserting Eq. (3.140) into Eq. (3.139) may even lead to still larger classes of higher derivative modified gravity for which all cosmological solutions can be obtained from second order field equations. Further generalizations are possible if one adds scalar, vector or tensor fields with no more than two derivatives to Eq. (3.139), and subsequently makes a field transformation of the type (3.140).

The field transformations (3.140) are a convenient way to construct effective actions (3.141) that only involve second order field equations for the scalar-graviton system. This does not mean that all models based on an action (3.141) are equivalent to those based on the action (3.139). The field transformations also affect the matter part $\mathscr{L}_m$. Consider a model where matter is minimally coupled to the metric $g_{\mu\nu}$ and particle masses are χ-independent. It becomes typically a model of coupled quintessence with non-minimal gravitational interactions once written in terms of $g'_{\mu\nu}$ and φ. The inverse of the transformation (3.140), which maps the action (3.141) onto (3.139), can induce in the matter and radiation sector a complicated dependence on φ and $g'_{\mu\nu}$. Even if we approximate $\mathscr{L}_m$ in the generalized Jordan frame (3.141) by free massive or massless particles, non-trivial interactions will appear in the Einstein frame (3.139). This is the way how the functions v, w, s_1, s_2 in Eq. (3.140) can affect the predictions for observations. Similar to $f(R)$-models also, the much more general class of models (3.141) encounters often problems with too large effective couplings $\sim \beta$ in the Einstein frame.

Conclusions

Can one distinguish modified gravity from dark energy by observation? In view of the equivalence of a large class of modified gravity models with coupled quintessence an answer to this question is not straightforward. Statements that modified gravity and quintessence lead to different growth factors for cosmic structures apply only to quintessence models without coupling to matter. We have seen, however, that the quintessence models that are equivalent to modified gravity typically have a nonzero coupling β between the cosmon and different forms of matter. (This coupling needs not to be the same for all species of massive particles.) In this view precision measurements of the growth rate can differentiate between uncoupled and coupled quintessence and determine bounds on β. The issue if there are modified gravity models that can be distinguished observationally from coupled quintessence is much harder to answer.

Modified gravity models almost always involve new degrees of freedom besides the graviton. This is a consequence of the fact that models for a massless spin two particle are severely constrained by consistency requirements. The conjecture that consistency requires diffeomorphism symmetry (more precisely its unimodular subgroup) has never been proven, but no counter examples are known either. A model containing a massless spin two particle as the only degree of freedom is then rather close to general relativity. Modifications of gravity therefore typically involve additional degrees of freedom, as scalars or massive spin two particles.

(continued)

The field description of the additional degrees of freedom is not unique. For example, a scalar may be described as a component of the metric (modified gravity) or by a separate field (quintessence). Very large classes of models can be mapped onto each other by non-linear field transformations. Field relativity states that observables cannot depend on the choice of fields. For models related by field transformations no observational distinction is possible. We have seen that this holds for variable gravity models where the Planck mass is field dependent. It also applies to $f(R)$-models and large classes of non-local gravity. Very general models equivalent to coupled quintessence models have been discussed in the preceding section.

For all these models modified gravity and coupled quintessence should merely be seen as two different pictures describing the same reality, in analogy to the Jordan frame and Einstein frame for the metric. For practical computations of the evolution of homogeneous cosmology and fluctuations around this background the simplest way uses the Einstein frame. This holds both for the linear treatment of fluctuations and for numerical simulations in the non-linear regime. The physical effects of the cosmon-matter coupling β are intuitively accessible in the Einstein frame.

For modified gravity models that are equivalent to coupled quintessence one may ask: why then discuss them at all? If there is no observational distinction, the discussion of such modifications of gravity may at first sight look like a redundant exercise. A deeper answer concerns questions of simplicity and naturalness. Models of modified gravity can be very simple and involve no unnatural parameters. Nevertheless, the equivalent description in the Einstein frame by coupled quintessence may hide simplicity and naturalness in the complexity of the field transformation. An example is the big bang singularity. We have presented in Sect. 3.6 a modified gravity model for which the "beginning" of the universe is very slow and cold. It has no big bang singularity, the cosmological solution can be continued to the infinite past $t \to -\infty$. In the Einstein frame the same model is described as a hot big bang. Models may be regular in the Jordan frame and show a big bang singularity in the Einstein frame. This singularity is then due to a singularity in the field transformation [32], in close analogy to a coordinate singularity.

The question of naturalness is often closely linked to symmetries. Scale symmetry is explicitly visible in the modified gravity description of the models in Sects. 3.5 and 3.6. It is realized by a multiplicative rescaling of the metric and the scalar field χ. In the presence of quantum fluctuations scale symmetry is violated by χ-dependent ("running") dimensionless couplings. For fixed points of the running exact (quantum-) scale symmetry is restored. For the quantum effective action (3.49) such fixed points are present for $\chi \to 0$ and $\chi \to \infty$ [17].

(continued)

In our model in Sect. 3.6 the asymptotic value

$$\lambda_\infty = \lim_{\chi\to\infty} V(\chi)/\chi^4 \tag{3.150}$$

vanishes for the fixed point at $\chi \to \infty$. This can be motivated by properties of a possible ultraviolet fixed point in dilaton quantum gravity [89] or by dilatation symmetry in higher dimensions [90,91]. The fixed point with $\lambda_\infty = 0$ is the deeper reason why the cosmological constant vanishes asymptotically in the Einstein frame, $\lim_{\varphi\to\infty} V'(\varphi) \to 0$. Without this understanding of naturalness as a consequence of fixed point properties one would argue in the Einstein frame that naturalness suggests the addition of a constant to Eq. (3.63). Apparently convincing qualitative arguments on the induction of a cosmological constant by quantum fluctuations in the Einstein frame yield very different results when applied in the Jordan frame. A constant term in $V(\chi)$ yields a term $V'(\varphi) \sim \exp(-2\alpha\varphi/M)$ in the Einstein frame which vanishes for $\varphi \to \infty$. This is one more example how modified gravity can shed new light on questions of naturalness.

The possibility of field transformations from modified gravity theories to coupled quintessence models in the Einstein frame is an extremely useful tool for the discussion of observational consequences of a model. It should not prevent us, however, to look for modified gravity theories distinguished by simplicity and naturalness.

Acknowledgements The author would like to thank L. Amendola and V. Pettorino for pointing out several useful references.

References

1. A.G. Riess et al. (Supernova Search Team), Observational evidence from supernovae for an accelerating universe and a cosmological constant. Astron. J. **116**, 1009–1038 (1998). arXiv:astro-ph/9805201 [astro-ph]
2. S. Perlmutter et al. (Supernova Cosmology Project), Measurements of Omega and Lambda from 42 high redshift supernovae. Astrophys. J. **517**, 565–586 (1999). arXiv:astro-ph/9812133 [astro-ph]
3. C. Wetterich, Cosmology and the fate of dilatation symmetry. Nucl. Phys. **B302**, 668 (1988)
4. C. Wetterich, The Cosmon model for an asymptotically vanishing time dependent cosmological 'constant'. Astron. Astrophys. **301**, 321–328 (1995). arXiv:hep-th/9408025 [hep-th]
5. L. Amendola, Coupled quintessence. Phys. Rev. **D62**, 043511 (2000). arXiv:astro-ph/9908023 [astro-ph]
6. S. 'i. Nojiri, S.D. Odintsov, Introduction to modified gravity and gravitational alternative for dark energy. eConf C **0602061**, 06 (2006) [Int. J. Geom. Meth. Mod. Phys. **4**, 115 (2007)] [hep-th/0601213]

7. S. Capozziello, M. Francaviglia, Extended theories of gravity and their cosmological and astrophysical applications. Gen. Rel. Grav. **40**. 357 (2008). [arXiv:0706.1146 [astro-ph]]
8. T.P. Sotiriou, V. Faraoni, f(R) theories of gravity. Rev. Mod. Phys. **82**, 451 (2010). [arXiv:0805.1726 [gr-qc]]
9. A. De Felice, S. Tsujikawa, f(R) theories. Living Rev. Rel. **13**, 3 (2010). [arXiv:1002.4928 [gr-qc]]
10. T. Clifton, P. G. Ferreira, A. Padilla, C. Skordis, Modified gravity and cosmology. Phys. Rept. **513**, 1 (2012). [arXiv:1106.2476 [astro-ph.CO]]
11. L. Amendola et al. [Euclid Theory Working Group Collaboration], Cosmology and fundamental physics with the Euclid satellite. Living Rev. Rel. **16**, 6 (2013). [arXiv:1206.1225 [astro-ph.CO]]
12. S. Capozziello, S. Nojiri, S.D. Odintsov, Dark energy: the equation of state description versus scalar-tensor or modified gravity. Phys. Lett. B **634**, 93 (2006). [hep-th/0512118]
13. W. Hu, I. Sawicki, A parameterized post-Friedmann framework for modified gravity. Phys. Rev. **D76**, 104043 (2007). arXiv:0708.1190 [astro-ph]
14. M. Kunz, L. Amendola, D. Sapone, Dark energy phenomenology, (2008). arXiv:0806.1323 [astro-ph]
15. G.W. Horndeski, Second-order scalar-tensor field equations in a four-dimensional space. Int. J. Theor. Phys. **10**, 363 (1974)
16. C. Wetterich, Variable gravity Universe. Phys. Rev. D **89**, 024005 (2014). [arXiv:1308.1019 [astro-ph.CO]]
17. C. Wetterich, Hot big bang or slow freeze? (2014). arXiv:1401.5313 [astro-ph.CO]
18. A.A. Starobinsky, A new type of isotropic cosmological models without singularity. Phys. Lett. **B91**, 99–102 (1980)
19. C. Wetterich, Spontaneous compactification in higher dimensional gravity. Phys. Lett. B **113**, 377 (1982)
20. Q. Shafi, C. Wetterich, Cosmology from higher dimensional gravity. Phys. Lett. B **129**, 387 (1983)
21. Q. Shafi, C. Wetterich, Inflation with higher dimensional gravity. Phys. Lett. B **152**, 51 (1985) [Conf. Proc. C **841031**, 454 (1984)]
22. H.A. Buchdahl, Non-linear Lagrangians and cosmological theory.' Mon. Not. Roy. Astron. Soc. **150**, 1 (1970)
23. G.R. Dvali, G. Gabadadze, M. Porrati, 4-D gravity on a brane in 5-D Minkowski space. Phys. Lett. B **485**, 208 (2000). [hep-th/0005016]
24. C. Brans, R.H. Dicke, Mach's principle and a relativistic theory of gravitation. Phys. Rev. **124**, 925–935 (1961)
25. H. Weyl, Gravitation and electricity. Sitzungsber. Preuss. Akad. Wiss. Berlin (Math. Phys.) **1918**, 465 (1918)
26. R.H. Dicke, Mach's principle and invariance under transformation of units. Phys. Rev. **125**, 2163–2167 (1962)
27. C. Wetterich, Cosmologies with variable Newton's 'Constant'. Nucl. Phys. B **302**, 645 (1988)
28. T. Damour, G. Esposito-Farese, Tensor multiscalar theories of gravitation. Class. Quant. Grav. **9**, 2093 (1992)
29. E. E. Flanagan, The Conformal frame freedom in theories of gravitation. Class. Quant. Grav. **21**, 3817 (2004). [gr-qc/0403063]
30. R. Catena, M. Pietroni, L. Scarabello, Einstein and Jordan reconciled: a frame-invariant approach to scalar-tensor cosmology. Phys. Rev. D **76**, 084039 (2007). [astro-ph/0604492]
31. N. Deruelle, M. Sasaki, Conformal equivalence in classical gravity: the example of 'Veiled' General Relativity. Springer Proc. Phys. **137**, 247 (2011). [arXiv:1007.3563 [gr-qc]]
32. C. Wetterich, A Universe without expansion. Dark Universe **2**, 184 (2013). arXiv:1303.6878 [astro-ph.CO]
33. T. Dent, S. Stern, C. Wetterich, Primordial nucleosynthesis as a probe of fundamental physics parameters. Phys. Rev. D **76**, 063513 (2007). [arXiv:0705.0696 [astro-ph]]

34. B. Bertotti, L. Iess, P. Tortora, A test of general relativity using radio links with the Cassini spacecraft. Nature **425**, 374 (2003)
35. J.R. Ellis, S. Kalara, K.A. Olive, C. Wetterich, Density dependent couplings and astrophysical bounds on light scalar particles. Phys. Lett. B **228**, 264 (1989)
36. J. Khoury, A. Weltman, Chameleon cosmology. Phys. Rev. D **69**, 044026 (2004) [astro-ph/0309411]
37. Y. Fujii, Origin of the gravitational constant and particle masses in scale invariant scalar-tensor theory. Phys. Rev. **D26**, 2580 (1982)
38. O. Bertolami, Time dependent cosmological term. Nuovo Cim. B **93**, 36 (1986)
39. L.H. Ford, Cosmological constant damping by unstable scalar fields. Phys. Rev. D **35**, 2339 (1987)
40. L. Amendola, M. Baldi, C. Wetterich, Quintessence cosmologies with a growing matter component. Phys. Rev. D **78**, 023015 (2008). [arXiv:0706.3064 [astro-ph]]
41. C. Wetterich, Growing neutrinos and cosmological selection. Phys. Lett. B **655**, 201 (2007). [arXiv:0706.4427 [hep-ph]]
42. T. Chiba, Quintessence, the gravitational constant, and gravity. Phys. Rev. D **60**, 083508 (1999). [gr-qc/9903094]
43. J.-P. Uzan, Cosmological scaling solutions of nonminimally coupled scalar fields. Phys. Rev. D **59**, 123510 (1999). [gr-qc/9903004]
44. F. Perrotta, C. Baccigalupi, S. Matarrese, Extended quintessence. Phys. Rev. D **61**, 023507 (1999). [astro-ph/9906066]
45. V. Faraoni, Inflation and quintessence with nonminimal coupling. Phys. Rev. D **62**, 023504 (2000). [gr-qc/0002091]
46. C. Baccigalupi, S. Matarrese, F. Perrotta, Tracking extended quintessence. Phys. Rev. D **62**, 123510 (2000). [astro-ph/0005543]
47. V. Pettorino, C. Baccigalupi, G. Mangano, Extended quintessence with an exponential coupling. JCAP **0501**, 014 (2005). [astro-ph/0412334]
48. L. Amendola, D. Tocchini-Valentini, Stationary dark energy: the present universe as a global attractor. Phys. Rev. D **64**, 043509 (2001). [astro-ph/0011243]
49. L. Amendola, D. Tocchini-Valentini, Baryon bias and structure formation in an accelerating universe. Phys. Rev. D **66**, 043528 (2002). [astro-ph/0111535]
50. D. Comelli, M. Pietroni, A. Riotto, Dark energy and dark matter. Phys. Lett. B **571**, 115 (2003). [hep-ph/0302080]
51. X. Zhang, Coupled quintessence in a power-law case and the cosmic coincidence problem. Mod. Phys. Lett. A **20**, 2575 (2005). [astro-ph/0503072]
52. V. Pettorino, C. Baccigalupi, Coupled and extended quintessence: theoretical differences and structure formation. Phys. Rev. D **77**, 103003 (2008). [arXiv:0802.1086 [astro-ph]]
53. M. Baldi, V. Pettorino, G. Robbers, V. Springel, Hydrodynamical N-body simulations of coupled dark energy cosmologies. Mon. Not. Roy. Astron. Soc. **403**, 1684 (2010). [arXiv:0812.3901 [astro-ph]]
54. F. Saracco, M. Pietroni, N. Tetradis, V. Pettorino, G. Robbers, Non-linear matter spectra in coupled quintessence. Phys. Rev. D **82**, 023528 (2010). [arXiv:0911.5396 [astro-ph.CO]]
55. H. Wei, J. Liu, Z.-C. Chen, X.-P. Yan, Indistinguishability of warm dark matter, modified gravity, and coupled cold dark matter. Phys. Rev. D **88**, 043510 (2013). [arXiv:1306.1364 [astro-ph.CO]]
56. H. Wei, S.N. Zhang, How to distinguish dark energy and modified gravity? Phys. Rev. D **78**, 023011 (2008). [arXiv:0803.3292 [astro-ph]]
57. V. Pettorino, Testing modified gravity with Planck: the case of coupled dark energy. Phys. Rev. D 88, **063519** (2013). [arXiv:1305.7457 [astro-ph.CO]]
58. C. Wetterich, Phenomenological parameterization of quintessence. Phys. Lett. B **594**, 17 (2004). [astro-ph/0403289]
59. M. Doran, G. Robbers, Early dark energy cosmologies. JCAP **0606**, 026 (2006). [astro-ph/0601544]

60. M. Birkel, S. Sarkar, Nucleosynthesis bounds on a time varying cosmological ‘constant’. Astropart. Phys. **6**,197 (1997). [astro-ph/9605055]
61. R. Bean, S.H. Hansen, A. Melchiorri, Early universe constraints on a primordial scaling field. Phys. Rev. D **64**, 103508 (2001). [astro-ph/0104162]
62. M. Doran, G. Robbers, C. Wetterich, Impact of three years of data from the Wilkinson Microwave Anisotropy Probe on cosmological models with dynamical dark energy. Phys. Rev. D **75**, 023003 (2007). [astro-ph/0609814]
63. E. Calabrese, R. de Putter, D. Huterer, E.V. Linder, A. Melchiorri, Future CMB constraints on early, cold, or stressed dark energy. Phys. Rev. D **83**,023011 (2011). [arXiv:1010.5612 [astro-ph.CO]]
64. C.L. Reichardt, R. de Putter, O. Zahn, Z. Hou, New limits on early dark energy from the south pole telescope. Astrophys. J. **749**, L9 (2012). [arXiv:1110.5328 [astro-ph.CO]]
65. J.L. Sievers et al. [Atacama Cosmology Telescope Collaboration], The Atacama Cosmology Telescope: cosmological parameters from three seasons of data. JCAP **1310**, 060 (2013). [arXiv:1301.0824 [astro-ph.CO]]
66. V. Pettorino, L. Amendola, C. Wetterich, How early is early dark energy? Phys. Rev. D **87**, 083009 (2013). [arXiv:1301.5279 [astro-ph.CO]]
67. P.A.R. Ade et al. [Planck Collaboration], Planck 2013 results. XVI. Cosmological parameters. arXiv:1303.5076 [astro-ph.CO]
68. J.V. Narlikar, H. Arp, Flat spacetime cosmology: a unified framework for extragalactic redshifts. Astrophys. J. **405**, 51–56 (1993)
69. J.V. Narlikar, Two astrophysical applications of conformal gravity. Ann. Phys. **107**, 325 (1977)
70. F. Hoyle, J.V. Narlikar, A conformal theory of gravitation. Proc. R. Soc. London **A294**, 138–148 (1966)
71. Y. Ayaita, M. Weber, C. Wetterich, Neutrino lump fluid in growing neutrino quintessence. Phys. Rev. D **87**(4), 043519 (2013). [arXiv:1211.6589 [astro-ph.CO]]
72. N.J. Nunes, L. Schrempp, C. Wetterich, Mass freezing in growing neutrino quintessence. Phys. Rev. D **83**, 083523 (2011). [arXiv:1102.1664 [astro-ph.CO]]
73. D.F. Mota, V. Pettorino, G. Robbers, C. Wetterich, Neutrino clustering in growing neutrino quintessence. Phys. Lett. B **663**, 160 (2008). [arXiv:0802.1515 [astro-ph]]
74. M. Baldi, V. Pettorino, L. Amendola, C. Wetterich, Oscillating nonlinear large scale structure in growing neutrino quintessence. arXiv:1106.2161 [astro-ph.CO]
75. Y. Ayaita, M. Weber, C. Wetterich, Structure formation and backreaction in growing neutrino quintessence. Phys. Rev. D **85**, 123010 (2012). [arXiv:1112.4762 [astro-ph.CO]]
76. J.C. Hwang, Cosmological perturbations in generalized gravity theories: Formulation. Class. Quant. Grav. **7**, 1613 (1990)
77. S. 'i. Nojiri, S.D. Odintsov, Modified gravity with negative and positive powers of the curvature: unification of the inflation and of the cosmic acceleration. Phys. Rev. D **68**, 123512 (2003). [hep-th/0307288]
78. S. Capozziello, S. Carloni, A. Troisi, Quintessence without scalar fields. Recent Res. Dev. Astron. Astrophys. **1**, 625 (2003). [astro-ph/0303041]
79. S.M. Carroll, V. Duvvuri, M. Trodden, M.S. Turner, Is cosmic speed - up due to new gravitational physics?. Phys. Rev. D **70**, 043528 (2004). [astro-ph/0306438]
80. V. Faraoni, de Sitter space and the equivalence between f(R) and scalar-tensor gravity. Phys. Rev. D **75**, 067302 (2007). [gr-qc/0703044 [GR-QC]]
81. S. Tsujikawa, Matter density perturbations and effective gravitational constant in modified gravity models of dark energy. Phys. Rev. D **76**, 023514 (2007). [arXiv:0705.1032 [astro-ph]]
82. W. Hu, I. Sawicki, Models of $f(R)$ cosmic acceleration that evade solar-system tests. Phys. Rev. D **76**, 064004 (2007). [arXiv:0705.1158 [astro-ph]]
83. L. Amendola, D. Polarski, S. Tsujikawa, Are $f(R)$ dark energy models cosmologically viable?. Phys. Rev. Lett. **98**,131302 (2007). [astro-ph/0603703]
84. L. Amendola, R. Gannouji, D. Polarski, S. Tsujikawa, Conditions for the cosmological viability of f(R) dark energy models. Phys. Rev. D **75**, 083504 (2007). [gr-qc/0612180]
85. R. P. Woodard, Nonlocal models of cosmic acceleration. arXiv:1401.0254 [astro-ph.CO]

86. C. Wetterich, Effective nonlocal Euclidean gravity. Gen. Rel. Grav. **30**, 159 (1998). [gr-qc/9704052]
87. D. Bettoni, S. Liberati, Disformal invariance of second order tensor-scalar theories: framing the Horndeski action. Phys. Rev. D **88**, 084020 (2013). [arXiv:1306.6724 [gr-qc]]
88. M. Zumalacárregui, J. García-Bellido, Transforming gravity: from derivative couplings to matter to second-order scalar-tensor theories beyond the Horndeski Lagrangian. arXiv:1308.4685 [gr-qc]
89. T. Henz, J.M. Pawlowski, A. Rodigast, C. Wetterich, Dilaton quantum gravity. Phys. Lett. B **727**, 298 (2013). [arXiv:1304.7743 [hep-th]]
90. C. Wetterich, Dilatation symmetry in higher dimensions and the vanishing of the cosmological constant. Phys. Rev. Lett. **102**, 141303 (2009). [arXiv:0806.0741 [hep-th]]
91. C. Wetterich, The cosmological constant and higher dimensional dilatation symmetry. Phys. Rev. D **81**, 103507 (2010). [arXiv:0911.1063 [hep-th]]

Chapter 4
The Effective Field Theory of Inflation/Dark Energy and the Horndeski Theory

Shinji Tsujikawa

Abstract The effective field theory (EFT) of cosmological perturbations is a useful framework to deal with the low-energy degrees of freedom present for inflation and dark energy. We review the EFT for modified gravitational theories by starting from the most general action in unitary gauge that involves the lapse function and the three-dimensional geometric scalar quantities appearing in the Arnowitt-Deser-Misner (ADM) formalism. Expanding the action up to quadratic order in the perturbations and imposing conditions for the elimination of spatial derivatives higher than second order, we obtain the Lagrangian of curvature perturbations and gravitational waves with a single scalar degree of freedom. The resulting second-order Lagrangian is exploited for computing the scalar and tensor power spectra generated during inflation. We also show that the most general scalar-tensor theory with second-order equations of motion—Horndeski theory—belongs to the action of our general EFT framework and that the background equations of motion in Horndeski theory can be conveniently expressed in terms of three EFT parameters. Finally we study the equations of matter density perturbations and the effective gravitational coupling for dark energy models based on Horndeski theory, to confront the models with the observations of large-scale structures and weak lensing.

4.1 Introduction

The inflationary paradigm, which was originally proposed to solve a number of cosmological problems in the standard Big Bang cosmology [1, 2], is now widely accepted as a viable phenomenological framework describing the accelerated expansion in the early Universe. In particular, the Cosmic Microwave Background (CMB) temperature anisotropies measured by COBE [3], WMAP [4], and Planck [5] satellites support the slow-roll inflationary scenario driven by a single scalar degree of freedom. Inflation generally predicts the nearly scale-invariant primordial

S. Tsujikawa (✉)
Department of Physics, Faculty of Science, Tokyo University of Science, 1-3 Kagurazaka, Shinjuku-ku, Tokyo 162-8601, Japan
e-mail: shinji@rs.kagu.tus.ac.jp

E. Papantonopoulos (ed.), *Modifications of Einstein's Theory of Gravity at Large Distances*, Lecture Notes in Physics 892, DOI 10.1007/978-3-319-10070-8_4

power spectrum of curvature perturbations [6], whose property is consistent with the observed CMB anisotropies. In spite of its great success, we do not yet know the origin of the scalar field responsible for inflation (dubbed "inflaton").

The observations of the type Ia Supernovae (SN Ia) [7, 8] showed that the Universe entered the phase of another accelerated expansion after the matter-dominated epoch. This has been also supported by other independent observations such as CMB [4] and Baryon Acoustic Oscillations (BAO) [9]. The origin of the late-time cosmic acceleration (dubbed "dark energy") is not identified yet. The simplest candidate for dark energy is the cosmological constant Λ, but if it originates from the vacuum energy appearing in particle physics, the theoretical value is enormously larger than the observed dark energy scale [10, 11]. There is a possibility that some scalar degree of freedom (like inflaton) is responsible for dark energy [12].

Although many models of inflation and dark energy have been constructed in the framework of General Relativity (GR), the modification of gravity from GR can also give rise to the epoch of cosmic acceleration. For example, the Starobinsky model characterized by the Lagrangian $f(R) = R + R^2/(6M^2)$ [1], where R is a Ricci scalar and M is a constant, leads to the quasi de Sitter expansion of the Universe. The recent observational constraints on the dark energy equation of state $w_{\rm DE} = P_{\rm DE}/\rho_{\rm DE}$ (where $P_{\rm DE}$ and $\rho_{\rm DE}$ is the pressure and the energy density of dark energy respectively) imply that the region $w_{\rm DE} < -1$ is favored from the joint data analysis of SN Ia, CMB, and BAO [5, 13, 14]. If we modify gravity from GR, it is possible to realize $w_{\rm DE} < -1$ without having a problematic ghost state (see [15] for reviews).

Given that the origins of inflation and dark energy have not been identified yet, it is convenient to construct a general framework dealing with gravitational degrees of freedom beyond GR. In fact, the EFT of inflation and dark energy provides a systematic parametrization that accommodates possible low-energy degrees of freedom by employing cosmological perturbations as small expansion parameters about the Friedmann-Lemaître-Robertson-Walker (FLRW) background [16–18]. This EFT approach allows one to facilitate the confrontation of models with the cosmological data.

Originally, the EFT of inflation was developed to quantify high-energy corrections to the standard slow-roll inflationary scenario [19]. Expanding the action up to third order in the cosmological perturbations, it is also possible to estimate higher-order correlation functions associated with primordial non-Gaussianities [20]. The EFT formalism was applied to dark energy in connection to the large-distance modification of gravity [21–33]. The advantage of this approach is that practically all the single-field models of inflation and dark energy can be accommodated in a unified way.

Starting from the most general action that depends on the lapse function and other geometric three-dimensional scalar quantities present in the ADM formalism, Gleyzes et al. [28] expanded the action up to quadratic order in cosmological perturbations of the ADM variables. In doing so, the perturbation $\delta\phi$ of a scalar field ϕ can be generally present, but the choice of unitary gauge ($\delta\phi = 0$) allows one to absorb the field perturbation in the gravitational sector. Once we fix the gauge in this way, introducing another scalar-field perturbation implies that the system possesses at least two-scalar degrees of freedom. In fact, such a multi-field scenario was studied in [33] to describe both dark energy and dark matter.

By construction, the EFT formalism developed in [16, 17, 28] keeps the time derivatives under control, while the spatial derivatives higher than second order are generally present. Imposing conditions to eliminate these higher-order spatial derivatives for the general theory mentioned above, Gleyzes et al. [28] derived the quadratic Lagrangian of cosmological perturbations with one scalar degree of freedom. If the scalar degree of freedom is responsible for inflation, for example, the resulting power spectrum of curvature perturbations can be computed on the quasi de Sitter background (along the same lines in [34–38]). In this review, we evaluate the inflationary power spectra of both scalar and tensor perturbations expressed in terms of the ADM variables.

In 1973, Horndeski derived the action of the most general scalar-tensor theories with second-order equations of motion [39]. This theory recently received much attention as an extension of (covariant) Galileons [40–42]. One can show that the four-dimensional action of "generalized Galileons" derived by Deffayet et al. [43] is equivalent to the Horndeski action after a suitable field redefinition [35]. Gleyzes et al. [28] expressed the Horndeski Lagrangian in terms of the ADM variables appearing in the EFT formalism. This allows one to have a connection between the Horndeski theory and the EFF of inflation/dark energy. In fact, it was shown that Horndeski theory belongs to a sub-class of the general EFT action [28].

For the background cosmology, the EFT of inflation/dark energy is characterized by three time-dependent parameters f, Λ, and c [16–18]. This property is useful to perform general analysis for the dynamics of dark energy [30]. In the EFT of dark energy, Gleyzes et al. [28] obtained the equations of linear cosmological perturbations in the presence of non-relativistic matter (dark matter, baryons). This result reproduces the perturbation equations in Horndeski theory previously derived in [44]. We note that the perturbation equations in the presence of another scalar field (playing the role of dark matter) were also derived in [33]. These results are useful to confront modified gravitational models of dark energy with the observations of large-scale structures, weak lensing, and CMB.

In this lecture note, we review the EFT of inflation/dark energy following the recent works of [28, 33].

In Sect. 4.2 we start from a general gravitational action in unitary gauge and derive the background equations of motion on the flat FLRW background.

In Sect. 4.3 we obtain the linear perturbation equations of motion and discuss conditions for avoiding ghosts and Laplacian instabilities of scalar and tensor perturbations.

In Sect. 4.4 the inflationary power spectra of scalar and tensor perturbations are derived for general single-field theories with second-order linear perturbation equations of motion.

In Sect. 4.5 we introduce the action of Horndeski theory and express it in terms of the ADM variables appearing in the EFT formalism.

In Sect. 4.6 we discuss how the second-order EFT action accommodates Horndeski theory as specific cases and provide the correspondence between them.

In Sect. 4.7 we apply the EFT formalism to dark energy and obtain the background equations of motion in a generic way. In Horndeski theory, the equations of matter density perturbations and the effective gravitational coupling are derived in the presence of non-relativistic matter.

The final section is devoted to conclusions.

Throughout the paper we use units such that $c = \hbar = 1$, where c is the speed of light and $\hbar$ is reduced Planck constant. The gravitational constant G is related to the reduced Planck mass $M_{\rm pl} = 2.4357 \times 10^{18}$ GeV via $8\pi G = 1/M_{\rm pl}^2$. The Greek and Latin indices represent components in space-time and in a three-dimensional space-adapted basis, respectively. For the covariant derivative of some physical quantity Y, we use the notation $Y_{;\mu}$ or $\nabla_\mu Y$. We adopt the metric signature $(-,+,+,+)$.

4.2 The General Gravitational Action in Unitary Gauge and the Background Equations of Motion

The EFT of cosmological perturbations allows one to deal with the low-energy degree of freedom appearing for inflation and dark energy. In particular, we are interested in the minimal extension of GR to modified gravitational theories with a single scalar degree of freedom ϕ. The EFT approach is based on the choice of unitary gauge in which the constant time hypersurface coincides with the constant ϕ hypersurface. In other words, this corresponds to the gauge choice

$$\delta\phi = 0\,, \tag{4.1}$$

where $\delta\phi$ is the field perturbation. In this gauge the dynamics of $\delta\phi$ is "eaten" by the metric, so the Lagrangian does not have explicit ϕ dependence about the flat FLRW background.

The EFT of cosmological perturbations is based on the 3+1 decomposition of the ADM formalism [45]. In particular, the $3+1$ splitting in unitary gauge allows one to keep the number of time derivatives under control, while higher spatial derivatives can be generally present. As we will see later, this property is especially useful for constructing theories with second-order time and spatial derivatives. The ADM line element is given by

$$ds^2 = g_{\mu\nu}dx^\mu dx^\nu = -N^2 dt^2 + h_{ij}(dx^i + N^i dt)(dx^j + N^j dt)\,, \tag{4.2}$$

where N is the lapse function, N^i is the shift vector, and h_{ij} is the three-dimensional metric. Then, the four-dimensional metric $g_{\mu\nu}$ can be expressed as $g_{00} = -N^2 + N^i N_i$, $g_{0i} = g_{i0} = N_i$, and $g_{ij} = h_{ij}$. A unit vector orthogonal to the constant t hypersurface Σ_t is given by $n_\mu = -N t_{;\mu} = (-N, 0, 0, 0)$, and hence $n^\mu =$

$(1/N, -N^i/N)$ with $n_\mu n^\mu = -1$. The induced metric $h_{\mu\nu}$ on Σ_t can be expressed as $h_{\mu\nu} = g_{\mu\nu} + n_\mu n_\nu$, so that it satisfies the orthogonal relation $n^\mu h_{\mu\nu} = 0$.

The extrinsic curvature is defined by

$$K_{\mu\nu} = h^\lambda_\mu n_{\nu;\lambda} = n_{\nu;\mu} + n_\mu a_\nu \,, \tag{4.3}$$

where $a^\nu = n^\lambda n^\nu_{;\lambda}$ is the acceleration (curvature) of the normal congruence n^ν. Since there is the relation $n^\mu K_{\mu\nu} = 0$, the extrinsic curvature is the quantity on Σ_t. The internal geometry of Σ_t can be quantified by the three-dimensional Ricci tensor $\mathcal{R}_{\mu\nu} \equiv {}^{(3)}R_{\mu\nu}$ associated with the metric $h_{\mu\nu}$. The three-dimensional Ricci scalar $\mathcal{R} = \mathcal{R}^\mu{}_\mu$ is related to the four-dimensional Ricci scalar R, as

$$R = \mathcal{R} + K_{\mu\nu}K^{\mu\nu} - K^2 + 2(Kn^\mu - a^\mu)_{;\mu} \,, \tag{4.4}$$

where $K \equiv K^\mu{}_\mu$ is the trace of the extrinsic curvature.

In the following we study general gravitational theories that depend on scalar quantities appearing in the ADM formalism. In addition to the lapse N, we have the following scalars

$$K \equiv K^\mu{}_\mu \,, \quad \mathcal{S} \equiv K_{\mu\nu}K^{\mu\nu} \,, \quad \mathcal{R} \equiv \mathcal{R}^\mu{}_\mu \,, \quad \mathcal{Z} \equiv \mathcal{R}_{\mu\nu}\mathcal{R}^{\mu\nu} \,, \quad \mathcal{U} \equiv \mathcal{R}_{\mu\nu}K^{\mu\nu} \,. \tag{4.5}$$

The Lagrangian L of general gravitational theories depends on these scalars, so that the action is given by

$$S = \int d^4x \sqrt{-g}\, L(N, K, \mathcal{S}, \mathcal{R}, \mathcal{Z}, \mathcal{U}; t) \,. \tag{4.6}$$

We do not include the dependence of the scalar quantity $\mathcal{N} = N^\mu N_\mu$ coming from the shift vector, since such a term does not appear even in the most general scalar-tensor theories with second-order equations of motion (see Sect. 4.5). In the action (4.6), the time dependence is also explicitly included because in unitary gauge its dependence is directly related to the scalar degree of freedom, such that $\phi = \phi(t)$. The field kinetic term[1]

$$X \equiv g^{\mu\nu}\partial_\mu\phi\partial_\nu\phi \tag{4.7}$$

depends on the lapse N and the time t. The field ϕ enters the equations of motion through the partial derivatives $L_N \equiv \partial L/\partial N$ and $L_{NN} \equiv \partial^2 L/\partial N^2$.

[1] We caution that the notation of the field kinetic energy is the same as that used in [28, 33], but the notation of X used in [35–38, 44] is $-1/2$ times different.

Let us consider four scalar metric perturbations A, ψ, ζ, and E about the flat FLRW background with the scale factor $a(t)$. The general perturbed metric is given by

$$ds^2 = -e^{2A}dt^2 + 2\psi_{|i}dx^i dt + a^2(t)(e^{2\zeta}\delta_{ij} + \partial_{ij}E)dx^i dx^j \,, \tag{4.8}$$

where $|i$ represents a covariant derivative with respect to h_{ij}, and $\partial_{ij} \equiv \nabla_i\nabla_j - \delta_{ij}\nabla^2/3$. Under the transformation $t \to t + \delta t$ and $x^i \to x^i + \delta^{ij}\partial_j\delta x$, the perturbations $\delta\phi$ and E transform as

$$\delta\phi \to \delta\phi - \dot{\phi}\,\delta t \,, \qquad E \to E - \delta x \,, \tag{4.9}$$

where a dot represents a derivative with respect to t. Choosing the unitary gauge (4.1), the time slicing δt is fixed. The spatial threading δx can be fixed with the gauge choice

$$E = 0 \,. \tag{4.10}$$

On the flat FLRW background with the line element $ds^2 = -dt^2 + a^2(t)\delta_{ij}dx^i dx^j$, the three-dimensional geometric quantities are given by

$$\bar{K}_{\mu\nu} = H\bar{h}_{\mu\nu} \,, \quad \bar{K} = 3H \,, \quad \bar{\mathcal{S}} = 3H^2 \,, \quad \bar{\mathcal{R}}_{\mu\nu} = 0 \,, \quad \bar{\mathcal{R}} = \bar{\mathcal{Z}} = \bar{\mathcal{U}} = 0 \,, \tag{4.11}$$

where a bar represents background values and $H \equiv \dot{a}/a$ is the Hubble parameter. We define the following perturbed quantities

$$\delta K^{\mu}_{\nu} = K^{\mu}_{\nu} - Hh^{\mu}_{\nu} \,, \quad \delta K = K - 3H \,, \quad \delta\mathcal{S} = \mathcal{S} - 3H^2 = 2H\delta K + \delta K^{\mu}_{\nu}\delta K^{\nu}_{\mu} \,, \tag{4.12}$$

where the last equation arises from the first equation and the definition of $\mathcal{S}$. Since $\mathcal{R}$ and $\mathcal{Z}$ vanish on the background, they appear only as perturbations. Up to quadratic order in perturbations, they can be expressed as

$$\delta\mathcal{R} = \delta_1\mathcal{R} + \delta_2\mathcal{R} \,, \qquad \delta\mathcal{Z} = \delta\mathcal{R}^{\mu}_{\nu}\delta\mathcal{R}^{\nu}_{\mu} \,, \tag{4.13}$$

where $\delta_1\mathcal{R}$ and $\delta_2\mathcal{R}$ are first-order and second-order perturbations in $\delta\mathcal{R}$, respectively. The perturbation $\mathcal{Z}$ is higher than first order. The first equality (4.12) implies

$$\mathcal{U} = H\mathcal{R} + \mathcal{R}^{\mu}_{\nu}\delta K^{\nu}_{\mu} \,, \tag{4.14}$$

where the last term is a second-order quantity.

In order to derive the background and perturbation equations of motion, we expand the action (4.6) up to quadratic order in perturbations, as

$$L = \bar{L} + L_N \delta N + L_K \delta K + L_{\mathcal{S}} \delta \mathcal{S} + L_{\mathcal{R}} \delta \mathcal{R} + L_{\mathcal{Z}} \delta \mathcal{Z} + L_{\mathcal{U}} \delta \mathcal{U}$$
$$+ \frac{1}{2} \left(\delta N \frac{\partial}{\partial N} + \delta K \frac{\partial}{\partial K} + \delta \mathcal{S} \frac{\partial}{\partial \mathcal{S}} + \delta \mathcal{R} \frac{\partial}{\partial \mathcal{R}} + \delta \mathcal{Z} \frac{\partial}{\partial \mathcal{Z}} + \delta \mathcal{U} \frac{\partial}{\partial \mathcal{U}} \right)^2 L, \tag{4.15}$$

where a lower index of the Lagrangian L denotes the partial derivatives with respect to the scalar quantities represented in the index. From the second and third relations of Eq. (4.12), the expansion of the term $L_K \delta K + L_{\mathcal{S}} \delta \mathcal{S}$ up to second order reads

$$L_K \delta K + L_{\mathcal{S}} \delta \mathcal{S} = \mathcal{F}(K - 3H) + L_{\mathcal{S}} \delta K^{\mu}_{\nu} \delta K^{\nu}_{\mu}$$
$$\simeq -\dot{\mathcal{F}} - 3H\mathcal{F} + \dot{\mathcal{F}} \delta N + L_{\mathcal{S}} \delta K^{\mu}_{\nu} \delta K^{\nu}_{\mu} - \dot{\mathcal{F}} \delta N^2 \,, \tag{4.16}$$

where

$$\mathcal{F} \equiv L_K + 2HL_{\mathcal{S}} \,. \tag{4.17}$$

In the second line of Eq. (4.16), the term $\mathcal{F}K$ has been integrated by using the relation $K = n^{\mu}{}_{;\mu}$, as

$$\int d^4x \sqrt{-g}\, \mathcal{F} K = -\int d^4x \sqrt{-g}\, n^{\mu} \mathcal{F}_{;\mu} = -\int d^4x \sqrt{-g} \frac{\dot{\mathcal{F}}}{N} \,, \tag{4.18}$$

where the boundary term is dropped. Note that we have also expanded the term $N^{-1} = (1 + \delta N)^{-1}$ up to second order in Eq. (4.16).

The term $\mathcal{U}$ satisfies the relation

$$\alpha(t) \mathcal{U} = \frac{1}{2} \alpha(t) \mathcal{R} K + \frac{1}{2N} \dot{\alpha}(t) \mathcal{R} \,, \tag{4.19}$$

where $\alpha(t)$ is an arbitrary function of t. Using this relation and the fact that $\mathcal{U}$ is a perturbed quantity, it follows that

$$L_{\mathcal{U}} \delta \mathcal{U} = \frac{1}{2} \left(\dot{L_{\mathcal{U}}} + 3HL_{\mathcal{U}} \right) \delta_1 \mathcal{R} + \frac{1}{2} \left(L_{\mathcal{U}} \delta K - \dot{L_{\mathcal{U}}} \delta N \right) \delta_1 \mathcal{R}$$
$$+ \frac{1}{2} \left(\dot{L_{\mathcal{U}}} + 3HL_{\mathcal{U}} \right) \delta_2 \mathcal{R} \,, \tag{4.20}$$

where the first term on the r.h.s. is the first-order quantity, whereas the rest is second-order.

Summing up the terms discussed above, the zeroth-order and first-order Lagrangians of (4.15) are given, respectively, by

$$L_0 = \bar{L} - \dot{\mathcal{F}} - 3H\mathcal{F} \,, \tag{4.21}$$
$$L_1 = (\dot{\mathcal{F}} + L_N) \delta N + \mathcal{E} \delta_1 \mathcal{R} \,, \tag{4.22}$$

where

$$\mathcal{E} \equiv L_{\mathcal{R}} + \frac{1}{2}\dot{L_{\mathcal{U}}} + \frac{3}{2}HL_{\mathcal{U}}\,. \tag{4.23}$$

Defining the Lagrangian density as $\mathcal{L} = \sqrt{-g}L = N\sqrt{h}\,L$, where h is the determinant of the three-dimensional metric h_{ij}, the zeroth-order and first-order terms read

$$\mathcal{L}_0 = a^3\left(\bar{L} - \dot{\mathcal{F}} - 3H\mathcal{F}\right), \tag{4.24}$$

$$\mathcal{L}_1 = a^3\left(\bar{L} + L_N - 3H\mathcal{F}\right)\delta N + \left(\bar{L} - \dot{\mathcal{F}} - 3H\mathcal{F}\right)\delta\sqrt{h} + a^3\mathcal{E}\delta_1\mathcal{R}\,. \tag{4.25}$$

The last term is a total derivative, so it can be dropped. Varying the first-order Lagrangian (4.25) with respect to δN and $\delta\sqrt{h}$, we can derive the following equations of motion respectively:

$$\bar{L} + L_N - 3H\mathcal{F} = 0\,, \tag{4.26}$$

$$\bar{L} - \dot{\mathcal{F}} - 3H\mathcal{F} = 0\,. \tag{4.27}$$

On using Eq. (4.27), the zero-th order Lagrangian density (4.24) vanishes. Subtracting Eq. (4.26) from Eq. (4.27), we obtain

$$L_N + \dot{\mathcal{F}} = 0\,. \tag{4.28}$$

Two of Eqs. (4.26)–(4.28) determine the cosmological dynamics on the flat FLRW background.

As an example, let us consider the non-canonical scalar-field model given by [46, 47]

$$L = \frac{M_{\rm pl}^2}{2}R + P(\phi, X)\,, \tag{4.29}$$

where P is an arbitrary function with respect to ϕ and X. Using Eq. (4.4) and dropping the total divergence term, it follows that

$$L = \frac{M_{\rm pl}^2}{2}\left(\mathcal{R} + \mathcal{S} - K^2\right) + P(\phi, X)\,, \tag{4.30}$$

where $X = -N^{-2}\dot{\phi}^2$. Since $\bar{L} = -3M_{\rm pl}^2H^2 + P$, $L_N = 2\dot{\phi}^2P_X$, and $\mathcal{F} = -2M_{\rm pl}^2H$ on the flat FLRW background, Eqs. (4.26) and (4.28) read

$$3M_{\rm pl}^2H^2 = -2P_X\dot{\phi}^2 - P\,, \tag{4.31}$$

$$M_{\rm pl}^2\dot{H} = \dot{\phi}^2P_X\,, \tag{4.32}$$

which match with those derived in [46]. Taking the time derivative of Eq. (4.31) and using Eq. (4.32), we obtain the field equation of motion

$$\frac{d}{dt}\left(a^3 P_X \dot{\phi}^2\right) + \frac{1}{2} a^3 \dot{P} = 0\,, \tag{4.33}$$

which is equivalent to $\frac{d}{dt}\left(a^3 P_X \dot{\phi}\right) + \frac{1}{2} a^3 P_\phi = 0$. For a canonical field characterized by the Lagrangian $P = -X/2 - V(\phi)$, this reduces to the well-known equation $\ddot{\phi} + 3H\dot{\phi} + V_\phi = 0$.

4.3 Second-Order Action for Cosmological Perturbations

In order to derive the equations of motion for linear cosmological perturbations, we need to expand the action (4.6) up to quadratic order. The Lagrangian (4.15) reads

$$\begin{aligned} L = {} & \bar{L} - \dot{\mathcal{F}} - 3H\mathcal{F} + (\dot{\mathcal{F}} + L_N)\delta N + \mathcal{E}\delta_1\mathcal{R} \\ & + \left(\frac{1}{2}L_{NN} - \dot{\mathcal{F}}\right)\delta N^2 + \frac{1}{2}\mathcal{A}\delta K^2 + \mathcal{B}\delta K\delta N + \mathcal{C}\delta K\delta_1\mathcal{R} + \mathcal{D}\delta N\delta_1\mathcal{R} \\ & + \mathcal{E}\delta_2\mathcal{R} + \frac{1}{2}\mathcal{G}\delta_1\mathcal{R}^2 + L_{\mathcal{S}}\delta K^\mu_\nu \delta K^\nu_\mu + L_{\mathcal{Z}}\delta\mathcal{R}^\mu_\nu\delta\mathcal{R}^\nu_\mu\,, \end{aligned} \tag{4.34}$$

where

$$\mathcal{A} = L_{KK} + 4HL_{\mathcal{S}K} + 4H^2 L_{\mathcal{S}\mathcal{S}}\,, \tag{4.35}$$

$$\mathcal{B} = L_{KN} + 2HL_{\mathcal{S}N}\,, \tag{4.36}$$

$$\mathcal{C} = L_{K\mathcal{R}} + 2HL_{\mathcal{S}\mathcal{R}} + \frac{1}{2}L_{\mathcal{U}} + HL_{K\mathcal{U}} + 2H^2 L_{\mathcal{S}\mathcal{U}}\,, \tag{4.37}$$

$$\mathcal{D} = L_{N\mathcal{R}} - \frac{1}{2}\dot{L}_{\mathcal{U}} + HL_{N\mathcal{U}}\,, \tag{4.38}$$

$$\mathcal{G} = L_{\mathcal{R}\mathcal{R}} + 2HL_{\mathcal{R}\mathcal{U}} + H^2 L_{\mathcal{U}\mathcal{U}}\,. \tag{4.39}$$

Then, we obtain the second-order Lagrangian density, as

$$\begin{aligned} \mathcal{L}_2 = {} & \delta\sqrt{h}\left[(\dot{\mathcal{F}} + L_N)\delta N + \mathcal{E}\delta_1\mathcal{R}\right] \\ & + a^3\Bigg[\left(L_N + \frac{1}{2}L_{NN}\right)\delta N^2 + \mathcal{E}\delta_2\mathcal{R} + \frac{1}{2}\mathcal{A}\delta K^2 + \mathcal{B}\delta K\delta N + \mathcal{C}\delta K\delta_1\mathcal{R} \\ & + (\mathcal{D} + \mathcal{E})\delta N\delta_1\mathcal{R} + \frac{1}{2}\mathcal{G}\delta_1\mathcal{R}^2 + L_{\mathcal{S}}\delta K^\mu_\nu\delta K^\nu_\mu + L_{\mathcal{Z}}\delta\mathcal{R}^\mu_\nu\delta\mathcal{R}^\nu_\mu\Bigg]\,. \end{aligned} \tag{4.40}$$

For the gauge choice (4.10), the three-dimensional metric following from the metric (4.8) is $h_{ij} = a^2(t)e^{2\zeta}\delta_{ij}$. Then, several perturbed quantities appearing in Eq. (4.40) can be expressed as

$$\delta\sqrt{h} = 3a^3\zeta\,, \qquad \delta\mathcal{R}_{ij} = -\left(\delta_{ij}\partial^2\zeta + \partial_i\partial_j\zeta\right)\,,$$
$$\delta_1\mathcal{R} = -4a^{-2}\partial^2\zeta\,, \qquad \delta_2\mathcal{R} = -2a^{-2}\left[(\partial\zeta)^2 - 4\zeta\partial^2\zeta\right]\,, \tag{4.41}$$

where $\partial^2\zeta \equiv \partial_i\partial_i\zeta = \sum_{i=1}^3 \partial^2/\partial(x^i)^2$ and $(\partial\zeta)^2 = (\partial_i\zeta)(\partial_i\zeta) = \sum_{i=1}^3(\partial_i\zeta)^2$.

From Eq. (4.3) the extrinsic curvature can be expressed in the form

$$K_{ij} = \frac{1}{2N}\left(\dot{h}_{ij} - N_{i|j} - N_{j|i}\right)\,. \tag{4.42}$$

For the perturbed metric (4.8), the first-order extrinsic curvature reads

$$\delta K^i_j = \left(\dot{\zeta} - H\delta N\right)\delta^i_j - \frac{1}{2a^2}\delta^{ik}(\partial_k N_j + \partial_j N_k)\,, \tag{4.43}$$

where we have used the fact that the Christoffel symbols Γ^k_{ij} are the first-order perturbations for non-zero values of k, i, j. Since the shift N_i is related to the metric perturbation ψ via $N_i = \partial_i\psi$, the trace of δK_{ij} can be expressed as

$$\delta K = 3\left(\dot{\zeta} - H\delta N\right) - \frac{1}{a^2}\partial^2\psi\,. \tag{4.44}$$

On using the relations (4.41), (4.43), and (4.44), the second-order Lagrangian density (4.40), up to boundary terms, reduces to

$$\begin{aligned}
\mathcal{L}_2 = a^3\Bigg\{&\frac{1}{2}(2L_N + L_{NN} + 9\mathcal{A}H^2 - 6\mathcal{B}H + 6L_{\mathcal{S}}H^2)\delta N^2 \\
&+ \left[(\mathcal{B} - 3\mathcal{A}H - 2L_{\mathcal{S}}H)\left(3\dot{\zeta} - \frac{\partial^2\psi}{a^2}\right) + 4(3H\mathcal{C} - \mathcal{D} - \mathcal{E})\frac{\partial^2\zeta}{a^2}\right]\delta N \\
&- (3\mathcal{A} + 2L_{\mathcal{S}})\dot{\zeta}\frac{\partial^2\psi}{a^2} - 12\mathcal{C}\dot{\zeta}\frac{\partial^2\zeta}{a^2} + \left(\frac{9}{2}\mathcal{A} + 3L_{\mathcal{S}}\right)\dot{\zeta}^2 + 2\mathcal{E}\frac{(\partial\zeta)^2}{a^2} \\
&+ \frac{1}{2}(\mathcal{A} + 2L_{\mathcal{S}})\frac{(\partial^2\psi)^2}{a^4} + 4\mathcal{C}\frac{(\partial^2\psi)(\partial^2\zeta)}{a^4} + 2(4\mathcal{G} + 3L_{\mathcal{Z}})\frac{(\partial^2\zeta)^2}{a^4}\Bigg\}\,,
\end{aligned} \tag{4.45}$$

where we have used the background equation (4.28) to eliminate the term $3a^3(L_N + \dot{\mathcal{F}})\zeta\delta N$. Variations of the second-order action $S_2 = \int d^4x\,\mathcal{L}_2$ with respect to δN and $\partial^2\psi$ lead to the following Hamiltonian and momentum constraints, respectively:

$$\left[2L_N + L_{NN} - 6H\mathcal{W} - 3H^2(3\mathcal{A} + 2L_{\mathcal{S}})\right]\delta N$$
$$-\mathcal{W}\frac{\partial^2\psi}{a^2} + 3\mathcal{W}\dot{\zeta} + 4\left(3H\mathcal{C} - \mathcal{D} - \mathcal{E}\right)\frac{\partial^2\zeta}{a^2} = 0\,, \tag{4.46}$$

$$\mathcal{W}\delta N - (\mathcal{A} + 2L_{\mathcal{S}})\frac{\partial^2\psi}{a^2} + (3\mathcal{A} + 2L_{\mathcal{S}})\dot{\zeta} - 4\mathcal{C}\frac{\partial^2\zeta}{a^2} = 0\,, \tag{4.47}$$

where

$$\mathcal{W} \equiv \mathcal{B} - 3\mathcal{A}H - 2L_{\mathcal{S}}H\,. \tag{4.48}$$

From Eqs. (4.46) and (4.47) one can express δN and $\partial^2\psi/a^2$ in terms of $\dot{\zeta}$ and $\partial^2\zeta/a^2$. The last three terms in Eq. (4.45) give rise to the equations of motion containing spatial derivatives higher than second order. If we impose the three conditions

$$\mathcal{A} + 2L_{\mathcal{S}} = 0\,, \tag{4.49}$$

$$\mathcal{C} = 0\,, \tag{4.50}$$

$$4\mathcal{G} + 3L_{\mathcal{Z}} = 0\,, \tag{4.51}$$

then such higher-order spatial derivatives are absent. Under the conditions (4.49)–(4.51), we obtain the following relations from Eqs. (4.46) and (4.47):

$$\frac{\partial^2\psi}{a^2} = \frac{3\mathcal{W}^2 + 4L_{\mathcal{S}}(2L_N + L_{NN} - 6H\mathcal{W} + 12H^2L_{\mathcal{S}})}{\mathcal{W}^2}\dot{\zeta}$$
$$-\frac{4(\mathcal{D} + \mathcal{E})}{\mathcal{W}}\frac{\partial^2\zeta}{a^2}, \tag{4.52}$$

$$\delta N = \frac{4L_{\mathcal{S}}}{\mathcal{W}}\dot{\zeta}\,, \tag{4.53}$$

where $\mathcal{W} = L_{KN} + 2HL_{\mathcal{S}N} + 4HL_{\mathcal{S}}$. Substituting these relations into Eq. (4.45), we find that the second-order Lagrangian density can be written in the form $\mathcal{L}_2 = c_1(t)\dot{\zeta}^2 + c_2(t)\dot{\zeta}\partial^2\zeta + c_3(t)(\partial\zeta)^2$, where $c_{1,2,3}(t)$ are time-dependent coefficients. After integration by parts, the term $c_2(t)\dot{\zeta}\partial^2\zeta$ reduces to $\dot{c}_2(t)(\partial\zeta)^2/2$ up to a boundary term. Then, the second-order Lagrangian density reads [28, 33]

$$\mathcal{L}_2 = a^3 Q_s\left[\dot{\zeta}^2 - \frac{c_s^2}{a^2}(\partial\zeta)^2\right], \tag{4.54}$$

where

$$Q_s \equiv \frac{2L_{\mathcal{S}}[3\mathcal{B}^2 + 4L_{\mathcal{S}}(2L_N + L_{NN})]}{\mathcal{W}^2}\,, \tag{4.55}$$

$$c_s^2 \equiv \frac{2}{Q_s}\left(\dot{\mathcal{M}} + H\mathcal{M} - \mathcal{E}\right), \tag{4.56}$$

and

$$\mathcal{M} \equiv \frac{4L_{\mathcal{S}}(\mathcal{D}+\mathcal{E})}{\mathcal{W}} = \frac{4L_{\mathcal{S}}}{\mathcal{W}}\left(L_{\mathcal{R}} + L_{N\mathcal{R}} + HL_{N\mathcal{U}} + \frac{3}{2}HL_{\mathcal{U}}\right). \tag{4.57}$$

Varying the action $S_2 = \int d^4x\, \mathcal{L}_2$ with respect to the curvature perturbation ζ, we obtain the equation of motion for ζ:

$$\frac{d}{dt}\left(a^3 Q_s \dot{\zeta}\right) - a Q_s c_s^2 \partial^2 \zeta = 0 \,. \tag{4.58}$$

This is the second-order equation of motion with a single scalar degree of freedom. Provided that the conditions (4.49)–(4.51) are satisfied, the gravitational theory described by the action (4.6) does not involve derivatives higher than quadratic order at the level of *linear* cosmological perturbations. As we will see in Sect. 4.5, Horndeski theory satisfies the conditions (4.49)–(4.51).

While we have focused on scalar perturbations so far, we can also perform a similar expansion for tensor perturbations. The three-dimensional metric including tensor modes γ_{ij} can expressed as

$$h_{ij} = a^2(t) e^{2\zeta} \hat{h}_{ij} \,, \qquad \hat{h}_{ij} = \delta_{ij} + \gamma_{ij} + \frac{1}{2}\gamma_{il}\gamma_{lj} \,, \qquad \det \hat{h} = 1 \,, \tag{4.59}$$

where γ_{ij} is traceless and divergence-free such that $\gamma_{ii} = \partial_i \gamma_{ij} = 0$. We have introduced the second-order term $\gamma_{il}\gamma_{lj}/2$ for the simplification of calculations [48]. On using the property that tensor modes decouple from scalar modes, we substitute Eq. (4.59) into the action (4.6) and then set scalar perturbations 0. We note that tensor perturbations satisfy the relations $\delta K = 0$, $\delta K_{ij}^2 = \dot{\gamma}_{ij}^2/4$, $\delta_1 \mathcal{R} = 0$, and $\delta_2 \mathcal{R} = -(\partial_k \gamma_{ij})^2/(4a^2)$. The second-order action for tensor perturbations reads

$$\begin{aligned} S_2^{(h)} &= \int d^4x\, a^3 \left[L_{\mathcal{S}}\left(\delta K^{\nu}_{\mu}\delta K^{\mu}_{\nu} - \delta K^2\right) + \mathcal{E}\delta_2\mathcal{R}\right] \\ &= \int d^4x\, \frac{a^3}{4}\left[L_{\mathcal{S}}\dot{\gamma}_{ij}^2 - \frac{\mathcal{E}}{a^2}(\partial_k \gamma_{ij})^2\right]. \end{aligned} \tag{4.60}$$

One can express γ_{ij} in terms of two polarization modes, as $\gamma_{ij} = h_+ e_{ij}^+ + h_\times e_{ij}^\times$. In Fourier space, the transverse and traceless tensors e_{ij}^+ and $e_{ij}^\times$ satisfy the normalization condition $e_{ij}(\boldsymbol{k})\, e_{ij}(-\boldsymbol{k})^* = 2$ for each polarization ($\boldsymbol{k}$ is a comoving wavenumber), whereas $e_{ij}^+(\boldsymbol{k})\, e_{ij}^\times(-\boldsymbol{k})^* = 0$. The second-order Lagrangian (4.60) can be written as the sum of two polarizations, as

$$S_2^{(h)} = \sum_{\lambda=+,\times} \int d^4x\, a^3 Q_t \left[\dot{h}_\lambda^2 - \frac{c_t^2}{a^2}(\partial h_\lambda)^2\right], \tag{4.61}$$

where

$$Q_t \equiv \frac{L_{\mathcal{S}}}{2}, \tag{4.62}$$

$$c_t^2 \equiv \frac{\mathcal{E}}{L_{\mathcal{S}}}. \tag{4.63}$$

Each mode h_λ ($\lambda = +, \times$) obeys the second-order equation of motion

$$\frac{d}{dt}\left(a^3 Q_t \dot{h}_\lambda\right) - a Q_t c_t^2 \partial^2 h_\lambda = 0. \tag{4.64}$$

In order to avoid the appearance of ghosts, the coefficient in front of the term $\dot{h}_\lambda$ needs to be positive and hence $Q_t > 0$. The small-scale instability associated with the Laplacian term $c_t^2 \partial^2 h_\lambda$ is absent for $c_t^2 > 0$. Then, the conditions for avoidance of the ghost and the Laplacian instability associated with tensor perturbations are given, respectively, by Gleyzes et al. [28] and Gergely and Tsujikawa [33]

$$L_{\mathcal{S}} > 0, \tag{4.65}$$

$$\mathcal{E} = L_{\mathcal{R}} + \frac{1}{2}\dot{L_{\mathcal{U}}} + \frac{3}{2}HL_{\mathcal{U}} > 0. \tag{4.66}$$

Similarly, the ghost and the Laplacian instability of scalar perturbations can be avoided for $Q_s > 0$ and $c_s^2 > 0$, respectively, i.e.,

$$3\left(L_{KN} + 2HL_{\mathcal{S}N}\right)^2 + 4L_{\mathcal{S}}(2L_N + L_{NN}) > 0, \tag{4.67}$$

$$\dot{\mathcal{M}} + H\mathcal{M} - \mathcal{E} > 0, \tag{4.68}$$

where we have used the condition (4.65). The four conditions (4.65)–(4.68) need to be satisfied for theoretical consistency.

4.4 Inflationary Power Spectra

The scalar degree of freedom discussed in the previous section can give rise to inflation in the early Universe. Moreover, the curvature perturbation ζ generated during inflation can be responsible for the origin of observed CMB temperature anisotropies [6]. The tensor perturbation not only contributes to the CMB power spectrum but also leaves an imprint for the B-mode polarization of photons.

In this section we derive the inflationary power spectra of scalar and tensor perturbations for the general action (4.6). We focus on the theory satisfying the conditions (4.49)–(4.51). In this case, the equations of linear cosmological perturbations do not involve time and spatial derivatives higher than second order.

Since the Hubble parameter H is nearly constant during inflation, the terms that do not contain the scale factor a slowly vary in time. Let us then assume that variations of the terms Q_s, Q_t, c_s, and c_t are small, such that the quantities

$$\delta_{Q_s} \equiv \frac{\dot{Q}_s}{HQ_s}, \quad \delta_{Q_t} \equiv \frac{\dot{Q}_t}{HQ_t}, \quad \delta_{c_s} \equiv \frac{\dot{c}_s}{Hc_s}, \quad \delta_{c_t} \equiv \frac{\dot{c}_t}{Hc_t} \tag{4.69}$$

are much smaller than unity.

We first study the evolution of the curvature perturbation ζ during inflation. In doing so, we express ζ in Fourier space, as

$$\zeta(\tau, \boldsymbol{x}) = \frac{1}{(2\pi)^3} \int d^3\boldsymbol{k}\, \hat{\zeta}(\tau, \boldsymbol{k}) e^{i\boldsymbol{k}\cdot\boldsymbol{x}}, \tag{4.70}$$

where

$$\hat{\zeta}(\tau, \boldsymbol{k}) = u(\tau, \boldsymbol{k}) a(\boldsymbol{k}) + u^*(\tau, -\boldsymbol{k}) a^\dagger(-\boldsymbol{k}). \tag{4.71}$$

Here, $\tau \equiv \int a^{-1}\, dt$ is the conformal time, $\boldsymbol{k}$ is the comoving wavenumber, $a(\boldsymbol{k})$ and $a^\dagger(\boldsymbol{k})$ are the annihilation and creation operators, respectively, satisfying the commutation relations

$$\begin{aligned} \left[a(\boldsymbol{k}_1), a^\dagger(\boldsymbol{k}_2)\right] &= (2\pi)^3 \delta^{(3)}(\boldsymbol{k}_1 - \boldsymbol{k}_2), \\ \left[a(\boldsymbol{k}_1), a(\boldsymbol{k}_2)\right] &= \left[a^\dagger(\boldsymbol{k}_1), a^\dagger(\boldsymbol{k}_2)\right] = 0. \end{aligned} \tag{4.72}$$

On the de Sitter background where H is constant, we have $a \propto e^{Ht}$ and hence $\tau = -1/(aH)$. Here, we have set the integration constant 0, such that the asymptotic past corresponds to $\tau \to -\infty$.

Using the equation of motion (4.58) for ζ, we find that each Fourier mode u obeys

$$\ddot{u} + \frac{(a^3 Q_s)^{\cdot}}{a^3 Q_s} \dot{u} + c_s^2 \frac{k^2}{a^2} u = 0. \tag{4.73}$$

For large k, the second term on the l.h.s. of Eq. (4.73) is negligible relative to the third one, so that the field u oscillates according to the approximate equation $\ddot{u} + c_s^2(k^2/a^2)u \simeq 0$. After the onset of inflation, the $c_s^2(k^2/a^2)u$ term starts to decrease quickly. Since the second term on the l.h.s. of Eq. (4.73) is of the order of $H^2 u$, the third term becomes negligible relative to the other terms for $c_s k < aH$. In the large-scale limit ($k \to 0$), the solution to Eq. (4.73) is given by

$$u = c_1 + c_2 \int \frac{1}{a^3 Q_s}\, dt, \tag{4.74}$$

where c_1 and c_2 are integration constants. As long as the variable Q_s changes slowly in time, u approaches a constant value c_1. The field u starts to be frozen once the perturbations with the wavenumber k cross $c_s k = aH$ [6, 49, 50].

We recall that the second-order Lagrangian for the curvature perturbation ζ is given by Eq. (4.54). Introducing a rescaled field $v = zu$ with $z = a\sqrt{2Q_s}$, the kinetic term in the second-order action $S_2 = \int d^4x\, \mathcal{L}_2$ can be rewritten as $\int d\tau d^3x\, v'^2/2$, where a prime represents a derivative with respect to τ. This means that v is a canonical field that should be quantized [34, 36]. Equation (4.73) can be written as

$$v'' + \left(c_s^2 k^2 - \frac{z''}{z}\right) v = 0\,. \tag{4.75}$$

On the de Sitter background with a slow variation of the quantity Q_s, we can approximate $z''/z \simeq 2/\tau^2$. In the asymptotic past ($k\tau \to -\infty$), we choose the Bunch-Davies vacuum characterized by the mode function $v = e^{-ic_s k\tau}/\sqrt{2c_s k}$. Then the solution to Eq. (4.75) is given by

$$u(\tau, k) = \frac{i\, H\, e^{-ic_s k\tau}}{2(c_s k)^{3/2}\sqrt{Q_s}}\,(1 + ic_s k\tau)\,. \tag{4.76}$$

The deviation from the exact de Sitter background gives rise to a small modification to the solution (4.76), but this difference appears as a next-order slow-roll correction to the power spectrum [51, 52].

In the regime $c_s k \ll aH$, the two-point correlation function of ζ is given by the vacuum expectation value $\langle 0|\hat{\zeta}(\tau, \boldsymbol{k}_1)\hat{\zeta}(\tau, \boldsymbol{k}_2)|0\rangle$ at $\tau \approx 0$. We define the scalar power spectrum $\mathcal{P}_\zeta(k_1)$, as

$$\langle 0|\hat{\zeta}(0, \boldsymbol{k}_1)\hat{\zeta}(0, \boldsymbol{k}_2)|0\rangle = \frac{2\pi^2}{k_1^3}\mathcal{P}_\zeta(k_1)\,(2\pi)^3\delta^{(3)}(\boldsymbol{k}_1 + \boldsymbol{k}_2)\,. \tag{4.77}$$

Using the solution (4.76) in the $\tau \to 0$ limit, it follows that

$$\mathcal{P}_\zeta = \frac{H^2}{8\pi^2 Q_s c_s^3}\,. \tag{4.78}$$

Since the curvature perturbation soon approaches a constant for $c_s k < aH$, it is a good approximation to evaluate the power spectrum (4.78) at $c_s k = aH$ during inflation. From the Planck data, the scalar amplitude is constrained as $\mathcal{P}_\zeta \simeq 2.2 \times 10^{-9}$ at the pivot wavenumber $k_0 = 0.002\,\mathrm{Mpc}^{-1}$ [5].

The spectral index of $\mathcal{P}_\zeta$ is defined by

$$n_s - 1 \equiv \left.\frac{d\ln\mathcal{P}_\zeta}{d\ln k}\right|_{c_s k = aH} = -2\epsilon - \delta_{Q_s} - 3\delta_{c_s}\,, \tag{4.79}$$

where δ_{Q_s} and δ_{c_s} are given by Eq. (4.69), and

$$\epsilon \equiv -\frac{\dot{H}}{H^2}\,. \tag{4.80}$$

The slow-roll parameter ϵ is much smaller than 1 on the quasi de Sitter background. Given that the variations of H and c_s are small during inflation, we can approximate the variation of $\ln k$ at $c_s k = aH$, as $d\ln k = d\ln a = H dt$. Since we are considering the situation with $|\delta_{Q_s}| \ll 1$ and $|\delta_{c_s}| \ll 1$, the power spectrum is close to scale-invariant ($n_s \simeq 1$).

We also define the running of the spectral index, as

$$\alpha_s \equiv \left.\frac{dn_s}{d\ln k}\right|_{c_s k = aH}, \tag{4.81}$$

which is of the order of ϵ^2 from Eq. (4.79). With the prior $\alpha_s = 0$, the scalar spectral index is constrained as $n_s = 0.9603 \pm 0.0073$ at 68 % confidence level (CL) from the Planck data [5]. Since ϵ is at most of the order 10^{-2}, it is a good approximation to neglect the running α_s in standard slow-roll inflation.

Let us also derive the spectrum of gravitational waves generated during inflation. The second-order action for tensor perturbations is given by Eq. (4.61), where h_λ obeys Eq. (4.64). A canonical field associated with h_λ ($\lambda = +, \times$) corresponds to $v_t = z_t h_\lambda$ and $z_t = a\sqrt{2Q_t}$. Following a same procedure as that for scalar perturbations, the solution to the Fourier-transformed mode h_λ, which recovers the Bunch-Davies vacuum in the asymptotic past, reads

$$h_\lambda(\tau, k) = \frac{i\,H\,e^{-ic_t k\tau}}{2(c_t k)^{3/2}\sqrt{Q_t}}(1 + ic_t k\tau)\,. \tag{4.82}$$

This solution approaches $h_\lambda \to iH/[2(c_t k)^{3/2}\sqrt{Q_t}]$ in the $\tau \to 0$ limit.

We also define the tensor power spectrum $\mathcal{P}_h$ in a similar way to (4.77). According to the chosen normalization for the tensors e_{ij}^λ explained in Sect. 4.3, we obtain $\mathcal{P}_h = 4 \cdot k^3 |h_\lambda(0,k)|^2/(2\pi^2)$, where $h_\lambda(0,k) = iH/[2(c_t k)^{3/2}\sqrt{Q_t}]$. It then follows that

$$\mathcal{P}_h = \frac{H^2}{2\pi^2 Q_t c_t^3}\,. \tag{4.83}$$

The tensor spectral index, which is evaluated at $c_t k = aH$, reads

$$n_t \equiv \left.\frac{d\ln \mathcal{P}_h}{d\ln k}\right|_{c_t k = aH} = -2\epsilon - \delta_{Q_t} - 3\delta_{c_t}\,, \tag{4.84}$$

where δ_{Q_t} and δ_{c_t} are given by Eq. (4.69). The tensor power spectrum is close to scale-invariant ($n_t \simeq 0$) provided that $\epsilon \ll 1$, $|\delta_{Q_t}| \ll 1$, and $|\delta_{c_t}| \ll 1$. The difference between the scalar and tensor spectral indices comes from the difference between (Q_s, c_s) and (Q_t, c_t).

For those times before the end of inflation ($\epsilon \ll 1$) when both $\mathcal{P}_\zeta$ and $\mathcal{P}_h$ are approximately constant, the tensor-to-scalar ratio can be estimated as

$$r \equiv \frac{\mathcal{P}_h}{\mathcal{P}_\zeta} = 4\frac{Q_s c_s^3}{Q_t c_t^3}\,. \tag{4.85}$$

The Planck data [5], combined with the WMAP large-angle polarization measurement [13] and ACT/SPT temperature data [53], showed that r is constrained as $r < 0.11$ (95 % CL). Recently, the Background Imaging of Cosmic Extragalactic Polarization (BICEP2) group [54] reported the first evidence for the primordial B-mode polarization of CMB photons and they derived the bound $r = 0.20^{+0.07}_{-0.05}$ (68 % CL) with $r = 0$ disfavored at 7σ. There is a tension between the data of Planck and BICEP2, but future measurements of the B-mode polarization will place more precise bounds on r.

The inflationary scalar and tensor power spectra (4.78) and (4.83) are valid for general theories given by the action (4.6), provided that the conditions (4.49)–(4.51) are satisfied. The quantities like Q_s and c_s^2 are written in terms of the partial derivatives of L with respect to the ADM variables such as K and N. For a given theory, we need to express the Lagrangian L in terms of the three-dimensional quantities and the lapse N to derive concrete forms of the inflationary power spectra. In the next section, we will perform this procedure for the most general scalar-tensor theories with second-order equations of motion.

4.5 Horndeski Theory

4.5.1 The Lagrangian of Horndeski Theory

In this section we apply the EFT formalism advocated in Sects. 4.2 and 4.3 to the most general scalar-tensor theories with second-order equations of motion–Horndeski theory [39]. This theory is described by the action $S = \int d^4x \sqrt{-g}\, L$, with the Lagrangian [43]

$$L = \sum_{i=2}^{5} L_i\,, \tag{4.86}$$

where

$$L_2 = G_2(\phi, X), \tag{4.87}$$

$$L_3 = G_3(\phi, X)\Box\phi, \tag{4.88}$$

$$L_4 = G_4(\phi, X)\, R - 2G_{4X}(\phi, X) \left[(\Box\phi)^2 - \phi^{;\mu\nu}\phi_{;\mu\nu} \right], \tag{4.89}$$

$$L_5 = G_5(\phi, X) G_{\mu\nu}\phi^{;\mu\nu} + \frac{1}{3} G_{5X}(\phi, X) \left[(\Box\phi)^3 - 3(\Box\phi)\, \phi_{;\mu\nu}\phi^{;\mu\nu} + 2\phi_{;\mu\nu}\phi^{;\mu\sigma}\phi^{;\nu}{}_{;\sigma} \right]. \tag{4.90}$$

Here G_i $(i = 2, 3, 4, 5)$ are functions in terms of a scalar field ϕ and its kinetic energy $X = g^{\mu\nu}\partial_\mu\phi\partial_\nu\phi$ with the partial derivatives $G_{iX} \equiv \partial G_i/\partial X$ and $G_{i\phi} \equiv \partial G_i/\partial\phi$, R is the Ricci scalar, and $G_{\mu\nu}$ is the Einstein tensor. In 1973, Horndeski derived the Lagrangian of the most general scalar-tensor theories in a different form [39], but as shown in [35], it is equivalent to the above form. The Horndeski's paper[2] has not been recognized much for a long time, but it was revived recently in connection to covariant Galileons [40,41] and generalized Galileon theories [42,43].

The Lagrangian (4.86) covers a wide variety of gravitational theories listed below.

- (1) General Relativity with a minimally coupled scalar field

  The minimally coupled scalar-field theory (4.29) is characterized by the functions [46]

$$G_2 = P(\phi, X), \qquad G_3 = 0, \qquad G_4 = M_{\rm pl}^2/2, \qquad G_5 = 0. \tag{4.91}$$

  The canonical scalar field with a potential $V(\phi)$ corresponds to the particular choice

$$G_2 = -X/2 - V(\phi). \tag{4.92}$$

- (2) Brans-Dicke theory

  The Lagrangian of Brans-Dicke (BD) theory is given by

$$G_2 = -\frac{M_{\rm pl}\omega_{\rm BD} X}{2\phi} - V(\phi), \qquad G_3 = 0, \qquad G_4 = \frac{1}{2} M_{\rm pl}\phi, \qquad G_5 = 0, \tag{4.93}$$

  where $\omega_{\rm BD}$ is the so-called BD parameter. In the original BD theory [55], the field potential $V(\phi)$ is absent. Dilaton gravity [56] corresponds to $\omega_{\rm BD} = -1$.

[2]When Horndeski wrote this paper, he was the Ph.D. student of David Lovelock. In 1981, he was taking a sabbatical year in Netherlands as a tenured professor of applied mathematics at the University of Waterloo. When he saw a van Gogh exhibition, he was deeply moved. He stated "I was never that interested in art. Then I stumbled onto van Gogh. I never knew art could be like that. I had always thought of it as very representational and not very interesting. But then I thought, 'This is something I eventually want to do.' When I saw van Gogh I was sure I could paint." After this, Horndeski left physics and became an artist.

- (3) $f(R)$ gravity

  This theory is characterized by the action

$$S = \int d^4x \sqrt{-g} \frac{M_{\rm pl}^2}{2} f(R)\,, \tag{4.94}$$

where $f(R)$ is an arbitrary function of the Ricci scalar R. The metric $f(R)$ gravity corresponds to the case in which the action (4.94) is varied with respect to $g_{\mu\nu}$. This can be accommodated by the Lagrangian (4.86) for the choice

$$G_2 = -\frac{M_{\rm pl}^2}{2}(RF - f), \qquad G_3 = 0\,, \qquad G_4 = \frac{1}{2} M_{\rm pl}^2 F\,, \qquad G_5 = 0\,, \tag{4.95}$$

where $F \equiv \partial f/\partial R$. There is a scalar degree of freedom $\phi = M_{\rm pl} F(R)$ with a gravitational origin. Comparing Eq. (4.93) with Eq. (4.95), we find that metric $f(R)$ gravity is equivalent to BD theory with $\omega_{\rm BD} = 0$ and the potential $V = (M_{\rm pl}^2/2)(RF - f)$.

In the Palatini formalism where the metric $g_{\mu\nu}$ and the connection $\Gamma^{\alpha}_{\beta\gamma}$ are treated as independent variables, the Ricci scalar is different from that in metric $f(R)$ gravity. The Palatini $f(R)$ gravity is equivalent to BD theory with the parameter $\omega_{\rm BD} = -3/2$ [15].

- (4) Non-minimally coupled theory

  This theory is described by the functions

$$G_2 = \omega(\phi) X - V(\phi)\,, \qquad G_3 = 0\,, \qquad G_4 = \frac{M_{\rm pl}^2}{2} - \frac{1}{2}\xi\phi^2\,, \qquad G_5 = 0\,. \tag{4.96}$$

where $\omega(\phi)$ and $V(\phi)$ are functions of ϕ. Higgs inflation [57] corresponds to a canonical field ($\omega(\phi) = -1/2$) with the potential $V(\phi) = (\lambda/4)(\phi^2 - v^2)^2$ (see also [58]).

- (5) Covariant Galileons

  The covariant Galileons [41], in the absence of the field potential, are described by the functions

$$G_2 = c_2 X\,, \qquad G_3 = c_3 X\,, \qquad G_4 = \frac{M_{\rm pl}^2}{2} + c_4 X^2\,, \qquad G_5 = c_5 X^2\,, \tag{4.97}$$

where c_i ($i = 2, 3, 4, 5$) are constants. The field equations of motion are invariant under the Galilean transformation $\partial_\mu \phi \to \partial_\mu \phi + b_\mu$ in the limit of Minkowski space-time [40].

- (6) Derivative couplings
  A scalar field whose derivative couples to the Einstein tensor in the form $G_{\mu\nu}\partial^\mu\phi\partial^\nu\phi$ [59, 60] corresponds to the choice

$$G_2 = -X/2 - V(\phi)\,, \qquad G_3 = 0\,, \qquad G_4 = 0\,, \qquad G_5 = c\phi\,, \tag{4.98}$$

  where c is a constant and $V(\phi)$ is the field potential. In fact, integration of the term $c\phi G_{\mu\nu}\phi^{;\mu\nu}$ by parts gives rise to the coupling $-c\,G_{\mu\nu}\partial^\mu\phi\partial^\nu\phi$.
- (7) Gauss-Bonnet couplings
  The Gauss-Bonnet couplings of the from $-\xi(\phi)R_{\rm GB}^2$, where $R_{\rm GB}^2 = R^2 - 4R_{\alpha\beta}R^{\alpha\beta} + R_{\alpha\beta\gamma\delta}R^{\alpha\beta\gamma\delta}$, can be accommodated by the choice [35]

$$\begin{aligned} G_2 &= -2\xi^{(4)}(\phi)X^2[3 - \ln(-X/2)]\,, \quad & G_3 &= 2\xi^{(3)}(\phi)X[7 - 3\ln(-X/2)]\,, \\ G_4 &= 2\xi^{(2)}(\phi)X[2 - \ln(-X/2)]\,, \quad & G_5 &= 4\xi^{(1)}(\phi)\ln(-X/2)\,, \end{aligned} \tag{4.99}$$

  where $\xi^{(n)}(\phi) = \partial^n\xi(\phi)/\partial\phi^n$.

4.5.2 Horndeski Lagrangian in Terms of ADM Variables

Let us express the Horndeski Lagrangians (4.87)–(4.90) in terms of the lapse N and the three-dimensional quantities introduced in Sect. 4.2. In unitary gauge, the unit vector n_μ orthogonal to the constant ϕ-hypersurface is given by Gleyzes et al. [28]

$$n_\mu = -\gamma\phi_{;\mu}\,, \qquad \gamma = \frac{1}{\sqrt{-X}}\,. \tag{4.100}$$

Taking the covariant derivative of Eq. (4.100) and using the relation (4.4), we obtain

$$\phi_{;\mu\nu} = -\frac{1}{\gamma}\left(K_{\mu\nu} - n_\mu a_\nu - n_\nu a_\mu\right) + \frac{\gamma^2}{2}\phi^{;\sigma}X_{;\sigma}n_\mu n_\nu\,. \tag{4.101}$$

The trace of Eq. (4.101) gives

$$\Box\phi = -\frac{1}{\gamma}K + \frac{\phi^{;\sigma}X_{;\sigma}}{2X}\,. \tag{4.102}$$

First of all, the Lagrangian L_2 depends on N through the field kinetic energy, i.e.,

$$L_2 = G_2(\phi, X(N))\,. \tag{4.103}$$

On using the property $X(N) = -\dot\phi^2/N^2$ on the flat FLRW background, the quantity like L_{2N} can be evaluated as $L_{2N} = 2\dot\phi^2 G_{2X}$.

For the computation of $L_3 = G_3\Box\phi$, it is convenient to introduce an auxiliary function $F_3(\phi, X)$, as

$$G_3 = F_3 + 2XF_{3X}\,. \tag{4.104}$$

After integration by parts, the term $F_3\Box\phi$ reduces to $-(F_{3\phi}\phi_{;\mu} + F_{3X}X_{;\mu})\phi^{;\mu}$ up to a boundary term. On using the relation (4.102) for the term $2XF_{3X}\Box\phi$, it follows that

$$L_3 = 2(-X)^{3/2}F_{3X}K - XF_{3\phi}\,. \tag{4.105}$$

Although the auxiliary function F_3 is present in the expression of L_3, the combination of quantities appearing in the background and linear perturbation equations of motion can be expressed in terms of G_3.

Substituting Eqs. (4.101) and (4.102) into Eq. (4.89), the term L_4 reads

$$L_4 = G_4R + 2XG_{4X}(K^2 - \mathcal{S}) + 2G_{4X}X_{;\mu}(Kn^\mu - a^\mu)\,, \tag{4.106}$$

where we have used the property $a_\mu = -h^\nu_\mu X_{;\nu}/(2X)$. Substituting Eq. (4.4) into Eq. (4.106) and employing the relations $G_{4X}X_{;\mu} = G_{4;\mu} + \gamma^{-1}G_{4\phi}n_\mu$ and $n_\mu a^\mu = 0$, we obtain

$$L_4 = G_4\mathcal{R} + (2XG_{4X} - G_4)(K^2 - \mathcal{S}) - 2\sqrt{-X}G_{4\phi}K\,. \tag{4.107}$$

The Lagrangian L_5 is most complicated to be dealt with. We refer readers to [28] for detailed calculations. Introducing an auxiliary function $F_5(\phi, X)$ such that

$$G_{5X} \equiv \frac{F_5}{2X} + F_{5X}\,, \tag{4.108}$$

the final expression of L_5 is given by

$$L_5 = \sqrt{-X}F_5\left(\frac{1}{2}K\mathcal{R} - \mathcal{U}\right) - H(-X)^{3/2}G_{5X}(2H^2 - 2KH + K^2 - \mathcal{S})$$
$$+\frac{1}{2}X(G_{5\phi} - F_{5\phi})\mathcal{R} + \frac{1}{2}XG_{5\phi}(K^2 - \mathcal{S})\,, \tag{4.109}$$

which is valid up to quadratic order in the perturbations.

Summing up the contributions (4.103), (4.105), (4.107), and (4.109), the Lagrangian (4.86) can be expressed as

$$L = G_2 + 2(-X)^{3/2}F_{3X}K - XF_{3\phi}$$
$$+G_4\mathcal{R} + (2XG_{4X} - G_4)(K^2 - \mathcal{S}) - 2\sqrt{-X}G_{4\phi}K$$

$$+\sqrt{-X}F_5\left(\frac{1}{2}K\mathcal{R}-\mathcal{U}\right)-H(-X)^{3/2}G_{5X}(2H^2-2KH+K^2-\mathcal{S})$$
$$+\frac{1}{2}X(G_{5\phi}-F_{5\phi})\mathcal{R}+\frac{1}{2}XG_{5\phi}(K^2-\mathcal{S})\,, \tag{4.110}$$

where $G_{2,3,4,5}$ and $F_{3,5}$ are functions of ϕ and $X(N)$. The Lagrangian (4.110) depends on N, K, $\mathcal{S}$, $\mathcal{R}$, $\mathcal{U}$, but not on $\mathcal{Z}$. We evaluate the partial derivatives of the Lagrangian (4.110) with respect to N, K etc. and finally set $N=1$, $K=3H$, $\mathcal{S}=3H^2$, $\mathcal{R}=0, \mathcal{U}=0$.

Among the terms appearing in Eqs. (4.49)–(4.51), the non-vanishing ones are given by

$$L_{KK}=-2L_{\mathcal{S}}=2(2XG_{4X}-G_4)-2H(-X)^{3/2}G_{5X}+XG_{5\phi}\,, \tag{4.111}$$
$$L_{K\mathcal{R}}=-\frac{1}{2}L_{\mathcal{U}}=\frac{1}{2}\sqrt{-X}F_5\,, \tag{4.112}$$

so that all the three conditions (4.49)–(4.51) are satisfied. In Horndeski theory, there are no spatial derivatives higher than second order.

4.5.3 Conditions for the Avoidance of Ghosts and Laplacian Instabilities

The conditions (4.65) and (4.66) for avoiding the ghost and the Laplacian instability of tensor perturbations translate to

$$L_{\mathcal{S}}=G_4-2XG_{4X}-H\dot{\phi}XG_{5X}-\frac{1}{2}XG_{5\phi}>0\,, \tag{4.113}$$
$$\mathcal{E}=G_4+\frac{1}{2}XG_{5\phi}-XG_{5X}\ddot{\phi}>0\,, \tag{4.114}$$

respectively. In the presence of the terms $G_4(X)$ and $G_5(\phi,X)$, the tensor propagation speed square $c_t^2=\mathcal{E}/L_{\mathcal{S}}$ is generally different from 1.

On using the properties $\mathcal{B}=L_{KN}+2HL_{\mathcal{S}N}$ and $\mathcal{W}=L_{KN}+2HL_{\mathcal{S}N}+4HL_{\mathcal{S}}$, the quantity Q_s in Eq. (4.55) can be expressed as

$$Q_s=\frac{2L_{\mathcal{S}}}{3\mathcal{W}^2}\left(9\mathcal{W}^2+8L_{\mathcal{S}}w\right)\,, \tag{4.115}$$

where[3]

$$
\begin{aligned}
w \equiv\ & 3L_N + 3L_{NN}/2 - 9H(L_{KN} + 2HL_{\mathcal{S}N}) - 18L_{\mathcal{S}}H^2 \\
=\ & -18H^2G_4 + 3(XG_{2X} + 2X^2G_{2XX}) - 18H\dot{\phi}(2XG_{3X} + X^2G_{3XX}) \\
& -3X(G_{3\phi} + XG_{3\phi X}) + 18H^2(7XG_{4X} + 16X^2G_{4XX} + 4X^3G_{4XXX}) \\
& -18H\dot{\phi}(G_{4\phi} + 5XG_{4\phi X} + 2X^2G_{4\phi XX}) + 6H^3\dot{\phi}(15XG_{5X} + 13X^2G_{5XX} \\
& +2X^3G_{5XXX}) + 9H^2X(6G_{5\phi} + 9XG_{5\phi X} + 2X^2G_{5\phi XX})\,, \qquad (4.116) \\
\mathcal{W} =\ & 4HG_4 + 2\dot{\phi}XG_{3X} - 16H(XG_{4X} + X^2G_{4XX}) + 2\dot{\phi}(G_{4\phi} + 2XG_{4\phi X}) \\
& -2H^2\dot{\phi}(5XG_{5X} + 2X^2G_{5XX}) - 2HX(3G_{5\phi} + 2XG_{5\phi X})\,. \qquad (4.117)
\end{aligned}
$$

Taking into account the requirement (4.113), the no-ghost condition for scalar perturbations reads

$$9\mathcal{W}^2 + 8L_{\mathcal{S}}w > 0\,. \qquad (4.118)$$

In Horndeski theory (4.110), we notice that there is the following relation

$$L_{\mathcal{S}} = \mathcal{D} + \mathcal{E} = L_{\mathcal{R}} + L_{N\mathcal{R}} + \frac{3}{2}HL_{\mathcal{U}} + HL_{N\mathcal{U}}\,, \qquad (4.119)$$

so that the quantity (4.57) reduces to

$$\mathcal{M} = \frac{4L_{\mathcal{S}}^2}{\mathcal{W}}\,. \qquad (4.120)$$

Then, the condition (4.68) for avoiding the Laplacian instability of scalar perturbations reads

$$\frac{d}{dt}\left(\frac{4L_{\mathcal{S}}^2}{\mathcal{W}}\right) + \frac{4HL_{\mathcal{S}}^2}{\mathcal{W}} - \mathcal{E} > 0\,, \qquad (4.121)$$

where $L_{\mathcal{S}}$, $\mathcal{E}$, and $\mathcal{W}$ are given by Eqs. (4.113), (4.114), and (4.117) respectively.

As an example, let us consider BD theory described by the functions (4.93). Since $L_{\mathcal{S}} = \mathcal{E} = G_4 = M_{\rm pl}\phi/2$ in this case, the conditions (4.113) and (4.114) are satisfied for

$$\phi > 0\,, \qquad (4.122)$$

[3]The four quantities $w_{1,2,3,4}$ introduced in [38] are related to $L_{\mathcal{S}}$, $\mathcal{W}$, w, and $\mathcal{E}$, as $w_1 = 2L_{\mathcal{S}}$, $w_2 = \mathcal{W}$, $w_3 = w$, and $w_4 = 2\mathcal{E}$.

with the tensor propagation speed square $c_t^2 = 1$. Since $\mathcal{W} = M_{\rm pl}(\dot{\phi} + 2H\phi)$ and $w = -3M_{\rm pl}(6H^2\phi^2 - \omega_{\rm BD}\dot{\phi}^2 + 6H\phi\dot{\phi})/(2\phi)$, the quantity (4.115) reads

$$\mathcal{Q}_s = \frac{(3 + 2\omega_{\rm BD})M_{\rm pl}\phi\dot{\phi}^2}{(\dot{\phi} + 2H\phi)^2}\,. \tag{4.123}$$

On using the condition (4.122), we find that the scalar ghost is absent for

$$\omega_{\rm BD} > -3/2\,. \tag{4.124}$$

The quantity $\mathcal{M}$ can be expressed as

$$\mathcal{M} = -\frac{M_{\rm pl}^2\phi^2}{\mathcal{F}}\,, \tag{4.125}$$

where we have used the fact that the term $\mathcal{F}$ in Eq. (4.17) is given by $\mathcal{F} = -M_{\rm pl}(\dot{\phi} + 2H\phi)$. From the background equation (4.28), it follows that

$$\dot{\mathcal{F}} = -L_N = -M_{\rm pl}\dot{\phi}(3H\phi - \omega_{\rm BD}\dot{\phi})/\phi\,. \tag{4.126}$$

Then, the condition (4.68) for avoiding the Laplacian instability of scalar perturbations translates to

$$\dot{\mathcal{M}} + H\mathcal{M} - \mathcal{E} = \frac{(3 + 2\omega_{\rm BD})M_{\rm pl}\phi\dot{\phi}^2}{2(\dot{\phi} + 2H\phi)^2} > 0\,, \tag{4.127}$$

which is satisfied under (4.122) and (4.124). In fact, from Eq. (4.56), the scalar propagation speed square c_s^2 is equivalent to 1 in BD theory.

4.5.4 Primordial Power Spectra in k-Inflation

Let us consider a non-canonical scalar-field theory described by the Lagrangian (4.29). This theory can be expressed in terms of the ADM variables as Eq. (4.30). Since $L_{\mathcal{S}} = \mathcal{E} = G_4 = M_{\rm pl}^2/2$, $\mathcal{Q}_t = M_{\rm pl}^2/4$ and $c_t^2 = 1$, the tensor mode is not plagued by any ghosts and Laplacian instabilities. From Eq. (4.83), the tensor power spectrum is given by

$$\mathcal{P}_h = \frac{2H^2}{\pi^2 M_{\rm pl}^2}\,, \tag{4.128}$$

which depends only on H. Therefore, if the amplitude of primordial gravitational waves is measured, the energy scale of inflation can be explicitly known.

We also have the relations $\mathcal{W} = 2HM_{\rm pl}^2$, $w = -9H^2M_{\rm pl}^2 + 3X(P_X + 2XP_{XX})$, and

$$\mathcal{Q}_s = -\frac{\dot{\phi}^2(P_X + 2XP_{XX})}{H^2}, \tag{4.129}$$

so the scalar ghost is absent for $P_X + 2XP_{XX} < 0$. Since $\mathcal{F} = -2M_{\rm pl}^2\dot{H}$ and $L_N = 2\dot{\phi}^2 P_X$, the background equation of motion (4.28) gives $M_{\rm pl}^2\dot{H} = \dot{\phi}^2 P_X$. Taking the time derivative of the quantity $\mathcal{M} = M_{\rm pl}^2/(2H)$, it follows that

$$\dot{\mathcal{M}} + H\mathcal{M} - \mathcal{E} = -\frac{M_{\rm pl}^2\dot{H}}{2H^2} = -\frac{\dot{\phi}^2 P_X}{2H^2}. \tag{4.130}$$

To avoid the instability of scalar perturbations, we require that $P_X < 0$. Substituting Eqs. (4.129) and (4.130) into Eq. (4.56), we obtain

$$c_s^2 = \frac{P_X}{P_X + 2XP_{XX}}. \tag{4.131}$$

In standard slow-roll inflation driven by the potential energy $V(\phi)$ of a canonical scalar field ($P = -X/2 - V(\phi)$), c_s^2 is equivalent to 1. If the Lagrangian P contains a non-linear term in X, the scalar propagation speed is generally different from 1.

From Eqs. (4.129) and (4.131), we find that the slow-roll parameter $\epsilon = -\dot{H}/H^2$ is related to $\mathcal{Q}_s$ and c_s^2, as

$$\epsilon = \frac{\mathcal{Q}_s c_s^2}{M_{\rm pl}^2}. \tag{4.132}$$

Then, the scalar power spectrum (4.78) reads

$$\mathcal{P}_\zeta = \frac{H^2}{8\pi^2 M_{\rm pl}^2 \epsilon c_s}. \tag{4.133}$$

From Eqs. (4.128) and (4.133), the tensor-to-scalar ratio is given by Garriga and Mukhanov [49]

$$r = 16 c_s \epsilon. \tag{4.134}$$

Since $\epsilon \ll 1$ during inflation, it follows that $r \ll 1$ for $c_s \le 1$.

4.6 Horndeski Theory in the Language of EFT

In this section, we relate the variables introduced in Sect. 4.2 with those employed in the EFT language of [17, 25, 26]. The action expanded up to quadratic order in the perturbations can be written in the following form

$$S = \int d^4x\sqrt{-g}\Big[\frac{M_*^2}{2}fR - \Lambda - c\,g^{00} + \frac{M_2^4}{2}(\delta g^{00})^2 - \frac{\bar{m}_1^3}{2}\delta K\delta g^{00} - \frac{\bar{M}_2^2}{2}\delta K^2$$
$$-\frac{\bar{M}_3^2}{2}\delta K^\mu_\nu \delta K^\nu_\mu + \frac{\mu_1^2}{2}\mathcal{R}\delta g^{00} + \frac{\bar{m}_5}{2}\mathcal{R}\delta K + \frac{\lambda_1}{2}\mathcal{R}^2 + \frac{\lambda_2}{2}\mathcal{R}^\mu_\nu \mathcal{R}^\nu_\mu\Big], \tag{4.135}$$

where $g^{00} = -1/N^2$, M_* is a constant, and other coefficients such as f, Λ, c, M_2^4 depend on time. We note that the four-dimensional Ricci scalar R can be written in terms of the three-dimensional quantities as Eq. (4.4). After integration by parts, the first term in Eq. (4.135) reads

$$\frac{M_*^2}{2}fR = \frac{M_*^2}{2}\left(f\mathcal{R} + f\mathcal{S} - fK^2 - 2\dot{f}\frac{K}{N}\right). \tag{4.136}$$

Now we substitute $\mathcal{R} = \delta_1\mathcal{R} + \delta_2\mathcal{R}$, $K = 3H^2 + \delta K$, and $\mathcal{S} = 3H^2 + 2H\delta K + \delta K^\mu_\nu \delta K^\nu_\mu$ into Eq. (4.136) and then expand the action (4.135) up to quadratic order in the perturbations. In doing so, we use the similar property to Eq. (4.18), i.e., $\int d^4x\sqrt{-g}\,\beta(t)\delta K = \int d^4x\sqrt{-g}(-\dot{\beta} - 3H\beta + \dot{\beta}\delta N - \dot{\beta}\delta N^2)$, where $\beta(t)$ is an arbitrary function in terms of t. Then, the resulting Lagrangian reads

$$L = M_*^2(\ddot{f} + 2H\dot{f} + 2\dot{H}f + 3H^2 f) - \Lambda + c$$
$$+[M_*^2(-\ddot{f} + H\dot{f} - 2\dot{H}f) - 2c]\delta N + \frac{M_*^2}{2}f\delta_1\mathcal{R}$$
$$+\left[M_*^2(\ddot{f} - H\dot{f} + 2\dot{H}f) + 3c + 2M_2^4\right]\delta N^2 - \left(\frac{M_*^2}{2}f + \frac{\bar{M}_2^2}{2}\right)\delta K^2$$
$$+(M_*^2\dot{f} - \bar{m}_1^3)\delta K\delta N + \frac{\bar{m}_5}{2}\delta K\delta_1\mathcal{R} + \mu_1^2\delta N\delta_1\mathcal{R} + \frac{M_*^2}{2}f\delta_2\mathcal{R}$$
$$+\left(\frac{M_*^2}{2}f - \frac{\bar{M}_3^2}{2}\right)\delta K^\mu_\nu\delta K^\nu_\mu + \frac{\lambda_1}{2}\mathcal{R}^2 + \frac{\lambda_2}{2}\delta\mathcal{R}^\mu_\nu\delta\mathcal{R}^\nu_\mu. \tag{4.137}$$

Comparing the terms up to the second line of Eq. (4.137) with those in Eq. (4.22), it follows that

$$M_*^2(\ddot{f} + 2H\dot{f} + 2\dot{H}f + 3H^2 f) - \Lambda + c = \bar{L} - \dot{\mathcal{F}} - 3H\mathcal{F}, \tag{4.138}$$
$$M_*^2(-\ddot{f} + H\dot{f} - 2\dot{H}f) - 2c = \dot{\mathcal{F}} + L_N, \tag{4.139}$$
$$f = \frac{2}{M_*^2}\mathcal{E} = \frac{1}{M_*^2}\left(2L_{\mathcal{R}} + \dot{L_{\mathcal{U}}} + 3HL_{\mathcal{U}}\right). \tag{4.140}$$

From Eqs. (4.27) and (4.28), the r.h.s. of Eq. (4.138) and (4.139) vanish in the absence of matter. The background equations of motion are characterized by the three parameters f, Λ, and c. Comparing the second-order terms in Eq. (4.137) with those in Eq. (4.34), we obtain the following relations

$$
\begin{aligned}
M_2^4 &= \frac{1}{4}(2L_N + L_{NN} - 2c)\,, \qquad \bar{m}_1^3 = 2\dot{\mathcal{E}} - L_{KN} - 2HL_{\mathcal{S}N}\,,\\
\bar{M}_2^2 &= -2\mathcal{E} - L_{KK} - 4HL_{\mathcal{S}K} - 4H^2 L_{\mathcal{S}\mathcal{S}}\,, \qquad \bar{M}_3^2 = 2\mathcal{E} - 2L_{\mathcal{S}}\,,\\
\mu_1^2 &= L_{N\mathcal{R}} - \frac{1}{2}\dot{L}_{\mathcal{U}} + HL_{N\mathcal{U}}\,,\\
\bar{m}_5 &= 2L_{K\mathcal{R}} + 4HL_{\mathcal{S}\mathcal{R}} + L_{\mathcal{U}} + 2HL_{K\mathcal{U}} + 4H^2 L_{\mathcal{S}\mathcal{U}}\,,\\
\lambda_1 &= L_{\mathcal{R}\mathcal{R}} + 2HL_{\mathcal{R}\mathcal{U}} + H^2 L_{\mathcal{U}\mathcal{U}}\,, \qquad \lambda_2 = 2L_{\mathcal{Z}}\,,
\end{aligned}
\tag{4.141}
$$

where we have used Eq. (4.139) to derive M_2^4. In Horndeski theory, the r.h.s. of Eq. (4.141) can be evaluated by taking partial derivatives of the Lagrangian (4.110) in terms of the scalar variables.

The conditions (4.49)–(4.51) reduce, respectively, to

$$
\bar{M}_2^2 + \bar{M}_3^2 = 0\,, \qquad \bar{m}_5 = 0\,, \qquad 8\lambda_1 + 3\lambda_2 = 0\,, \tag{4.142}
$$

under which the spatial derivatives higher than second order are absent. On using these conditions, the Lagrangian (4.135) can be expressed as

$$
\begin{aligned}
S = \int d^4x\sqrt{-g}\Big[&\frac{M_*^2}{2}fR - \Lambda - c\,g^{00} + \frac{M_2^4}{2}(\delta g^{00})^2 - \frac{\bar{m}_1^3}{2}\delta K\delta g^{00}\\
&- m_4^2\left(\delta K^2 - \delta K^{\mu}_{\nu}\delta K^{\nu}_{\mu}\right) + \frac{\mu_1^2}{2}\mathcal{R}\delta g^{00}\Big]\,,
\end{aligned}
\tag{4.143}
$$

where

$$
m_4^2 \equiv \frac{1}{4}\left(\bar{M}_2^2 - \bar{M}_3^2\right) = \frac{1}{4}\left(-4\mathcal{E} + 2L_{\mathcal{S}} - L_{KK} - 4HL_{\mathcal{S}K} - 4H^2 L_{\mathcal{S}\mathcal{S}}\right)\,. \tag{4.144}
$$

The terms containing $\mathcal{R}^2 = 16(\partial^2\zeta)^2/a^4$ and $\mathcal{R}_{ij}\mathcal{R}^{ij} = [5(\partial^2\zeta)^2 + (\partial_i\partial_j\zeta)^2]/a^4$ are absent in Eq. (4.143) because they only involve spatial derivatives of ζ higher than second order.

In Horndeski theory described by the action (4.110), the coefficients in the action (4.143) can be computed by using Eqs. (4.138)–(4.141). They are given by

$$
M_*^2 f = 2G_4 - G_{5\phi}\dot{\phi}^2 + 2G_{5X}\dot{\phi}^2\ddot{\phi}\,, \tag{4.145}
$$

$$
\begin{aligned}
\Lambda = {} & XG_{2X} - G_2 + \dot{\phi}^2(\ddot{\phi} + 3H\dot{\phi})G_{3X} + \dot{\mathcal{F}}_4/2 + 3H\dot{X}G_{4X} - 18H^2G_{4X}\dot{\phi}^2\\
&+6HG_{4\phi X}\dot{\phi}^3 + 12H^2G_{4XX}\dot{\phi}^4 + \dot{\mathcal{F}}_5/2 + 3M_*^2H^2 f_5 + 3M_*^2H\dot{f}_5/2\\
&-6H^2G_{5\phi}\dot{\phi}^2 - 7H^3G_{5X}\dot{\phi}^3 + 3H^2G_{5\phi X}\dot{\phi}^4 + 2H^3G_{5XX}\dot{\phi}^5\,,
\end{aligned}
\tag{4.146}
$$

$$
c = XG_{2X} + \dot{\phi}^2(-\ddot{\phi} + 3H\dot{\phi})G_{3X} + \dot{\phi}^2G_{3\phi} - \dot{\mathcal{F}}_4/2 + 3H\dot{X}G_{4X}
$$

$$-6H^2G_{4X}\dot{\phi}^2+6HG_{4\phi X}\dot{\phi}^3+12H^2G_{4XX}\dot{\phi}^4-\dot{\mathcal{F}}_5/2+3M_*^2H\dot{f}_5/2$$
$$-3H^2G_{5\phi}\dot{\phi}^2-3H^3G_{5X}\dot{\phi}^3+3H^2G_{5\phi X}\dot{\phi}^4+2H^3G_{5XX}\dot{\phi}^5\,, \tag{4.147}$$

$$\begin{aligned}M_2^4 =\ & X^2G_{2XX}+(\ddot{\phi}+3H\dot{\phi})G_{3X}\dot{\phi}^2/2-3HG_{3XX}\dot{\phi}^5-G_{3\phi X}\dot{\phi}^4/2\\ &+\dot{\mathcal{F}}_4/4-3H\dot{X}G_{4X}/2+6HG_{4\phi X}\dot{\phi}^3+18H^2G_{4XX}\dot{\phi}^4-6HG_{4\phi XX}\dot{\phi}^5\\ &-12H^2G_{4XXX}\dot{\phi}^6+\dot{\mathcal{F}}_5/4-3M_*^2H\dot{f}_5/4-3H^3G_{5X}\dot{\phi}^3/2\\ &+6H^2G_{5\phi X}\dot{\phi}^4+6H^3G_{5XX}\dot{\phi}^5-3H^2G_{5\phi XX}\dot{\phi}^6-2H^3G_{5XXX}\dot{\phi}^7\,,\end{aligned} \tag{4.148}$$

$$\begin{aligned}\bar{m}_1^3 =\ & 2G_{3X}\dot{\phi}^3+2\dot{X}G_{4X}-8HG_{4X}\dot{\phi}^2+4G_{4\phi X}\dot{\phi}^3+16HG_{4XX}\dot{\phi}^4\,,\\ &+M_*^2\dot{f}_5-4HG_{5\phi}\dot{\phi}^2-6H^2G_{5X}\dot{\phi}^3+4HG_{5\phi X}\dot{\phi}^4+4H^2G_{5XX}\dot{\phi}^5\,,\end{aligned} \tag{4.149}$$

$$m_4^2=\mu_1^2=2G_{4X}\dot{\phi}^2+G_{5\phi}\dot{\phi}^2+HG_{5X}\dot{\phi}^3-G_{5X}\dot{\phi}^2\ddot{\phi}\,, \tag{4.150}$$

where

$$\mathcal{F}_4=2\dot{X}G_{4X}-8HG_{4X}\dot{\phi}^2\,, \tag{4.151}$$
$$\mathcal{F}_5=2M_*^2Hf_5+M_*^2\dot{f}_5-2HG_{5\phi}\dot{\phi}^2-2H^2G_{5X}\dot{\phi}^3\,, \tag{4.152}$$
$$M_*^2f_5=-G_{5\phi}\dot{\phi}^2+2G_{5X}\dot{\phi}^2\ddot{\phi}\,. \tag{4.153}$$

We stress that Horndeski theory satisfies the additional relation $m_4^2 = \mu_1^2$. The time and spatial derivatives for the theory (4.143) are kept up to second order for *linear* cosmological perturbations. If $m_4^2 \neq \mu_1^2$, then higher-order spatial derivatives should appear beyond linear order. For the computation of primordial non-Gaussianities of curvature perturbations generated during inflation, we need to expand the action (4.6) higher than quadratic order. In such cases, the presence of higher-order spatial derivatives can modify the shape of non-Gaussianities [20, 61] relative to that derived for Horndeski theory [37, 38, 52].

4.7 Application to Dark Energy

In this section, we study the dynamics of dark energy based on Horndeski theory in the presence of matter (cold dark matter, baryons, photons etc.). The action in such a theory is given by

$$S=\int d^4x\sqrt{-g}\sum_{i=2}^{5}L_i+\int d^4x\,L_m\,, \tag{4.154}$$

where $L_{2,3,4,5}$ are given by Eqs. (4.87)–(4.90) and L_m is the matter Lagrangian of a barotropic perfect fluid. The scalar degree of freedom is responsible for the late-time cosmic acceleration. We assume that matter does not have a direct coupling to ϕ.

4.7.1 Background Equations of Motion

On the flat FLRW background, the energy-momentum tensor of the barotropic perfect fluid is given by $T^0_0 = -\rho_m$ and $T^i_j = P_m \delta^i_j$, where ρ_m is the energy density and P_m is the pressure. This satisfies the continuity equation $T^{\mu}_{0;\mu} = 0$, i.e.,

$$\dot{\rho}_m + 3H(\rho_m + P_m) = 0\,. \tag{4.155}$$

In the presence of matter, the background equations of motion (4.26) and (4.28) are modified to

$$\bar{L} + L_N - 3H\mathcal{F} = \rho_m\,, \tag{4.156}$$

$$\dot{\mathcal{F}} + L_N = \rho_m + P_m\,. \tag{4.157}$$

Substituting Eqs. (4.156)–(4.157) into Eqs. (4.138)–(4.139), we obtain

$$\Lambda + c = 3M_*^2(fH^2 + \dot{f}H) - \rho_m\,, \tag{4.158}$$

$$\Lambda - c = M_*^2(2f\dot{H} + 3fH^2 + 2\dot{f}H + \ddot{f}) + P_m\,. \tag{4.159}$$

In Horndeski theory, the functions f, Λ, c are given, respectively, by Eqs. (4.145), (4.146), and (4.147). Among the four functions $G_{2,3,4,5}$, the three combinations of them (i.e., f, Λ, c) determine the cosmological dynamics.

Taking the time derivative of Eq. (4.158) and using Eqs. (4.155) and (4.159), we obtain

$$\dot{\Lambda} + \dot{c} + 6Hc = 3M_*^2\dot{f}(2H^2 + \dot{H})\,. \tag{4.160}$$

The background equations of motion (4.158) and (4.159) can be expressed as

$$3M_{\rm pl}^2H^2 = \rho_{\rm DE} + \rho_m\,, \tag{4.161}$$

$$M_{\rm pl}^2(2\dot{H} + 3H^2) = -P_{\rm DE} - P_m\,, \tag{4.162}$$

where

$$\rho_{\rm DE} = c + \Lambda + 3H^2(M_{\rm pl}^2 - M_*^2 f) - 3M_*^2\dot{f}H\,, \tag{4.163}$$

$$P_{\rm DE} = c - \Lambda - (2\dot{H} + 3H^2)(M_{\rm pl}^2 - M_*^2 f) + M_*^2(2H\dot{f} + \ddot{f})\,. \tag{4.164}$$

On using Eq. (4.160), we find that the "dark" component satisfies the standard continuity equation

$$\dot{\rho}_{\rm DE} + 3H(\rho_{\rm DE} + P_{\rm DE}) = 0\,. \tag{4.165}$$

Then, we can define the equation of state of dark energy, as

$$w_{\rm DE} = \frac{P_{\rm DE}}{\rho_{\rm DE}} = -1 + \frac{2c - 2\dot{H}(M_{\rm pl}^2 - M_*^2 f) - M_*^2(H\dot{f} - \ddot{f})}{c + \Lambda + 3H^2(M_{\rm pl}^2 - M_*^2 f) - 3M_*^2 \dot{f} H} . \qquad (4.166)$$

For quintessence described by the Lagrangian $G_2 = P(\phi, X)$, $G_3 = 0$, $G_4 = M_{\rm pl}^2/2$, and $G_5 = 0$, we have $M_*^2 f = M_{\rm pl}^2$, $\Lambda = V(\phi)$, and $c = \dot{\phi}^2/2$. Since $w_{\rm DE} = [\dot{\phi}^2/2 - V(\phi)]/[\dot{\phi}^2/2 + V(\phi)]$ in this case, it follows that $w_{\rm DE} > -1$. For a non-canonical scalar field with the Lagrangian (4.29) we have $w_{\rm DE} < -1$ for $P_X > 0$, but the scalar ghost is present. For the theories in which the quantity f varies in time (i.e., G_4 or G_5 varies), it is possible to realize $w_{\rm DE} < -1$ under the condition

$$2c - 2\dot{H}(M_{\rm pl}^2 - M_*^2 f) - M_*^2(H\dot{f} - \ddot{f}) < 0 , \qquad (4.167)$$

where we have assumed $\rho_{\rm DE} > 0$. In $f(R)$ gravity [62–66] and Galileons [67], the dark energy equation of state can be smaller than -1, while avoiding the appearance of ghosts.

4.7.2 *Matter Density Perturbations and Effective Gravitational Couplings*

Let us proceed to discuss the equations of motion for linear cosmological perturbations. The discussion in Sect. 4.2 is based on unitary gauge, but for the study of dark energy, the Newtonian gauge is commonly used. The general metric in the presence of scalar perturbations Ψ, ψ, Φ, and E can be written as

$$ds^2 = -(1+2\Psi)dt^2 + 2\psi_{|i}dx^i dt + a^2(t)\left[(1+2\Phi)\delta_{ij} + \partial_{ij}E\right]dx^i dx^j . \qquad (4.168)$$

The Newtonian gauge corresponds to $\psi = 0$ and $E = 0$.

Since the Horndeski action is equivalent to the EFT action (4.143) in unitary gauge with $m_4 = \mu_1^2$ (up to second order), it is possible to derive the perturbation equations in general gauge by reintroducing the scalar perturbation $\delta\phi$ via the Stueckelberg trick [16, 17, 28]. The quantities appearing in the action (4.143) transform under the time coordinate change $t \to t + \delta\phi(t, \boldsymbol{x})$, e.g., $\delta K_{ij} \to \delta K_{ij} - \dot{H}\delta\phi h_{ij} - \partial_i\partial_j\delta\phi$, ${}^{(3)}R_{ij} \to {}^{(3)}R_{ij} + H(\partial_i\partial_j\delta\phi + \delta_{ij}\partial^2\delta\phi)$. This transformation allows one to write the action (4.6) up to quadratic order in the perturbations for the general metric (4.168). Varying the resulting action S with respect to Ψ, ψ, Φ, E, $\delta\phi$ and finally setting $\psi = 0 = E$, we can derive the perturbation equations in the Newtonian gauge. This is the approach taken in [28].

As performed in [44], the perturbation equations can be also derived by directly expanding the Horndeski action (4.154) for the metric (4.168). In the following we assume that the matter Lagrangian L_m is described by a barotropic perfect fluid of non-relativistic matter with the energy-momentum tensor

$$T_0^0 = -(\rho_m + \delta\rho_m)\,, \qquad T_i^0 = -\rho_m \partial_i v_m\,, \qquad T_j^i = 0\,. \tag{4.169}$$

Since there is no direct coupling between matter and the field ϕ, the perturbed energy-momentum tensor obeys the continuity equation

$$\delta T^{\mu\nu}{}_{;\mu} = 0\,. \tag{4.170}$$

From the $\nu = 0$ and $\nu = i$ components of Eq. (4.170), we obtain the following equations in Fourier space respectively,

$$\delta\dot{\rho}_m + 3H\delta\rho_m + 3\rho_m\dot{\Phi} + \frac{k^2}{a^2}\rho_m v_m = 0\,, \tag{4.171}$$

$$\dot{v}_m = \Psi\,, \tag{4.172}$$

where k is a comoving wavenumber. We introduce the gauge-invariant density contrast

$$\delta_m \equiv \frac{\delta\rho_m}{\rho_m} + 3Hv_m\,. \tag{4.173}$$

Taking the time derivative of (4.171) and using Eq. (4.172), the density contrast satisfies

$$\ddot{\delta}_m + 2H\dot{\delta}_m + \frac{k^2}{a^2}\Psi = 3\ddot{Q} + 6H\dot{Q}\,, \tag{4.174}$$

where $Q \equiv Hv_m - \Phi$.

Expanding the action (4.154) for the metric (4.168) up to quadratic order in the perturbations, varying the resulting action with respect to E, Ψ, $\delta\phi$, and finally setting $\psi = E = 0$, we obtain the following perturbation equations respectively:

$$B_6\Phi + B_7\delta\phi + B_8\Psi = 0\,, \tag{4.175}$$

$$A_1\dot{\Phi} + A_2\dot{\delta\phi} - \rho_m\Psi + B_8\frac{k^2}{a^2}\Phi + A_4\Psi + \left(A_6\frac{k^2}{a^2} - \mu\right)\delta\phi - \delta\rho_m = 0\,, \tag{4.176}$$

$$D_1\ddot{\Phi} + D_2\ddot{\delta\phi} + D_3\dot{\Phi} + D_4\dot{\delta\phi} + D_5\dot{\Psi} + \left(B_7\frac{k^2}{a^2} + D_8\right)\Phi$$
$$+ \left(D_9\frac{k^2}{a^2} - M^2\right)\delta\phi + \left(A_6\frac{k^2}{a^2} + D_{11}\right)\Psi = 0\,, \tag{4.177}$$

where

$$B_6 = 4\mathcal{E} = 4G_4 + 2XG_{5\phi} - 4XG_{5X}\ddot{\phi}\,, \tag{4.178}$$

$$\begin{aligned} B_7 &= \frac{4}{\dot{\phi}}\left[\dot{L_{\mathcal{S}}} + H(L_{\mathcal{S}} - \mathcal{E})\right]\,, \\ &= 8G_{4X}H\dot{\phi} + 8(G_{4X} + 2XG_{4XX})\ddot{\phi} + 4G_{4\phi} - 8XG_{4\phi X} \\ &\quad +4(G_{5\phi} + XG_{5\phi X})\ddot{\phi} + 4H\left[2(G_{5X} + XG_{5XX})\ddot{\phi} + G_{5\phi} - XG_{5\phi X}\right]\dot{\phi} \\ &\quad -2XG_{5\phi\phi} - 4(H^2 + \dot{H})XG_{5X}\,, \end{aligned} \tag{4.179}$$

$$B_8 = 4L_{\mathcal{S}} = 4G_4 - 8XG_{4X} - 4H\dot{\phi}XG_{5X} - 2XG_{5\phi}\,. \tag{4.180}$$

Explicit forms of the time-dependent coefficients A_i and D_i as well as other perturbations equations (derived by the variations of Φ and ψ) are given in [44]. The definition of the term μ in Eq. (4.176) is $\mu = \mathcal{H}_{\phi}$, where $\mathcal{H} \equiv -(\bar{L} + L_N - 3H\mathcal{F})$. The term M in Eq. (4.177) is defined by

$$M^2 \equiv [\dot{\mu} + 3H(\mu + \nu)]/\dot{\phi}\,, \tag{4.181}$$

where $\nu = \mathcal{P}_{\phi}$ with $\mathcal{P} \equiv \bar{L} - \dot{\mathcal{F}} - 3H\mathcal{F}$. The mass square M^2 involves the second derivative of $-G_2$ with respect to ϕ [44]. For a canonical field with the potential $V(\phi)$, this means that the second derivative $V_{\phi\phi}$ is present in the expression of M^2. For dark energy models in which the so-called chameleon mechanism [68] works to suppress the fifth force mediated by the field ϕ, the models are designed to have a large mass M in the region of high density [62–66, 69]. In the low-energy regime where the late-time cosmic acceleration comes into play, the mass M should be as small as H_0.

The perturbations related to the observations of large-scale structures and weak lensing have been deep inside the Hubble radius in the low-redshift regime. In the following we use the quasi-static approximation on sub-horizon scales, under which the dominant contributions to Eqs. (4.176) and (4.177) are those involving the terms k^2/a^2, $\delta\rho_m$, and M^2 [70]. In doing so, we neglect the contribution of the oscillating term of the field perturbation $\delta\phi$ relative to the one induced from the matter perturbation $\delta\rho_m$. Under this approximation scheme, the variations of the gravitational potentials Φ and Ψ are small such that $|\dot{\Phi}| < |H\Phi|$ and $|\dot{\Psi}| < |H\Psi|$. Then, Eqs. (4.176) and (4.177) read

$$B_8\frac{k^2}{a^2}\Phi + A_6\frac{k^2}{a^2}\delta\phi - \delta\rho_m \simeq 0\,, \tag{4.182}$$

$$B_7\frac{k^2}{a^2}\Phi + \left(D_9\frac{k^2}{a^2} - M^2\right)\delta\phi + A_6\frac{k^2}{a^2}\Psi \simeq 0\,, \tag{4.183}$$

where

$$
\begin{aligned}
A_6 &= 2XG_{3X} + 8H(G_{4X} + 2XG_{4XX})\dot{\phi} + 2G_{4\phi} + 4XG_{4\phi X} \\
&\quad + 4H\left(G_{5\phi} + XG_{5\phi X}\right)\dot{\phi} - 2H^2X\left(3G_{5X} + 2XG_{5XX}\right), \qquad (4.184)
\end{aligned}
$$

$$
\begin{aligned}
D_9 &= 2G_{2X} - 4\left(G_{3X} + XG_{3XX}\right)\ddot{\phi} - 8HG_{3X}\dot{\phi} - 2G_{3\phi} + 2XG_{3\phi X} \\
&\quad + [-16H(3\,G_{4XX} + 2XG_{4XXX})\ddot{\phi} - 8H(3G_{4\phi X} - 2XG_{4\phi XX})]\dot{\phi} \\
&\quad - 4(3G_{4\phi X} + 2XG_{4\phi XX})\ddot{\phi} + 40H^2XG_{4XX} + 4XG_{4\phi\phi X} \\
&\quad + 8\dot{H}(G_{4X} + 2XG_{4XX}) + 12H^2G_{4X} + \{-8H(2G_{5\phi X} + XG_{5\phi XX})\ddot{\phi} \\
&\quad + 8H(H^2 + \dot{H})(G_{5X} + XG_{5XX}) + 4HXG_{5\phi\phi X}\}\dot{\phi} - 4H^2X^2G_{5\phi XX} \\
&\quad + 4H^2(G_{5X} + 5XG_{5XX} + 2X^2G_{5XXX})\ddot{\phi} + 2(3H^2 + 2\dot{H})G_{5\phi} \\
&\quad + 4\dot{H}XG_{5\phi X} + 10H^2XG_{5\phi X}\,. \qquad (4.185)
\end{aligned}
$$

Solving Eqs. (4.175), (4.182), and (4.183) for Ψ and Φ, it follows that

$$
\frac{k^2}{a^2}\Psi \simeq -\frac{(B_6D_9 - B_7^2)\,(k/a)^2 - B_6M^2}{(A_6^2B_6 + B_8^2D_9 - 2A_6B_7B_8)\,(k/a)^2 - B_8^2M^2}\delta\rho_m\,, \qquad (4.186)
$$

$$
\frac{k^2}{a^2}\Phi \simeq -\frac{(A_6B_7 - B_8D_9)\,(k/a)^2 + B_8M^2}{(A_6^2B_6 + B_8^2D_9 - 2A_6B_7B_8)\,(k/a)^2 - B_8^2M^2}\delta\rho_m\,. \qquad (4.187)
$$

From Eq. (4.171), we find that the term Hv_m is at most of the order of $(aH/k)^2\delta\rho_m/\rho_m$. For the modes deep inside the Hubble radius ($k \gg aH$), we then have $\delta_m \simeq \delta\rho_m/\rho_m$ in Eq. (4.173). Under the quasi-static approximation on sub-horizon scales, the r.h.s. of Eq. (4.174) is negligible relative to the l.h.s. of it. On using Eq. (4.186), the linear matter perturbation obeys

$$
\ddot{\delta}_m + 2H\dot{\delta}_m - 4\pi G_{\rm eff}\rho_m\delta_m \simeq 0\,, \qquad (4.188)
$$

where

$$
G_{\rm eff} = \frac{2M_{\rm pl}^2[(B_6D_9 - B_7^2)\,(k/a)^2 - B_6M^2]}{(A_6^2B_6 + B_8^2D_9 - 2A_6B_7B_8)\,(k/a)^2 - B_8^2M^2}G\,. \qquad (4.189)
$$

Note that G is the bare gravitational constant related with the reduced Planck mass $M_{\rm pl}$ via the relation $8\pi G = M_{\rm pl}^{-2}$. Since the effective gravitational coupling $G_{\rm eff}$ is different depending on gravitational theories, it is possible to discriminate between different modified gravity models from the growth of matter perturbations.

In order to quantify the difference between the two gravitational potentials Ψ and Φ, we define

$$
\eta \equiv -\Phi/\Psi\,. \qquad (4.190)
$$

On using the solutions (4.186) and (4.187), the anisotropy parameter reads

$$\eta \simeq \frac{(B_8 D_9 - A_6 B_7)(k/a)^2 - B_8 M^2}{(B_6 D_9 - B_7^2)(k/a)^2 - B_6 M^2}\,. \tag{4.191}$$

The effective gravitational potential associated with deviation of the light rays in CMB and weak lensing observations is defined by [71]

$$\Phi_{\rm eff} \equiv (\Psi - \Phi)/2\,, \tag{4.192}$$

From Eqs. (4.186), (4.189), and (4.190), we obtain

$$\Phi_{\rm eff} \simeq -4\pi G_{\rm eff} \frac{1+\eta}{2} \left(\frac{a}{k}\right)^2 \rho_m \delta_m\,, \tag{4.193}$$

which is related to both δ_m and η.

4.7.3 Growth of Matter Perturbations

Introducing the matter density parameter $\Omega_m = \rho_m/(3M_{\rm pl}^2 H^2)$, we can write the matter perturbation equation (4.188) in the form

$$\delta_m'' + \left(2 + \frac{H'}{H}\right)\delta_m' - \frac{3}{2}\frac{G_{\rm eff}}{G}\Omega_m \delta_m \simeq 0\,, \tag{4.194}$$

where a prime represents a derivative with respect to $\ln a$.

Let us first consider a non-canonical scalar field described by the Lagrangian

$$L = \frac{M_{\rm pl}^2}{2} R + P(\phi, X)\,, \tag{4.195}$$

in which case $G_2 = P(\phi, X)$, $G_3 = 0$, $G_4 = M_{\rm pl}^2/2$, and $G_5 = 0$. Since $B_6 = B_8 = 2M_{\rm pl}^2$, $B_7 = A_6 = 0$, and $D_9 = 2P_X$, it follows that $G_{\rm eff} = G$ and $\eta = 1$ from Eqs. (4.189) and (4.191). During the matter-dominated epoch characterized by $\Omega_m = 1$ and $H'/H = -3/2$, there is the growing-mode solution to Eq. (4.194):

$$\delta_m \propto t^{2/3}\,. \tag{4.196}$$

In this regime, the effective gravitational potential (4.193) is constant. After the Universe enters the epoch of cosmic acceleration, the growth rate of δ_m becomes smaller than that given in Eq. (4.196), so $\Phi_{\rm eff}$ starts to decay. Since $G_{\rm eff}$ is equivalent to G for the models in the framework of GR, the difference of the growth rate between the models comes from the different background expansion history. In the

ΛCDM model characterized by $P = -\Lambda$, the growth rate $f \equiv \dot{\delta}_m/(H\delta_m)$ can be estimated as $f = (\Omega_m)^\gamma$ with $\gamma \simeq 0.55$ in the low-redshift regime ($z < 1$) [72]. As long as the dark energy equation of state does not significantly deviate from -1, γ is close to the value 0.55 for the models in the framework of GR [73, 74].

As an example of modified gravity models, we consider BD theory described by the action (4.93). Since $B_6 = 2M_{\rm pl}\phi$, $B_7 = 2M_{\rm pl}$, $B_8 = 2M_{\rm pl}\phi$, $A_6 = M_{\rm pl}$, and $D_9 = -M_{\rm pl}\omega_{\rm BD}/\phi$, Eqs. (4.189) and (4.191) reduce to

$$G_{\rm eff} = \frac{M_{\rm pl}}{\phi}\frac{4 + 2\omega_{\rm BD} + 2(\phi/M_{\rm pl})(Ma/k)^2}{3 + 2\omega_{\rm BD} + 2(\phi/M_{\rm pl})(Ma/k)^2}G\,, \tag{4.197}$$

$$\eta = \frac{1 + \omega_{\rm BD} + (\phi/M_{\rm pl})(Ma/k)^2}{2 + \omega_{\rm BD} + (\phi/M_{\rm pl})(Ma/k)^2}\,, \tag{4.198}$$

where

$$M^2 = V_{\phi\phi} + \frac{\omega_{\rm BD}M_{\rm pl}}{\phi^3}\left[\dot{\phi}^2 - \phi\left(\ddot{\phi} + 3H\dot{\phi}\right)\right]. \tag{4.199}$$

In the $\omega_{\rm BD} \to \infty$ limit with $\phi \to M_{\rm pl}$, we obtain $G_{\rm eff} \to G$ and $\eta \to 1$, so the General Relativistic behavior can be recovered. The same property also holds for $M \to \infty$, as the scalar field does not propagate.

In the massless limit $M^2 \to 0$, it follows that $G_{\rm eff} \simeq (M_{\rm pl}/\phi)(4+2\omega_{\rm BD})G/(3+2\omega_{\rm BD})$ and $\eta \simeq (1 + \omega_{\rm BD})/(2 + \omega_{\rm BD})$, so the growth rates of δ_m and $\Phi_{\rm eff}$ are different from those in GR. Since $\omega_{\rm BD} = 0$ in metric $f(R)$ gravity, we have $G_{\rm eff} \simeq (M_{\rm pl}/\phi)(4/3)G$ and $\eta \simeq 1/2$. The viable dark energy models based on $f(R)$ gravity [62–66] are constructed in a way that the mass M is large for $R \gg H_0^2$ and that M decreases to the similar order to H_0 by today. There is a transition from the "massive" regime $M > k/a$ to the "massless" regime $M < k/a$, depending on the wavenumber k [64, 65, 75]. If this transition happens in the deep matter era characterized by $H'/H \simeq -3/2$ and $\tilde{\Omega}_m = \rho_m/(3M_{\rm pl}\phi H^2) \simeq 1$, the growing-mode solution to Eq. (4.194) during the "massless" regime of metric $f(R)$ gravity is given by

$$\delta_m \propto t^{(\sqrt{33}-1)/6}\,, \tag{4.200}$$

whose growth rate is larger than that in GR. This leaves an imprint for the measurement of red-shift space distortions in the galaxy power spectrum [76]. From Eq. (4.193), the effective gravitational coupling evolves as $\Phi_{\rm eff} \propto t^{(\sqrt{33}-5)/6}$. This modification affects the weak lensing power spectrum as well as the ISW effect in CMB [77, 78].

In other modified gravity models like covariant Galileons [79], the growth rate of perturbations is different from that in GR and $f(R)$ gravity. Although the current observations are not enough to discriminate between different models precisely, we hope that future observations will allow us to do so.

Conclusions

We have reviewed a framework for studying the most general four-dimensional gravitational theories with a single scalar degree of freedom. The EFT of cosmological perturbations is useful for the unified description of modified gravitational theories in that it can be describe practically all single-field models proposed in the literature. This unified scheme can allow one to provide model-independent constraints on the properties of inflation/dark energy and to put constraints on individual models consistent with observations.

Starting from the general action (4.6) that depends on the lapse N and other three-dimensional scalar ADM variables, we have expanded the action up to quadratic order in cosmological perturbations about the FLRW background. The choice of unitary gauge allows one to absorb dynamics of the field perturbation $\delta\phi$ into the gravitational sector. Provided that the three conditions (4.49)–(4.51) are satisfied, the second-order Lagrangian density reduces to the simple form (4.54) with a single scalar degree of freedom characterized by the curvature perturbation ζ. We have also shown that the quadratic action for tensor perturbations is given by Eq. (4.60). In order to avoid ghosts and Laplacian instabilities of scalar and tensor perturbations, we require the conditions $Q_s > 0$, $c_s^2 > 0$, $Q_t > 0$, and $c_t^2 > 0$.

The most general scalar-tensor theories with second-order equations of motion–Horndeski theory–belong to a sub-class of the action (4.6) in the framework of EFT. The Horndeski Lagrangian can be expressed in terms of the ADM scalar quantities in the form (4.110). Using the relations (4.138)–(4.141) between the EFT variables appearing in the action (4.135) and the partial derivatives of the Lagrangian L with respect to the ADM variables, we have shown that, up to quadratic order in perturbations, Horndeski theory corresponds to the action (4.143) with the additional condition $m_4^2 = \mu_1^2$. The dictionary between the EFT variables and the functions $G_i(\phi, X)$ in Horndeski theory is given by Eqs. (4.145)–(4.150).

In Sect. 4.4 we have also derived the power spectra of scalar and tensor perturbations generated during inflation for general second-order theories satisfying the conditions (4.49)–(4.51). The formulas (4.78) and (4.83) cover a wide variety of modified gravitational theories presented in Sect. 4.5.1, so they can be used for constraining each inflationary model from the CMB observations (along the lines of [80]). In particular, it will be of interest to discriminate between a host of single-field inflationary models from the precise B-mode polarization data available in the future.

In Sect. 4.7 we have applied the EFT of cosmological perturbations to dark energy in the presence of a barotropic perfect fluid. The background cosmology is described by three time-dependent functions f, Λ, and c,

(continued)

with which different models can be distinguished from the evolution of the dark energy equation of state. In Horndeski theory, we have obtained the effective gravitational coupling (4.189) appearing in the matter perturbation equation (4.188) under the quasi-static approximation on sub-horizon scales. Together with the effective gravitational potential given in Eq. (4.193), it will be possible to discriminate between different modified gravity models from the observations of large-scale structures, weak lensing, and CMB.

While we have studied the effective single-field scenario in unitary gauge, another scalar degree of freedom can be also taken into account in the action (4.6) [33]. Such a second scalar field can be potentially responsible for dark matter. It will be of interest to provide a unified framework for understanding the origins of inflation, dark energy, and dark matter.

Acknowledgements The author is grateful to the organizers of the 7th Aegean Summer School for wonderful hospitality. The author thanks Antonio De Felice, Laszlo Arpad Gergely, and Federico Piazza for useful discussions. This work was supported by Grant-in-Aid for Scientific Research Fund of the JSPS (No. 30318802) and Grant-in-Aid for Scientific Research on Innovative Areas (No. 21111006).

References

1. A.A. Starobinsky, Phys. Lett. B **91**, 99 (1980)
2. D. Kazanas, Astrophys. J. **241**, L59 (1980); K. Sato, Mon. Not. R. Astron. Soc. **195**, 467 (1981); Phys. Lett. **99B**, 66 (1981); A.H. Guth, Phys. Rev. D **23**, 347 (1981)
3. G.F. Smoot et al., Astrophys. J. **396**, L1 (1992)
4. D.N. Spergel et al. [WMAP Collaboration], Astrophys. J. Suppl. **148**, 175 (2003)
5. P.A.R. Ade et al. [Planck Collaboration], (2014) [arXiv:1303.5076 [astro-ph.CO]]
6. V.F. Mukhanov, G.V. Chibisov, JETP Lett. **33**, 532 (1981); A.H. Guth, S.Y. Pi, Phys. Rev. Lett. **49** (1982) 1110; S.W. Hawking, Phys. Lett. B **115**, 295 (1982); A.A. Starobinsky, Phys. Lett. B **117** (1982) 175; J.M. Bardeen, P.J. Steinhardt, M.S. Turner, Phys. Rev. D **28**, 679 (1983).
7. A.G. Riess et al. [Supernova Search Team Collaboration], Astron. J. **116**, 1009 (1998) [astro-ph/9805201]
8. S. Perlmutter et al. [Supernova Cosmology Project Collaboration], Astrophys. J. **517**, 565 (1999) [astro-ph/9812133]
9. D.J. Eisenstein et al. [SDSS Collaboration], Astrophys. J. **633**, 560 (2005) [astro-ph/0501171]
10. S. Weinberg, Rev. Mod. Phys. **61**, 1 (1989)
11. E.J. Copeland, M. Sami, S. Tsujikawa, Int. J. Mod. Phys. D **15**, 1753 (2006) [hep-th/0603057]
12. Y. Fujii, Phys. Rev. D **26**, 2580 (1982); L.H. Ford, Phys. Rev. D **35**, 2339 (1987); C. Wetterich, Nucl. Phys B. **302**, 668 (1988); T. Chiba, N. Sugiyama, T. Nakamura, Mon. Not. R. Astron. Soc. **289**, L5 (1997) [astro-ph/9704199]; P.G. Ferreira, M. Joyce, Phys. Rev. Lett. **79**, 4740 (1997) [astro-ph/9707286]; R.R. Caldwell, R. Dave, P.J. Steinhardt, Phys. Rev. Lett. **80**, 1582 (1998) [astro-ph/9708069]; T. Chiba, N. Sugiyama, T. Nakamura, Mon. Not. R. Astron. Soc. **301**, 72 (1998) [astro-ph/9806332]; I. Zlatev, L.-M. Wang, P.J. Steinhardt, Phys. Rev. Lett. **82**, 896 (1999) [astro-ph/9807002]

13. G. Hinshaw et al. [WMAP Collaboration], Astrophys. J. Suppl. **208**, 19 (2013) [arXiv:1212.5226 [astro-ph.CO]]
14. T. Chiba, A. De Felice, S. Tsujikawa, Phys. Rev. D **87**, 083505 (2013) [arXiv:1210.3859 [astro-ph.CO]]
15. T.P. Sotiriou, V. Faraoni, Rev. Mod. Phys. **82**, 451 (2010) [arXiv:0805.1726 [gr-qc]]; A. De Felice, S. Tsujikawa, Living Rev. Rel. **13**, 3 (2010) [arXiv:1002.4928 [gr-qc]]; S. Tsujikawa, Lect. Notes Phys. **800**, 99 (2010) [arXiv:1101.0191 [gr-qc]]; T. Clifton, P.G. Ferreira, A. Padilla, C. Skordis, Phys. Rept. **513**, 1 (2012) [arXiv:1106.2476 [astro-ph.CO]]
16. P. Creminelli, M.A. Luty, A. Nicolis, L. Senatore, J. High Energy Phys. **0612**, 080 (2006) [hep-th/0606090]
17. C. Cheung, P. Creminelli, A.L. Fitzpatrick, J. Kaplan, L. Senatore, J. High Energy Phys. **0803**, 014 (2008) [arXiv:0709.0293 [hep-th]]
18. P. Creminelli, G. D'Amico, J. Norena, F. Vernizzi, JCAP **0902**, 018 (2009) [arXiv:0811.0827 [astro-ph]]
19. S. Weinberg, Phys. Rev. D **72**, 043514 (2005) [hep-th/0506236]
20. L. Senatore, K.M. Smith, M. Zaldarriaga, JCAP **1001**, 028 (2010) [arXiv:0905.3746 [astro-ph.CO]]; N. Bartolo, M. Fasiello, S. Matarrese, A. Riotto, JCAP **1009**, 035 (2010) [arXiv:1006.5411 [astro-ph.CO]]; N. Agarwal, R. Holman, A.J. Tolley, J. Lin, J. High Energy Phys. **1305**, 085 (2013) [arXiv:1212.1172 [hep-th]]
21. M. Park, K.M. Zurek, S. Watson, Phys. Rev. D **81**, 124008 (2010) [arXiv:1003.1722 [hep-th]]
22. J.K. Bloomfield, E.E. Flanagan, JCAP **1210**, 039 (2012) [arXiv:1112.0303 [gr-qc]]
23. R.A. Battye, J.A. Pearson, JCAP **1207**, 019 (2012) [arXiv:1203.0398 [hep-th]]
24. E.-M. Mueller, R. Bean, S. Watson, Phys. Rev. D **87**, 083504 (2013) [arXiv:1209.2706 [astro-ph.CO]]
25. G. Gubitosi, F. Piazza, F. Vernizzi, JCAP **1302**, 032 (2013) [arXiv:1210.0201 [hep-th]]
26. J.K. Bloomfield, E. Flanagan, M. Park, S. Watson, JCAP **1308**, 010 (2013) [arXiv:1211.7054 [astro-ph.CO]]
27. J. Bloomfield, JCAP **1312**, 044 (2013) [arXiv:1304.6712 [astro-ph.CO]]
28. J. Gleyzes, D. Langlois, F. Piazza, F. Vernizzi, JCAP **1308**, 025 (2013) [arXiv:1304.4840 [hep-th]]
29. F. Piazza, F. Vernizzi, Class. Quant. Grav. **30**, 214007 (2013) [arXiv:1307.4350]
30. N. Frusciante, M. Raveri, A. Silvestri, JCAP **1402**, 026 (2014) [arXiv:1310.6026 [astro-ph.CO]]
31. B. Hu, M. Raveri, N. Frusciante, A. Silvestri, Phys. Rev. D **89**, 103530 (2014) [arXiv:1312.5742 [astro-ph.CO]]
32. F. Piazza, H. Steigerwald, C. Marinoni, JCAP **1405**, 043 (2014) [arXiv:1312.6111 [astro-ph.CO]]
33. L. Gergely, S. Tsujikawa, Phys. Rev. D **89**, 064059 (2014) [arXiv:1402.0553 [hep-th]]
34. V.F. Mukhanov, H.A. Feldman, R.H. Brandenberger, Phys. Rept. **215**, 203 (1992)
35. T. Kobayashi, M. Yamaguchi, J.'I. Yokoyama, Prog. Theor. Phys. **126**, 511 (2011) [arXiv:1105.5723 [hep-th]]
36. A. De Felice, S. Tsujikawa, JCAP **1104**, 029 (2011) [arXiv:1103.1172 [astro-ph.CO]]
37. X. Gao, D.A. Steer, JCAP **1112**, 019 (2011) [arXiv:1107.2642 [astro-ph.CO]]
38. A. De Felice, S. Tsujikawa, Phys. Rev. D **84**, 083504 (2011) [arXiv:1107.3917 [gr-qc]]
39. G.W. Horndeski, Int. J. Theor. Phys. **10**, 363–384 (1974)
40. A. Nicolis, R. Rattazzi, E. Trincherini, Phys. Rev. D **79**, 064036 (2009) [arXiv:0811.2197 [hep-th]]
41. C. Deffayet, G. Esposito-Farese, A. Vikman, Phys. Rev. D **79**, 084003 (2009) [arXiv:0901.1314 [hep-th]]; C. Deffayet, S. Deser, G. Esposito-Farese, Phys. Rev. D **80**, 064015 (2009) [arXiv:0906.1967 [gr-qc]]
42. C. Charmousis, E.J. Copeland, A. Padilla, P.M. Saffin, Phys. Rev. Lett. **108**, 051101 (2012) [arXiv:1106.2000 [hep-th]]
43. C. Deffayet, X. Gao, D.A. Steer, G. Zahariade, Phys. Rev. D **84**, 064039 (2011) [arXiv:1103.3260 [hep-th]]

44. A. De Felice, T. Kobayashi, S. Tsujikawa, Phys. Lett. B **706**, 123 (2011) [arXiv:1108.4242 [gr-qc]]
45. R.L. Arnowitt, S. Deser, C.W. Misner, Phys. Rev. **117**, 1595 (1960)
46. C. Armendariz-Picon, T. Damour, V.F. Mukhanov, Phys. Lett. B **458**, 209 (1999) [hep-th/9904075]
47. T. Chiba, T. Okabe, M. Yamaguchi, Phys. Rev. D **62**, 023511 (2000) [astro-ph/9912463]; C. Armendariz-Picon, V.F. Mukhanov, P.J. Steinhardt, Phys. Rev. Lett. **85**, 4438 (2000) [astro-ph/0004134]
48. J.M. Maldacena, J. High Energy Phys. **0305**, 013 (2003) [astro-ph/0210603]
49. J. Garriga, V.F. Mukhanov, Phys. Lett. B **458**, 219 (1999) [hep-th/9904176]
50. B.A. Bassett, S. Tsujikawa, D. Wands, Rev. Mod. Phys. **78**, 537 (2006) [astro-ph/0507632]
51. X. Chen, M.-X. Huang, S. Kachru, G. Shiu, JCAP **0701**, 002 (2007) [hep-th/0605045]
52. A. De Felice, S. Tsujikawa, JCAP **1303**, 030 (2013) [arXiv:1301.5721 [hep-th]]
53. S. Das et al., arXiv:1301.1037 [astro-ph.CO]; C.L. Reichardt et al., Astrophys. J. **755**, 70 (2012) [arXiv:1111.0932 [astro-ph.CO]]
54. P.A.R. Ade et al. [BICEP2 Collaboration], Phys. Rev. Lett. **112**, 241101 (2014) [arXiv:1403.3985 [astro-ph.CO]]
55. C. Brans, R.H. Dicke, Phys. Rev. **124**, 925 (1961)
56. M. Gasperini, G. Veneziano, Astropart. Phys. **1**, 317 (1993) [hep-th/9211021]; M. Gasperini, G. Veneziano, Phys. Rept. **373**, 1 (2003) [hep-th/0207130]
57. F.L. Bezrukov, M. Shaposhnikov, Phys. Lett. B **659**, 703 (2008) [arXiv:0710.3755 [hep-th]]
58. T. Futamase, K.-I. Maeda, Phys. Rev. D **39**, 399 (1989); R. Fakir, W.G. Unruh, Phys. Rev. D **41**, 1783 (1990); J.L. Cervantes-Cota, H. Dehnen, Phys. Rev. D **51**, 395 (1995) [astro-ph/9412032]; Nucl. Phys. B **442**, 391 (1995) [astro-ph/9505069]
59. L. Amendola, Phys. Lett. B **301**, 175 (1993) [gr-qc/9302010]
60. C. Germani, A. Kehagias, Phys. Rev. Lett. **105**, 011302 (2010) [arXiv:1003.2635 [hep-ph]]; C. Germani, A. Kehagias, Phys. Rev. Lett. **106**, 161302 (2011) [arXiv:1012.0853 [hep-ph]]; S. Tsujikawa, Phys. Rev. D **85**, 083518 (2012) [arXiv:1201.5926 [astro-ph.CO]]
61. P. Creminelli, G. D'Amico, M. Musso, J. Norena, E. Trincherini, JCAP **1102**, 006 (2011) [arXiv:1011.3004 [hep-th]]
62. W. Hu, I. Sawicki, Phys. Rev. D **76**, 064004 (2007) [arXiv:0705.1158 [astro-ph]]
63. S.A. Appleby, R.A. Battye, Phys. Lett. B **654**, 7 (2007) [arXiv:0705.3199 [astro-ph]]
64. A.A. Starobinsky, JETP Lett. **86**, 157 (2007) [arXiv:0706.2041 [astro-ph]]
65. S. Tsujikawa, Phys. Rev. D **77**, 023507 (2008) [arXiv:0709.1391 [astro-ph]]
66. E.V. Linder, Phys. Rev. D **80**, 123528 (2009) [arXiv:0905.2962 [astro-ph.CO]]
67. F. Silva, K. Koyama, Phys. Rev. D **80**, 121301 (2009) [arXiv:0909.4538 [astro-ph.CO]]; T. Kobayashi, Phys. Rev. D **81**, 103533 (2010) [arXiv:1003.3281 [astro-ph.CO]]; C. Deffayet, O. Pujolas, I. Sawicki, A. Vikman, JCAP **1010**, 026 (2010) [arXiv:1008.0048 [hep-th]]; R. Gannouji, M. Sami, Phys. Rev. D **82**, 024011 (2010) [arXiv:1004.2808 [gr-qc]]; A. De Felice, S. Tsujikawa, Phys. Rev. Lett. **105**, 111301 (2010) [arXiv:1007.2700 [astro-ph.CO]]; A. De Felice, S. Tsujikawa, Phys. Rev. D **84**, 124029 (2011) [arXiv:1008.4236 [hep-th]]
68. J. Khoury, A. Weltman, Phys. Rev. Lett. **93**, 171104 (2004) [astro-ph/0309300]; J. Khoury, A. Weltman, Phys. Rev. D **69**, 044026 (2004) [astro-ph/0309411]
69. S. Tsujikawa, K. Uddin, S. Mizuno, R. Tavakol, J.'I. Yokoyama, Phys. Rev. D **77**, 103009 (2008) [arXiv:0803.1106 [astro-ph]]; R. Gannouji, B. Moraes, D.F. Mota, D. Polarski, S. Tsujikawa, H.A. Winther, Phys. Rev. D **82**, 124006 (2010) [arXiv:1010.3769 [astro-ph.CO]]
70. A.A. Starobinsky, JETP Lett. **68**, 757 (1998) [Pisma Zh. Eksp. Teor. Fiz. **68**, 721 (1998)] [astro-ph/9810431]; B. Boisseau, G. Esposito-Farese, D. Polarski, A.A. Starobinsky, Phys. Rev. Lett. **85**, 2236 (2000) [gr-qc/0001066]; S. Tsujikawa, Phys. Rev. D **76**, 023514 (2007) [arXiv:0705.1032 [astro-ph]]; S. Nesseris, Phys. Rev. D **79**, 044015 (2009) [arXiv:0811.4292 [astro-ph]]
71. C. Schimd, J.-P. Uzan, A. Riazuelo, Phys. Rev. D **71**, 083512 (2005) [astro-ph/0412120]; L. Amendola, M. Kunz, D. Sapone, JCAP **0804**, 013 (2008) [arXiv:0704.2421 [astro-ph]]
72. L.-M. Wang, P.J. Steinhardt, Astrophys. J. **508**, 483 (1998) [astro-ph/9804015]

73. E.V. Linder, Phys. Rev. D **72**, 043529 (2005) [astro-ph/0507263]
74. S. Tsujikawa, A. De Felice, J. Alcaniz, JCAP **1301**, 030 (2013) [arXiv:1210.4239 [astro-ph.CO]]
75. S. Tsujikawa, R. Gannouji, B. Moraes, D. Polarski, Phys. Rev. D **80**, 084044 (2009) [arXiv:0908.2669 [astro-ph.CO]]; H. Okada, T. Totani, S. Tsujikawa, Phys. Rev. D **87**, 103002 (2013) [arXiv:1208.4681 [astro-ph.CO]]
76. M. Tegmark et al. [SDSS Collaboration], Phys. Rev. D **74**, 123507 (2006) [astro-ph/0608632]; C. Blake et al., Mon. Not. R. Astron. Soc. **415**, 2876 (2011) [arXiv:1104.2948 [astro-ph.CO]]; L. Samushia, W.J. Percival, A. Raccanelli, Mon. Not. R. Astron. Soc. **420**, 2102 (2012) [arXiv:1102.1014 [astro-ph.CO]]
77. S. Tsujikawa, T. Tatekawa, Phys. Lett. B **665**, 325 (2008) [arXiv:0804.4343 [astro-ph]]; F. Schmidt, Phys. Rev. D **78**, 043002 (2008) [arXiv:0805.4812 [astro-ph]]; E. Beynon, D.J. Bacon, K. Koyama, Mon. Not. R. Astron. Soc. **403**, 353 (2010) [arXiv:0910.1480 [astro-ph.CO]]
78. Y.-S. Song, W. Hu, I. Sawicki, Phys. Rev. D **75**, 044004 (2007) [astro-ph/0610532]; Y.-S. Song, H. Peiris, W. Hu, Phys. Rev. D **76**, 063517 (2007) [arXiv:0706.2399 [astro-ph]]
79. A. De Felice, R. Kase, S. Tsujikawa, Phys. Rev. D **83**, 043515 (2011) [arXiv:1011.6132 [astro-ph.CO]]
80. S. Tsujikawa, J. Ohashi, S. Kuroyanagi, A. De Felice, Phys. Rev. D **88**, 023529 (2013) [arXiv:1305.3044 [astro-ph.CO]]

Part II
Massive Gravity

Chapter 5
Introduction to Massive Gravity

Claudia de Rham

Abstract We review recent progress on massive gravity. We first show how extra dimensions prove to be a useful tool in building theories of modified gravity, including Galileon theories and their DBI extensions. DGP arises from an infinite size extra dimension, and we show how massive gravity arises from 'deconstructing' the extra dimension in the vielbein formalism. We then explain how the ghost issue is resolved in that special theory of massive gravity. The viability of such models relies on the Vainshtein mechanism which is best described in terms of Galileons. While its implementation is successful in most of these models it also comes hand in hand with superluminalities and strong coupling which are reviewed and their real consequences are discussed.

5.1 Gravitational Waves and Degrees of Freedom

5.1.1 Polarizations

One of the genuine predictions of General Relativity is the existence of a graviton or massless spin-2 field under the Poincaré group which mediates the gravitational force. The existence of this particle implies the presence of Gravitational Waves (GWs). Whilst advanced LIGO and other interferometer [1] are expected to be on the edge of discovering GWs, the indirect detection of GWs has been confirmed for forty years via the spin-down of binary pulsars and particularly the Hulse Taylor pulsar [2]. The spin-down is in perfect agreement with the emission of gravitational radiation and the prediction that in GR gravitational waves have two polarizations. Nevertheless this does not necessarily rule out the existence of additional polarizations which could be screened for instance via the Vainshtein mechanism see [3] and [4].

C. de Rham (✉)
CERCA & Physics Department, Case Western Reserve University, OH, USA

Perimeter Institute for Theoretical Physics, 31 Caroline St N, Waterloo, ON, Canada N2L 6B9, Emmy Noether Visiting Fellow
e-mail: Claudia.deRham@case.edu

E. Papantonopoulos (ed.), *Modifications of Einstein's Theory of Gravity at Large Distances*, Lecture Notes in Physics 892, DOI 10.1007/978-3-319-10070-8_5

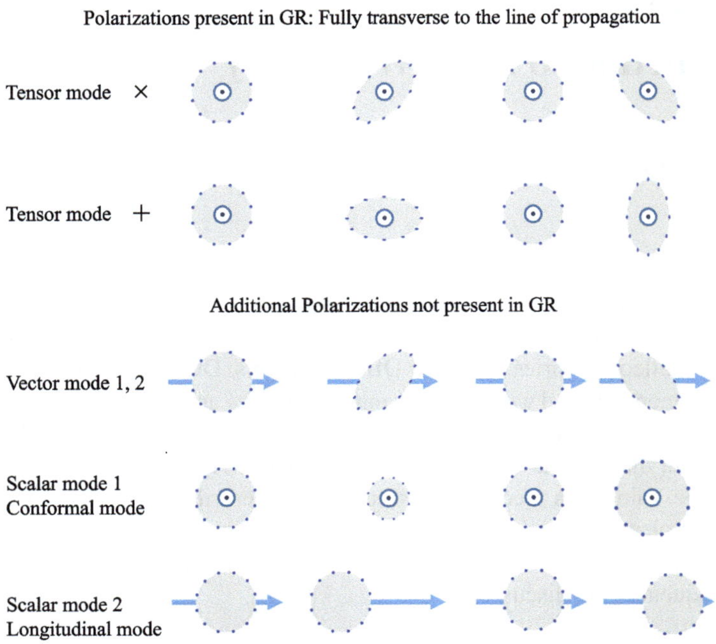

Fig. 5.1 Polarizations of Gravitational Waves in General Relativity and potential additional polarizations in modified gravity. From [6]

In modified theories of gravity GWs could have up to four additional polarizations: two 'vector' polarizations which mix the longitudinal and the transverse directions, as well as two 'scalar' polarizations, one of each being a conformal or breathing mode and the other one a purely longitudinal mode as depicted in Fig. 5.1.

These last four polarizations are absent in GR. However in theories of modified gravity one could in principle excite them. For instance in massive gravity the graviton is instead seen as a massive spin-2 field. In four dimensions, a massive spin-s fields is known to propagate $2s + 1$ dofs (degrees of freedom), so a massive spin-2 field should propagate five dofs.

At the same time, massive gravity breaks diffeomorphism invariance, corresponding to four symmetries in four dimensions. This means that we expect massive gravity to propagate four dofs more than in GR, this would correspond to the four additional polarizations depicted in Fig. 5.1. This corresponds to one additional polarization compared to what a massive spin-2 field should have. If present, this additional fourth new polarization is always pathological and enters as a ghost, now commonly known as the Boulware–Deser (BD) ghost [5]. This BD ghost correspond to the last polarization depicted in Fig. 5.1, namely the longitudinal scalar mode. So for a theory of massive gravity to be free of the BD ghost it should only have at most the three first additional polarizations of Fig. 5.1 and should not excite the last one. In what follows we explain why the presence of a BD ghost would always invalidate the theory and then proceed by constructing explicit models of massive gravity which are free from this pathology. We refer to [6] for a recent review on massive gravity.

5.1.2 *Implications of the BD Ghost*

To understand the implications of the BD ghost, we consider a simple but representative example of how this ghost can present itself. Let us consider a free scalar field ϕ with kinetic term[1] $-1/2(\partial\phi)^2$. For definiteness, one way the BD ghost can manifest itself is via a new operator of the form $(\Box\phi)^3$ arising at a scale Λ,

$$\mathscr{L} = -\frac{1}{2}(\partial\phi)^2 + \frac{1}{6\Lambda^5}(\Box\phi)^3 . \tag{5.1}$$

Considering the fluctuations about a non-trivial background $\phi = \phi_0 + \delta\phi$, with say $\phi_0 = \Lambda^3/8B_0\eta_{\mu\nu}x^\mu x^\nu$, the Lagrangian for the fluctuations is

$$\mathscr{L} = -\frac{1}{2}\delta\phi\left(1 + \frac{B_0}{\Lambda^2}\Box\right)\Box\delta\phi . \tag{5.2}$$

The associated propagator has two poles signaling the presence of two dofs

$$\mathscr{G} = \frac{1}{\left(1 + \frac{B_0}{\Lambda^2}\Box\right)\Box} = \frac{1}{\Box} - \frac{1}{\Box + \Lambda^2/B_0^2} . \tag{5.3}$$

The pole at zero mass represents the standard degree of freedom associated with ϕ, but we see a new pole with (tachyonic) mass square Λ^2/B_0^2 which always enters with the wrong sign. So the new degree of freedom at Λ/B_0 is a ghost.

We emphasize that a ghost represents a degree of freedom with the wrong sign kinetic term and should be distinguished from a tachyon which corresponds to a degree of freedom with the wrong sign mass term or an instability in the potential. For a tachyon the scale of the instability is governed by the mass of the mode and we can thus survive with small mass tachyonic modes as the time scale of the instability is long compared to other process that may be taking place. For a ghost on the other hand, the scale associated with the instability is the momentum of the field and so the instability scale is always at least of the order of the cutoff of the theory. This implies that if a ghost is present at a scale μ then one cannot trust the theory beyond the scale μ. In the case of the BD ghost, the ghost enters at the background dependent scale Λ/B_0. By choosing an arbitrarily large background B_0, one can brings the scale at which the theory breaks down arbitrarily low, which would mean that one can

[1] In this contribution we use a mainly + convention and so $-1/2(\partial\phi)^2$ represents the correct sign kinetic term.

never trust this theory, neither at the classical level nor at the quantum level. New physics has to enter at the cutoff scale or at the scale Λ/B_0 to help making sense of the theory. This is distinct from having a low strong coupling scale where classical predictions break down at that scale but not the quantum ones. New physics does not need to enter at the strong coupling scale.

To summarize, a ghost leads to an arbitrarily fast instability already at the classical level and signals the fact that the theory cannot be trusted neither classically nor quantum mechanically at and above the mass of the ghost. However as we shall see below the Vainshtein mechanism relies crucially on classical configurations at the low scale Λ. It is therefore essential to be able to trust the theory at the scale at which the first interactions enter (i.e., at the strong coupling scale). To get some intuition on how to obtain a ghost-free theory of massive gravity and other modifications of gravity a useful tool is to rely on a higher dimensional theory of gravity. In some cases this higher-dimensional theory is merely a 'mathematical trick' but it will show to provide useful insights.

5.2 Consistent Modifications of Gravity From Extra Dimensions

One of the most straight-forward way to derive a sensible and theoretical consistency theory of modified gravity is to start with General Relativity in higher dimensions. Higher dimensional gravity is known to lead to consistent high energy modifications of gravity. Here we shall focus on infrared (IR) modifications and see how it can lead to different interconnected models like Galileon theories of DGP and massive gravity which behave as Galileons in some limit. In the rest of this contribution we will use the notation that y represents the fifth extra dimension and x^μ are the 4d space-time coordinates. The 5d coordinates are given by $\{x^\alpha\}_{\alpha=0}^4 = \{x^\mu, y\}$.

5.2.1 DBI–Galileon

5.2.1.1 Five-Dimensional Minkowski

Starting with five dimensional GR, we can consider all the Lovelock invariants namely a cosmological constant (CC), a five dimensional scalar curvature $R^{(5)}$ and a Gauss-Bonnet (GB) invariant $\mathscr{L}_{\rm GB}$. The presence of a cosmological constant leads to a non-flat maximally symmetric 5d spacetime (AdS) and will be mentioned in what follows. To start with we stick to a flat Minkowski 5d spacetime $\mathrm{d}s^2 = \mathrm{d}y^2 + \eta_{\mu\nu}\,\mathrm{d}x^\mu\,\mathrm{d}x^\nu$ and set the CC to zero. In order to recover 4d gravity in some regime we consider a probe brane located at $y = \pi(x^\mu)$ and consider the boundary terms induced by the Lovelock invariants. The scalar curvature leads to an extrinsic boundary term K on the brane and the GB to a related term $K_{\rm GB}$. Furthermore induced on the brane one can consider a tension or a 4d CC λ and a 4d induced scalar curvature $R^{(4)}$. If the brane is localized at $y = \pi(x^\mu)$, the induced metric on the brane is $g_{\mu\nu} = \eta_{\mu\nu} + \partial_\mu\pi\partial_\nu\pi$ leading to what will represent a disformal coupling to matter. In the weak field limit, these invariants lead to a generalized Galileon-DBI set of interactions in 4d [7],

$$
\begin{array}{ccccl}
\text{5d} & & \text{4d} & & \text{DBI} - \text{Galileon} \\
\hline
 & & \lambda & \to & \mathcal{L}_2 \sim -\lambda\sqrt{1-(\partial\pi)^2} \\
R^{(5)} & \to & K & \to & \mathcal{L}_3 \sim (\partial_\mu\pi\partial_\nu\pi\Pi^{\mu\nu})/(1-(\partial\pi)^2) \\
 & & R^{(4)} & \to & \mathcal{L}_4 \sim \left([\Pi]^2-[\Pi]^2\right)/\sqrt{1-(\partial\pi)^2}+\cdots \\
\mathcal{L}_{\rm GB} & \to & K_{\rm GB} & \to & \mathcal{L}_5 \sim \left([\Pi]^2-3[\Pi][\Pi^2]+2[\Pi^3]\right)/(1-(\partial\pi)^2)+\cdots\,,
\end{array}
\tag{5.4}
$$

where here and in what follows $\Pi_{\mu\nu} = \partial_\mu\partial_\nu\pi$ and square brackets represent the trace of a tensor with respect to $\eta_{\mu\nu}$, $[\Pi] = \eta^{\mu\nu}\Pi_{\mu\nu}$, etc. In the weak field limit, these invariants lead to the Galileon terms on the brane [8]

$$
\begin{aligned}
L_2 &= (\partial\pi)^2 \\
L_3 &= (\partial\pi)^2\,[\Pi] \\
L_4 &= (\partial\pi)^2\left([\Pi]^2-[\Pi^2]\right) \\
L_5 &= (\partial\pi)^2\left([\Pi]^3-3[\Pi][\Pi^2]+2[\Pi^3]\right)\,.
\end{aligned}
\tag{5.5}
$$

This is a finite set of interactions and the fact that these terms be it in their exact form (5.4) or in their weak field limit (5.5) derive from Lovelock invariants in five dimensions ensures that they are ghost free. Furthermore Poincaré invariance in five dimensions leads to the following four-dimensional global symmetry [7]

$$
\pi \to \pi + c + v_\mu x^\mu + \pi v^\mu\partial_\mu\pi\,, \tag{5.6}
$$

for the interactions (5.4) and the *Galilean* symmetry for the Galileon interactions (5.5)

$$
\pi \to \pi + c + v_\mu x^\mu\,. \tag{5.7}
$$

In addition they also satisfy a non-renormalization theorem [9] which means that the coefficient governing any of these interactions can be set to any desired value without the loops of the field itself destabilizing it.

5.2.1.2 Curved Five Dimensions

As mentioned previously, one can also consider a CC in five dimensions, leading to 5d AdS rather than Minkowski. Since this is still a maximally symmetric spacetime, there is an equivalent to the symmetries presented in (5.6) or (5.7) for Minkowski, simply involving the AdS curvature [7]. The results sets of interactions are a Galileon generalization of the warped DBI and satisfy the same properties as previously namely the absence of ghost and radiative stability.

One can also extend the setup to arbitrary matter in five dimensions leading to an arbitrary five dimensional metric $q_{\mu\nu}$. The induced metric on the brane is then $g_{\mu\nu} = q_{\mu\nu} + \partial_\mu \pi \partial_\nu \pi$ and the resulting Galileon field π leaves on a curved metric $q_{\mu\nu}$. This leads to the covariant set of Galileon interactions first proposed in [10] which remains free of ghost but does satisfy the Galileon symmetry nor a generalized one. The reason is clear: the five dimensional spacetime is no longer maximally symmetric and there is therefore no reason to expect any resulting global symmetry.

These Galileon scalar fields can play an important role on cosmological scales (for instance they can be a good candidate for dark energy) and yet remain frozen on short distance scales thanks to a Vainshtein mechanism. Before describing this mechanism in Sect. 5.5 (see also other contributions), we show how theories of modified gravity are derived from extra dimensions.

5.2.2 *Massive Gravity*

5.2.2.1 Infinite Extra Dimension: DGP

If one is to start with five dimensional gravity to derive theories of IR modifications of gravity one first needs to confine gravity in four dimensions. This can be performed in two ways: Either by compactifying the extra dimension, which is performed in Sects. 5.2.2.2 and 5.3 or by considering an large (even infinite) extra dimension and inducing a four-dimensional curvature on a four-dimensional brane. This is the idea behind the DGP (Dvali–Gabadadze–Porrati) model where we start with five-dimensional gravity with a five-dimensional Planck scale M_5 and induce a four-dimensional curvature with Planck scale $M_{\rm Pl}$ in four dimensions [11]. The effective Friedman equation on the brane is then [12]

$$M_{\rm Pl}^2 H^2 \pm M_5^3 H = \rho \,, \tag{5.8}$$

where H is the Hubble parameter and ρ the energy density of fields localized on the four-dimensional brane. This modified Friedman equation has lead to a wealth of new directions for testing cosmology.

The brane-bending mode on the brane behaves as a cubic Galileon [9], given by $\mathcal{L}_3$ in (5.5). From a four-dimensional view point, the graviton is effectively massive and at the linearized level it satisfies (symbolically) the following equation

$$\left(\Box - m\sqrt{-\Box}\right) h_{\mu\nu} = M_{\rm Pl}^{-1} T_{\mu\nu} \,, \tag{5.9}$$

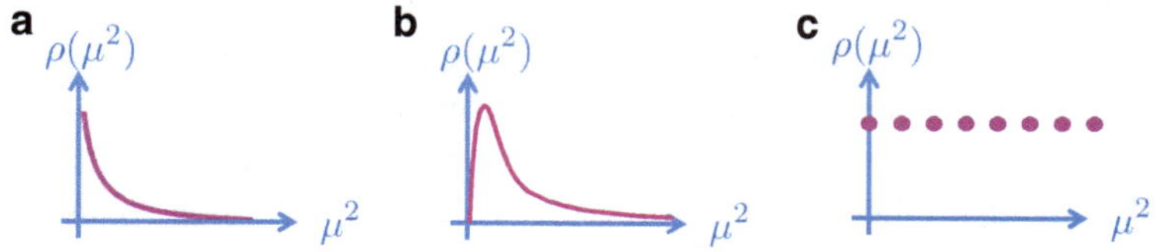

Fig. 5.2 Spectral representation of different models. (**a**) DGP, (**b**) higher-dimensional cascading gravity and (**c**) multi-gravity. Bi-gravity is the special case of multi-gravity with one massless mode and one massive mode. Massive gravity is the special case where only one massive mode couples to the rest of the standard model and the other modes decouple. (**a**) and (**b**) are models of soft massive gravity where the graviton mass can be thought of as a resonance. From [6]

where $m = M_5^3/M_{\rm Pl}^2$ and the effective mass of the graviton is momentum-dependent $m_{\rm eff}^2(k) = mk$. So rather than having a fixed pole at the scale m, the propagator has rather a resonance. In this sense DGP is a model of 'soft-massive gravity'.

For DGP, the peak of the spectral distribution still occurs at zero mass as can be seen in Fig. 5.2. However extensions of DGP to higher dimensions (known as Cascading gravity [13–15]) can lead to a peak in the spectral representation as depicted in Fig. 5.2 and are possibly closer to models of a hard mass graviton. In what follows, we discuss an alternative way to derive a theory of massive gravity from five dimensional GR, via Kaluza–Klein reduction or deconstruction.

5.2.2.2 Compact Extra Dimension

An alternative to the DGP model and its extensions is to consider a compact extra dimension of size R. A Kaluza-Klein decomposition (discretization in the momentum along the extra dimension) leads to a massless mode and an infinite tower of massive Kaluza-Klein modes, with mass gap $m = 1/R$. Rather than performing a Kaluza-Klein decomposition, one can also consider a deconstruction of the extra dimension which is a discretization of the extra dimension directly in real space rather than in momentum. Rather than considering a smooth extra dimension $0 < y < R$, we replace that direction by a series of N points y_n. In the large N limit one should in principle recover 5d GR but as we shall see below this does not occur in some special gauge choices.

The deconstruction framework will be explained in more detail below and as we shall see, for a finite number of site N one obtains a four-dimensional theory of N interacting gravitons (multi-gravity), with one massless graviton and $N-1$ massive ones. Moreover this theory is identical to a truncated Kaluza-Klein decomposition after a non-trivial field redefinition.

As we have seen before, in the case of an infinite extra dimension à la DGP, we obtain a theory of gravity where the graviton acquires a soft mass or resonance. In the case of a compact extra dimension, the deconstruction framework leads to a finite number N of discrete graviton(s) with mass $\sim n/R$ as can be seen from the spectral representation in Fig. 5.2.

In both cases starting from five-dimensional GR ensures (to some extend[2]) a consistent resulting four-dimensional theory of modified gravity. Indeed a massless graviton in 5d propagates 5 dofs which is precisely the right number of dofs that a massive spin-2 field should propagate in 4d without the BD pathology discussed in Sect. 5.1.2. In what follows we thus proceed by showing how 5d gravity can lead to a consistent theory of 4d massive gravity free of the BD ghost.

5.3 Deconstruction and Massive Gravity

We now present how to deconstruct 5d GR and recover 4d multi-gravity. We will then specialize to bi-gravity and to massive gravity as special cases. We follow the formalism derived in [17].

Starting with 5d gravity in the Einstein–Cartan form, the 5d metric is given by

$$g_{\alpha\beta}(x,y) = e^A_\alpha(x,y) e^B_\beta(x,y) \eta_{AB} \,. \tag{5.10}$$

The connection is set of the torsionless condition,

$$\omega^{AB}_\alpha = \frac{1}{2} e^C_\alpha \left(O^{AB}{}_C - O_C{}^{AB} - O^B{}_C{}^A \right) , \tag{5.11}$$

with $O^{AB}{}_C = 2 e^{A\alpha} e^{B\beta} \partial_{[\alpha} e_{\beta]C}$. The 5d curvature 2-form is then

$$\mathscr{R}^{AB} = \mathrm{d}\omega^{AB} + \omega^A{}_C \wedge \omega^{CB} \,. \tag{5.12}$$

The 5d Einstein-Hilbert action is then

$$S^{(5)}_{\rm EH} = \frac{M_5^3}{2} \int \mathrm{d}^4x\, \mathrm{d}y \sqrt{-g} R^{(5)}[g] \tag{5.13}$$

$$= \frac{M_5^3}{2\times 3!} \int \varepsilon_{ABCDE} \mathscr{R}^{AB} \wedge e^C \wedge e^D \wedge e^E \,. \tag{5.14}$$

Before we proceed with discretizing this action we first briefly discuss the gauge choice we use.

5.3.1 Gauge-Fixing

The theory has 5 spacetime symmetries associated with 5d diffeomorphism invariance. In addition in the veilbein language there are 10 Lorentz symmetries. As a

[2] There are some exceptions to the rule, see for instance [16].

result one can make 15 gauge choices. We chose the following conditions on the vielbein and the connection

$$\begin{aligned} e^a_y = 0,\ e^5_\mu = 0,\ e^5_y = 1 \quad &\longrightarrow 9 \text{ gauge fixing} \\ \omega_y^{ab} = e^{\mu[a}\partial_y e_\mu^{b]} = 0 \quad &\longrightarrow \quad 6 \text{ gauge fixing}, \end{aligned} \tag{5.15}$$

which fully fixes all the gauge freedom. The condition on the vielbein implies $e^A = \left(e^a_\mu\, \mathrm{d}x^\mu,\ \mathrm{d}y\right)$ and the condition on the connection implies the symmetric vielbein condition. Interestingly this condition ensures that the theory can be written back in terms of the metric. Here it appears as a simple consequence of our gauge choice. In the metric language this gauge choice implies that the lapse is unity and the shift vanishes, $\mathrm{d}s^2 = \mathrm{d}y^2 + g_{\mu\nu}(x,y)\,\mathrm{d}x^\mu\,\mathrm{d}x^\nu$. We now proceed with discretizing 5d GR in this gauge.

5.3.2 *From 5d Gravity to 4d Multi-Gravity*

In the gauge chosen previously, 5d GR can be written as

$$S^{(5)}_{\rm EH} = \frac{M_5^3}{2}\int \mathrm{d}^4x\,\mathrm{d}y\sqrt{-g}\left(R^{(4)}[g] + [K]^2 - [K^2]\right), \tag{5.16}$$

where K is the extrinsic curvature, in the metric language

$$K_{\mu\nu} = \frac{1}{2}g^{\mu\alpha}(x,y)\partial_y g_{\alpha\nu}(x,y)\,. \tag{5.17}$$

We now discretize the extra dimension as follows:

$$\begin{aligned} &y \longrightarrow y_n \\ &e^a_A(x,y) \longrightarrow e_n{}^a_A(x) \\ &g_{\mu\nu}(x,y) = \eta_{ab}e^a_\mu(x,y)e^b_\nu(x,y) \longrightarrow g^{(n)}_{\mu\nu}(x) = \eta_{ab}e_n{}^a_\mu(x)e_n{}^b_\nu(x) \\ &\partial_y e(x,y) \longrightarrow m\left(e_{n+1}(x) - e_n(x)\right), \end{aligned} \tag{5.18}$$

with $m = N/R$. Applying this discretization procedure on the extrinsic curvature,

$$K^\mu_\nu \sim g^{\mu\alpha}\partial_y g_{\alpha\nu} \sim e^{-1}\partial_y e\,, \tag{5.19}$$

and using the symmetric vielbein condition we obtain

$$\begin{aligned} K^\mu_\nu \to m e_n^{-1}\left(e_{n+1} - e_n\right) &= -m\left(\delta^\mu_\nu - \sqrt{\left(g^{(n)}\right)^{\mu\alpha} g^{(n+1)}_{\alpha\nu}}\right) \\ &\equiv -m\mathcal{K}^\mu_\nu[g^{(n)}, g^{(n+1)}] \equiv -m\mathcal{K}_{n,n+1}{}^\mu_\nu\,. \end{aligned} \tag{5.20}$$

Using this expression into the 5d Einstein-Hilbert action (5.16) we obtain [17]

$$\mathcal{L} = M_5^3 \int \mathrm{d}y \left(R^{(4)}[g] + [K]^2 - [K^2]\right) \tag{5.21}$$

$$\to \frac{M_5^3}{m} \sum_{n=1}^{N} \left(R_n^{(4)} + m^2 \left([\mathcal{K}_{n,n+1}]^2 - [\mathcal{K}_{n,n+1}^2]\right)\right) . \tag{5.22}$$

This is a 4d theory of multi-gravity as presented in [18] with the specific interactions governed by $\mathcal{K}_{n,m}$ derived in [19,20]. The 4d fundamental Planck scale is then given by $M_{\rm Pl}^2 = M_5^3/(mN) = M_5^3 R$.

5.3.3 Generalized Mass Term

The multi-gravity theory derived previously has only one of the possible sets of allowed interactions derived in [19,20]. In the previous derivation we have applied the most straightforward discretization procedure but there is some freedom on how one wishes to define a field or its derivative at a point. To see the most general discretization procedure it is convenient to return to the vielbein language where rather than using

$$e^a \to e_n^a , \tag{5.23}$$

we may use the more general procedure

$$e^a \to \left(w e_n^a + (1-w) e_{n+1}^a\right) . \tag{5.24}$$

The mass term then gets generalized to

$$\begin{aligned} \sqrt{-g}\left([K]^2 - [K^2]\right) &= \varepsilon_{abcd} e^a \wedge e^b \wedge \partial_y e^c \wedge \partial_y e^d \\ &\to m^2 \varepsilon_{abcd} \left(w_1 e_n^a + (1-w_1) e_{n+1}^a\right) \wedge \left(w_2 e_n^b + (1-w_2) e_{n+1}^b\right) \\ &\quad \wedge \left(e_n^c - e_{n+1}^c\right) \wedge \left(e_n^d - e_{n+1}^d\right) \\ &\equiv m^2 \sqrt{-g}\left(\mathcal{L}_2(\mathcal{K}) + (w_1 + w_2)\mathcal{L}_3(\mathcal{K}) + w_1 w_2 \mathcal{L}_4(\mathcal{K})\right) , \end{aligned} \tag{5.25}$$

where we recover the ghost-free interaction terms $L_{2,3,4}$ first derived in [20] (sometimes known as 'dRGT' mass terms or interactions),

$$2\mathcal{L}_2[\mathcal{K}] = \varepsilon^{abcd} \varepsilon_{a'b'cd} \mathcal{K}_a^{a'} \mathcal{K}_b^{b'} \tag{5.26}$$

$$\mathcal{L}_3[\mathcal{K}] = \varepsilon^{abcd} \varepsilon_{a'b'c'd} \mathcal{K}_a^{a'} \mathcal{K}_b^{b'} \mathcal{K}_c^{c'} \tag{5.27}$$

$$\mathcal{L}_4[\mathcal{K}] = \varepsilon^{abcd} \varepsilon_{a'b'c'd'} \mathcal{K}_a^{a'} \mathcal{K}_b^{b'} \mathcal{K}_c^{c'} \mathcal{K}_d^{d'} , \tag{5.28}$$

or equivalently,

$$\mathcal{L}_2[\mathcal{K}] = \left([\mathcal{K}^2] - [\mathcal{K}]^2\right) \tag{5.29}$$

$$\mathcal{L}_3[\mathcal{K}] = \left([\mathcal{K}]^2 - 3[\mathcal{K}][\mathcal{K}^2] + 2[\mathcal{K}^3]\right) \tag{5.30}$$

$$\mathcal{L}_4[\mathcal{K}] = \left([\mathcal{K}]^4 - 6[\mathcal{K}^2][\mathcal{K}]^2 + 3[\mathcal{K}^2]^2 + 8[\mathcal{K}][\mathcal{K}^3] - 6[\mathcal{K}^4]\right) . \tag{5.31}$$

This structure is very similar to that of the Galileons [8] and as we shall see they are indeed very closely related and are the essence of the absence of BD ghost.

5.3.4 Strong Coupling Scale

This theory of multi-gravity has one massless mode with 2 dofs and $(N-1)$ massive modes with 5 dofs each, meaning that there is no BD ghost for any mode. The lightest mode has a mass $m_1 = 1/R = m/N$, while the heaviest mode has a mass set by $m \sim Nm_1$ (in the large N limit.)

The strong coupling scale for this theory (the scale at which the lowest interactions arise) is the same as for a normal (ghost-free) theory of massive gravity and is given by de Rham and Gabadadze [19]

$$\Lambda = (M_{\rm Pl} m_1^2)^{1/3} , \tag{5.32}$$

where m_1 is the mass of the lightest mode. Interestingly in what should be the continuum limit $R \to \infty$ or $m_1 \to 0$ the degree of freedom that interact at the scale Λ (namely the helicity-0 mode of the lightest mode), as well as all the other helicity-0 modes entirely decouple in that limit. This means that in this specific theory, we do not recover 5d GR in the limit $R \to \infty$ or $m_1 \to 0$ but rather N decoupled massless spin-2 fields, $(N-1)$ decoupled spin-0 fields and $(N-1)$ decoupled spin-1 fields. This decoupling is ensured by the low strong coupling scale (5.32) and is responsible for the Vainshtein mechanism [3] and the absence of vDVZ (van Dam–Veltman–Zakharov) discontinuity in the massless limit [21, 22]. In this sense the strong coupling scale (5.32) is a desirable (and even required) feature of the theory if one would like to be able to consider it as a truncated theory in its own right.

There is an alternative to the low strong coupling scale (5.32) which implies choosing a different gauge choice that what was performed here. If instead we keep the lapse dynamical, the presence of low strong coupling scale is avoided but at the price of introducing a ghost at the scale of the heaviest mode. This means that the truncated theory is not consistent, and one should keep an infinite number of modes or work at energy scales well below the mass of the heaviest mode. In that case one recovers 5d GR in the continuum limit $R \to \infty$ or $m_1 \to 0$.

5.3.5 Bi-Gravity

Focusing on a discretization with two sites only with respective metrics $g_{\mu\nu}$ and $f_{\mu\nu}$, we obtain the following bi-gravity theory [23] with the same ghost-free interactions [20]

$$\mathcal{L}_{g,f} = M_g^2\sqrt{-g}R[g] + M_f^2\sqrt{-f}R[f] \tag{5.33}$$

$$+ m^2 M_f M_g \sqrt{-g}\sum_{n=0}^{4}\alpha_n \mathcal{L}_n\left(\mathcal{K}[g,f]\right), \tag{5.34}$$

with

$$\mathcal{K}^\mu_\nu[g,f] = \delta^\mu_\nu - \left(\sqrt{g^{-1}f}\right)^\mu_\nu. \tag{5.35}$$

In the absence of the interaction governed by m, this would the theory of two non-interactive massless spin-2 fields bearing $2 \times 2 = 4$ dofs. This theory would have two copies of diffeomorphism invariance.

Including the interaction breaks one copy of diffeomorphism invariance which excites three new dofs in the theory leading to a total of $4+3 = 7$ dofs, which is the correct counting for one massless mode and one massive mode which carry a total of $2+5 = 7$ dofs without any BD ghost.

It is sometimes stated that unlike massive gravity bi-gravity does not break diffeomorphism invariance. This statement is quite incorrect, just like massive gravity, bi-gravity breaks one copy of diffeomorphism invariance and just like in massive gravity four Stückelberg fields (only three of which are independent) should be included in bi-gravity to restore that homeomorphism invariance.

5.3.6 Massive Gravity

We can now easily see how to obtain a theory of massive gravity and a decoupled massless spin-2 field out of massive gravity.[3] From simplicity let us imagine that no matter couples directly to the metric $f_{\mu\nu}$ (such a coupling does not affect the argument it simply allows to generalize massive gravity on arbitrary reference metrics [24]) and we set $\alpha_0 = \alpha_1 = 0$. In that case it is useful to split the metric $f_{\mu\nu}$ as follows

[3]In reality multi-gravity was obtained out bi-gravity which was obtained out of massive gravity but for pedagogical reasons it is more intuitive to derive bi-gravity from multi-gravity and massive gravity from bi-gravity.

$$f_{\mu\nu} = \eta_{\mu\nu} + \frac{1}{M_f}\chi_{\mu\nu} \,. \tag{5.36}$$

Taking the scaling limit $M_f \to \infty$ while keeping $\chi_{\mu\nu}$ fixed does not change the number of dofs in the theory but simply decouples some of them. In this limit, we obtain a theory of massive gravity and a decoupled massless-spin-2 field,

$$\begin{aligned} \mathcal{L}_{M_f\to\infty} = & \, M_g^2\sqrt{-g}\left(R[g] + m^2 \sum_{n=2}^{4} \alpha_n \mathcal{L}_n(\mathcal{K}[g,\eta])\right) \\ & -\frac{1}{2}\chi^{\mu\nu}\hat{\mathcal{E}}^{\alpha\beta}_{\mu\nu}\chi_{\alpha\beta} \,, \end{aligned} \tag{5.37}$$

where $\hat{\mathcal{E}}$ is the Lichnerowicz operator which is the healthy linearized kinetic term for a massless spin-2 field. Notice that the second line is exact to all orders in χ, so the massless sector of the theory is not interacting at all, not even with itself. Nevertheless it still carries the two standard dofs of a massless spin-2 field, and the massive graviton carried in $g_{\mu\nu}$ carries five dofs, leading once again to the same number of dofs as any other healthy bi-gravity theory.

As already mentioned, one could generalize this procedure to allow for a non-trivial background metric, $f_{\mu\nu} = \bar{f}_{\mu\nu} + \frac{1}{M_f}\chi_{\mu\nu}$ before taking the limit $M_f \to \infty$. In that case, the resulting theory is massive gravity on the reference metric $\bar{f}_{\mu\nu}$ and a decoupled non-interacting massless spin-2 field.

The fact that this theory emerges from 5d GR which carries the correct number of dofs for a massive graviton is suggestive that the theory of massive gravity we have derived here does not suffer from the BD ghost. We shall prove this more explicitly in what follows working both in the ADM language and in the decoupling limit.

5.4 Absence of Boulware–Deser Ghost

5.4.1 ADM Language

The presence of a BD ghost in a large class of massive gravity theories was originally presented in the ADM language [5]. Starting with the ADM decomposition,

$$\mathrm{d}s^2 = -N_0^2\,\mathrm{d}t^2 + \gamma_{ij}\left(\mathrm{d}x^i + N^i\,\mathrm{d}t\right)\left(\mathrm{d}x^j + N^j\,\mathrm{d}t\right) \,, \tag{5.38}$$

GR is special in that both the lapse and the shift are Lagrange multipliers, propagating 4 first class constraints. This means that the phase space has a priori 6×2 dofs in γ_{ij} and its conjugate momentum but 8 of them are removed by the 4 first class constraints, leading to a total of $4 = 2\times 2$ dofs in phase space or 2 dofs in field space which is the correct counting for GR (leading to the two first polarizations presented in Fig. 5.1.)

Now focusing on massive gravity (without the decoupled linearized massless spin-2 field), neither the lapse nor the shift remain linear. A priori this means that one looses four first class constraints, and one is left with a priori 6 degrees of freedom in γ_{ij} in field space, which would correspond to the five expected dofs and an additional sixth BD ghost, which as we have seen would always signal a disaster (see Sect. 5.4.)

However this naive estimation does not account from the fact that not all the shift and lapse are necessarily independent. As first explained in [19] and then carried out in [20], the real criteria for determining the number of degrees of freedom in field space in d spacetime dimensions is

$$\#\,\text{field space dof} = \frac{1}{2}d(d-1) - (d - \text{rank}\,(L))\,, \tag{5.39}$$

where the Hessian $L_{\mu\nu}$ is given by the second derivative of the potential $\mathscr{U} = \sqrt{-g}\sum_n \alpha_n \mathscr{L}_n$,

$$L_{\mu\nu} = \frac{\partial^2 \mathscr{U}}{\partial N^\mu \partial N^\nu}\,. \tag{5.40}$$

In $d = 2$ dimensions, it was shown in [20] that the rank of L was 1 and so the number of physical dofs in 2 dimensions is zero, as it should be for a healthy spin-2 field without BD pathology. The counting carries through to any number of dimensions and in $d = 4$ is was shown in [19] for special cases and then in [25] in all generality that rank$L = 3$, for the special form of the potential given in (5.37) and so in the theory given in (5.37) has only 5 and not 6 dofs in the massive spin-2 field. This theory is thus free of the BD ghost.

5.4.2 Decoupling Limit

The theory of multi-gravity presented previously breaks $(N - 1)$ copies of diffeomorphism invariance. To restore them one can introduce $(N - 1)$ Stückelberg fields. The same counting remains for bi-gravity and massive gravity. In what follows we shall focus on the case of massive gravity bearing in mind that the same derivation follows for bi- and multi-gravity as well as for New Massive Gravity (NMG) [26].

When formulating the theory of massive gravity, we made use of a reference metric $\bar{f}_{\mu\nu}$ which can be chosen to be Minkowski or other. We focus the discussion on a Minkowski reference metric $\bar{f}_{\mu\nu} = \eta_{\mu\nu}$ but the essence of the argument remains the same for other reference metrics. See for instance [27] for the decoupling limit of a de Sitter reference metric. The existence of a reference metric breaks diff invariance, but it can be restored by introducing four Stückelberg fields ϕ^a which transform as scalar under local diffs

$$\eta_{\mu\nu} \to \tilde{\eta}_{\mu\nu} = \partial_\mu \phi^a \partial_\nu \phi^b \eta_{ab} \,, \tag{5.41}$$

where $\tilde{\eta}_{\mu\nu}$ now transforms as a tensor under local diffs. Even if in bi-gravity the two metrics are dynamical it does not change the fact that the interaction between the two metrics breaks one copy of diff and the theory is not fully diff invariant unless the same four Stückelberg fields are introduced. The same remain valid for NMG and multi-gravity.

We can further split the Stückelberg fields into a helicity-0 and -1 modes:

$$\phi^a = x^a + \frac{1}{m M_{\rm Pl}} A^a + \frac{1}{m^2 M_{\rm Pl}} \eta^{ab} \partial_b \pi \,, \tag{5.42}$$

where the scales are introduced for later convenience and in what follows we only focus on the helicity-0 mode π. The full decoupling limit including the vector A^a was derived in [28].

Using the expression (5.41) into (5.35) with $f_{\mu\nu} \to \tilde{\eta}_{\mu\nu}$, we see directly that

$$\mathcal{K}^\mu_\nu = \frac{1}{m^2 M_{\rm Pl}} \eta^{\mu\alpha} \Pi_{\alpha\nu} + \mathcal{O}\left(\frac{1}{M_{\rm Pl}} h\right) \,, \tag{5.43}$$

where we write the metric $g_{\mu\nu}$ as

$$g_{\mu\nu} = \eta_{\mu\nu} + \frac{1}{M_{\rm Pl}} h_{\mu\nu} \,. \tag{5.44}$$

We now take the decoupling limit where $M_{\rm Pl} \to \infty$ and $m \to 0$ while keeping the scale $\Lambda = (m^2 M_{\rm Pl})^{1/3}$ fixed. Clearly in this decoupling limit $\mathcal{K} \to \Pi$ and the mass terms for massive gravity given in ((5.26)–(5.28)) or equivalently ((5.29)–(5.31)) reduce to total derivatives. As a result to zeroth order in $h/M_{\rm Pl}$ the theory has no ghost.

We now proceed to first order in $h/M_{\rm Pl}$, to that order the mass term becomes

$$\mathcal{L}_{\rm mGR} = M_{\rm Pl}^2 \sqrt{-g} \left(\frac{1}{2} R + m^2 \left(\mathcal{L}_2[\mathcal{K}] + \alpha_3 \mathcal{L}_3[\mathcal{K}] + \alpha_4 \mathcal{L}_4[\mathcal{K}]\right)\right) \tag{5.45}$$

$$\mathcal{L}^{\rm (dec)}_{\rm mGR} = -\frac{1}{2} h^{\mu\nu} \hat{\mathcal{E}}^{\alpha\beta}_{\mu\nu} h_{\alpha\beta} - h^{\mu\nu} \left(X^{(1)}_{\mu\nu} + \frac{1+3\alpha_3}{\Lambda^3} X^{(2)}_{\mu\nu} + \frac{\alpha_3 + 4\alpha_4}{\Lambda^6} X^{(3)}_{\mu\nu} \right), \tag{5.46}$$

where as $\hat{\mathcal{E}}$ is the Lichnerowicz operator and

$$X^{(1)\mu}{}_{\mu'} = \varepsilon^{\mu\nu\alpha\beta} \varepsilon_{\mu'\nu'\alpha\beta} \Pi^{\nu}_{\nu'} \tag{5.47}$$

$$X^{(2)\mu}{}_{\mu'} = \varepsilon^{\mu\nu\alpha\beta} \varepsilon_{\mu'\nu'\alpha'\beta} \Pi^{\nu}_{\nu'} \Pi^{\alpha}_{\alpha'} \tag{5.48}$$

$$X^{(3)\mu}{}_{\mu'} = \varepsilon^{\mu\nu\alpha\beta} \varepsilon_{\mu'\nu'\alpha'\beta'} \Pi^{\nu}_{\nu'} \Pi^{\alpha}_{\alpha'} \Pi^{\beta}_{\beta'} \,. \tag{5.49}$$

This set of tensors satisfies some remarkable properties: First they are identically conserved. Second they share a similar structure as Galileon interactions and are indeed closely related. Third, they trivially satisfy the Galileon symmetry by construction (and this already at the level of the Lagrangian unlike Galileon interactions). Finally and most importantly, these interactions can be proven to have no ghost. The reason for that is that their respective equations of motion never bear more than two derivatives and the X^{00} bears no time derivative, while X^{0i} carries at most a single time derivative.

The helicity-0 and -2 modes can be 'semi-diagonalized' by performing a field redefinition,

$$h_{\mu\nu} = \bar{h}_{\mu\nu} + \pi \eta_{\mu\nu} + \frac{1+3\alpha_3}{\Lambda^3} \partial_\mu \pi \partial_\nu \pi \,, \tag{5.50}$$

leading to a Galileon theory

$$\mathcal{L}^{(\mathrm{dec})}_{\mathrm{mGR}} = -\frac{1}{2} \bar{h}^{\mu\nu} \hat{\mathcal{E}}^{\alpha\beta}_{\mu\nu} \bar{h}_{\alpha\beta} + \sum_{n=2}^{5} \frac{c_n}{\Lambda^{3(n-2)}} \mathcal{L}^{(n)}_{\mathrm{Gal}} + \frac{\alpha_3 + 4\alpha_4}{\Lambda^6} \bar{h}^{\mu\nu} X^{(3)}_{\mu\nu} \,, \tag{5.51}$$

where $\mathcal{L}^{(n)}_{\mathrm{Gal}}$ are the Galileon Lagrangian, $\mathcal{L}^{(n)}_{\mathrm{Gal}} = \pi X^{(n-1)\mu}{}_{\mu}$ and the c_n are dimensionless coefficients related to the α_n. We see that when $\alpha_3 + 4\alpha_4 = 0$, the helicity-2 and -0 modes fully decouple in this limit and the interactions for the helicity-0 mode are pure Galileon interactions. The only two differences with a standard Galileon model is that it only has one free parameter (namely α_3) and the coupling to matter includes a disformal contribution

$$\mathcal{L}_{\mathrm{matter}} = \frac{1}{M_{\mathrm{Pl}}} h_{\mu\nu} T^{\mu\nu} = \frac{1}{M_{\mathrm{Pl}}} \bar{h}_{\mu\nu} T^{\mu\nu} + \frac{1}{M_{\mathrm{Pl}}} \pi T \frac{1+3\alpha_3}{M_{\mathrm{Pl}} \Lambda^3} \partial_\mu \pi \partial_\nu \pi T^{\mu\nu} \,, \tag{5.52}$$

which can lead to specific observational signatures as the field now also couple to radiation.

In [29, 30] the BD ghost was connected to the existence of an Ostrogradsky instability in the decoupling limit. The fact the decoupling limit of this theory is a Galileon which is known to be free of Ostrogradsky instability was therefore the first indication that the theory was in fact free of the BD ghost. As explained earlier this was later confirmed by a multitude of independent studies.

In what follows we will introduce the Vainshtein mechanism using the cubic Galileon as a toy model and discuss the existence of superluminalities.

5.5 Vainshtein Mechanism

The essence of the Vainshtein mechanism, and its subtleties is already manifest in the cubic Galileon

$$\mathscr{L}_{\text{cub}} = -\frac{1}{2}(\partial\pi)^2 + \frac{1}{\Lambda^3}(\partial\pi)^2\Box\pi + \frac{1}{M_{\text{Pl}}}\pi T\,. \tag{5.53}$$

In the absence of the cubic interaction, the field π would always couple to matter with gravitational strength and would be incompatible with observations. In what follows we show how the cubic interaction at the low scale $\Lambda \ll M_{\text{Pl}}$ is key in screening this scalar field.

5.5.1 *Redressed Coupling*

Let us consider a macroscopic source $\bar{T}$ and smaller perturbations on top of it, $T = \bar{T} + \delta T$. Similarly we may split the field as the configuration $\bar{\pi}$ soured by $\bar{T}$ and its fluctuations $\pi = \bar{\pi} + \delta\pi$. For definiteness we consider a constant source $\bar{T}$ although the argument is relatively unaffected by the precise form of source, so long as there is a regime where $\bar{T} \gg M_{\text{Pl}}\Lambda^3$. The background configuration is then given by

$$\bar{\pi} = -\Lambda^3 A x^2 \tag{5.54}$$

$$\text{with} \quad A = -\frac{1}{24}\left(1 - \left(1 + \frac{6\bar{T}}{M_{\text{Pl}}\Lambda^3}\right)^{1/2}\right) \simeq \frac{1}{4\sqrt{6}}\sqrt{\frac{\bar{T}}{M_{\text{Pl}}\Lambda^3}} \gg 1\,, \tag{5.55}$$

$$\text{so} \quad \bar{\Pi} \sim \partial^2\bar{\pi} \sim \Lambda^3 A \gg \Lambda^3 \tag{5.56}$$

On top of these background configuration, the effective Lagrangian for the fluctuations is

$$\mathscr{L}_{\delta\pi} = -\frac{Z}{2}(\partial\delta\pi)^2 + \frac{1}{\Lambda^3}(\partial\delta\pi)^2\Box\delta\pi + \frac{1}{M_{\text{Pl}}}\delta\pi\delta T\,, \tag{5.57}$$

with

$$Z = 1 + 24A\,. \tag{5.58}$$

When $\bar{T} \gg M_{\text{Pl}}\Lambda^3$ then $A \gg 1$ and it follows that $Z \gg 1$. Next we canonically normalize the field,

$$\delta\pi = Z^{-1/2}\delta\hat{\pi}\,, \tag{5.59}$$

so that the properly canonically normalized field sees the effective Lagrangian

$$\mathscr{L}_{\delta\hat{\pi}} = -\frac{1}{2}(\partial\delta\hat{\pi})^2 + \frac{1}{\Lambda_\star^3}(\partial\delta\hat{\pi})^2\Box\delta\hat{\pi} + \frac{1}{M_{\text{Pl}}\sqrt{Z}}\delta\hat{\pi}\delta T\,, \tag{5.60}$$

with the new 'redressed' scale $\Lambda_* = \Lambda Z^{1/2} \gg \Lambda$. As a result on the background of the source T_0 the field is no longer strongly coupled at the scale Λ but rather at the much scale Λ_*. Notice that at no point do we consider the scale Λ or Λ_* to be the cutoff, as it would simply not make sense to have a cutoff which is background dependent unless some very peculiar mixing with high energy physics occurs. Instead Λ (resp. Λ_*) are the scales at which tree-level unitarity breaks down. This scale differs from the cutoff which is the scale at which new physics enters (see [31] for other examples in physics where the strong coupling scale which dictates the breakdown of tree-level unitarity is distinct from the cutoff scale at which new physics enters.)

Moreover we see that the coupling to matter occurs at the new scale $M_{\rm Pl}\sqrt{Z} \gg M_{\rm Pl}$. This means that in the vicinity of large sources T_0 (for instance the Sun), the coupling to other sources (for instance the planets of the solar system) is very much suppressed. This is precisely how the Vainshtein mechanism succeeds at screening the field π. In what follows we will show how this Vainshtein mechanism comes at the price of allowing superluminal classical velocities. After reviewing a simple example we shall see why the presence of these superluminalities do not imply acausality.

5.5.2 *Superluminalities*

5.5.2.1 Classical Superluminalities

Similarly as seen previously, if we split the field into a background configuration $\bar{\pi}$ and a fluctuation $\delta\pi$, with $\bar{\Pi}_{\mu\nu} = \partial_\mu\partial_\nu\bar{\pi} \gg \Lambda^3$ (by that we mean, that at least some of the eigenvalues of $\bar{\Pi}_{\mu\nu}$ are larger than Λ^3), then the fluctuations $\delta\pi$ see the effective second order Lagrangian

$$\mathcal{L}^{(2)} = -\frac{1}{2} Z^{\mu\nu}\partial_\mu\delta\pi\partial_\nu\delta\pi\,, \tag{5.61}$$

with the effective metric

$$Z^{\mu\nu} = \eta^{\mu\nu} + \frac{4}{\Lambda^3}\left(\bar{\Pi}^{\mu\nu} - [\bar{\Pi}]\eta^{\mu\nu}\right)\,. \tag{5.62}$$

Now without loss of generality, at any point x one can perform a global Lorentz transformation to a frame where Z^μ_ν is diagonal. In that frame the speed of propagation along the direction x^1 is

$$c^2_{s,1} = \frac{Z^1_1}{Z^0_0} = \frac{1 - \frac{4}{\Lambda^3}\left(\bar{\Pi}^0_0 + \bar{\Pi}^2_2 + \bar{\Pi}^3_3\right)}{1 - \frac{4}{\Lambda^3}\left(\bar{\Pi}^1_1 + \bar{\Pi}^2_2 + \bar{\Pi}^3_3\right)}\,. \tag{5.63}$$

As a result, the field $\delta\pi$ propagates with superluminal classical (group and phase) velocity along the direction x^1 for any configuration admitting $\bar{\Pi}^0_0 > \bar{\Pi}^1_1$ at least at one point. This is easily achieved, at least locally, for instance considering a plane wave $\bar{\pi} = F(x^1 - t)$ which satisfies the background equations of motion in the absence of any source. Then the fluctuations travel with classical superluminal group and phase velocity as soon as $F'' > 0$ [32, 33].

5.5.2.2 Front Velocity and Causality

The existence of these classical superluminalities has been the object of much concern and claims connecting them to acausality have been made. However it is important to emphasize that causality is not determined from the classical group or phase velocity but rather from the front velocity which is the high frequency limit of the phase velocity. As a consequence quantum corrections ought to be included in order to compute the front velocity and before any claims may be made on the causality of the theory. This is especially important in the context of these theories since we have seen that the strong coupling scale, or scale at which tree-level calculations can no longer be trusted depend on the background. As a result the tree-level (or classical) calculation presented above of the front velocity are only valid at low energy and break down precisely in the regime where one would want to connect it with causality. Consequently there has been so far no evidence that massive gravity or other theories that exhibit the Vainshtein mechanism are causal or acausal.

5.5.2.3 Galileon Duality

To emphasize further how the notion of classical group or front velocity can be misleading we perform a coordinate and field redefinition to specific example of quintic Galileon. Consider the following quintic Galileon [33],

$$S_{\text{quintic}} = \int \mathrm{d}^4x \left(-\frac{1}{12}\mathcal{L}^{(2)}_{\text{Gal}} + \frac{1}{6\Lambda^3}\mathcal{L}^{(3)}_{\text{Gal}} - \frac{1}{8\Lambda^6}\mathcal{L}^{(4)}_{\text{Gal}} + \frac{1}{30\Lambda^9}\mathcal{L}^{(5)}_{\text{Gal}} \right), \tag{5.64}$$

where the Galileon Lagrangian are given below Eq. (5.51). The same analysis as for the cubic Galileon applies here and similarly it is straightforward to find exact solutions in the absence of matter which exhibits superluminal propagation along any direction for the field fluctuation $\delta\pi$.

Now performing the following combined field and coordinate transformation

$$x^\mu \to \tilde{x}^\mu = x^\mu + \frac{1}{\Lambda^3}\partial\pi(x) \tag{5.65}$$

$$\pi(x) \to \rho(\tilde{x}) = \pi(x) + \frac{1}{\Lambda^3}(\partial\pi(x))^2\,, \tag{5.66}$$

the quintic Galileon theory introduced in (5.64) simplifies to a free theory for ρ [33]

$$S_{\text{quintic}} \to \tilde{S} = \int \mathrm{d}^4x \left(-\frac{1}{2} (\partial\rho)^2 \right) , \tag{5.67}$$

which can never exhibit any superluminalities and is manifestly causal. This does not mean that the causal structure between the two representations is different, quite the opposite the causal structure is the same but is distinct from the notion of superluminalities. This comes to show how the notion of classical superluminality can be misleading and one ought to keep track of the front velocity (with in this case its full quantum corrections) in order to infer whether or not the theory is causal.

5.6 Summary and Outlook

In this proceedings we have reviewed how to derive consistent and ghost-free four-dimensional theories of massive gravity using five-dimensional General Relativity as our starting point. In the case of an infinite extra dimension, gravity may be localized in four dimensions by inducing a four-dimensional Einstein Hilbert term on a four-dimensional brane. Depending on the setup, this leads to a general DBI-Galileon model or to a soft theory of massive gravity known as DGP. Alternatively for finite size-extra dimensions, a discretization of this extra dimension (either in real space or in Fourier space) leads to a ghost-free theory of massive gravity (sometimes known as dRGT) provided the discretization is performed in the vielbein formalism. Galileons are ubiquitous to all these theories of massive gravity and provide a simple way to understand the Vainshtein mechanism whereby the helicity-0 mode of the graviton is screened in the vicinity of large matter sources. This Vainshtein mechanism is also shown to come hand in hand with classical superluminalities. While superluminalities in the front velocity would indeed imply acausality superluminal classical group and front velocities do not have the same implications and have been observed in nature. In order to comment on the causality of the theory it is therefore essential to find a prescription which allows us to compute the front velocity with all its quantum corrections. This is where the Vainshtein mechanism and its implementation at the quantum level could come in useful.

Acknowledgements CdR is supported by a Department of Energy grant DE-SC0009946. CdR wishes to thank the organizers of the 7th Aegean Summer School for a very productive and interactive meeting.

References

1. G.M. Harry [LIGO Scientific Collaboration], Class. Quant. Grav. **27**, 084006 (2010)
2. R.A. Hulse, J.H. Taylor, Astrophys. J. **195**, L51 (1975)

3. A.I. Vainshtein, Phys. Lett. B **39**, 393 (1972)
4. C. de Rham, A.J. Tolley, D.H. Wesley, Phys. Rev. D **87**(4), 044025 (2013) [arXiv:1208.0580 [gr-qc]]
5. D.G. Boulware, S. Deser, Ann. Phys. **89**, 193 (1975)
6. C. de Rham, Living Rev. Relativity **17**, 7 (2014), http://www.livingreviews.org/lrr-2014-7 [arXiv:1401.4173 [hep-th]]
7. C. de Rham, A.J. Tolley, JCAP **1005**, 015 (2010) [arXiv:1003.5917 [hep-th]]
8. A. Nicolis, R. Rattazzi, E. Trincherini, Phys. Rev. D **79**, 064036 (2009) [arXiv:0811.2197 [hep-th]]
9. M.A. Luty, M. Porrati, R. Rattazzi, J. High Energy Phys. **0309**, 029 (2003) [hep-th/0303116]
10. C. Deffayet, G. Esposito-Farese, A. Vikman, Phys. Rev. D **79**, 084003 (2009) [arXiv:0901.1314 [hep-th]]
11. G.R. Dvali, G. Gabadadze, M. Porrati, Phys. Lett. B **485**, 208 (2000) [hep-th/0005016]
12. C. Deffayet, Phys. Lett. B **502**, 199 (2001) [hep-th/0010186]
13. C. de Rham, S. Hofmann, J. Khoury, A.J. Tolley, JCAP **0802**, 011 (2008) [arXiv:0712.2821 [hep-th]]
14. C. de Rham, G. Dvali, S. Hofmann, J. Khoury, O. Pujolas, M. Redi, A.J. Tolley, Phys. Rev. Lett. **100**, 251603 (2008) [arXiv:0711.2072 [hep-th]]
15. C. de Rham, J. Khoury, A.J. Tolley, Phys. Rev. Lett. **103**, 161601 (2009) [arXiv:0907.0473 [hep-th]]
16. C. de Rham, A. Matas, A.J. Tolley, Class. Quant. Grav. **31**, 165004 (2014) [arXiv:1311.6485 [hep-th]]
17. C. de Rham, A. Matas, A.J. Tolley, Class. Quant. Grav. **31**, 025004 (2014) [arXiv:1308.4136 [hep-th]]
18. K. Hinterbichler, R.A. Rosen, J. High Energy Phys. **1207**, 047 (2012) [arXiv:1203.5783 [hep-th]]
19. C. de Rham, G. Gabadadze, Phys. Rev. D **82**, 044020 (2010) [arXiv:1007.0443 [hep-th]]
20. C. de Rham, G. Gabadadze, A.J. Tolley, Phys. Rev. Lett. **106**, 231101 (2011) [arXiv:1011.1232 [hep-th]]
21. H. van Dam, M.J.G. Veltman, Nucl. Phys. B **22**, 397 (1970)
22. V.I. Zakharov, JETP Lett. **12**, 312 (1970) [Pisma Zh. Eksp. Teor. Fiz. **12**, 447 (1970)]
23. S.F. Hassan, R.A. Rosen, J. High Energy Phys. **1202**, 126 (2012) [arXiv:1109.3515 [hep-th]]
24. S.F. Hassan, R.A. Rosen, A. Schmidt-May, J. High Energy Phys. **1202**, 026 (2012) [arXiv:1109.3230 [hep-th]]
25. S.F. Hassan, R.A. Rosen, Phys. Rev. Lett. **108**, 041101 (2012) [arXiv:1106.3344 [hep-th]]
26. E.A. Bergshoeff, O. Hohm, P.K. Townsend, Phys. Rev. Lett. **102**, 201301 (2009) [arXiv:0901.1766 [hep-th]]
27. C. de Rham, S. Renaux-Petel, JCAP **1301**, 035 (2013) [arXiv:1206.3482 [hep-th]]
28. N.A. Ondo, A.J. Tolley, J. High Energy Phys. **1311**, 059 (2013) [arXiv:1307.4769 [hep-th]]
29. P. Creminelli, A. Nicolis, M. Papucci, E. Trincherini, J. High Energy Phys. **0509**, 003 (2005) [arXiv:hep-th/0505147]
30. C. Deffayet, J.W. Rombouts, Phys. Rev. **D72**, 044003 (2005) [arXiv:gr-qc/0505134]
31. U. Aydemir, M.M. Anber, J.F. Donoghue, Phys. Rev. **D86**, 014025 (2012) [arXiv:1203.5153 [hep-th]]
32. C. Burrage, C. de Rham, L. Heisenberg, A.J. Tolley, JCAP **1207**, 004 (2012) [arXiv:1111.5549 [hep-th]]
33. C. de Rham, M. Fasiello, A.J. Tolley, Phys. Lett. B **733**, 46 (2014) [arXiv:1308.2702 [hep-th]]

Chapter 6
Hairy Black Holes in Theories with Massive Gravitons

Mikhail S. Volkov

Abstract This is a brief survey of the known black hole solutions in the theories of ghost-free bigravity and massive gravity. Various black holes exist in these theories, in particular those supporting a massive graviton hair. However, it seems that solutions which could be astrophysically relevant are the same as in General Relativity, or very close to them. Therefore, the no-hair conjecture essentially applies, and so it would be hard to detect the graviton mass by observing black holes.

6.1 Black Holes and the No-Hair Conjecture

More than 40 year ago J.A. Wheeler summarized the progress in the area of black hole physics at the time by his famous phrase: *black holes have no hair* [1]. More precisely, this means that

- All stationary black holes are completely characterized by their mass, angular momentum, and electric charge measurable from infinity.
- Black holes cannot support hair = external fields distributed close to the horizon but not seen from infinity.

Therefore, according to the ho-hair conjecture, the only allowed characteristics of stationary black holes are those associated with the Gauss law. The logic behind this is the following. Black holes are formed in the gravitational collapse, which is so violent a process that it breaks all the usual conservation laws not related to the exact symmetries. For example, the chemical content, the baryon number, etc. are not conserved during the collapse—the black hole 'swallows' all the memory

M.S. Volkov (✉)
Laboratoire de Mathématiques et Physique Théorique CNRS-UMR 7350, Université de Tours, Parc de Grandmont, 37200 Tours, France

Institut des Hautes Etudes Scientifiques (IHES), 91440 Bures-sur-Yvette, France

Department of General Relativity and Gravitation, Institute of Physics, Kazan Federal University, Kremlevskaya str. 18, 420008 Kazan, Russia
e-mail: volkov@lmpt.univ-tours.fr

E. Papantonopoulos (ed.), *Modifications of Einstein's Theory of Gravity at Large Distances*, Lecture Notes in Physics 892, DOI 10.1007/978-3-319-10070-8_6

of them. Everything that can be absorbed by the black hole gets absorbed. Only few exact local symmetries, such as the local Lorentz or local U(1), can survive the gravitational collapse. Associated to them conserved quantities—the mass, angular momentum, and electric charge—cannot be absorbed by the black hole and remain attached to it as parameters. They give rise to the Gaussian fluxes that can be measured at infinity.

The no-hair conjecture essentially implies that the only asymptotically flat black holes in Nature should be those described by the Kerr-Newman solutions. And indeed, a number of *the uniqueness theorems* [2–4] confirm that all stationary and asymptotically flat *electrovacuum* black holes with a non-degenerate horizon should belong to the Kerr-Newman family.

The electrovacuum uniqueness theorems do not directly apply to systems with matter fields other than the electromagnetic field. The field equations for such systems read schematically

$$G_{\mu\nu} = 8\pi G T_{\mu\nu}(\Psi), \qquad \Box\Psi = V(\Psi), \tag{6.1}$$

where Ψ denotes the matter field, or several interacting matter fields. One can wonder if these equations admit asymptotically flat black hole solutions with the curvature bounded everywhere outside the black hole horizon. According to the no-hair conjecture, the answer should be negative, but to prove this requires considering each matter type separately. In view of this, a number of *the no-hair theorems* have been proven to confirm the absence of static black hole solutions of Eq. (6.1) in the cases where Ψ denotes scalar, spinor, etc. fields [5–10]. The common feature in all these cases is that if Ψ does not vanish, then the field equations require that it should diverge at the black hole horizon, where the curvature diverges too. Therefore, to get regular black holes one is bound to set $\Psi = 0$, but then the solution is a vacuum black hole belonging to the Kerr-Newman family.[1] All this confirms the non-existence of *hairy* black holes.

The first explicit evidence against the no-hair conjecture was found 20 years after its formulation, in the context of the Einstein-Yang-Mills theory with gauge group SU(2). This theory contains all the electrovacuum solutions, hence all Kerr-Newman black holes [13], because the electromagnetic U(1) gauge group is contained in SU(2). However, it also admits static black holes supporting a non-trivial Yang-Mills field which asymptotically decays as $1/r^3$, so that the corresponding Gaussian flux is zero [14, 15]. Close to the horizon the geometry deviates from the Schwarzschild one, but the deviations rapidly decay with distance and cannot be seen from infinity. Therefore, such black holes support a *hair*.

Subsequent developments have revealed that the Einstein-Yang-Mills black holes can be generalized to include scalar fields, as for example a Higgs field, which leads

[1]It has recently been shown that these arguments can be circumvented for fine-tuned black hole mass and angular momentum [11]. This allows one to construct spinning hairy black holes which do not admit a static limit [12].

to a variety of new solutions describing hairy black holes [16]. In particular, it turns out that regular gravitating solitons, as for example gravitating magnetic monopoles or gravitating Skyrmions, can be generalized to contain inside a small black hole. This gives rise to black holes with a 'solitonic hair'. However, when the black hole size exceeds a certain critical value, the black hole 'swallows the soliton' and 'looses its hair', becoming a Kerr-Newman black hole [16].

Yet more hairy black holes can be obtained in models inspired by string theory and including a dilaton [17], the curvature corrections and so on [16]. Adding a cosmological term, positive or negative, gives asymptotically (anti)-de Sitter hairy black holes [18]. Summarizing, one can say that hairy black holes arise generically in physical models. However, large hairy black holes are typically unstable and loose the hair when perturbed, whereas the stable ones are typically very small [16]. As a result, despite a large number of solutions describing hairy black holes in various systems, it seems that the no-hair conjecture essentially holds for the astrophysical black holes, all of which should be of the Kerr-Newman type.

In what follows we shall be considering black holes in theories with massive gravitons—the ghost-free bigravity and massive gravity. Some of these black holes are of the known Kerr-Newman(-de Sitter) type, but there are also black holes supporting a massive graviton hair. However, the hairy black holes turn out to be either asymptotically anti-de Sitter (AdS), or cosmologically large, which contradicts the observations. Therefore, the astrophysical black holes should be described by the Kerr-Newman(-de Sitter) metrics, possibly with small corrections in the near-horizon region, so that the no-hair conjecture essentially holds.

6.2 Theories with Massive Gravitons

The idea that gravitons could have a tiny mass was proposed long ago [19], but it attracted a particular interest after the recent discovery of the special massive gravity theory by de Rham, Gabadadze, and Tolley (dRGT) [20] (see [21, 22] for a review). Before this discovery it had been known that the massive gravity theory generically had six propagating degrees of freedom (Dof). Five of them could be associated with the polarizations of the massive graviton, while the sixth one, usually called Boulware-Deser (BD) ghost, is unphysical, because it has a negative kinetic energy and renders the whole theory unstable [23]. The specialty of the dRGT theory is that it contains two Hamiltonian constraints which eliminate one of the six Dof [24–28]. Therefore, there remain just the right number of Dof to describe massive gravitons and so the theory is referred to as ghost-free. This does not mean that all solutions are stable in this theory, since there could be other instabilities, which should be checked in each particular case. However, since the most dangerous BD ghost instability is absent, the theory of [20] and its bigravity generalization [29] can be considered as healthy physical models for interpreting the observational data.

These theories can be used to explain the current cosmic acceleration [30, 31]. This acceleration could be accounted for by introducing a cosmological term in

Einstein equations, however, this would pose the problem of explaining the origin and value of this term. An alternative possibility is to consider modifications of General Relativity (GR), and theories with massive gravitons are natural candidates for this, since the graviton mass can effectively manifest itself as a small cosmological term [32].

Theories with massive gravitons are described by two metrics, $g_{\mu\nu}$ and $f_{\mu\nu}$. In massive gravity theories the f-metric is non-dynamical and is usually chosen to be flat, although other choices are also possible, while the dynamical g-metric describes massive gravitons. In bigravity theories [29] both metrics are dynamical and describe together two gravitons, one massive and one massless. The theory contains two gravitational couplings, κ_g and κ_f, and in the $\kappa_f \to 0$ limit the f-metric decouples and can be chosen to be flat. Therefore, the bigravity theory is more general, while the massive gravity theory can be viewed as its special case.

All known bigravity black holes were obtained in [33] (see also [34]), with the exception of special solutions discovered in [35]. These black holes can be divided into three types. First, there are solutions for which the two metrics are proportional, $f_{\mu\nu} = C^2 g_{\mu\nu}$ with a constant C, where $g_{\mu\nu}$ fulfills the Einstein equations with a cosmological term $\Lambda(C) \propto m^2$. If $C = 1$ then $\Lambda = 0$ and one obtains all solutions of the vacuum GR, in particular the vacuum black holes. For other values of C one has $\Lambda(C) \neq 0$, which gives rise to black holes with a cosmological term. None of these solutions fulfill equations of the massive gravity theory with a flat f.

Secondly, imposing spherical symmetry, there are black holes described by two metrics which are not simultaneously diagonal. They formally decouple one from the other and each of them fulfills its own set of Einstein equations with its own cosmological term. The g-metric is Schwarzschild-de Sitter, whereas the f-metric can be chosen to be AdS, with $\Lambda_f \sim \kappa_f^2$, and it becomes flat when $\kappa_f \to 0$, in which limit the dRGT massive gravity is naturally recovered. Therefore, these solutions exist both in the bigravity and dRGT massive gravity theories. In the latter case they exhaust all known black hole solutions.

Solutions of the third type are obtained when the two metrics are both diagonal but not proportional. One obtains in this case more complex solutions describing static black holes with a massive graviton hair, which can be either asymptotically AdS [33], or asymptotically flat [35], although in the latter case their size should be comparable with the Hubble radius.

A more detailed description of the currently known bigravity and massive gravity black holes is given below.

6.3 Ghost-Free Bigravity

The theory of the ghost-free bigravity [29] is defined on a four-dimensional spacetime manifold equipped with two metrics, $g_{\mu\nu}$ and $f_{\mu\nu}$, which describe two interacting gravitons, one massive and one massless. The kinetic term for each metric is chosen to be of the standard Einstein-Hilbert form, while the interaction between them is described by a local potential $\mathcal{U}[g, f]$ which does not contain derivatives and is expressed by a scalar function of the tensor

$$\gamma^{\mu}_{\ \nu} = \sqrt{g^{\mu\alpha} f_{\alpha\nu}}. \tag{6.2}$$

Here $g^{\mu\nu}$ is the inverse of $g_{\mu\nu}$ and the square root is understood in the matrix sense, i.e.

$$(\gamma^2)^{\mu}_{\ \nu} \equiv \gamma^{\mu}_{\ \alpha}\gamma^{\alpha}_{\ \nu} = g^{\mu\alpha} f_{\alpha\nu}. \tag{6.3}$$

The action is (with the metric signature $-+++$)

$$S[g, f] = \frac{1}{2\kappa_g^2}\int d^4x\,\sqrt{-g}\,R(g) + \frac{1}{2\kappa_f^2}\int d^4x\,\sqrt{-f}\,\mathcal{R}(f)$$
$$-\frac{m^2}{\kappa^2}\int d^4x\,\sqrt{-g}\,\mathcal{U}[g, f]\,, \tag{6.4}$$

where R and $\mathcal{R}$ are the Ricci scalars for $g_{\mu\nu}$ and $f_{\mu\nu}$, respectively, $\kappa_g^2 = 8\pi G$ and $\kappa_f^2 = 8\pi\mathcal{G}$ are the corresponding gravitational couplings, while $\kappa^2 = \kappa_g^2 + \kappa_f^2$ and m is the graviton mass. The interaction between the two metrics is given by

$$\mathcal{U} = \sum_{k=0}^{4} b_k\,\mathcal{U}_k(\gamma), \tag{6.5}$$

where b_k are parameters, while $\mathcal{U}_k(\gamma)$ are defined by the relations

$$\mathcal{U}_0(\gamma) = 1, \quad \mathcal{U}_1(\gamma) = \sum_A \lambda_A = [\gamma],$$
$$\mathcal{U}_2(\gamma) = \sum_{A<B} \lambda_A\lambda_B = \frac{1}{2!}([\gamma]^2 - [\gamma^2]),$$
$$\mathcal{U}_3(\gamma) = \sum_{A<B<C} \lambda_A\lambda_B\lambda_C = \frac{1}{3!}([\gamma]^3 - 3[\gamma][\gamma^2] + 2[\gamma^3]),$$
$$\mathcal{U}_4(\gamma) = \lambda_0\lambda_1\lambda_2\lambda_3 = \frac{1}{4!}([\gamma]^4 - 6[\gamma]^2[\gamma^2] + 8[\gamma][\gamma^3] + 3[\gamma^2]^2 - 6[\gamma^4])\,. \tag{6.6}$$

Here λ_A ($A = 0, 1, 2, 3$) are the eigenvalues of $\gamma^{\mu}_{\ \nu}$, and, using the hat to denote matrices, one has defined $[\gamma] = \mathrm{tr}(\hat{\gamma}) \equiv \gamma^{\mu}_{\ \mu}$, $[\gamma^k] = \mathrm{tr}(\hat{\gamma}^k) \equiv (\gamma^k)^{\mu}_{\ \mu}$. The (real) parameters b_k could be arbitrary, however, if one requires flat space to be a solution of the theory, and m to be the Fierz-Pauli mass of the graviton [19], then the five b_k's are expressed in terms of *two* free parameters c_3, c_4 as follows:

$$b_0 = 4c_3 + c_4 - 6, \;\; b_1 = 3 - 3c_3 - c_4, \;\; b_2 = 2c_3 + c_4 - 1,$$
$$b_3 = -(c_3 + c_4), \;\; b_4 = c_4. \tag{6.7}$$

The theory (6.4) propagates $7 = 5+2$ Dof corresponding to the polarizations of two gravitons, one massive and one massless. Before this theory was discovered [29], more general bigravity models, sometimes called f-g theories, had been considered [36]. In these models the potential $\mathscr{U}$ is a scalar function of $H^\mu_\nu = \delta^\mu_\nu - g^{\mu\alpha} f_{\alpha\nu}$ of the form

$$\mathscr{U} = \frac{1}{8}\,(H^\mu_\nu H^\nu_\mu - (H^\mu_\mu)^2) + \ldots, \tag{6.8}$$

where the dots denote all possible higher order scalars made of H^μ_ν. A particular choice of these terms leads to (6.5). The generic f-g theories propagate $7 + 1$ Dof, the additional one being the BD ghost [23].

Introducing the mixing angle η such that $\kappa_g = \kappa \cos\eta$, $\kappa_f = \kappa \sin\eta$ and varying the action (6.4) gives the field equations

$$G^\mu_\nu = m^2 \cos^2\eta\, T^\mu_\nu\,, \tag{6.9}$$
$$\mathscr{G}^\mu_\nu = m^2 \sin^2\eta\, \mathscr{T}^\mu_\nu\,, \tag{6.10}$$

where G^μ_ν and $\mathscr{G}^\mu_\nu$ are the Einstein tensors for $g_{\mu\nu}$ and $f_{\mu\nu}$. The graviton energy-momentum tensors obtained by varying the interaction $\mathscr{U}$ are

$$T^\mu_\nu = \tau^\mu_\nu - \mathscr{U}\,\delta^\mu_\nu, \qquad \mathscr{T}^\mu_\nu = -\frac{\sqrt{-g}}{\sqrt{-f}}\,\tau^\mu_\nu\,, \tag{6.11}$$

where

$$\begin{aligned}\tau^\mu_\nu = &\{b_1\,\mathscr{U}_0 + b_2\,\mathscr{U}_1 + b_3\,\mathscr{U}_2 + b_4\,\mathscr{U}_3\}\gamma^\mu_\nu \\ &-\{b_2\,\mathscr{U}_0 + b_3\,\mathscr{U}_1 + b_4\,\mathscr{U}_2\}(\gamma^2)^\mu_\nu \\ &+\{b_3\,\mathscr{U}_0 + b_4\,\mathscr{U}_1\}(\gamma^3)^\mu_\nu \\ &-b_4\,\mathscr{U}_0\,(\gamma^4)^\mu_\nu\,,\end{aligned} \tag{6.12}$$

with $\mathscr{U}_k \equiv \mathscr{U}_k(\gamma)$. The Bianchi identities for (6.9) and (6.10) imply that

$$\overset{(g)}{\nabla}_\mu T^\mu_\nu = 0, \qquad \overset{(f)}{\nabla}_\mu \mathscr{T}^\mu_\nu = 0, \tag{6.13}$$

where $\overset{(g)}{\nabla}$ and $\overset{(f)}{\nabla}$ are the covariant derivatives with respect to $g_{\mu\nu}$ and $f_{\mu\nu}$. In fact, the latter of these conditions is not independent and follows from the former one in view of the diffeomorphism invariance of the interaction term.

If $\eta \to 0$ and $\sin^2\eta\,\mathscr{T}^\mu_\nu \to 0$, then Eq. (6.10) for the f-metric decouple and their solution enters the g-equations (6.9) as a fixed reference metric. The g-equations describe in this case a massive gravity theory. If f becomes flat for $\eta \to 0$, then one recovers the dRGT theory [20]. Therefore, the massive gravity theory is contained in the bigravity.

6.4 Proportional Backgrounds

The simplest solutions of the bigravity equations are obtained by assuming the two metrics to be proportional [33, 37],

$$f_{\mu\nu} = C^2 g_{\mu\nu}. \tag{6.14}$$

The energy-momentum tensors (6.11) then become

$$T^\mu_\nu = -\Lambda_g(C)\delta^\mu_\nu, \quad \mathscr{T}^\mu_\nu = -\Lambda_f(C)\delta^\mu_\nu, \tag{6.15}$$

with

$$\begin{aligned} \Lambda_g(C) &= m^2\cos^2\eta\left(b_0 + 3b_1\,C + 3b_2\,C^2 + b_3\,C^3\right), \\ \Lambda_f(C) &= m^2\frac{\sin^2\eta}{C^3}\left(b_1 + 3b_2 C + 3b_3 C^2 + b_4 C^3\right). \end{aligned} \tag{6.16}$$

Since the energy-momentum tensors should be conserved, it follows that C is a constant. As a result, one obtains two sets of Einstein equations,

$$G^\nu_\mu + \Lambda_g(C)\delta^\nu_\mu = 0, \quad \mathscr{G}^\nu_\mu + \Lambda_f(C)\delta^\nu_\mu = 0. \tag{6.17}$$

Since one has $\mathscr{G}^\nu_\mu = G^\nu_\mu/C^2$, it follows that $\Lambda_f = \Lambda_g/C^2$, which gives an algebraic equation for C. If the parameters b_k are chosen according to Eq. (6.7), then this equation reads

$$\begin{aligned} 0 = (C-1)[&(c_3+c_4)C^3 + (3-5c_3+(\chi-2)c_4)C^2 \\ &+((4-3\chi)c_3+(1-2\chi)c_4-6)C + (3c_3+c_4-1)\chi], \end{aligned} \tag{6.18}$$

with $\chi = \tan^2\eta$, while the cosmological constant is

$$\frac{\Lambda_g}{m^2\cos^2\eta} = (1-C)((c_3+c_4)C^2 + (3-5c_3-2c_4)C + 4c_3 + c_4 - 6). \tag{6.19}$$

Depending on values of c_3, c_4, η, Eq. (6.18) can have up to four real roots, so that there can be solutions with four different values of the cosmological constant, which can be positive, negative, or zero.

One solution of (6.18) is $C = 1$, in which case the two metrics coincide, $g_{\mu\nu} = f_{\mu\nu}$, while $\Lambda_g = 0$, so that the vacuum GR is recovered. Therefore, the black hole solutions obtained in this case are either Kerr, or Kerr-de Sitter, or Kerr-AdS. None of these solutions admit the massive gravity limit with a flat f-metric.

6.5 Solutions with Non-Bidiagonal Metrics

Let us assume both metrics to be invariant under spatial SO(3) rotations. Since the theory is invariant under diffeomorphisms, one can choose the spacetime coordinates such that the g-metric is diagonal. However, the f-metric will in general contain an off-diagonal term, so that the two metrics can be parameterized as

$$ds_g^2 = -N^2 dt^2 + \frac{dr^2}{\Delta^2} + R^2 d\Omega^2\,,$$
$$ds_f^2 = -\left(aNdt + \frac{c}{\Delta}\,dr\right)^2 + \left(cNdt - \frac{b}{\Delta}\,dr\right)^2 + u^2 R^2 d\Omega^2\,, \tag{6.20}$$

with $d\Omega^2 = d\vartheta^2 + \sin^2\vartheta d\varphi^2$. The amplitudes N, Δ, R depend on r, while a, b, c, u can in general depend on t, r. It is straightforward to check that the matrix square root is

$$\gamma^\mu_\nu = \sqrt{g^{\mu\alpha} f_{\alpha\nu}} = \begin{pmatrix} a & c/(\Delta N) & 0 & 0 \\ -c\Delta N & b & 0 & 0 \\ 0 & 0 & u & 0 \\ 0 & 0 & 0 & u \end{pmatrix}, \tag{6.21}$$

whose eigenvalues are

$$\lambda_{0,1} = \frac{1}{2}\left(a + b \pm \sqrt{(a-b)^2 - 4c^2}\right), \quad \lambda_2 = \lambda_3 = u. \tag{6.22}$$

Inserting this to (6.6) gives

$$\mathscr{U}_1 = a + b + 2u, \qquad \mathscr{U}_2 = u(u + 2a + 2b) + ab + c^2\,,$$
$$\mathscr{U}_3 = u\,(au + bu + 2ab + 2c^2), \qquad \mathscr{U}_4 = u^2(ab + c^2). \tag{6.23}$$

Although the eigenvalues (6.22) can be complex-valued, the $\mathscr{U}_k$'s are always real. It is straightforward to compute the energy-momentum tensors T^μ_ν and $\mathscr{T}^\mu_\nu$ defined by Eqs .(6.11),(6.12). In particular, one finds

$$T^0_r = \frac{c}{\Delta N}\,[b_1 + 2b_2 u + b_3 u^2]. \tag{6.24}$$

Since the g-metric is static, there is no radial energy flux, and so T^0_r should be zero. Therefore, either c should vanish, or the expression in brackets in (6.24) vanishes. The former option will be considered in the next section, while presently let us assume that $c \neq 0$ and

$$b_1 + 2b_2 u + b_3 u^2 = 0. \tag{6.25}$$

This yields

$$u = \frac{1}{b_3}\left(-b_2 \pm \sqrt{b_2^2 - b_1 b_3}\right). \tag{6.26}$$

Notice that u was a priori a function of t, r, but now it is restricted to be a constant. Using this, one finds that $T^0_0 = T^r_r = -\lambda_g$ and $\mathscr{T}^0_0 = \mathscr{T}^r_r = -\lambda_f$ where

$$\lambda_g = b_0 + 2b_1 u + b_2 u^2, \qquad \lambda_f = \frac{b_2 + 2b_3 u + b_4 u^2}{u^2}. \tag{6.27}$$

The conditions $\overset{(g)}{\nabla}_\rho\, T^\rho_\lambda = 0$ reduce in this case to the requirement that $T^0_0 - T^\vartheta_\vartheta$ should vanish. On the other hand, one finds

$$T^0_0 - T^\vartheta_\vartheta = (b_2 + b_3 u)[(u-a)(u-b) + c^2], \tag{6.28}$$

and since this has to vanish, either the first or the second factor on the right should be zero. Let us assume that one of these conditions is fulfilled. Then one has $T^0_0 = T^\vartheta_\vartheta$ and $\mathscr{T}^0_0 = \mathscr{T}^\vartheta_\vartheta$, hence both energy-momentum tensors are proportional to the unit tensor, $T^\mu_\nu = -\lambda_g \delta^\mu_\nu$ and $\mathscr{T}^\mu_\nu = -\lambda_f \delta^\mu_\nu$. The field equations (6.9) then reduce to

$$G^\rho_\lambda + \Lambda_g \delta^\rho_\lambda = 0\,, \tag{6.29}$$

$$\mathscr{G}^\rho_\lambda + \Lambda_f \delta^\rho_\lambda = 0\,, \tag{6.30}$$

where

$$\Lambda_g = m^2 \cos^2\eta\, \lambda_g, \qquad \Lambda_f = m^2 \sin^2\eta\, \lambda_f. \tag{6.31}$$

As a result, the two metrics decouple one from the other, and the graviton mass gives rise to the two cosmological terms. If the parameters b_k are chosen according to (6.7), then $\lambda_g + u^2\lambda_f = -(u-1)^2$, therefore, if $\Lambda_g > 0$ then $\Lambda_f < 0$.

Since we want the g-metric to describe a black hole geometry, the solution of (6.29) is the Schwarzschild-de Sitter metric. On the other hand, as the cosmological term for the f-metric is negative, the solution of (6.30) can be chosen to be AdS. Therefore,

$$ds_g^2 = -\Sigma(r)\, dt^2 + \frac{dr^2}{\Sigma(r)} + r^2 d\Omega^2\,, \qquad \Sigma(r) = 1 - \frac{2M}{r} - \frac{\Lambda_g}{3} r^2,$$

$$ds_f^2 = -\mathscr{D}(U)\, dT^2 + \frac{dU^2}{\mathscr{D}(U)} + U^2 d\Omega^2\,, \qquad \mathscr{D}(U) = 1 - \frac{\Lambda_f}{3} U^2, \tag{6.32}$$

with $U = ur$. It is worth noting that, since $\Lambda_f \sim \sin^2\eta \to 0$ when $\eta \to 0$, the f-metric becomes flat in this limit. Therefore, the solutions apply both in the bigravity theory and in the dRGT massive gravity.

6.5.1 Imposing the Consistency Condition

The solution (6.32) is not yet complete, since the two metrics are expressed in two different coordinate systems, t, r and T, U, whose relation to each other is not known. One has $U = ur$ but the function $T(t, r)$ is still undetermined. We therefore remember that up to now we have not considered the consistency condition, which requires that the expression in (6.28) should vanish. This condition will be fulfilled in either of the following two cases:

$$\text{I:} \qquad (b_2 + b_3 u) = 0; \tag{6.33}$$

$$\text{II:} \qquad (u-a)(u-b) + c^2 = 0. \tag{6.34}$$

In case I, since u is already expressed in terms of b_1, b_2, b_3 by Eq. (6.26), the condition (6.33) imposes a constraint on values of these parameters. Therefore, this condition is possible only for the special subclass of the theory characterized by the restricted values of b_k. Within this subclass the consistency condition is fulfilled without specifying $T(t, r)$. Therefore, the function $T(t, r)$ in (6.32) remains arbitrary, which can probably be traced to a some kind of hidden gauge invariance.

In case II no restrictions on the coefficients b_k arise, so that this case is generic. The coefficients a, b, c can be obtained by comparing the line element ds_f^2 in (6.20) with that in (6.32), which gives

$$a^2 - c^2 = \frac{\mathscr{D}\dot{T}^2}{\Sigma}, \quad b^2 - c^2 = \Sigma\left(\frac{u^2}{\mathscr{D}} - \mathscr{D}T'^2\right), \quad c(a+b) = \mathscr{D}\dot{T}T'. \tag{6.35}$$

Resolving these relations with respect to a, b, c and inserting the result to (6.34) yields the equation,

$$\frac{\mathscr{D}}{\Sigma}\dot{\mathscr{T}}^2 + \frac{\Sigma\mathscr{D}}{\Sigma - \mathscr{D}}\mathscr{T}'^2 = 1\,, \tag{6.36}$$

with $T = u\mathscr{T}$. A simple solution can be obtained by separating the variables,

$$\mathscr{T} = t + \int \frac{dr}{\Sigma} - \int \frac{dr}{\mathscr{D}} \equiv t + r^*_\Sigma - r^*_\mathscr{D}. \tag{6.37}$$

One can think that this solution is singular, since the tortoise coordinate r^*_Σ diverges at the black hole and cosmological horizons, where Σ vanishes. However, introducing the light-like coordinate

$$V = t + r^*_\Sigma = \mathscr{T} + r^*_\mathscr{D}\,, \tag{6.38}$$

both metrics can be written in the Eddington-Finkelstein form

$$\begin{aligned} ds_g^2 &= -\Sigma dV^2 + 2dVdr + r^2 d\Omega^2\,, \\ \frac{1}{u^2}ds_f^2 &= -\mathscr{D}dV^2 + 2dVdr + r^2 d\Omega^2\,, \end{aligned} \tag{6.39}$$

from where it is obvious that the solution is regular. This solution is valid for all values of the parameters b_k. All the above solutions have been obtained in the ghost-free bigravity context in [33] (see also [34]), but in fact solutions of this type were considered already long ago in the generic f-g bigravity theories [38–40]. The generalization for a nonzero electric charge was considered in [41].

Since the f-metric becomes flat for $\eta \to 0$, the solutions describe in this limit black holes in the dRGT massive gravity. In this context they were studied in [42,43] for the special case I, and in [44,45] for the generic case II. These solutions and their generalization for a nonzero electric charge [42,43,46] exhaust all static, spherically symmetric black holes in the dRGT theory.

6.6 Hairy Black Holes, Lumps, and Stars

Black holes considered in the previous two sections are described by the known GR metrics. New black holes are obtained in the case where the two metrics are simultaneously diagonal [33],

$$ds_g^2 = N^2 dt^2 - \frac{dr^2}{\Delta^2} - r^2 d\Omega^2, \quad ds_f^2 = A^2 dt^2 - \frac{U'^2}{Y^2}dr^2 - U^2 d\Omega^2\,. \tag{6.40}$$

Here N, Δ, Y, U, A are five functions of r which fulfill the equations

$$G_0^0 = m^2\cos^2\eta\, T_0^0, \qquad G_r^r = m^2\cos^2\eta\, T_r^r,$$
$$\mathscr{G}_0^0 = m^2\sin^2\eta\, \mathscr{T}_0^0, \qquad \mathscr{G}_r^r = m^2\sin^2\eta\, \mathscr{T}_r^r,$$
$$T_r^{r\prime} + \frac{N'}{N}(T_r^r - T_0^0) + \frac{2}{r}(T_\vartheta^\vartheta - T_r^r) = 0. \tag{6.41}$$

The simplest solutions are obtained if $f_{\mu\nu} = C^2 g_{\mu\nu}$, where $g_{\mu\nu}$ fulfills (6.17) while $C, \Lambda_g(C)$ are defined by (6.16),(6.18). Since Λ_g can be positive, negative, or zero, there are the Schwarzschild, Schwarzschild-de Sitter, and Schwarzschild-AdS black holes. Let us call them background black holes.

More general solutions are obtained by numerically integrating Eq. (6.41). It turns out [33] that the equations for the three amplitudes Δ, Y, U comprise a closed system. Its local solution near the horizon,

$$\Delta^2 = \sum_{n\geq 1} a_n (r - r_h)^n, \quad Y^2 = \sum_{n\geq 1} b_n (r - r_h)^n, \quad U = u r_h + \sum_{n\geq 1} c_n (r - r_h)^n,$$

contains only one free parameter $u = U(r_h)/r_h$, which is the ratio of the horizon radius measured by $f_{\mu\nu}$ to that measured by $g_{\mu\nu}$. The horizon is common for both metrics, in addition, its surface gravities and temperatures determined with respect to both metrics are the same [47].

Choosing a value of u and integrating numerically the equations from $r = r_h$ towards large r, the result is as follows [33]. If $u = C$ where C is a root of the algebraic equation (6.18), then the solution is one of the background black holes. If $u = C + \delta u$ then one can expect the solution to be the background black hole slightly deformed by a massive graviton 'hair' localized in the horizon vicinity. This is indeed confirmed for the Schwarzschild-AdS type solutions ($\Lambda_g < 0$), which can support a short massive hair and show deviations from the pure Schwarzschild-AdS in the horizon vicinity, but far away from the horizon the deviations tend to zero (see Fig. 6.1). Therefore, there are asymptotically AdS hairy black holes in the theory.

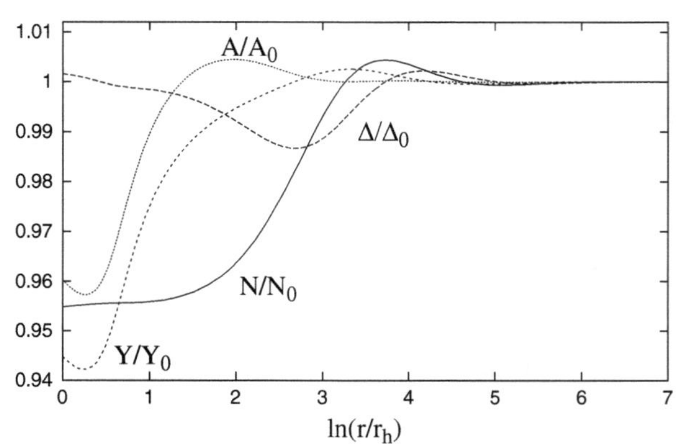

Fig. 6.1 Hairy deformations of the Schwarzschild-AdS background, where A_0, N_0, Δ_0, Y_0 correspond to the undeformed solution

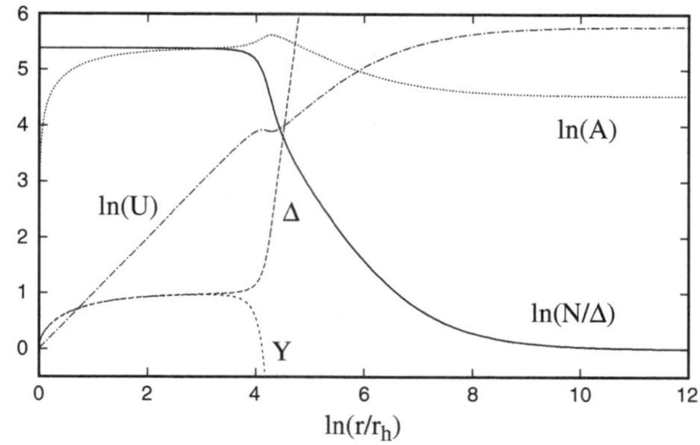

Fig. 6.2 Hairy deformations of the Schwarzschild background

However, the procedure goes differently for $\Lambda_g \geq 0$. When one deforms the Schwarzschild background by setting $u = r_h + \delta u$, then the solutions first stay very close to Schwarzschild. However, at large r they deviate away and show a completely different asymptotic behavior at infinity (Fig. 6.2), characterized by a quasi-AdS g-metric and a compact f-metric [33]. Therefore, the only asymptotically flat black hole one finds is the pure Schwarzschild, while its hairy deformations loose the asymptotic flatness. Similarly, trying to deform the Schwarzschild-de Sitter background produces a curvature singularity at a finite proper distance away from the black hole horizon, hence the only asymptotically de Sitter black hole is the pure Schwarzschild-de Sitter.

The conclusion is that there are hairy black holes in the theory, but they are not asymptotically flat. The following argument helps to understand this. Let us *require* the solution to be asymptotically flat. Then one should have at large r

$$\begin{aligned} \Delta &= 1 - \frac{A\sin^2\eta}{r} + B\cos^2\eta\,\frac{mr+1}{r}\,e^{-mr} + \ldots, \\ U &= r + B\,\frac{m^2r^2+mr+1}{m^2r^2}\,e^{-mr} + \ldots, \\ Y &= 1 - \frac{A\sin^2\eta}{r} - B\,\sin^2\eta\,\frac{1+mr}{r}\,e^{-mr} + \ldots, \end{aligned} \tag{6.42}$$

where A, B are integration constants. Suppose that one wants to find black hole solutions with this asymptotic behavior using the multiple shooting method. In this method one tries to match the asymptotics (6.42) and (6.42) by integrating the equations starting from the horizon towards large r, and at the same time starting from infinity towards small r. The two solutions should match at some intermediate point, which gives three matching conditions for Δ, Y, U. These conditions should be fulfilled by adjusting the free parameters A, B, u in Eqs. (6.42),(6.42). Solutions of this problem may exist at most for discrete sets of values of A, B, u, hence one cannot vary continuously the horizon parameter u. Therefore, there could be no continuous, asymptotically flat hairy deformations of the Schwarzschild solution. However, this does not exclude isolated solutions, and in fact they exist, but to find them requires a good initial guess for A, B, u.

It is interesting to see what happens to the hairy black holes when one changes the horizon radius r_h. It turns out that in the $r_h \to 0$ limit, where the black hole disappears, its 'hair' survives and becomes a static 'lump' made of massive field modes. Such lumps are described by globally regular solutions for which the event horizon is replaced by a regular center at $r = 0$, while at infinity the asymptotic behavior is the same as for the black holes [33]. None of the lumps are asymptotically flat. Neither lumps no hairy black holes admit the dRGT limit, they exist only in the bigravity theory.

It is worth mentioning in this context that there are asymptotically flat solutions with a matter [33]. Such solutions describe regular stars, and for them one can take limits where one of the two metrics becomes flat. Suppose that the f-sector is empty, while the g-sector contains $T^{[\mathrm{m}]\mu}{}_{\nu} = \mathrm{diag}[-\rho(r), P(r), P(r), P(r)]$ with $\rho = \rho_\star \theta(r_\star - r)$, corresponding to a 'star' with a constant density $\rho_\star$ and a radius $r_\star$. Adding this source to the field equations (6.41) and assuming a regular center at $r = 0$, one finds solutions for which both metrics approach Minkowski metric at infinity according to (6.42). Introducing the mass functions M_g, M_f via $g^{rr} = \Delta^2 = 1 - 2M_g(r)/r$ and $f^{rr} = Y^2/U'^2 = 1 - 2M_f(r)/r$, one finds that M_g, M_f rapidly increase inside the star, while outside they approach the same asymptotic value $M_g(\infty) = M_f(\infty) \sim \sin^2\eta$ (see Fig. 6.3). For $\eta = \pi/2$ the g-metric is coupled only to the matter and is described by the GR Schwarzschild solution, $M_g(r) = \rho_\star r^3/6$ for $r < r_\star$ and $M_g(r) = \rho_\star r_\star^3/6 \equiv M_{\mathrm{ADM}}$ for $r > r_\star$. For $\eta < \pi/2$ the star mass M_{ADM} is partially screened by the negative graviton energy. For $\eta = 0$ (dRGT theory) the f-metric becomes flat, so that $M_f = 0$, while M_g asymptotically approaches zero and the star mass is totally screened, because the massless graviton decouples and there could be no $1/r$ terms in the metric.

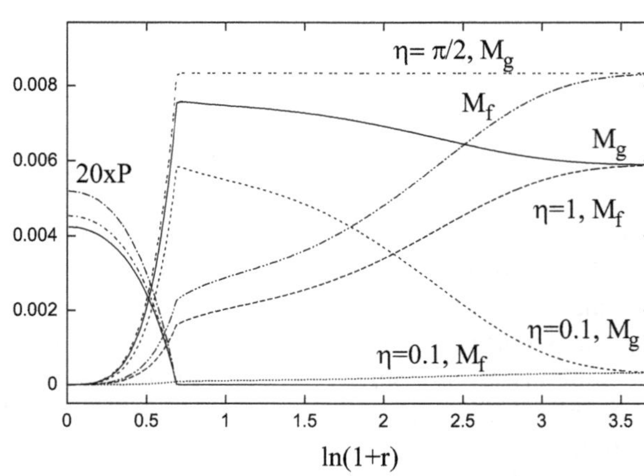

Fig. 6.3 Profiles of the asymptotically flat star solution sourced by a regular matter distribution

If the graviton mass is very small, then the $m^2 T^{\mu}_{\ \nu}$ contribution to the equations is small as compared to $T^{[\mathrm{m}]\mu}_{\ \nu}$, and for this reason M_g rests approximately constant for $r_\star < r < r_{\mathrm{V}} \sim (M_{\mathrm{ADM}}/m^2)^{1/3}$. This illustrates the Vainshtein mechanism of recovery of General Relativity in a finite region [48]. This mechanism has also been confirmed by the numerical analysis within the generic massive gravity theory with the BD ghost [49, 50], and also in the dRGT theory [51]. The approximate analytical solutions in the weak field limit were considered in [44,45,52] within the dRGT theory and in [53] within the bigravity theory.

6.7 Black Hole Stability and New Hairy Black Holes

As discussed in Sect. 6.4, if the two metrics coincide, $g_{\mu\nu} = f_{\mu\nu}$, then the bigravity theory reduces to the vacuum GR, hence one can choose the Schwarzschild metric as a solution. This solution is known to be linearly stable in the GR context, but one can wonder if it is stable also within the bigravity theory. Let us consider small perturbations around this solution,

$$g_{\mu\nu} = g^{(0)}_{\mu\nu} + \delta g_{\mu\nu}, \quad f_{\mu\nu} = g^{(0)}_{\mu\nu} + \delta f_{\mu\nu}\,, \tag{6.43}$$

where $g^{(0)}_{\mu\nu}$ is the Schwarzschild metric. If one sets $\delta g_{\mu\nu} = \delta f_{\mu\nu}$, then the GR result will be recovered. However, the perturbations of the two metrics need not be the same in general. Linearizing the bigravity field equations with respect to the perturbations, it turns out that the linear combinations

$$h_{\mu\nu} = \cos\eta\,\delta g_{\mu\nu} + \sin\eta\,\delta f_{\mu\nu}, \quad h^{(0)}_{\mu\nu} = \cos\eta\,\delta f_{\mu\nu} - \sin\eta\,\delta g_{\mu\nu} \tag{6.44}$$

decouple from each other and can be identified with the massive and massless gravitons, respectively. Equations for the massless graviton are the same as in GR, while for the massive graviton one obtains [54]

$$\begin{aligned} &\overset{(0)}{\Box}\, h_{\mu\nu} + 2\,\overset{(0)}{R}_{\mu\alpha\nu\beta}\, h^{\alpha\beta} = m^2 h_{\mu\nu}, \\ &\overset{(0)}{\nabla}_{\mu}\, h^{\mu}_{\nu} = 0, \quad h^{\mu}_{\mu} = 0. \end{aligned} \tag{6.45}$$

An interesting observation [54] is that these equations have exactly the same structure as those describing perturbations of the black strings—Schwarzschild black holes uplifted to five spacetime dimensions. At the same time, it is known that the black strings are prone to the Gregory-Laflamme instability [55]. Specifically, setting $h_{\mu\nu} = e^{i\omega t} H_{\mu\nu}(r,\vartheta,\varphi)$, it turns out that Eq. (6.45) admit a bound state solution with $\omega^2 < 0$ in the spherically-symmetric sector, provided that [56]

$$m r_h = \frac{\text{black hole radius}}{\text{graviton's Compton length}} < 0.86. \tag{6.46}$$

It follows that small black holes are unstable, since the frequency ω is imaginary and so the perturbations grow in time [54]. The condition of smallness is not crucial, since all usual black holes are small compared to the Hubble radius and so fulfill the bound (6.46), so that all of them should be unstable. On the other hand, since the frequency $|\omega| \propto m$, this instability is very mild, as it needs a Hubble time $\sim 1/m$ to develop. Therefore, even if real astrophysical black hole were described by the bigravity theory, their instability would be largely irrelevant and they would actually be stable for all practical purposes over a cosmologically long period of time.

A similar instability was found also for the Kerr black holes [56] and for the Schwarzschild-de Sitter black holes with proportional metrics described in Sect. 6.4 [57]. Interestingly, it was found in the latter case that the instability disappears in the partially massless limit, where the graviton mass is related to the cosmological constant as $m^2 = 2\Lambda/3$ [57].

As discussed in Sect. 6.5 above, the Schwarzschild-de Sitter solution in the bigravity theory can exist also in a different version, for which the two metrics are not simultaneously diagonal. The linear stability of this solution was studied with respect to all possible perturbations, but only in the restricted case (6.33) [58], and also in the generic case (6.34), but only with respect to spherically symmetric perturbations [59]. In both cases the solution was found to be stable.

Getting back to the unstable Schwarzschild black holes, it turns out that their instability can be used to find new black holes which support hair and which are asymptotically flat. As was explained above, asymptotically flat solutions subject to the boundary conditions (6.42),(6.42) may exist, but to find them requires to fine-tune the parameters A, B, u in (6.42),(6.42), for which an additional information is needed. Now, the existence of the black hole instability provides such an information [35].

Indeed, Eq. (6.45) admit solutions with $\omega^2 < 0$ only for $mr_h < 0.86$, while for $mr_h > 0.86$ all solutions have $\omega^2 > 0$. This means that for $mr_h \approx 0.86$ there is a zero mode: a static solution of (6.45) with $\omega = 0$. This zero mode can be viewed as approximating a new black hole solution which exists for $mr_h < 0.86$ and which merges with the Schwarzschild solution for $mr_h \approx 0.86$. Close to the merging point the deviations of the new solution from the Schwarzschild are small and can be described by the linear theory. Therefore, one can use the linear zero mode to read-off the values of the parameters A, B, u in (6.42),(6.42), after which one can iteratively decrease r_h to obtain the 'fully-fledged' non-perturbative hairy black holes. This was done in [35].

The conclusion is that there are asymptotically flat black holes with a massive hair in the bigravity theory. However, it seems that their parameter mr_h cannot be too small (unless for $c_3 = -c_4 = 2$) [35], which means that these black holes are cosmologically large, their size being comparable with the Hubble radius. Such solutions are unlikely to be relevant.

All described above black holes have been obtained in the theory either without a matter source or in the theory with an electromagnetic field. At the same time, the perturbative analysis of [60] predicts that hairy black holes should generically exist in the massive gravity theory coupled to a matter with a non-vanishing trace of the energy-momentum tensor. It would be very interesting to test this prediction by fully non-perturbative calculations.

Concluding Remarks
Summarizing the above discussion, all possible static, spherically symmetric black holes in the dRGT massive gravity theory are described by the Schwarzschild-de Sitter metrics. They belong to the type studied in Sect. 6.5 and they are probably stable. One may wonder why one does not find asymptotically flat black holes. However, our universe is actually in the de Sitter phase, and the main motivation for considering theories with massive gravitons is to describe this fact. Hence, it is not astonishing that the solutions are not asymptotically flat.

One finds more solutions in the bigravity theory, as for example the hairy black holes. However, these seem to be not very relevant, since they are either asymptotically AdS, which contradicts the observations, or they are too large. There are also asymptotically flat or asymptotically de Sitter black hole solutions, but they are unstable. However, they can describe astrophysical black holes, since the instability takes cosmologically long times to develop. One can also wonder what these black holes decay to, and one possibility is that their instability actually implies that there is a slow accretion of massive graviton modes to the horizon [61]. If this is true, then the black holes should be almost exactly Kerr (Kerr-de Sitter), apart from small corrections in the near-horizon region where the accretion takes place.

Some aspects of the graviton mass can be captured within a simplified description in the context of the Galileon theory [62]. This is essentially the General Relativity coupled to a self-interacting scalar field that mimics the scalar polarization mode of the massive graviton. It turns out that black holes in these theory are described by the GR metrics [63], and a no-hair theorem can be proven in this case [64].

To recapitulate, even if the gravitons are indeed massive, this would be hard to detect by observing black holes.

Acknowledgements This work was partly supported by the Russian Government Program of Competitive Growth of the Kazan Federal University.

References

1. R. Ruffini, J.A. Wheeler, Introducing the black hole. Phys. Today **24**, 30–41 (1971)
2. W. Israel, Event horizons in static vacuum space-times. Phys. Rev. **164**, 1776–1779 (1967)
3. B. Carter, Black hole equilibrium states, in *Black Holes*, ed. by C. DeWitt, B.S. DeWit (Gordon and Breach, New York, 1973)

4. P.O. Mazur, Proof of uniqueness of the Kerr-Newman black hole solution. J. Phys. **A15**, 3173–3180 (1982)
5. J.D. Bekenstein, Transcendence of the law of baryon-number conservation in black hole physics. Phys. Rev. Lett. **28**, 452–455 (1972)
6. J.D. Bekenstein, Nonexistence of baryon number for static black holes. Phys. Rev. **D5**, 1239–1246 (1972)
7. J.D. Bekenstein, Nonexistence of baryon number for black holes. II. Phys. Rev. **D5**, 2403–2412 (1972)
8. J.D. Bekenstein, Novel 'no scalar hair' theorem for black holes. Phys. Rev. **D51**, 6608–6611 (1995)
9. A.E. Mayo, J.D. Bekenstein, No hair for spherical black holes: charged and nonminimally coupled scalar field with selfinteraction. Phys. Rev. **D54**, 5059–5069 (1996)
10. J.D. Bekenstein, Black hole hair: 25 - years after (1996) [arXiv:gr-qc/9605059]
11. S. Hod, Stationary scalar clouds around rotating black holes. Phys. Rev. **D86**, 104026 (2012)
12. C.A.R. Herdeiro, E. Radu, Kerr black holes with scalar hair. Phys. Rev. Lett. **112**, 221101 (2014)
13. P.B. Yasskin, Solutions for gravity coupled to massless gauge fields. Phys. Rev. **D12**, 2212–2217 (1975)
14. M.S. Volkov, D.V. Galtsov, Non-Abelian Einstein Yang-Mills black holes. JETP Lett. **50**, 346–350 (1989)
15. M.S. Volkov, D.V. Galtsov, Black holes in Einstein Yang-Mills theory. Sov. J. Nucl. Phys. **51**, 747–753 (1990)
16. M.S. Volkov, D.V. Gal'tsov, Gravitating non-Abelian solitons and black holes with Yang-Mills fields. Phys. Rept. **319**, 1–83 (1999)
17. B. Kleihaus, J. Kunz, Static axially symmetric Einstein Yang-Mills dilaton solutions. 2. Black hole solutions. Phys. Rev. **D57**, 6138–6157 (1998)
18. S.S. Gubser, S.S. Pufu, The Gravity dual of a p-wave superconductor. J. High Energy Phys. **0811**, 033 (2008)
19. M. Fierz, W. Pauli, On relativistic wave equations for particles of arbitrary spin in an electromagnetic field. Proc. R. Soc. Lond. **A173**, 211–232 (1939)
20. C. de Rham, G. Gabadadze, A.J. Tolley, Resummation of massive gravity. Phys. Rev. Lett. **106**, 231101 (2011)
21. K. Hinterbichler, Theoretical aspects of massive gravity. Rev. Mod. Phys. **84**, 671–710 (2012)
22. C. de Rham, Massive gravity (2014) [arXiv:1401.4173]
23. D.G. Boulware, S. Deser, Can gravitation have a finite range? Phys. Rev. **D6**, 3368–3382 (1972)
24. S.F. Hassan, R.A. Rosen, Resolving the ghost problem in non-linear massive gravity. Phys. Rev. Lett. **108**, 041101 (2012)
25. S.F. Hassan, R.A. Rosen, Confirmation of the secondary constraint and absence of ghost in massive gravity and bimetric gravity. J. High Energy Phys. **1204**, 123 (2012)
26. J. Kluson, Non-linear massive gravity with additional primary constraint and absence of ghosts. Phys. Rev. **D86**, 044024 (2012)
27. D. Comelli, M. Crisostomi, F. Nesti, L. Pilo, Degrees of freedom in massive gravity. Phys. Rev. **D86**, 101502 (2012)
28. D. Comelli, F. Nesti, L. Pilo, Massive gravity: a general analysis (2013) [arXiv:1305.0236]
29. S.F. Hassan, R.A. Rosen, Bimetric gravity from ghost-free massive gravity. J. High Energy Phys. **1202**, 126 (2012)
30. A.G. Riess et al., Observational evidence from supernovae for an accelerating universe and a cosmological constant. Astron. J. **116**(3), 1009 (1998)
31. S. Perlmutter et al., Measurements of Omega and Lambda from 42 high-redshift supernovae. Astrophys. J. **517**(2), 565 (1999)
32. T. Damour, I.I. Kogan, A. Papazoglou, Nonlinear bigravity and cosmic acceleration. Phys. Rev. D **66**, 104025 (2002)

33. M.S. Volkov, Hairy black holes in the ghost-free bigravity theory. Phys. Rev. **D85**, 124043 (2012)
34. D. Comelli, M. Crisostomi, F. Nesti, L. Pilo, Spherically symmetric solutions in ghost-free massive gravity. Phys. Rev. **D85**, 024044 (2012)
35. R. Brito, V. Cardoso, P. Pani, Black holes with massive graviton hair. Phys. Rev. **D88**, 064006 (2013)
36. C.J. Isham, A. Salam, J.A. Strathdee, F-dominance of gravity. Phys. Rev. **D3**, 867–873 (1971)
37. M.S. Volkov, Self-accelerating cosmologies and hairy black holes in ghost-free bigravity and massive gravity. Class. Quant. Grav. **30**, 184009 (2013)
38. A. Salam, J.A. Strathdee, A class of solutions for the strong gravity equations. Phys. Rev. **D16**, 2668 (1977)
39. C.J. Isham, D. Storey, Exact spherically symmetric classical solutions for the f-g theory of gravity. Phys. Rev. **D18**, 1047 (1978)
40. Z. Berezhiani, D. Comelli, F. Nesti, L. Pilo, Exact spherically symmetric solutions in massive gravity. J. High Energy Phys. **0807**, 130 (2008)
41. E. Babichev, A. Fabbri, A class of charged black hole solutions in massive (bi)gravity. J. High. Energy Phys. **1407**, 016 (2014)
42. Th.M. Nieuwenhuizen, Exact Schwarzschild-de Sitter black holes in a family of massive gravity models. Phys. Rev. **D84**, 024038 (2011)
43. L. Berezhiani, G. Chkareuli, C. de Rham, G. Gabadadze, A.J. Tolley, On black holes in massive gravity. Phys. Rev. **D85**, 044024 (2012)
44. K. Koyama, G. Niz, G. Tasinato, Analytic solutions in non-linear massive gravity. Phys. Rev. Lett. **107**, 131101 (2011)
45. K. Koyama, G. Niz, G. Tasinato, Strong interactions and exact solutions in non-linear massive gravity. Phys. Rev. **D84**, 064033 (2011)
46. Y.-F. Cai, D.A. Easson, C. Gao, E.N. Saridakis, Charged black holes in nonlinear massive gravity. Phys. Rev. **D87**(6), 064001 (2013)
47. C. Deffayet, T. Jacobson, On horizon structure of bimetric spacetimes. Class. Quant. Grav. **29**, 065009 (2012)
48. A.I. Vainshtein, To the problem of nonvanishing gravitation mass. Phys. Lett. **B39**, 393–394 (1972)
49. E. Babichev, C. Deffayet, R. Ziour, Recovering general relativity from massive gravity. Phys. Rev. Lett. **103**, 201102 (2009)
50. E. Babichev, C. Deffayet, R. Ziour, The recovery of general relativity in massive gravity via the Vainshtein mechanism. Phys. Rev. **D82**, 104008 (2010)
51. A. Gruzinov, M. Mirbabayi, Stars and black holes in massive gravity. Phys. Rev. **D84**, 124019 (2011)
52. F. Sbisa, G. Niz, K. Koyama, G. Tasinato, Characterising Vainshtein solutions in massive gravity. Phys. Rev. **D86**, 024033 (2012)
53. E. Babichev, M. Crisostomi, Restoring general relativity in massive bi-gravity theory. Phys. Rev. **D88**, 084002 (2013)
54. E. Babichev, A. Fabbri, Instability of black holes in massive gravity. Class. Quant. Grav. **30**, 152001 (2013)
55. R. Gregory, R. Laflamme, Black strings and p-branes are unstable. Phys. Rev. Lett. **70**, 2837–2840 (1993)
56. R. Brito, V. Cardoso, P. Pani, Massive spin-2 fields on black hole spacetimes: instability of the Schwarzschild and Kerr solutions and bounds on the graviton mass. Phys. Rev. **D88**(2), 023514 (2013)
57. R. Brito, V. Cardoso, P. Pani, Partially massless gravitons do not destroy general relativity black holes. Phys. Rev. **D87**(12), 124024 (2013)
58. H. Kodama, I. Arraut, Stability of the Schwarzschild-de Sitter black hole in the dRGT massive gravity theory. Progr. Theor. Exp. Phys. **2014**, 023E02 (2014)
59. E. Babichev, A. Fabbri, Stability analysis of black holes in massive gravity: a unified treatment. Phys. Rev. **D89**, 081502 (2014)

60. S. Deser, A. Waldron, Non-Einstein source effects in massive gravity. Phys. Rev. **D89**, 027503 (2014)
61. M. Mirbabayi, A. Gruzinov, Black hole discharge in massive electrodynamics and black hole disappearance in massive gravity. Phys. Rev. **D88**, 064008 (2013)
62. A. Nicolis, R. Rattazzi, E. Trincherini, The Galileon as a local modification of gravity. Phys. Rev. **D79**, 064036 (2009)
63. E. Babichev, C. Charmousis, Dressing a black hole with a time-dependent Galileon. J. High. Energy Phys. **1408**, 106 (2014)
64. L. Hui, A. Nicolis, No-hair theorem for the Galileon. Phys. Rev. Lett. **110**(24), 241104 (2013)

Chapter 7
Chern–Simons-Like Gravity Theories

Eric A. Bergshoeff, Olaf Hohm, Wout Merbis, Alasdair J. Routh, and Paul K. Townsend

Abstract A wide class of three-dimensional gravity models can be put into "Chern–Simons-like" form. We perform a Hamiltonian analysis of the general model and then specialise to Einstein-Cartan Gravity, General Massive Gravity, the recently proposed Zwei-Dreibein Gravity and a further parity violating generalisation combining the latter two.

7.1 Introduction: CS-Like Gravity Theories

In three space-time dimensions (3D), General Relativity (GR) can be interpreted as a Chern–Simons (CS) gauge theory of the 3D Poincaré, de Sitter (dS) or anti-de Sitter (AdS) group, depending on the value of the cosmological constant [1, 2]. The action is the integral of a Lagrangian three-form L constructed from the wedge products of Lorentz-vector valued one-form fields: the dreibein e^a and the dualised spin-connection ω^a. Using a notation in which the wedge product is implicit, and a "mostly plus" metric signature convention, we have

$$L = -e_a R^a + \frac{\Lambda}{6}\varepsilon^{abc} e_a e_b e_c \,, \tag{7.1}$$

E.A. Bergshoeff (✉) • W. Merbis
Centre for Theoretical Physics, University of Groningen, Nijenborgh 4, 9747 AG Groningen, The Netherlands
e-mail: e.a.bergshoeff@rug.nl; w.merbis@rug.nl

O. Hohm
Center for Theoretical Physics, Massachusetts Institute for Technology, Cambridge, MA 02139, USA
e-mail: ohohm@mit.edu

A.J. Routh • P.K. Townsend
Department of Applied and Mathematical Physics, Centre for Mathematical Sciences, University of Cambridge, Wilberforce Road, Cambridge CB3 0WA, UK
e-mail: A.J.Routh@damtp.cam.ac.uk; P.K.Townsend@damtp.cam.ac.uk

E. Papantonopoulos (ed.), *Modifications of Einstein's Theory of Gravity at Large Distances*, Lecture Notes in Physics 892, DOI 10.1007/978-3-319-10070-8_7

where R^a is the dualised Riemann 2-form:

$$R^a = d\omega^a + \frac{1}{2}\varepsilon^{abc}\omega_b\omega_c\,. \tag{7.2}$$

This action is manifestly local Lorentz invariant, in addition to its manifest invariance under diffeomorphisms, which are on-shell equivalent to local translations. The field equations are zero field strength conditions for the Poincaré or (A)dS group.

Strictly speaking, the CS gauge theory is equivalent to 3D GR only if one assumes invertibility of the dreibein; this is what allows the Einstein field equations to be written as zero field-strength conditions, and it is one way to see that 3D GR has no local degrees of freedom, and hence no gravitons. However, there are variants of 3D GR that do propagate gravitons. The simplest of these are 3D "massive gravity" theories found by including certain higher-derivative terms in the action.[1] The best known example is Topologically Massive Gravity (TMG), which includes the parity violating Lorentz–Chern–Simons term and leads to third-order field equations that propagate a single spin-2 mode [3]. A more recent example is New Massive Gravity (NMG) which includes certain curvature-squared terms; this leads to parity-preserving fourth-order equations that propagate a parity-doublet of massive spin-2 modes; combining TMG and NMG we get a parity violating fourth-order General Massive Gravity (GMG) theory that propagates two massive gravitons, but with different masses [4].

Although GMG is fourth order in derivatives, it is possible to introduce auxiliary tensor fields to get a set of equivalent first-order equations [5]; in this formulation the fields can all be taken to be Lorentz vector-valued 1-forms, and the action takes a form that is "CS-like" in the sense that it is the integral of a Lagrangian 3-form defined without an explicit space-time metric (which appears only on the assumption of an invertible dreibein). The general model of this type can be constructed as follows [5]. We start from a collection of N Lorentz-vector valued 1-forms $a^{r\,a} = a^{r\,a}_\mu dx^\mu$, where $r = 1,\ldots,N$ is a "flavour" index; the generic Lagrangian 3-form constructible from these 1-form fields is

$$L = \frac{1}{2}g_{rs}a^r\cdot da^s + \frac{1}{6}f_{rst}a^r\cdot(a^s\times a^t)\,, \tag{7.3}$$

where g_{rs} is a symmetric constant metric on the flavour space which we will require to be invertible, so it can be used to raise and lower flavour indices, and the coupling constants f_{rst} define a totally symmetric "flavour tensor". We now use a 3D-vector algebra notation for Lorentz vectors in which contractions with η_{ab} and ϵ_{abc} are represented by dots and crosses respectively. The 3-form (7.3) is a CS 3-form when the constants

[1]It is possible, at least in some cases, to take a massless limit but since "spin" is not defined for massless 3D particles, one cannot get a theory of "massless gravitons" this way, if by "graviton" we mean a particle of spin-2.

$$f^{ar}{}_{bs\;ct} \equiv \epsilon^{a}{}_{bc} f^{r}{}_{st} \quad \& \quad g_{ar\;bs} \equiv \eta_{ab} g_{rs} \tag{7.4}$$

are, respectively, the structure constants of a Lie algebra, and a group invariant symmetric tensor on this Lie algebra.[2] For example, with $N = 2$ we may choose $a_1^a = e^a$ and $a_2^a = \omega^a$, and then a choice of the flavour metric and coupling constants that ensures local Lorentz invariance will yield a CS 3-form equivalent, up to field redefinitions, to (7.1). For $N > 2$, we will continue to suppose that $a_1^a = e^a$ and $a_2^a = \omega^a$, and that the flavour metric and coupling constants are such that the action is local Lorentz invariant, but even with this restriction the generic $N > 2$ model will be only CS-like. In particular TMG has a CS-like formulation with $N = 3$ and both NMG and GMG have CS-like formulations with $N = 4$. Since these models have local degrees of freedom they are strictly CS-like, and not CS models.

The generic $N = 4$ CS-like gravity model also includes the recently analysed Zwei-Dreibein Gravity (ZDG) [6]. This is a parity preserving massive gravity model with the same local degrees of freedom as NMG (two propagating spin-2 modes of equal mass in a maximally-symmetric vacuum background) but has advantages in the context of the AdS/CFT correspondence since, in contrast to NMG, it leads to a positive central charge for a possible dual CFT at the AdS boundary.

We focus here on the Hamiltonian formulation of CS-like gravity models for a number of reasons. One is that the CS-like formulation allows us to discuss various 3D massive gravity models as special cases of a generic model, and this formulation is well-adapted to a Hamiltonian analysis. Another is that there are some unusual features of the Hamiltonian formulation of massive gravity models that are clarified by the CS-like formalism. One great advantage of the Hamiltonian approach is that it allows a determination of the number of local degrees of freedom independently of a linearised approximation (which can give misleading results). In particular, massive gravity models typically have an additional local degree of freedom, the Boulware-Deser ghost [7]. It is known that GMG has no Boulware-Deser ghost, and this is confirmed by its Hamiltonian analysis, but ZDG does have a Boulware-Deser ghost for generic parameters [8], even though it is ghost-free in a linearised approximation. Fortunately, this problem can be avoided by assuming invertibility of a linear combination of the two dreibeine of ZDG. A special case of this assumption imposes a restriction of the parameters; this point was made in an erratum to [6] and here we present a detailed substantiation of it. We also present a parity violating CS-like extension of ZDG, which we call "General Zwei-Dreibein Gravity" (GZDG), and we show that it has the same number of local degrees of freedom as ZDG.

[2] There are CS gauge theories for which the Lagrangian 3-form is *not* of the form (7.3) because not all of the generators of the Lie algebra of the gauge group are Lorentz vectors. If we wish the class of CS gravity theories to be a subclass of the class of CS-like gravity theories, we should define the latter by a larger class of 3-form Lagrangians, as in [5], but (7.3) will be sufficient for our purposes.

7.2 Hamiltonian Analysis

It is straightforward to put the CS-like model defined by (7.3) into Hamiltonian form. We perform the space-time split

$$a^{r\,a} = a_0^{r\,a} dt + a_i^{r\,a} dx^i \,, \tag{7.5}$$

which leads to the Lagrangian density

$$\mathscr{L} = -\frac{1}{2}\varepsilon^{ij} g_{rs} a_i^r \cdot \dot{a}_j^s + a_0^r \cdot \phi_r \,, \tag{7.6}$$

where $\varepsilon^{ij} = \varepsilon^{0ij}$. The time components of the fields, $a_0^{r\,a}$, become Lagrange multipliers for the primary constraints ϕ_r^a:

$$\phi_r^a = \varepsilon^{ij}\left(g_{rs}\partial_i a_j^{s\,a} + \frac{1}{2} f_{rst}\left(a_i{}^s \times a_j{}^t\right)^a \right) . \tag{7.7}$$

The Hamiltonian density is just the sum of the primary constraints, each with a Lagrange multiplier $a_0^{r\,a}$,

$$\mathscr{H} = -\frac{1}{2}\varepsilon^{ij} g_{rs} a_i^r \cdot \partial_0 a_j^s - \mathscr{L} = -a_0{}^r \cdot \phi_r \,. \tag{7.8}$$

We must now work out the Poisson brackets of the primary constraints. Then, following Dirac's procedure [9], we must consider any secondary constraints. We consider these two steps in turn.

7.2.1 Poisson Brackets and the Primary Constraints

The Lagrangian is first order in time derivatives, so the Poisson brackets of the canonical variables can be determined by inverting the first term of (7.6); this gives

$$\left\{ a_{i\,a}^r(x), a_{j\,b}^s(y) \right\}_{\text{P.B.}} = \varepsilon_{ij} g^{rs} \eta_{ab} \delta^{(2)}(x-y) \,. \tag{7.9}$$

Using this result we may calculate the Poisson brackets of the primary constraint functions. It will be convenient to first define the "smeared" functions $\phi[\xi]$ associated to the constraints (7.7) by integrating them against a test function $\xi_a^r(x)$ as follows

$$\phi[\xi] = \int_\Sigma d^2x \; \xi_a^r(x) \phi_r^a(x) \,, \tag{7.10}$$

where Σ is space-like hypersurface. In general, the functionals $\phi[\xi]$ will not be differentiable, but we can make them so by adding boundary terms. Varying (7.10) with respect to the fields $a_i{}^s$ gives

$$\delta\phi[\xi] = \int_\Sigma d^2x\, \xi^r_a \frac{\delta\phi^a_r}{\delta a^{s\,b}_i}\delta a_i{}^{s\,b} + \int_{\partial\Sigma} dx\, B[\xi, a, \delta a]\,. \tag{7.11}$$

A non-zero $B[\xi, a, \delta a]$ could lead to delta-function singularities in the brackets of the constraint functions. To remove these, we can choose boundary conditions which make B a total variation

$$\int_{\partial\Sigma} dx\, B[\xi, a, \delta a] = -\delta Q[\xi, a]\,. \tag{7.12}$$

We then work with the quantities

$$\varphi[\xi] = \phi[\xi] + Q[\xi, a]\,, \tag{7.13}$$

which have well-defined variations, with no boundary terms. In our case, after varying $\phi[\xi]$ with respect to the fields $a_i{}^s$, we find

$$B[\xi, a, \delta a] = \int_{\partial\Sigma} dx^i\, \xi^r_a g_{rs} \delta a_i{}^{s\,a}\,. \tag{7.14}$$

The Poisson brackets of the constraint functions can now be computed using Eq. (7.9). They are given by

$$\begin{aligned}\{\varphi(\xi), \varphi(\eta)\}_{\text{P.B.}} &= \varphi([\xi, \eta]) + \int_\Sigma d^2x\, \xi^r_a \eta^s_b\, \mathscr{P}^{ab}_{rs} \\ &\quad - \int_{\partial\Sigma} dx^i\, \xi^r \cdot \left[g_{rs}\partial_i \eta^s + f_{rst}(a_i{}^s \times \eta^t)\right],\end{aligned} \tag{7.15}$$

where

$$[\xi, \eta]^t_c = f_{rs}{}^t \epsilon^{ab}{}_c \xi^r_a \eta^s_b\,, \tag{7.16}$$

and

$$\mathscr{P}^{ab}_{rs} = f^t{}_{q[r} f_{s]pt} \eta^{ab} \Delta^{pq} + 2 f^t{}_{r[s} f_{q]pt} (V^{ab})^{pq}\,, \tag{7.17}$$

$$V^{pq}_{ab} = \varepsilon^{ij} a^p_{i\,a} a^q_{j\,b}\,, \qquad \Delta^{pq} = \varepsilon^{ij} a^p_i \cdot a^q_j\,. \tag{7.18}$$

In general, adopting non-trivial boundary conditions may lead to a (centrally extended) asymptotic symmetry algebra spanned by the first-class constraint functions if the corresponding test functions $\xi^r_a(x)$ are the gauge parameters of boundary condition preserving gauge transformations. Here we will focus on the bulk theory and assume that the test functions $\xi^r_a(x)$ do not give rise to boundary terms in (7.11) and (7.15).

The consistency conditions guaranteeing time-independence of the primary constraints are

$$\frac{d}{dt}\phi^b_s = \{\mathscr{H}, \phi^b_s\}_{\text{P.B.}} \approx -a^r_{0a}\mathscr{P}^{ab}_{rs} \approx 0\,. \tag{7.19}$$

This expression is equivalent to a set of "integrability conditions" which can be derived from the equations of motion. The field equations of (7.3) are

$$g_{rs}da^{s\,a} + \frac{1}{2}f_{rst}(a^s \times a^t)^a = 0\,. \tag{7.20}$$

Taking the exterior derivative of this equation and using $d^2 = 0$, we get the conditions

$$f^t{}_{q[r}f_{s]pt}a^{r\,a}a^p \cdot a^q = 0\,. \tag{7.21}$$

Using the space-time decomposition (7.5) we have

$$0 = f^t{}_{q[r}f_{s]pt}a^{r\,b}a^p \cdot a^q = a^r_{0a}\mathscr{P}^{ab}_{rs}\,, \tag{7.22}$$

the right hand side being exactly what is required to vanish for time-independence of the primary constraints. These conditions are 3-form equations in which each 3-form necessarily contains a Lagrange multiplier one-form factor, so they could imply that some linear combinations of the Lagrange multipliers is zero.

If the matrix $\mathscr{P}^{ab}_{rs}$ vanishes identically then all primary constraints are first-class and there is no constraint on any Lagrange multiplier. In this case the model is actually a Chern–Simons theory, that of the Lie algebra with structure constants $\epsilon^a{}_{bc}f^r{}_{st}$. In general, however, $\mathscr{P}^{ab}_{rs}$ will not vanish and rank($\mathscr{P}$) will be non-zero. We can pick a basis of constraint functions such that $3N$ − rank($\mathscr{P}$) have zero Poisson bracket with all constraints, while the remaining rank($\mathscr{P}$) constraint functions have non-zero Poisson brackets amongst themselves. At this point, it might appear that the Lagrange multipliers for the latter set of constraints will be set to zero by the conditions (7.22). However, when one of the fields is a dreibein, this may involve setting $e_0{}^a = 0$. This is not acceptable for a theory of gravity, as the dreibein must be invertible! When specifying a model, we must therefore be clear whether we are assuming invertibility of any fields as it affects the Hamiltonian analysis. In general, if we require invertibility of any one-form field then we may need to impose further, secondary, constraints.

In other words, the consistency of the primary constraints is equivalent to satisfying the integrability conditions (7.22). If some one-form is invertible, then some integrability condition may reduce to a two-form constraint on the canonical variables, which we must add as a secondary constraint in our theory. We now turn to an investigation of these secondary constraints.

7.2.2 Secondary Constraints

To be precise, consider for fixed s the expression $f^t{}_{q[r}f_{s]pt}a^{r\,a}$. If the sum over r is non-zero for only one value of r, say $a^{ra} = h^a$, and h^a is invertible, then the integrability conditions (7.21) imply that

$$f^t{}_{q[r}f_{s]pt}a^p \cdot a^q = 0\,. \tag{7.23}$$

In particular, taking the space-space part of this two-form, we find

$$\varepsilon^{ij} f^t{}_{q[r}f_{s]pt}a_i^p \cdot a_j^q = 0\,, \tag{7.24}$$

which depends only on the canonical variables and is therefore a new, secondary, constraint. One invertible field may lead to several constraints if the above equation holds for multiple values of s. The secondary constraints arising in this way[3] are therefore the inequivalent components of the field space vector $\psi_s = f^t{}_{q[r}f_{s]pt}\Delta^{pq}$. Let M be the number of these secondary constraints, and let us write them as

$$\psi_I = f_{I,pq}\Delta^{pq}\,, \qquad I = 1,\ldots,M\,. \tag{7.25}$$

We now have a total of $3N + M$ constraints.

According to Dirac, after finding the secondary constraints we should add them to the Hamiltonian with new Lagrange multipliers [9]. However, in general this can change the field equations. To see why let us suppose that we have a phase-space action $I[z]$ for some phase space coordinates z, and that the equations of motion imply the constraint $\phi(z) = 0$. If we add this constraint to the action with a Lagrange multiplier λ then we get a new action for which the equations of motion are

$$\frac{\delta I}{\delta z} = \lambda\frac{\partial\phi}{\partial z}\,, \qquad \phi(z) = 0\,. \tag{7.26}$$

Any solution of the original equations of motions, together with $\lambda = 0$, solves these equations, but there may be more solutions for which $\lambda \neq 0$. This is precisely what happens for NMG and GMG (although not for TMG) [5]; the field equations of these models lead to a (field-dependent) cubic equation for one of the secondary constraint Lagrange multipliers, leading to two possible non-zero solutions for this Lagrange multiplier.[4] In this case, Dirac's procedure would appear to lead us to a Hamiltonian

[3]Here we should issue a warning: a linear combination of invertible one-forms is not in general invertible, so if $f^t{}_{q[r}f_{s]pt}a^{r\,a}$ sums over multiple values of r with each corresponding one-form invertible, this does not in general imply a new constraint.

[4]This problem appears to be distinct from the problem of whether the "Dirac conjecture" is satisfied, since that concerns the values of Lagrange multipliers of first-class constraints. It may be related to the recently discussed "sectors" issue [10].

formulation of a theory that is more general than the one we started with (in that its solution space is larger). Perhaps more seriously, adding the secondary constraints to the Hamiltonian will generally lead to a violation of symmetries of the original model.

Because of this problem, we will omit the secondary constraints from the total Hamiltonian. This omission could lead to difficulties. The first-class constraints are found by consideration of the matrix of Poisson brackets of *all* constraints, so it could happen that some are linear combinations of primary with secondary constraints. We would then have a situation in which not all first-class constraints are imposed by Lagrange multipliers in the (now restricted) total Hamiltonian, and this would appear to lead to inconsistencies. Fortunately, this problem does not actually arise for any of the CS-like gravity models that we shall consider, as they satisfy conditions that we now spell out.

The Poisson brackets of the primary with the secondary constraint functions are

$$\{\phi[\xi], \psi_I\}_{\text{P.B.}} = \varepsilon^{ij}\left[f_{I,rp}\partial_i(\xi^r)\cdot a_j^p + f_{rs}{}^t f_{I,pt}\xi^r \cdot \left(a_i^s \times a_j^p\right)\right], \tag{7.27}$$

and the Poisson brackets of the secondary constraint functions amongst themselves are

$$\{\psi_I, \psi_J\}_{\text{P.B.}} = 4 f_{I,pq} f_{J,rs} \Delta^{pr} g^{qs}. \tag{7.28}$$

We now make the following two assumptions, *which hold for all our examples*:

- We assume that all Poisson brackets of secondary constraints with other secondary constraints vanish on the full constraint surface. It then follows that the total matrix of Poisson brackets of all $3N + M$ constraint functions takes the form

$$\mathbb{P} = \begin{pmatrix} \mathscr{P}' & -\{\phi, \psi\}^T \\ \{\phi, \psi\} & 0 \end{pmatrix}, \tag{7.29}$$

  where $\mathscr{P}'$ is the matrix of Poisson brackets of the $3N$ primary constraints evaluated on the new constraint surface defined by all $3N + M$ constraints.
- We assume that inclusion of the secondary constraints in the set of all constraints does not lead to new first-class constraints. This means that the secondary constraints must all be second-class, and any linear combination of secondary constraints and the rank($\mathscr{P}'$) primary constraints with non-vanishing Poisson brackets on the full constraint surface must be second-class.

The rank of $\mathbb{P}$, as given in (7.29), is the number of its linearly independent columns. By the second assumption, this is M plus the number of linearly independent columns of

$$\begin{pmatrix} \mathscr{P}' \\ \{\phi, \psi\} \end{pmatrix} . \tag{7.30}$$

The number of linearly independent columns of this matrix, as for any other matrix, is the same as the number of linearly independent rows, which by the second assumption is rank($\mathscr{P}'$) $+ M$. The rank of $\mathbb{P}$, and therefore the number of second-class constraints, is then rank($\mathscr{P}'$) $+ 2M$.

In principle one should now check for tertiary constraints. However, in this procedure the invertibility of certain fields will be guaranteed by the secondary constraints. The consistency of the primary constraints under time evolution can be guaranteed by fixing rank($\mathscr{P}'$) of the Lagrange multipliers. The consistency of the secondary constraints under time evolution, $a^r_{0a}\{\phi^a_r, \psi_I\} \approx 0$, can be guaranteed, under the second assumption, by fixing a further M of the Lagrange multipliers. The fact that these M multipliers are distinct from the rank($\mathscr{P}'$) multiplier fixed before follows from the second assumption. The remaining consistency condition, $\{\psi, \psi\} \approx 0$, is guaranteed by the first assumption.

We therefore have $3N$ - rank($\mathscr{P}'$) - M undetermined Lagrange multipliers, corresponding to the $3N$ - rank($\mathscr{P}'$) - M first-class constraints. The remaining rank($\mathscr{P}'$) $+ 2M$ constraints are second-class. The dimension of the physical phase space per space point is the number of canonical variables a^{ra}_i, minus twice the number of first-class constraints, minus the number of second-class constraints, or

$$D = 6N - 2 \times \big(3N - \text{rank}(\mathscr{P}') - M\big) - 1 \times \big(\text{rank}(\mathscr{P}') + 2M\big) = \text{rank}(\mathscr{P}') . \tag{7.31}$$

We will now apply this procedure to determine the number of local degrees of freedom of various 3D gravity models with a CS-like formulation.

7.3 Specific Examples

We will now derive the Hamiltonian form of a number of three-dimensional CS-like gravity models of increasing complexity following the above general procedure.

7.3.1 Einstein-Cartan Gravity

To illustrate our formalism we will start by using it to analyse 3D General Relativity with a cosmological constant Λ, in its first-order Einstein-Cartan form. There are two flavours of one-forms: the dreibein, $a^{e\,a} = e^a$, and the dualised spin-connection $a^{\omega\,a} = \omega^a = \frac{1}{2}\varepsilon^{abc}\omega_{bc}$. The Lagrangian 3-form is that of (7.1). This takes the general form of (7.3), with the flavour index $r, s, t, \ldots = \omega, e$. The first step is to read off g_{rs} and f_{rst}, and for later convenience we also determine the components of the inverse metric g^{rs} and the structure constants with one index raised, $f^r{}_{st}$. The non-zero components of these quantities are:

$$g_{\omega e} = -1\,, \qquad f_{eee} = \Lambda\,, \qquad f_{e\omega\omega} = -1\,, \qquad (7.32)$$
$$g^{\omega e} = -1\,, \qquad f^{\omega}{}_{ee} = -\Lambda\,, \qquad f^{\omega}{}_{\omega\omega} = 1\,, \qquad f^{e}{}_{e\omega} = 1\,.$$

These constants define a Chern–Simons 3-form, as mentioned in the introduction; the structure constants are $\varepsilon^a{}_{bc} f^r{}_{st}$. This algebra is spanned by the Hamiltonian constraint functions, which are all first-class. In three-dimensions, General Relativity, like any Chern–Simons theory, has no local degrees of freedom.

To see how the details work in our general formalism, we can work out the matrix (7.17) and find that it vanishes. Then, by equation (7.31) the dimension of the physical phase space is

$$D = 12 - 2 \times 6 = 0\,, \qquad (7.33)$$

as expected. Using (7.15) we can also verify that

$$\{\phi^a_\omega, \phi^b_\omega\}_{\text{P.B.}} = \epsilon^{ab}{}_c\, \phi^c_\omega\,, \quad \{\phi^a_e, \phi^b_\omega\}_{\text{P.B.}} = \epsilon^{ab}{}_c\, \phi^c_e\,, \quad \{\phi^a_e, \phi^b_e\}_{\text{P.B.}} = -\Lambda\epsilon^{ab}{}_c\, \phi^c_\omega\,,$$

which is the $SO(2,2)$ algebra for $\Lambda < 0$, $SO(3,1)$ for $\Lambda > 0$ and $ISO(2,1)$ for $\Lambda = 0$, as expected.

7.3.2 General Massive Gravity

General Relativity was a very simple application of our general formalism; as a Chern–Simons theory the Poisson brackets of the constraint functions formed a closed algebra, so it did not require our full analysis. We will now turn to a more complicated example, General Massive Gravity (GMG). This theory does have local degrees of freedom; it propagates two massive spin-2 modes at the linear level. It contains two well known theories of 3D massive gravity as limits: Topologically Massive Gravity (TMG) [3] and New Massive Gravity (NMG) [4].

We can write the Lagrangian 3-form of GMG in the general form (7.3). There are four flavours of one-form, $a^{r\,a} = (\omega^a, h^a, e^a, f^a)$, the dualised spin-connection and dreibein and two extra fields h^a and f^a, and the Lagrangian 3-form is

$$L_{\text{GMG}} = -\,\sigma e_a R^a + \frac{\Lambda_0}{6}\epsilon^{abc} e_a e_b e_c + h_a T^a + \frac{1}{2\mu}\left[\omega_a d\omega^a + \frac{1}{3}\epsilon^{abc}\omega_a\omega_b\omega_c\right]$$
$$- \frac{1}{m^2}\left[f_a R^a + \frac{1}{2}\epsilon^{abc} e_a f_b f_c\right], \qquad (7.34)$$

where we recall that R^a is the dualised Riemann 2-form. The flavour-space metric g_{rs} and the structure constants $f^r{}_{st}$ can again be read off:

$$\begin{aligned}
g_{\omega e} &= -\sigma\,, & g_{eh} &= 1\,, & g_{f\omega} &= -\frac{1}{m^2}\,, & g_{\omega\omega} &= \frac{1}{\mu}\,,\\
f_{e\omega\omega} &= -\sigma\,, & f_{eh\omega} &= 1\,, & f_{eff} &= -\frac{1}{m^2}\,, & f_{\omega\omega\omega} &= \frac{1}{\mu}\,,\\
f_{eee} &= \Lambda_0\,, & & & f_{\omega\omega f} &= -\frac{1}{m^2}\,. & &
\end{aligned} \tag{7.35}$$

The next step is to work out the integrability conditions (7.21). We find three inequivalent 3-form relations,

$$e^a e\cdot f = 0\,,\quad f^a\left(\frac{1}{\mu}e\cdot f + h\cdot e\right) - h^a e\cdot f = 0\,,\quad e^a\left(\frac{1}{\mu}e\cdot f + h\cdot e\right) = 0\,. \tag{7.36}$$

We will demand that the dreibein, e^a, is invertible. Following our general analysis, we find the two secondary constraints

$$\psi_1 = \Delta^{eh} = 0\,,\qquad \psi_2 = \Delta^{ef} = 0\,. \tag{7.37}$$

Next, we compute the matrix $\mathscr{P}^{ab}_{rs}$ as defined in (7.17). All the Δ^{pq} terms drop out because of the secondary constraints, and in the basis (ω, h, e, f) we get

$$(\mathscr{P}'_{ab})_{rs} = \begin{pmatrix} 0 & 0 & 0 & 0\\ 0 & 0 & V^{ef}_{ab} & -V^{ee}_{ab}\\ 0 & V^{fe}_{ab} & -2V^{hf}_{[ab]} + \frac{1}{\mu}V^{ff}_{ab} & V^{he}_{ab} - \frac{1}{\mu}V^{fe}_{ab}\\ 0 & -V^{ee}_{ab} & V^{eh}_{ab} - \frac{1}{\mu}V^{ef}_{ab} & \frac{1}{\mu}V^{ee}_{ab} \end{pmatrix}. \tag{7.38}$$

We must now determine the rank of this matrix at an arbitrary point in space-time. A Mathematica calculation shows that the rank of $\mathscr{P}'$ is 4. To complete the analysis and verify if the two assumptions stated in Sect. 7.2.2 are met, we need the Poisson brackets of the secondary constraint functions ψ_I $(I = 1, 2)$ with themselves and with the primary constraint functions. The Poisson bracket $\{\psi_1, \psi_2\}$ is zero on the constraint surface, which verifies the first assumption, and the Poisson brackets of ψ_I with the primary constraint functions are

$$\begin{aligned}
\{\phi[\xi], \psi_1\}_{\text{P.B.}} = \epsilon^{ij}\Big[&\partial_i\xi^h\cdot e_j - \xi^h\cdot(\omega_i\times e_j) - \partial_i\xi^e\cdot h_j + \xi^e\cdot(\omega_i\times h_j)\\
&+\left(\sigma\xi^e + \frac{1}{m^2}\xi^f\right)\cdot(e_i\times f_j) + \left(\sigma\xi^f + \Lambda_0\xi^e\right)(e_i\times e_j)\Big]\,,
\end{aligned} \tag{7.39}$$

$$\{\phi[\xi], \psi_2\}_{\text{P.B.}} = \epsilon^{ij}\Bigg[\partial_i \xi^f \cdot e_j - \xi^f \cdot (\omega_i \times e_j) - \partial_i \xi^e \cdot f_j + \xi^e \cdot (\omega_i \times f_j) + \left(m^2 \xi^h - \frac{m^2}{\mu}\xi^f\right)(e_i \times e_j) + m^2 \xi^e \cdot \left(e_i \times \left(h_j - \frac{1}{\mu} f_j\right)\right)\Bigg]. \tag{7.40}$$

The full matrix of Poisson brackets is a 14×14 matrix $\mathbb{P}$ given by

$$\mathbb{P} = \begin{pmatrix} \mathscr{P}' & v_1 & v_2 \\ -v_1^T & 0 & 0 \\ -v_2^T & 0 & 0 \end{pmatrix}, \tag{7.41}$$

where the v_I, $(I = 1, 2)$, are column vectors with components

$$v_I = \begin{pmatrix} \{\phi^a_\omega, \psi_I\}_{\text{P.B.}} \\ \{\phi^a_h, \psi_I\}_{\text{P.B.}} \\ \{\phi^a_e, \psi_I\}_{\text{P.B.}} \\ \{\phi^a_f, \psi_I\}_{\text{P.B.}} \end{pmatrix}. \tag{7.42}$$

These brackets can be read off from Eqs. (7.39) and (7.40). The vectors (7.42) are linearly independent from each other and from the columns of $\mathscr{P}'$, which verifies the second assumption of Sect. 7.2.2 and the rank of $\mathbb{P}$ is increased by 4. The full (14×14) matrix therefore has rank 8, so eight constraints are second-class and the remaining six are first-class. By Eq. (7.31), the dimension of the physical phase space per space point is

$$D = 24 - 8 - 2 \times 6 = 4. \tag{7.43}$$

This means there are two local degrees of freedom, and we conclude that the non-linear theory has the same degrees of freedom as the linearised theory, two massive states of helicity ± 2.

7.3.3 Zwei-Dreibein Gravity

We now turn our attention to another theory of massive 3D gravity, the recently proposed Zwei-Dreibein Gravity (ZDG) [6]. This is a theory of two interacting dreibeine, e_1^a and e_2^a, each with a corresponding spin-connection, $\omega_1{}^a$ and $\omega_2{}^a$. It also has a Lagrangian 3-form of our general CS-like form (7.3). Like NMG, ZDG preserves parity and has two massive spin-2 degrees of freedom when linearised about a maximally-symmetric vacuum background, but this does not exclude the possibility of additional local degrees of freedom appearing in other backgrounds. In fact, it was shown by Bañados et al. [8] that the generic ZDG model does have an additional local degree of freedom, the Boulware-Deser ghost. We will see why this is so, and also how it can be removed by assuming invertibility of a special linear combination of the two dreibein.

The Lagrangian 3-form is

$$L_{\mathrm{ZDG}} = -M_P\Big\{\sigma e_{1a}R_1{}^a + e_{2a}R_2{}^a + \frac{m^2}{6}\epsilon^{abc}\left(\alpha_1 e_{1a}e_{1b}e_{1c} + \alpha_2 e_{2a}e_{2b}e_{2c}\right) \\ -L_{12}(e_1,e_2)\Big\}\,, \tag{7.44}$$

where $R_1{}^a$ and $R_2{}^a$ are the dualised Riemann 2-forms constructed from $\omega_1{}^a$ and $\omega_2{}^a$ respectively, and the interaction Lagrangian 3-form L_{12} is given by

$$L_{12}(e_1,e_2) = \frac{1}{2}m^2\epsilon^{abc}\left(\beta_1 e_{1a}e_{1b}e_{2c} + \beta_2 e_{1a}e_{2b}e_{2c}\right)\,. \tag{7.45}$$

Here $\sigma = \pm 1$ is a sign parameter, α_1 and α_2 are two dimensionless cosmological parameters and β_1 and β_2 are two dimensionless coupling constants. The parameter m^2 is a redundant, but convenient, dimensionful parameter. For now these parameters are arbitrary, but we will soon see that some restrictions are necessary.

From (7.44) we can read off the components of g_{rs} and f_{rst}. We will ignore the overall factor M_P to simplify the analysis; after this step they become

$$\begin{aligned} g_{e_1\omega_1} = g_{\omega_1 e_1} &= -\sigma\,,\quad g_{e_2\omega_2} = g_{\omega_2 e_2} = -1\,,\\ f_{e_1\omega_1\omega_1} &= -\sigma\,,\quad f_{e_2\omega_2\omega_2} = -1\,,\\ f_{e_1e_1e_2} &= m^2\beta_1\,,\quad f_{e_1e_2e_2} = m^2\beta_2\,,\\ f_{e_1e_1e_1} &= -m^2\alpha_1\,,\quad f_{e_2e_2e_2} = -m^2\alpha_2\,. \end{aligned} \tag{7.46}$$

We also work out the inverse metric g^{rs} and the structure constants $f^r{}_{st}$,

$$\begin{aligned} g^{e_1\omega_1} &= g^{\omega_1 e_1} = -\frac{1}{\sigma}\,, g^{e_2\omega_2} = g^{\omega_2 e_2} = -1\,,\\ f^{\omega_1}{}_{\omega_1\omega_1} &= f^{e_1}{}_{\omega_1 e_1} = 1\,, f^{\omega_2}{}_{\omega_2\omega_2} = f^{e_2}{}_{\omega_2 e_2} = 1\,,\\ f^{\omega_1}{}_{e_1e_2} &= f^{\omega_1}{}_{e_2e_1} = -\frac{m^2\beta_1}{\sigma}\,, f^{\omega_1}{}_{e_2e_2} = -\frac{m^2\beta_2}{\sigma}\,,\\ f^{\omega_1}{}_{e_1e_1} &= \frac{m^2}{\sigma}\alpha_1\,, f^{\omega_2}{}_{e_2e_2} = m^2\alpha_2\,,\\ f^{\omega_2}{}_{e_1e_2} &= f^{\omega_2}{}_{e_2e_1} = -m^2\beta_2\,, f^{\omega_2}{}_{e_1e_1} = -m^2\beta_1\,. \end{aligned} \tag{7.47}$$

Equipped with these expressions, we can evaluate the 12×12 matrix of Poisson brackets (7.15), in the flavour basis $(\omega_1,\omega_2,e_1,e_2)$

$$
\begin{aligned}
(\mathscr{P}_{ab})_{rs} = m^2\eta_{ab} & \begin{pmatrix} 0 & 0 & -\beta_1\Delta^{e_1e_2} & -\beta_2\Delta^{e_1e_2} \\ 0 & 0 & \beta_1\Delta^{e_1e_2} & \beta_2\Delta^{e_1e_2} \\ \beta_1\Delta^{e_1e_2} & -\beta_1\Delta^{e_1e_2} & 0 & -\beta_1\Delta^{\omega-e_1} - \beta_2\Delta^{\omega-e_2} \\ \beta_2\Delta^{e_1e_2} & -\beta_2\Delta^{e_1e_2} & \beta_1\Delta^{\omega-e_1} + \beta_2\Delta^{\omega-e_2} & 0 \end{pmatrix} \\
+ m^2\beta_1 & \begin{pmatrix} 0 & 0 & V_{ab}^{e_1e_2} & -V_{ab}^{e_1e_1} \\ 0 & 0 & -V_{ab}^{e_1e_2} & V_{ab}^{e_1e_1} \\ V_{ab}^{e_2e_1} & -V_{ab}^{e_2e_1} & -(V_{[ab]}^{\omega_1e_2} - V_{[ab]}^{\omega_2e_2}) & (V_{ab}^{\omega_1e_1} - V_{ab}^{\omega_2e_1}) \\ -V_{ab}^{e_1e_1} & V_{ab}^{e_1e_1} & (V_{ab}^{e_1\omega_1} - V_{ab}^{e_1\omega_2}) & 0 \end{pmatrix} \\
+ m^2\beta_2 & \begin{pmatrix} 0 & 0 & V_{ab}^{e_2e_2} & -V_{ab}^{e_2e_1} \\ 0 & 0 & -V_{ab}^{e_2e_2} & V_{ab}^{e_2e_1} \\ V_{ab}^{e_2e_2} & -V_{ab}^{e_2e_2} & 0 & -(V_{ab}^{e_2\omega_1} - V_{ab}^{e_2\omega_2}) \\ -V_{ab}^{e_1e_2} & V_{ab}^{e_1e_2} & -(V_{ab}^{\omega_1e_2} - V_{ab}^{\omega_2e_2}) & (V_{[ab]}^{\omega_1e_1} - V_{[ab]}^{\omega_2e_1}) \end{pmatrix} .
\end{aligned}
\tag{7.48}
$$

Where $\omega_- \equiv \omega_1 - \omega_2$. We determine the rank of this matrix as before using Mathematica, and find it to be 6. This means that there are $12 - 6 = 6$ gauge symmetries in the theory.

To find the secondary constraints we must study the integrability conditions (7.21). There are three independent equations

$$(\beta_1 e_1{}^a + \beta_2 e_2{}^a) e_1 \cdot e_2 = 0\,, \tag{7.49}$$

$$e_2{}^a \omega_- \cdot (\beta_1 e_1 + \beta_2 e_2) - \beta_1 \omega_-{}^a e_1 \cdot e_2 = 0\,, \tag{7.50}$$

$$e_1{}^a \omega_- \cdot (\beta_1 e_1 + \beta_2 e_2) + \beta_2 \omega_-{}^a e_1 \cdot e_2 = 0\,. \tag{7.51}$$

Assuming invertibility of both dreibeine, e_1^a and e_2^a, is not sufficient to generate a secondary constraint; from (7.49) we need that $(\beta_1 e_1{}^a + \beta_2 e_2{}^a)$ is invertible. This does not follow from the invertibility of the two separate dreibeine. Without any secondary constraints, the dimension of the physical phase space, using Eq. (7.31), is 6. This corresponds to 3 local degrees of freedom, one massive graviton and the other presumably a scalar ghost.

We are interested in theories of massive gravity without ghosts, so we must restrict our general model to ensure secondary constraints. By analysing (7.49)-(7.51) we see that to derive two secondary constraints, we should assume the invertibility of the linear combination $\beta_1 e_1{}^a + \beta_2 e_2{}^a$. A special case of this assumption, where the ZDG parameter space is restricted to $\beta_1\beta_2 = 0$, but one of the β_i is non-zero and the corresponding dreibein is assumed to be invertible, was considered in an erratum to [6]. We will first analyse this special case in more detail and then move to the generic case.

7.3.3.1 The Case $\beta_1\beta_2 = 0$

In the case that we set to zero one of the two parameters β_i we may choose, without loss of generality, to set

$$\beta_2 = 0\,. \tag{7.52}$$

In this case the invertibility of ${e_1}^a$ alone implies the two secondary constraints.

$$\psi_1 = \Delta^{e_1 e_2} = 0\,, \qquad \psi_2 = \Delta^{\omega - e_1} = 0\,. \tag{7.53}$$

These constraints and parameter choices cause the first and last matrices in Eq. (7.48) to vanish. The remaining matrix $\mathscr{P}'$ has rank 4.

The secondary constraints (7.53) are in involution with each other, and their brackets with the primary constraint functions are given by

$$\begin{aligned}\{\phi[\xi], \psi_1\}_{\text{P.B.}} = \varepsilon^{ij}\Big[&\partial_i \xi^{e_1}\cdot e_{2\,j} - \xi^{e_1}\cdot \omega_{1\,i}\times e_{2\,j} - \partial_i\xi^{e_2}\cdot e_{1\,j} + \xi^{e_2}\cdot\omega_{2\,i}\times e_{1\,j}\\ &- (\xi^{\omega_1} - \xi^{\omega_2})\cdot e_{1\,i}\times e_{2\,j}\Big]\,,\end{aligned} \tag{7.54}$$

and

$$\begin{aligned}\{\phi[\xi], \psi_2\}_{\text{P.B.}} =\varepsilon^{ij}\Big[&(\partial_i\xi^{\omega_1} - \partial_i\xi^{\omega_2})\cdot e_{1\,j} - (\xi^{\omega_1} - \xi^{\omega_2})\cdot(\omega_{2\,i}\times e_{1\,j}) - \partial_i\xi^{e_1}\cdot\omega_{-\,j}\\ &+ \xi^{e_1}\cdot(\omega_{1\,i}\times\omega_{-\,j}) + m^2\,(\sigma\beta_1\xi^{e_1} + \alpha_2\xi^{e_2})\cdot(e_{1\,i}\times e_{2\,j})\\ &- m^2\,((\sigma\alpha_1 + \beta_1)\xi^{e_1} - \sigma\beta_1\xi^{e_2})\cdot(e_{1\,i}\times e_{1\,j})\Big]\,.\end{aligned} \tag{7.55}$$

The full matrix of Poisson brackets is again a 14×14 matrix $\mathbb{P}$ given by (7.41), where the v_I with $I = 1, 2$ are now

$$v_I = \begin{pmatrix} \{\phi^a_{\omega_1}, \psi_I\}_{\text{P.B.}} \\ \{\phi^a_{\omega_2}, \psi_I\}_{\text{P.B.}} \\ \{\phi^a_{e_1}, \psi_I\}_{\text{P.B.}} \\ \{\phi^a_{e_2}, \psi_I\}_{\text{P.B.}} \end{pmatrix}. \tag{7.56}$$

These brackets can be read off from equations (7.54) and (7.55). The vectors (7.56) are linearly independent from each other and with the columns of $\mathbb{P}$, so this increases the rank of $\mathbb{P}$ by 4. The total number of second-class constraints is 8, leaving 6 first-class constraints. Using (7.31) we find that for general values of the parameters

α_1, α_2 and β_1 the dimension of the physical phase space per space point is 4. This corresponds to the 2 local degrees of freedom of a massive graviton.

7.3.3.2 The Case of Invertible $\beta_1 e_1{}^a + \beta_2 e_2{}^a$

The more general case is to assume invertibility of the linear combination of the two dreibeine, $\beta_1 e_1{}^a + \beta_2 e_2{}^a$. In this case, to keep track of the invertible field, we make a field redefinition in the original Lagrangian (7.44). We define, for $\beta_1 + \sigma\beta_2 \neq 0$,

$$e^a = \frac{2}{\beta_1 + \sigma\beta_2}\left(\beta_1 e_1{}^a + \beta_2 e_2{}^a\right), \qquad f^a = \sigma e_1{}^a - e_2{}^a . \tag{7.57}$$

For convenience we will work with the sum and difference of the spin connections[5]

$$\omega^a = \frac{1}{2}\left(\omega_1{}^a + \omega_2{}^a\right), \qquad h^a = \frac{1}{2}\left(\omega_1{}^a - \omega_2{}^a\right). \tag{7.58}$$

In terms of these new fields, the ZDG Lagrangian 3-form is

$$\begin{aligned} L = &- M_P\Big\{\sigma e_a R^a(\omega) + c f_a R^a(\omega) + f_a \mathscr{D} h^a + \frac{1}{2}\epsilon_{abc}(\sigma e^a + c f^a)h^b h^c \\ &+ m^2\epsilon_{abc}\left(\frac{a_1}{6}e^a e^b e^c - \frac{b_1}{2}e^a e^b f^c - \frac{b_2}{2}e^a f^b f^c \right. \\ &\left. + \frac{(c^2-1)b_1 - 2c\sigma b_2}{6} f^a f^b f^c\right)\Big\}, \end{aligned} \tag{7.59}$$

where $\mathscr{D}$ is the covariant derivative with respect to ω. The new dimensionless constants (a_1, b_1, b_2, c) are given in terms of the old (α_I, β_I) parameters as follows

$$a_1 = \frac{1}{8}\left(\alpha_1 - 3\sigma\beta_1 - 3\beta_2 + \sigma\alpha_2\right), \qquad b_1 = \frac{\alpha_2\beta_1 + \beta_2^2 - \beta_1^2 - \alpha_1\beta_2}{4(\beta_1 + \sigma\beta_2)}, \tag{7.60}$$

$$b_2 = -\frac{\alpha_1\beta_2^2 + \sigma\beta_1\beta_2^2 + \beta_1^2\beta_2 + \sigma\alpha_2\beta_1^2}{2(\beta_1 + \sigma\beta_2)^2}, \qquad c = \frac{\sigma\beta_2 - \beta_1}{\sigma\beta_2 + \beta_1}.$$

By construction, this theory has two secondary constraints for invertible e^a. Indeed, when we calculate the integrability conditions (7.21) for this theory we find the three equations

[5] Note that the sum of the two connections also transforms as a connection, while the difference transforms as a tensor under the diagonal gauge symmetries.

$$\frac{1}{2}(\beta_1 + \sigma\beta_2)e^a f \cdot e = 0\,, \qquad \frac{1}{2}(\beta_1 + \sigma\beta_2)e^a h \cdot e = 0\,, \tag{7.61}$$

and

$$\frac{1}{2}(\beta_1 + \sigma\beta_2)\,(h^a f \cdot e + f^a h \cdot e) = 0\,. \tag{7.62}$$

From (7.61) we can derive two secondary constraints, since we assumed that e^a was invertible and that $\beta_1 + \sigma\beta_2 \neq 0$. The secondary constraints are

$$\psi_1 = \Delta^{fe} = 0\,, \qquad \psi_2 = \Delta^{he} = 0\,. \tag{7.63}$$

After imposing these constraints, the matrix of Poisson brackets in the basis (ω, h, f, e) reduces to

$$(\mathscr{P}'_{ab})_{rs} = \frac{1}{2}m^2(\beta_1 + \sigma\beta_2)\begin{pmatrix} 0 & 0 \\ 0 & Q \end{pmatrix}, \tag{7.64}$$

where

$$Q = \begin{pmatrix} 0 & V^{ee}_{ab} & -V^{ef}_{ab} \\ V^{ee}_{ab} & 0 & -V^{eh}_{ab} \\ -V^{fe}_{ab} & -V^{he}_{ab} & V^{hf}_{[ab]} \end{pmatrix}. \tag{7.65}$$

Using the same techniques as previously, we find that this matrix has rank 4.

The secondary constraints are again in involution with themselves, and their brackets with the primary constraint functions are given by

$$\begin{aligned}\{\phi[\xi], \psi_1\}_{\text{P.B.}} = \varepsilon^{ij}\Big[&\partial_i \xi^f \cdot e_j - \xi^f \cdot \omega_i \times e_j - \partial_i \xi^e \cdot f_j + \xi^e \cdot \omega_i \times f_j \\ &- \left(\sigma\xi^e + c\,\xi^f\right) \cdot e_i \times h_j - \left(c\,\xi^e + \sigma(c^2 - 1)\xi^f\right) \cdot f_i \times h_j \\ &- \xi^h \cdot \left(\sigma e_i \times e_j + 2c\, e_i \times f_j + \sigma(c^2 - 1) f_i \times f_j\right)\Big],\end{aligned} \tag{7.66}$$

and

$$\begin{aligned}\{\phi[\xi], \psi_2\}_{\text{P.B.}} = \varepsilon^{ij}\Big[&\partial_i \xi^h \cdot e_j - \xi^h \cdot \omega_i \times e_j - \partial_i \xi^e \cdot h_j + \xi^e \cdot \omega_i \times h_j \\ &+ m^2\left((c\,\sigma a_1 + b_1)\xi^e - (c\,\sigma b_1 - b_2)\xi^f\right) \cdot e_i \times e_j \\ &- m^2\left((c\,\sigma b_1 - b_2)\xi^e + ((c^2 - 1)b_1 - c\,\sigma b_2)\xi^f\right) \cdot e_i \times f_j\end{aligned} \tag{7.67}$$

$$- (c\,\xi^e + \sigma(c^2-1)\xi^f)\cdot h_i \times h_j - \xi^h\cdot$$
$$\left(c\,e_i \times h_j + \sigma(c^2-1) f_i \times h_j\right)\Big].$$

For generic values of the parameters these constraints increase the rank of the total matrix of Poisson brackets, $\mathbb{P}$, by 4, leading to a 14×14 matrix of rank 8. This implies that there are eight second-class constraints and six first-class constraints, leading to two degrees of freedom, those of two massive spin-2 modes in 3 dimensions.

To summarize, demanding the presence of secondary constraints in ZDG to remove unwanted degrees of freedom forces us to make an additional assumption about the theory. We must assume invertibility of a linear combination of the two dreibeine. With an additional restriction on the parameter space of the theory, the invertibility of one of the original dreibein is sufficient to remove the Boulware-Deser ghost. Note that only one dreibeine (or one combination of the two dreibeine) need be assumed invertible. This suggests that we identify its square as the "physical" metric with which distances are measured. In the case where $\beta_1\beta_2 = 0$, this suggestion is supported by the fact that the second dreibein may be solved for in terms of the invertible dreibein and its derivatives, leading to an equation of motion for a single dreibein containing an infinite sum of higher derivative contributions [11]. It would be interesting to investigate whether this is also possible in the generic case.

7.3.4 General Zwei-Dreibein Gravity

It is natural to look for a parity violating generalisation of ZDG, just as GMG is a parity violating version of NMG. To this end we add to the ghost-free, $\beta_2 = 0$, ZDG Lagrangian 3-form (7.44) a Lorentz–Chern–Simons (LCS) term for the spin-connection ${\omega_1}^a$.[6]

$$L_{\text{GZDG}} = L_{\text{ZDG}}(\beta_2 = 0) + \frac{M_P}{2\mu}\omega_{1\,a}\left(d{\omega_1}^a + \frac{1}{3}\epsilon^{abc}\omega_{1\,b}\omega_{1\,c}\right). \tag{7.68}$$

The introduction of the LCS term for ${\omega_1}^a$ introduces non-zero torsion for ${e_1}^a$. One might consider adding a torsion constraint for ${e_1}^a$, enforced by a Lagrange multiplier field h^a, but this introduces new degrees of freedom [6]. In any case, the equations of motion for General ZDG are such that the torsion constraint is not needed in order to solve for the spin-connections, and there exists a scaling limit similar to

[6]It is also possible to include a LCS term for ${\omega_2}^a$, in this case the expressions presented in this subsection are only slightly modified and lead to the same conclusion.

the NMG-limit presented in [6] where the General ZDG Lagrangian reduces to the GMG Lagrangian (7.34).

From the point of view of our general formalism, the addition of the LCS term adds the following non-zero components to g_{rs} and f_{rst}

$$g_{\omega_1\omega_1} = \frac{1}{\mu}, \qquad f_{\omega_1\omega_1\omega_1} = \frac{1}{\mu}. \tag{7.69}$$

The integrability conditions now read

$$e_1{}^a e_1 \cdot e_2 = 0, \tag{7.70}$$

$$e_1{}^a\left(\omega_- \cdot e_1 + \frac{\beta_1 m^2}{\mu} e_1 \cdot e_2\right) = 0, \tag{7.71}$$

$$e_2{}^a \omega_- \cdot e_1 + \left(\frac{\beta_1 m^2}{\mu} e_2{}^a - \omega_-{}^a\right) e_1 \cdot e_2 = 0. \tag{7.72}$$

Invertibility of $e_1{}^a$ implies the same secondary constraints as in Eq. (7.53), and the counting of degrees of freedom proceeds analogously. After a linear redefinition of the constraints to $\phi_{\omega'} = \phi_{\omega_1} + \phi_{\omega_2}$, the matrix of Poisson brackets reduces to

$$(\mathscr{P}'_{ab})_{rs} = m^2\beta_1 \begin{pmatrix} 0 & 0 \\ 0 & Q \end{pmatrix}, \tag{7.73}$$

where

$$Q = \begin{pmatrix} 0 & -V_{ab}^{e_1e_2} & V_{ab}^{e_1e_1} \\ -V_{ab}^{e_2e_1} & -(V_{[ab]}^{\omega_1e_2} - V_{[ab]}^{\omega_2e_2}) + \frac{\beta_1 m^2}{\mu} V_{ab}^{e_2e_2} & (V_{ab}^{\omega_1e_1} - V_{ab}^{\omega_2e_1}) - \frac{\beta_1 m^2}{\mu} V_{ab}^{e_2e_1} \\ V_{ab}^{e_1e_1} & (V_{ab}^{e_1\omega_1} - V_{ab}^{e_1\omega_2}) - \frac{\beta_1 m^2}{\mu} V_{ab}^{e_1e_2} & \frac{\beta_1 m^2}{\mu} V_{ab}^{e_1e_1} \end{pmatrix}. \tag{7.74}$$

We find that this matrix has rank 4. The Poisson brackets of the secondary constraints with the primary ones are now:

$$\begin{aligned} \{\phi[\xi], \psi_1\}_{\text{P.B.}} = \varepsilon^{ij}\Big[& \partial_i \xi^{e_1} \cdot e_{2j} - \xi^{e_1} \cdot \omega_{1i} \times e_{2j} - \partial_i \xi^{e_2} \cdot e_{1j} + \xi^{e_2} \cdot \omega_{2i} \times e_{1j} \\ & - \left(\xi^{\omega_1} - \xi^{\omega_2} + \frac{\alpha_1 m^2}{\mu}\xi^{e_1} - \frac{m^2\beta_1}{\mu}\xi^{e_2}\right) \cdot e_{1i} \times e_{2j} \\ & + \frac{\beta_1 m^2}{\mu}\xi^{e_1} \cdot e_{2i} \times e_{2j}\Big], \end{aligned} \tag{7.75}$$

and

$$
\begin{aligned}
\{\phi[\xi], \psi_2\}_{\text{P.B.}} = & \varepsilon^{ij}\Big[(\partial_i \xi^{\omega_1} - \partial_i \xi^{\omega_2}) \cdot e_{1j} - (\xi^{\omega_1} - \xi^{\omega_2}) \cdot (\omega_{2i} \times e_{1j}) - \partial_i \xi^{e_1} \cdot \omega_{-j} \\
& + \xi^{e_1} \cdot (\omega_{1i} \times \omega_{-j}) + m^2 (\sigma\beta_1 \xi^{e_1} + \alpha_2 \xi^{e_2}) \cdot (e_{1i} \times e_{2j}) \\
& - m^2 ((\sigma\alpha_1 + \beta_1)\xi^{e_1} - \sigma\beta_1 \xi^{e_2}) \cdot (e_{1i} \times e_{1j}) \\
& + m^2 \left(\frac{\alpha_1}{\mu}\xi^{e_1} - \frac{\beta_1}{\mu}\xi^{e_2}\right) \cdot (e_{1i} \times \omega_{-j}) - m^2 \frac{\beta_1}{\mu} \xi^{e_1} \cdot (e_{2i} \times \omega_{-j})\Big].
\end{aligned}
\tag{7.76}
$$

Again, the secondary constraints are in involution, and the new columns are linearly independent from each other and the original columns. The usual analysis shows that there are 8 second-class constraints and 6 first-class constraints. The total dimension of the physical phase space remains 4, and so the model has the same number of local degrees of freedom as GMG.

Conclusions

It is a remarkable fact that many of the 3D "massive gravity" models that have been found and analysed in recent years have a CS-like formulation in which the action is an integral over a Lagrangian 3-form constructed from wedge products of 1-forms that include an invertible dreibein. One example not considered here is Topologically Massive Supergravity [12].

Many of these CS-like models have an alternative formulation as a higher-derivative extension of 3D General Relativity, and it is certainly not the case that all such higher-derivative extensions can be recast as CS-like models. It appears that the unitary (ghost-free) 3D massive models are also special in this respect. Whatever the reason may be for this, it is fortunate because the CS-like formalism is well-adapted to a Hamiltonian analysis, which we have reviewed, and refined, extending the results of [5] for General Massive Gravity (GMG) to include the recently proposed Zwei-Dreibein Gravity (ZDG) [6].

This Hamiltonian analysis leads to a simple determination of the number of local degrees of freedom, independent of any linearisation about a particular background. This allows one to establish that a class of 3D massive gravity models is free of the Boulware-Deser ghost that typically afflicts massive gravity models [7]. This class includes ZDG, provided a linear combination of the dreibeine is assumed to be invertible. Conversely, the CS-like formulation of these models can be used as a starting point to find higher-derivative extensions of New Massive Gravity which are guaranteed to be free of scalar ghosts [13].

(continued)

We have also discussed a parity violating extension of ZDG; it has some similarities to GMG (and has a limit to GMG for a certain range of its parameters) so it could be called "General Zwei-Dreibein Gravity" (GZDG). We have shown that it has exactly the same number of local degrees of freedom as GMG. We know that ZDG propagates two spin-2 modes of equal mass in a maximally symmetric vacuum, so it seems that GZDG will propagate two spin-2 modes of different masses, like GMG. It would be interesting to see whether there is some limit of the parameters of GZDG that sends one mass to infinity keeping the other fixed, because we would then have a model similar to TMG but possibly with better behaviour in relation to the AdS/CFT correspondence.

Acknowledgements This paper is based upon lectures given by Eric Bergshoeff and Paul Townsend at the Seventh Aegean Summer School *Beyond Einstein's Theory of Gravity* in Paros, Greece. Eric A. Bergshoeff, Wout Merbis, Alasdair J. Routh and Paul K. Townsend thank the organizers of the Paros School for providing an inspiring environment. We are also grateful to Joaquim Gomis and Marc Henneaux for discussions and correspondence on Hamiltonian methods.

References

1. A. Achúcarro, P.K. Townsend, A Chern-Simons action for three-dimensional anti-De Sitter supergravity theories. Phys. Lett. B **180**, 89 (1986)
2. E. Witten, (2+1)-Dimensional gravity as an exactly soluble system. Nucl. Phys. B **311**, 46 (1988)
3. S. Deser, R. Jackiw, S. Templeton, Topologically massive Gauge theories. Ann. Phys. **140**, 372 (1982) [Erratum Ann. Phys. **185**, 406 (1988)]
4. E.A. Bergshoeff, O. Hohm, P.K. Townsend, Massive gravity in three dimensions. Phys. Rev. Lett. **102**, 201301 (2009) [arXiv:0901.1766]; More on massive 3D gravity. Phys. Rev. **D79**, 124042 (2009) [arXiv:0905.1259]
5. O. Hohm, A. Routh, P.K. Townsend, B. Zhang, On the Hamiltonian form of 3D massive gravity. Phys. Rev. D **86**, 084035 (2012) [arXiv:1208.0038]
6. E.A. Bergshoeff, S. de Haan, O. Hohm, W. Merbis, P.K. Townsend, Zwei-Dreibein gravity. Phys. Rev. Lett. **111**, 111102 (2013) [arXiv:1307.2774]
7. D.G. Boulware, S. Deser, Can gravitation have a finite range? Phys. Rev. D **6**, 3368 (1972)
8. M. Bañados, C. Deffayet, M. Pino, The Boulware-Deser mode in Zwei-Dreibein gravity. Phys. Rev. D **88**, 124016 (2013) [arXiv:1310.3249]
9. P.A.M. Dirac, *Lectures on Quantum Mechanics* (Dover, New York, 2001)
10. D. Dominici, J. Gomis, K. Kamimura, G. Longhi, Dynamical sectors of a relativistic two particle model. Phys. Rev. **D89**, 045001 (2014) [arXiv:1312.1000]
11. E.A. Bergshoeff, A.F. Goya, W. Merbis, J. Rosseel, Logarithmic AdS waves and Zwei-Dreibein gravity. JHEP **1404**, 012 (2014) [arXiv:1401.5386]
12. A. Routh, The Hamiltonian form of topologically massive supergravity. Phys. Rev. D **88**, 024022 (2013) [arXiv:1301.7671]
13. H.R. Afshar, E.A. Bergshoeff, W. Merbis, Extended massive gravity in three dimensions JHEP **1408**, 115 (2014) [arXiv:1405.6213]

Chapter 8
Cosmological Applications of Massive Gravity

Andrew J. Tolley

Abstract Models of modified gravity in the infrared are especially appealing for their late-time cosmology. We review different models before focusing on the cosmology of massive gravity. We start by information derived from its decoupling limit where a self-acceleration solution can be found but suffers from strong coupling issues in the vector modes. This feature is carried through for most FRW self-accelerating solutions in the full theory. We emphasize the role played by inhomogeneous solutions which reduce to a self-accelerating FRW solution on distances comparable to our current Universe but are inhomogeneous at larger distances. We also give an overview of cosmological solutions in extensions of massive gravity such as bi-gravity and quasi-dilaton massive gravity.

8.1 Introduction and Motivations

Most modifications of gravity change physics at high energies. Examples include string theory, Kaluza–Klein theories etc.... In gravity, *high energy* means *high curvatures* which means *early times*. Thus string theory/Kaluza–Klein modifications and other UV modifications of gravity have little impact on late-time cosmology. Ironically, it is late time cosmology that we least understand and particularly Cosmic Acceleration. In this proceedings we explore the effects of IR (low-energy) modifications of gravity on late-time cosmology.

We start by reviewing the Dvali–Gabadadze–Porrati (DGP) model and its cosmology. While DGP is well-known for admitting a self-acceleration branch, it is plagued by ghost which makes that solution unphysical. Nevertheless DGP has played a profound role in our understanding of IR modifications of gravity. While the "normal" branch (or ghost) of DGP does not self-accelerate it exhibits important features which remain for any modification of gravity. We then review the features behind degravitating/screening solutions and explore extensions of DGP including massive gravity and Cascading gravity. The rest of these proceedings are

A.J. Tolley (✉)
CERCA & Physics Department, Case Western Reserve University, Cleveland, OH, USA
e-mail: Andrew.J.Tolley@case.edu

E. Papantonopoulos (ed.), *Modifications of Einstein's Theory of Gravity at Large Distances*, Lecture Notes in Physics 892, DOI 10.1007/978-3-319-10070-8_8

then dedicated to the cosmology of massive gravity. Starting with its decoupling limit we emphasize the existence of several key features, including self-acceleration and strong-coupling. We then present a no-go for FRW solutions in the full spatially flat massive gravity theory on flat spacetime and different resolutions to get round this no-go theorem. We show how massive gravity with FRW reference metric is never an observationally consistent resolution due to the "Higuchi" problem but explain how bi-Gravity bypasses these issues. We then present the cosmology in the quasi-dilaton. Finally we discuss Partially Massless (Bi)Gravity and show that such a non-linear theory cannot exist.

8.1.1 DGP: The Quintessential IR Modification

8.1.1.1 Self-Acceleration

Imagine a brane in an infinite fifth dimension with a localized Einstein–Hilbert term on the brane [38]

$$S = \int d^4x \sqrt{-g_4} \frac{M_{\rm Pl}^2}{2} R_4 + \int d^4x \sqrt{-g-4} \mathscr{L}_M + \int d^5x \sqrt{-g_5} \frac{M_5^3}{2} R_5, \quad (8.1)$$

where the terms of the left are the most irrelevant and the ones that dominate in the UV while to the right are the most relevant operators which dominate in the IR. This means that at low energies we feel all five dimensions. As a result, the force of gravity in at large distances (or low-energy) falls as r^{-3}.

At high energies, the brane kinetic term dominates which forces gravity to behave four-dimensional. At short distances (or high-energies), one recovers the standard Newton's square law for the force of gravity falling as r^{-2}.

One of the main interests of DGP is that gravitons can condense to form a condensate whose energy density *sources* self-acceleration [36]. This corresponds to the self-accelerating solution of DGP that can be inferred from the Friedman equation [36]

$$H^2 \mp mH = \frac{1}{3M_{\rm Pl}^2} \rho_{\rm matter} , \quad (8.2)$$

where $m = M_5^3/M_{\rm Pl}^2$. The two different signs correspond to the two embeddings of the brane. In the '−'-branch, the Universe accelerates $H \sim m$ even in the absence of matter $\rho_{\rm matter} = 0$. The 'unfortunate' news is that this branch also has a ghost [7,45,63]. Furthermore the solution also sits at a strong coupling threshold which makes the question of the quantum stability particularly interesting [65,68].

A few years later, one of the motivations for Galileon models was to find self-accelerating solutions without the Ghost issue [69]. Depending on the context, Galileons may or may not be seen as scalar fields in their own right. They may arise as the brane-bedding mode of a probe brane in higher dimensions [24]. In DGP, they arise a remnants of higher spin fields and this also happens in Cascading gravity and massive gravity as we shall see later. In this case the Galileon symmetry is exact even in the presence of gravity.

8.1.1.2 Screening/Self-Tuning Mechanism

As we have seen, IR modifications (like DGP) can be used to weaker the strength of gravity at large (cosmological) distances. But this is not all. Rather than providing a self-accelerating solution, IR modifications of gravity can lead to screening or self/tuning mechanism whereby a large cosmological constant could be screened resulting in a small late-time acceleration of the Universe. If this screening happens dynamically for any value of the cosmological constant it could lead to a degravitation mechanism [4, 39–41].

Unlike for self-accelerating solutions, for a degravitating/screening solution, gravitons can condense to form a condensate whose energy density *compensate* the cosmological constant. This would mean that the Cosmological Constant could be *large* but the cosmic acceleration would be *small.*

As we have seen, the Friedman equation in DGP (8.2) is a completely local relation between the energy density and the Hubble rate. As long as the FRW equation is local we can never use IR modifications to resolve the OLD cosmological constant problem.

• **In higher than five dimensions**, the full evolution is expected to be non-local from a four-dimensional viewpoint

$$H^2 + F(H) \sim \frac{8\pi}{3} G(\Box)\rho \,. \tag{8.3}$$

• **In Massive gravity**, the effective Einstein equation in the presence of a Cosmological Constant is expected to be of the form

$$G_{\mu\nu} + m^2 \frac{\partial \mathscr{L}_m}{\partial g^{\mu\nu}} = -8\pi G\, \Lambda g_{\mu\nu} \,, \tag{8.4}$$

where m is the graviton mass and $\mathscr{L}_m$ the Lagrangian for the mass term. For a self-screening graviton condensate, we expect the spacetime to be Minkowski, for instance

$$g_{\mu\nu} = \left(1 + f\left(\frac{\Lambda}{m^2}\right)\right)\eta_{\mu\nu} \qquad G_{\mu\nu} = 0 \,, \tag{8.5}$$

in the presence of an arbitrary large Cosmological Constant Λ,

$$m^2 \frac{\partial \mathscr{L}_m}{\partial g^{\mu\nu}} = -8\pi G \, \Lambda g_{\mu\nu} \, . \tag{8.6}$$

This means that the cosmological constant can be reabsorbed into a *redefinition* of the metric and coupling constants—and is hence a *redundant* operator.

Independently of its explicit realization, the idea behind degravitation [4, 39–41] is to have a dynamical relaxation, meaning a dynamical evolution towards screened solutions. **Can we modify gravity in the IR such that at low energy sources couple more weakly to gravity?** In GR, a cosmological constant is the most relevant operator one can write down (the operator which dominates at low energy) since $\partial_\mu \Lambda = 0$. Modifications of gravity such as DGP provide one step in answering this question, but DGP is not sufficiently modified in the IR. The Friedman equation ought to be more 'non-local'. A possible solution would be too generalize DGP to higher dimensions, known as Cascading Gravity [26, 27]. Another possibility could be to work straight with a theory of massive gravity.

8.1.2 IR Modifications of Gravity

8.1.2.1 Extending DGP to Higher Dimensions

In $4 + n$ dimensions, the gravitational potential scales as

$$V(r) \sim \frac{1}{r^{(1+n)}} \, , \tag{8.7}$$

and so gravity is weaker in larger dimensions. We would like this behaviour in the IR while maintaining the standard Newton's square law in the UV,

$$\begin{array}{ccc} V(r) \sim \dfrac{1}{r} & \longrightarrow & V(r) \sim \dfrac{1}{r^{(1+n)}} \\ \text{UV, small } r & & \text{IR, large } r \, . \end{array} \tag{8.8}$$

The gravitational potential $V(r)$ can be expanded using the Kallen–Lehman spectral representation

$$V(r) = \frac{Z}{r} + \int_0^\infty \mathrm{d}s^2 \rho(s^2) \frac{e^{-sr}}{r} \, , \tag{8.9}$$

corresponding to the propagator[1] (in Fourier space)

[1] The tensor structure has been omitted in the expression for the propagator but can easily be restored: all the massive modes in the spectral representation have the massive Fierz–Pauli tensor

$$G_F(k) = \frac{Z}{k^2 - i\epsilon} + \int_0^\infty \mathrm{d}s^2 \rho(s^2) \frac{1}{k^2 + s^2 - i\epsilon} \tag{8.10}$$

$$= \frac{Z}{k^2 - i\epsilon} + \frac{1 - Z}{k^2 + m^2(k) - i\epsilon}, \tag{8.11}$$

and we can interpret $m^2(k)$ as an effective mass for the graviton. For an infinite extra dimension, $Z = 0$.

In higher dimensional theories, we find [27] that the mass scales as

$$m^2(k) \sim m_0^2 \left(-k^2 L^2\right)^\alpha \tag{8.12}$$

with $\alpha = 1/2$ in 5d, $\alpha \sim 0$ (up to logarithmical corrections) in 6d and $\alpha = 0$ in seven dimensions or more. This means that one should consider six dimensions or more to potentially obtain degravitation.

When dealing with higher dimensions the first guess would be to consider a 3-brane embedded in six or more dimensions. The 3-brane is then an object of codimension-2 or higher which suffers from classical UV divergences that ought to be renormalized already at the classical level [20, 44]. To avoid needing to address these issues one can consider instead a "cascading" setup where a codimension-one brane lies within a codimension-one brane etc.... In six dimensions this leads to the following Cascading gravity theory [21, 26, 27, 29]

$$S = \int \mathrm{d}^4x \sqrt{-g_4} \left(\frac{M_{\rm Pl}^2}{2} R_4 + \mathcal{L}_M \right) + \int \mathrm{d}^5x \sqrt{-g_5} \left(\frac{M_5^3}{2} R_5 \right)$$
$$+ \int \mathrm{d}^6x \sqrt{-g_6} \left(\frac{M_6^4}{2} R_4 \right),$$

where the terms on the left are the most irrelevant operators which dominate in the UV and terms to the right are the most relevant operators which dominate in the IR. In this model gravity transitions from a four-dimensional behaviour at short distances to a five-dimensional one and finally a six-dimensional one at large distances. See also [28] for considerations in higher dimensions, where the same type of behaviour occurs.

One advantage in going to seven or more dimensions is that there could be no maximal value for the cosmological constant to be carried by the 3-brane. To understand whether or not Cascading gravity realizes a dynamical relaxation, two criteria should be satisfied:

Criterion I: Screening/Self-Tuning Existence of a Minkowski vacuum solution in the presence of a cosmological constant on the 3-brane. In a six-dimensional

---

structure $G^{(m)}_{\mu\nu\alpha\beta} \sim (\tilde\eta_{\mu(\alpha}\tilde\eta_{\nu\beta)} - \frac{1}{3}\tilde\eta_{\mu\nu}\tilde\eta_{\alpha\beta})$ with $\tilde\eta_{\mu\nu} \sim \eta_{\mu\nu} - k_\mu k_\nu/m^2$, while the massless modes have the Einsteinian tensor structure $G^{(0)}_{\mu\nu\alpha\beta} \sim (\eta_{\mu(\alpha}\eta_{\nu\beta)} - \frac{1}{2}\eta_{\mu\nu}\eta_{\alpha\beta})$.

spacetime, tension on a 3-brane creates a deficit angle in the bulk rather than leading to an acceleration on the brane. Similar properties were found in seven dimensional Cascading gravity where a 3-brane lies in a 4-brane which lies in a 5-brane in 7d.

Criterion II: Dynamical Relaxation For a model to degravitate it should not only satisfy the previous criterion but also admit a dynamical and *causal* process by which one can relax to the solution found in criterion I. At the linearized level this was shown to work in [41]. Non-linearly, this criterion is much harder to check. As yet this has not been demonstrated non-linearly mainly due to the complexity of the problem.

One strong motivation for considering massive gravity is as a toy-model of higher dimensional gravity models (e.g. for cascading gravity) that potentially exhibit degravitation.

8.1.2.2 Why Massive Gravity

In many respects, massive gravity is simpler than large extra dimensions and cascading gravity. In massive gravity, the departure from GR is governed by essentially a single parameter: the Graviton Mass.

The *gain* is that the theory is easier to solve than the higher dimensional framework. One may worry that massive gravity looses diffeomorphism invariance. In practice this is not so: Massive gravity can be formulated in a perfectly covariant (or diff invariant way) at the price of introducing four Stückelberg fields. These fields lead to new degrees of freedom, but far less than one would have for gravity in six or more dimensions.

In massive gravity, the Vainshtein mechanism [74] is the screening mechanism which ensures the recovery of GR in the massless limit $m \to 0$. This ensures that massive gravity can be easily made to be consistent with most tests of GR (effectively placing an upper bound on the graviton mass) without spoiling its role as an IR modification of GR. We now turn to the formulation of massive gravity and its cosmological applications.

8.2 Ghost-Free Massive Gravity

8.2.1 The Model

Other proceedings are dedicated to the description of massive gravity (de Rham) and its Vainshtein Mechanism (so we only summarize its formulation in what follows. See [22] for a recent review. The Lagrangian for massive gravity takes the form [30]

$$\mathcal{L} = \frac{M_{\rm Pl}^2}{2}\sqrt{-g}\left(R + 2m^2\mathcal{U}(g, f)\right) + \mathcal{L}_M\,, \tag{8.13}$$

where all quantities are four-dimensional, $f_{\mu\nu}$ is the reference metric that be Minkowski or other [54]. The potential term has only a finite number of possible interactions in any dimensions. In four dimensions, it may be written as

$$\mathscr{U}(g,f) = \mathscr{U}_2 + \alpha_3\mathscr{U}_3 + \alpha_4\mathscr{U}_4\,, \tag{8.14}$$

with [30]

$$\mathscr{U}_2[\mathscr{K}] = \left([\mathscr{K}]^2 - [\mathscr{K}^2]\right) \tag{8.15}$$

$$\mathscr{U}_3[\mathscr{K}] = \left([\mathscr{K}]^3 - 3[\mathscr{K}][\mathscr{K}^2] + 2[\mathscr{K}^3]\right) \tag{8.16}$$

$$\mathscr{U}_4[\mathscr{K}] = \left([\mathscr{K}]^2 - 6[\mathscr{K}]^2[\mathscr{K}^2] + 8[\mathscr{K}^3][\mathscr{K}] + 3[\mathscr{K}^2]^2 - 6[\mathscr{K}^4]\right), \tag{8.17}$$

and

$$\mathscr{K}^\mu_\nu(g,f) = \delta^\mu_\nu - \sqrt{g^{\mu\alpha} f_{\alpha\nu}}. \tag{8.18}$$

This model is sometimes referred to as dRGT and we keep the same terminology to avoid confusion. The mass term can equivalently be written as characteristic polynomials [52]

$$\mathscr{L} = \frac{1}{2}\sqrt{-g}\left(M_{\rm Pl}^2 R - m^2 \sum_{n=0}^{4} \beta_n \mathscr{U}_n[X]\right) + \mathscr{L}_M\,, \tag{8.19}$$

with

$$X^\mu_\nu = \sqrt{g^{\mu\alpha} f_{\alpha\nu}}\,. \tag{8.20}$$

These interactions appear as very non-trivial, yet it can be shown that they are protected by a non-renormalization theorem [33].

8.2.1.1 Cosmology of Massive Gravity: A Basic Tension

The theory of massive gravity presented previously ensures the absence of a sixth degree of freedom in four dimensions, but it does not guarantee that all five remaining degree are ghost free.

The representation theory of the de Sitter group gives the Higuchi bound for massive spin 2 representations [55]

- $m^2 = 0$: Corresponds to GR and has two degrees of freedom
- $0 < m^2 < 2H^2$: Corresponds to massive gravity and has five degrees of freedom, one of them being a ghost (the Higuchi ghost)

- $m^2 = 2H^2$: Corresponds to Partially massless gravity and has four degrees of freedom
- $m^2 > 2H^2$: Corresponds to massive gravity and has five degrees of freedom without Higuchi ghost.

For every cosmological solution we need to check carefully whether or not the helicity-0 mode is unitary, since this is not guaranteed a priori by the theory. However this is not guaranteed to be a problem either, for instance in DGP the bound is always satisfied [7, 63]. This may not obviously be relevant for a Minkowski reference metric which breaks the de Sitter symmetry.

8.2.2 Decoupling Limit Cosmology

We can take a decoupling limit of massive gravity (and as we shall see later of bi-gravity as well) where after diagonalization massive gravity is equivalent to a free helicity-2 particle and a helicity-1 coupled to a helicity-0 particle. This limit sends $M_{\rm Pl} \to \infty$, $m \to 0$ while keeping the scale $\Lambda_3 = (M_{\rm Pl} m^2)^{1/3}$ fixed [23]

$$\mathscr{L} = -\frac{1}{2} h^{\mu\nu} \hat{\mathscr{E}}^{\alpha\beta}_{\mu\nu} h_{\alpha\beta} + h^{\mu\nu} \sum_{n=1}^{3} \frac{a_n}{\Lambda_3^{3(n-1)}} X^{(n)}_{\mu\nu}[\Pi] + \frac{1}{M_{\rm Pl}} h^{\mu\nu} T_{\mu\nu} , \tag{8.21}$$

with $\Pi_{\mu\nu} = \partial_\mu \partial_\nu \pi$ and $X^{(n)}_{\mu\nu}[\Pi] \sim \Pi^n_{\mu\nu}$ in such a way that the trace of X would be a total derivative. The coefficients a_n are related to the previous coefficients α_n.

The helicity-0 mode interactions are true Galileons and preserve the Galileon symmetry. Since the Galileon symmetry is EXACT, we only require that $\Pi_{\mu\nu}$ is homogeneous and isotropic to describe FRW. The generic solution for the helicity-0 mode near $x = 0$ which is isotropic in this limit is [31]

$$\pi \sim A(t) + B(t)\mathbf{x}^2 . \tag{8.22}$$

Interestingly there is no equivalent of this form in the covariant Galileon or Horndeski theory [37, 58] because the symmetry is broken in these cases but not in massive gravity. The resulting metric takes the form

$$\begin{aligned} ds^2 &= -\left[1 - (\dot{H} + H^2)\mathbf{x}^2\right] dt^2 + \left[1 - \frac{1}{2} H^2 \mathbf{x}^2\right] d\mathbf{x}^2 \\ &= \left(\eta_{\mu\nu} + h^{\rm FRW}_{\mu\nu}\right) dx^\mu \, dx^\nu , \end{aligned} \tag{8.23}$$

where the equations of motion fix A and B in (8.22) for instance for a pure Cosmological constant source B is constant and $A = -Bt^2$.

8.2.2.1 Self-Accelerating Solution

Considering the following background solution plus perturbation split [31]

$$\pi = \frac{1}{2} q_{\rm dS} \Lambda_3^3 x^2 + \phi \tag{8.24}$$

$$h_{\mu\nu} = -\frac{1}{2} H_{\rm dS}^2 x^2 \eta_{\mu\nu} + \chi_{\mu\nu} \tag{8.25}$$

$$T_{\mu\nu} = -\lambda \eta_{\mu\nu} + \tau_{\mu\nu} \,, \tag{8.26}$$

the background quantities satisfy the equations of motion for the self-accelerating branch,

$$a_1 + 2a_2 q_{\rm dS} + 3a_3 q_{\rm dS}^2 = 0 \tag{8.27}$$

$$H_{\rm dS}^2 = \frac{\lambda}{3M_{\rm Pl}^2} + \frac{2\Lambda_3^3}{M_{\rm Pl}} \left(a_1 q_{\rm dS} + a_2 q_{\rm dS}^2 + a_3 q_{\rm dS}^3 \right) \,, \tag{8.28}$$

where we see an acceleration $H_{\rm dS}^2 > 0$ even in the absence of a cosmological constant, $\lambda = 0$. Unlike for DGP, this self-accelerating solution admits no ghost for $a_2 + 3a_3 q_{\rm dS} > 0$

$$\mathcal{L}^{(2)} = -\frac{1}{2} \chi^{\mu\nu} \hat{\mathcal{E}}^{\alpha\beta}_{\mu\nu} \chi_{\alpha\beta} + \frac{6H_{\rm dS}^2 M_{\rm Pl}}{\Lambda_3^3} (a_2 + 3a_3 q_{\rm dS}) \phi \Box \phi + \frac{1}{M_{\rm Pl}} \chi^{\mu\nu} \tau_{\mu\nu} . \tag{8.29}$$

A remarkable feature is worth pointing out at this level: the fluctuation ϕ does not directly couple to matter. As a result there is no need for a Vainshtein mechanism to screen to field [31].

8.2.2.2 Screening/Self-Tuning (Degravitating) branch

Another background solution to classical equations of motion in the decoupling limit (8.21) is

$$\pi = \frac{1}{2} q_{\rm dS} \Lambda_3^3 x^2 + \phi \tag{8.30}$$

$$h_{\mu\nu} = 0 + \chi_{\mu\nu} \tag{8.31}$$

$$T_{\mu\nu} = -\lambda \eta_{\mu\nu} + \tau_{\mu\nu} \,, \tag{8.32}$$

where we obtain a Minkowski solution for any value of the cosmological constant !!! Perturbations are stable and present the Vainshtein mechanism

$$\mathcal{L}^{(2)} = -\frac{1}{2} \chi^{\mu\nu} \hat{\mathcal{E}}^{\alpha\beta}_{\mu\nu} \chi_{\alpha\beta} + \frac{3}{2} \phi \Box \phi + \frac{1}{M_{\rm Pl}} \left(\chi^{\mu\nu} + \phi \eta^{\mu\nu} \right) \tau_{\mu\nu} \tag{8.33}$$

however the scale of the strong coupling ends up being too large for being observationally viable. Nevertheless this still provides a proof of principle for how one could evade Weinberg's no-go theorem.

8.2.2.3 DL Cosmology Summary

More generally, the decoupling limit implies the existence of isotropic and **inhomogeneous** cosmological solutions for massive gravity which for suitable range of parameters are free from the Higuchi bound (no ghost in helicity-0 sector).

The absence of Higuchi bound opens up possibilities for background Vainshtein effects where the mass can be as small as desired leading to consistent results with the expansion history at early times.

All the solutions presented so far are in the decoupling limit. They will all map to solutions in the full non-linear theory but may be hard to find.

8.3 Cosmology of Massive Gravity

8.3.1 A No-Go and Ways Out

The simplest model (dRGT massive gravity in Minkowski) does not support spatially flat (or closed) cosmological solutions which are FRW meaning homogeneous and isotropic.

The argument is simple: as in GR we have a Friedman equation and a Raychaudhuri equation. In GR, the second follows from the first by diff invariance. In massive gravity diff invariance is broken and so the would-be Raychaudhuri equation no longer follows from the first equation. The consistency of both equations imposes a condition on the scale factor [10].

Indeed, assuming an FRW metric,

$$\mathrm{d}s^2 = -N^2(t)\,\mathrm{d}t^2 + a(t)^2\,\mathrm{d}\mathbf{x}^2\,, \tag{8.34}$$

the lagrangian for massive gravity becomes

$$\mathscr{L} = 3M_{\rm Pl}^2\left(-\frac{a\dot{a}^2}{N} - m^2(a^3-a^2) + m^2N(2Na^3-3a^2+a)\right)\,. \tag{8.35}$$

The constraint imposed on the scale factor by consistency of the would be Friedman and Raychaudhuri equation is then [10]

$$m^2\partial_t\left(a^3-a^2\right) = 0\,, \tag{8.36}$$

which is clearly uninteresting.

To bypass this no-go several options can and have been considered:

- **Resolution I: Accept Inhomogeneities**
  The most natural and certainly the most physical resolution to the previous no-go (although also probably the hardest to implement from a purely technical aspect) is to accept the existence of inhomogeneities [10]. While inhomogeneities may be important at large distances (beyond our observable Universe—which is the picture modern cosmology has in mind), the Vainshtein mechanism would guarantee that the inhomogeneities are unobservable at short distance scales (within the observable Universe) and before late times. The inhomogeneities would only appear on a scale set by the graviton mass (which is usually assumed to be close to the current Hubble parameter). Since observational constraints on inhomogeneities at the current Hubble scale are actually very weak, the presence of these inhomogeneities would thus have little observational effects and yet would resolve the previous no-go.

  Moreover, inhomogeneities and anisotropies can be hidden inside the Stückelberg fields which do not directly couple to matter but only indirectly though the Planck scale. Inhomogeneities in the Stückelberg fields are thus observationally very weak.

  To summarize, the metric could even remain perfectly homogeneous and isotropic at the price of introducing some inhomogeneities in the Stückelberg fields that would show up at the level of cosmological perturbations but could easily be small [15, 17, 18, 46, 50, 51, 60, 62, 72, 76, 77, 81].

- **Resolution II: Modify the Assumptions**
  The previous no-go had several underlying assumptions which can be bypassed to allow for FRW solutions:

  - Considering an open Universe rather than a flat or closed one allow for FRW solutions [49] which are however unstable [75].
  - Consider a de Sitter or FRW reference metric, however this also leads to instabilities [42] as we shall see later. (See also [25] for the decoupling limit of massive gravity on de Sitter).
  - Make the reference metric dynamical, leading to bi-gravity [53]. As we shall see later, this could prevent the instabilities [2, 5, 8, 61, 78, 79].

- **Resolution III: Extension or Modification of the Theory**
  Other more significant modifications of the theory allow for FRW solutions:

  - Quasi-Dilaton massive gravity which admits self-accelerating solutions but which appear to be unstable [11, 13, 19]
  - Generalized Quasi-Dilaton massive gravity which admits stable self-accelerating solutions [14, 16, 67]
  - Lorentz-violating massive gravity [9]
  - Varying mass gravity [19, 64, 80]
  - Multi-vierbeins gravity [71]

- Extended Massive gravity [3,57]
- Non-local Massive gravity [59,66]

In what follows we look at a few of these alternatives and show how massive gravity with FRW reference metric allows for an FRW solution but inevitably suffers from an Higuchi ghost at early times.

8.3.1.1 From Acceleration to Decceleration

Consider the spacetime metric

$$\mathrm{d}s^2 = -N^2(t)\,\mathrm{d}t^2 + a^2(t)\,\mathrm{d}\mathbf{x}^2\,, \tag{8.37}$$

and the reference metric,

$$\partial_\mu \phi^a \partial_\nu \phi^b f_{ab}\,\mathrm{d}x^\mu\,\mathrm{d}x^\nu = -\left(\dot{\phi}^0\right)^2 \mathrm{d}t^2 + b^2(\phi^0)\,\mathrm{d}\mathbf{x}^2\,. \tag{8.38}$$

For instance if the reference metric was de Sitter it could be written as previously with $b(\phi^0) = \exp\left(H_f \phi^0\right)$.

Before proceeding, let us address at this point what happens when the metric transits from an acceleration to a decceleration. In that case $\dot{\phi}^0$ changes sign and one of the eigenvalues of $\sqrt{g^{-1}f}$ vanishes. To better understand the physics at that point, let us move onto the vierbein formulation which can accommodate a change of sign.

The vierbein formulation is analytic in the Stückelberg fields ϕ^a [6,48,56] and the mass term takes the form

$$\det\left[e^a_\mu + \lambda \Lambda^a_b f^b_c \partial_\mu \phi^c\right]. \tag{8.39}$$

There is no singularity in the formulation as long as it is possible to solve the following equation for the Lorentz Stückelberg fields Λ^a_b ($\Lambda \eta \Lambda^T = \eta$)

$$e^{\mu[a} \Lambda^{b]}_c f^c_d \partial_\mu \phi^d = 0\,, \tag{8.40}$$

which corresponds to six equations for six unknown Lorentz transformations. The main point to notice is that even when $\dot{\phi}^0 = 0$, one can solve for $\delta\Lambda^a_b = \ldots\delta\phi_c$. This point originally made in [73] was later explained in [47].

8.3.1.2 Dressed Mass and Partially Massless

Using the ansate (8.37) and (8.38) into the Lagrangian for massive gravity, we obtain the following equations:

$$\left(\sum_{n=0}^{2}\frac{\hat{\beta}_{n+1}}{(2-n)!n!}\left(\frac{b}{a}\right)^{n+1}\right)\left(\frac{H}{b}-\frac{H_f}{a}\right) \tag{8.41}$$

$$H^2=\sum_{n=0}^{3}\frac{3m^2\beta_n}{(3-n)!n!}\left(\frac{H}{H_f}\right)^n+\frac{\rho}{3M_{\rm Pl}^2}\,, \tag{8.42}$$

where $H_f = b'(\phi^0)/\dot{\phi}^0 b(\phi^0)$ is the effective Hubble parameter of the FRW reference metric and $H = \dot{a}/aN$. The normal branch solution of (8.41) is given by

$$\frac{b}{a}=\frac{H}{H_f}\,. \tag{8.43}$$

The effective mass (governing the kinetics of the helicity-0 mode) is given by [42]

$$\tilde{m}^2(H)=\frac{m^2}{2M_{\rm Pl}^2}\frac{H}{H_f}\left[\beta_1+2\beta_2\frac{H}{H_f}+\beta_3\frac{H^2}{H_f^2}\right], \tag{8.44}$$

and the coefficient of the kinetic term for the helicity-0 mode is

$$\mathscr{L}_{\rm helicity-0}\propto -\tilde{m}^2(H)\left(\tilde{m}^2(H)-2H^2\right)(\partial\pi)^2\,, \tag{8.45}$$

so the generalized Higuchi bound is

$$\tilde{m}^2(H)>2H^2\,. \tag{8.46}$$

If we make the special choice $\beta_1 = \beta_3 = 0$ and $\beta_2 = 1$ and $m^2 = 2H_f^2$ then the effective mass term is simplify $\tilde{m}^2(H) = 2H^2$ and the kinetic term vanishes regardless of the source [25]. This corresponds to partially massless case. Unfortunately this theory keeps some interactions between the helicity-0 mode and the vectors, and the theory is thus infinitely strongly coupled. This happens in massive gravity as in bi-gravity.

8.3.1.3 Higuchi Versus Vainshtein

As seen before, considering massive gravity on a FRW reference metric leads to the effective mass term (8.44) and the Higuchi bound imposes the relation (8.46). In parallel, observations and the screening of the Helicity-0 mode impose an upper bound on the effective graviton mass [42]

$$\frac{m^2}{2M_{\rm Pl}^2}\left[3\beta_1\frac{H}{H_f}+3\beta_2\frac{H^2}{H_f^2}+\beta_3\frac{H^3}{H_f^3}\right]\ll 3H^2\,. \tag{8.47}$$

As result to satisfy both the Higuchi bound and the Vainshtein requirements one should satisfy

$$\left[\frac{3}{2}\beta_1\frac{H}{H_f}+3\beta_2\frac{H^2}{H_f^2}+\frac{3}{2}\beta_3\frac{H^3}{H_f^3}\right]\gg\left[3\beta_1\frac{H}{H_f}+3\beta_2\frac{H^2}{H_f^2}+\beta_3\frac{H^3}{H_f^3}\right], \tag{8.48}$$

which is impossible to satisfy. We shall see in what follows how bi-gravity resolves this tension.

8.3.2 *Extensions*

8.3.2.1 Bi-gravity

We now consider the theory of bi-gravity [53],

$$\mathscr{L}=\frac{1}{2}\sqrt{-g}\left[M_{\rm Pl}^2R[g]-m^2\sum_{n=0}^{4}\beta_n\mathscr{U}_n(\mathscr{K})\right]$$
$$+\frac{1}{2}\sqrt{-f}M_f^2R[f]+\mathscr{L}_m. \tag{8.49}$$

The analogue of the Higuchi bound in that case is [43]

$$\tilde{m}^2\times\left[1+\left(\frac{H_f/M_f}{H/M_{\rm Pl}}\right)^2\right]>2H^2. \tag{8.50}$$

We recover the massive gravity bound by taking the limit $M_f\to\infty$, while keeping $M_{\rm Pl}$ and H_f finite. In that massive gravity limit it was not possible to obtain

$$\frac{H_f}{M_f}\gg\frac{H}{M_{\rm Pl}}, \tag{8.51}$$

but in bi-gravity any solution which satisfies (8.51) at early times automatically satisfies the Higuchi bound and is thus free from this ghost.

The resulting Friedman equations are then

$$H^2=\frac{1}{3M_{\rm Pl}^2}\left[\rho(a)+\sum_{n=0}^{3}\frac{3m^2\beta_n}{(3-n)!n!}\left(\frac{H}{H_f}\right)^n\right] \tag{8.52}$$

$$H_f^2=\frac{1}{3M_f^2}\left[\sum_{n=0}^{3}\frac{3m^2\beta_{n+1}}{(3-n)!n!}\left(\frac{H}{H_f}\right)^{n-3}\right]. \tag{8.53}$$

When $H_f/M_f \gg H/M_{\rm Pl}$ these equations can be used to solve for $\tilde{m}^2$ and H_f. In the region where $\beta_1 \neq 0$, the resulting bound simplifies to

$$3H^2 > 2H^2 \tag{8.54}$$

which is always satisfied ! As a result the tension between the stability of the theory and the observations is resolved in bi-gravity [43].

As an example, one can set $\beta_2 = \beta_3 = 0$ and $\beta_1 = 2M_{\rm Pl}^2$ leading to

$$H^2 = \frac{1}{6M_{\rm Pl}^2}\left(\rho(a) + \sqrt{\rho(a)^2 + \frac{12m^4 M_{\rm Pl}^6}{M_f^2}}\right), \tag{8.55}$$

which has been shown to be observationally viable, [1]. Moreover in that case the stability bound reduces to [43]

$$\left(\frac{1}{M_{\rm Pl}^2} + \frac{12M_f^2}{m^4\beta_1^2}H^4\right) > 0\,, \tag{8.56}$$

which is also always satisfied.

8.3.2.2 Decoupling Limit of Bi-gravity

In massive gravity (without introducing the Stückelberg fields), the mass term breaks a single copy of the local diffeomorphism group down to a global Lorentz group

$$\text{Diff}(M) \quad \longrightarrow \quad \text{Global Lorentz}\,. \tag{8.57}$$

In Bi-gravity (without introducing the Stückelberg fields), the mass term (or interaction term between the two metrics) breaks two copies of local diffeomorphism group down to a single copy local diffeomorphism group

$$\text{Diff}(M)\times\text{Diff}(M) \quad \longrightarrow \quad \text{Diff}(M)_{\rm diagonal}\,. \tag{8.58}$$

As a result bi-gravity is also best understood with the Stückelberg fields for the broken diffs which in turn lead to a Galileon field in its decoupling limit—dominating the interactions of the bi-gravity model.[2]

[2]The Stückelberg fields are as necessary or unnecessary in bi-gravity than they are in massive gravity. The physics is easier to follow when the Stückelberg fields are introduced but even with the Stückelberg fields, both theories are still strongly coupled at the scale $\Lambda_3 = (M_{\rm Pl}m^2)^{1/3}$ or the equivalent in terms of M_f.

In bi-gravity we can work with the following metrics:

$$\begin{cases} \text{Dynamical Metric I}: & g_{\mu\nu}(x) \\ \text{Dynamical Metric II}: & F_{\mu\nu} = f_{AB}(\phi)\partial_\mu\phi^A\partial_\nu\phi^B \,, \end{cases} \tag{8.59}$$

where to start with we can express the Stückelberg fields in terms of the helicity-0 mode (and omitting for now the helicity-1 mode)

$$\phi^A = x^a + \frac{1}{m^2 M_{\rm Pl}}\partial^a\pi(x)\,. \tag{8.60}$$

Denoting by $h_{\mu\nu}$ the fluctuations of the metric $g_{\mu\nu}$ and by $v_{\mu\nu}$ the fluctuations of the metric $f_{\mu\nu}$, then the decoupling limit of bi-gravity is [43] (omitting the helicity-1 modes)

$$\begin{aligned} S_{\rm helicity-2/0} = \int {\rm d}^4x\Bigg[& -\frac{1}{4}h^{\mu\nu}\hat{\mathcal{E}}^{\alpha\beta}_{\mu\nu}h_{\alpha\beta} - \frac{1}{4}v^{\mu\nu}\hat{\mathcal{E}}^{\alpha\beta}_{\mu\nu}v_{\alpha\beta} + \frac{\Lambda_3^3}{2}h_{\mu\nu}(x)X^{\mu\nu} \\ & + \frac{\Lambda_3^3}{2}\frac{M_{\rm Pl}}{M_f}h_{\mu\nu}(x)X^{\mu\nu}v_{\mu A}[x^a + \Lambda_3^{-3}\partial^a\pi]\left(\eta^a_\nu + \hat{\Pi}^A_\nu\right)Y^{\mu\nu}\Bigg]\,, \end{aligned} \tag{8.61}$$

with $\hat{\Pi}_{\mu\nu} = \partial_\mu\partial_\nu\pi/\Lambda_3^3$ and

$$X^{\mu\nu} = -\frac{1}{2}\sum_{n=0}^{4}\frac{\hat{\beta}_n}{(3-n)!n!}\epsilon^{\mu\cdots}\epsilon^{\nu\cdots}\left(\eta+\hat{\Pi}\right)^n\eta^{3-n} \tag{8.62}$$

$$Y^{\mu\nu} = -\frac{1}{2}\sum_{n=0}^{4}\frac{\hat{\beta}_n}{(4-n)!(n-1)!}\epsilon^{\mu\cdots}\epsilon^{\nu\cdots}\left(\eta+\hat{\Pi}\right)^{n-1}\eta^{4-n}\,. \tag{8.63}$$

As a result, the two massless spin-two fields coupled to a Galileon in a highly non-minimal way. Now including the helicity-1 modes, the decoupling limit of bi-gravity and massive gravity gives the following helicity-1/helicity-0 interactions [70]

$$\begin{aligned} & S_{\rm helicity-1/0} \\ & = -\frac{1}{8}\delta^{\mu\nu\rho\sigma}_{abcd}\left(2G^a_\mu(\delta+\hat{\Pi})^b_\nu\omega^c_\rho\delta^d_\sigma + (\delta+\hat{\Pi})^a_\mu(\delta+\hat{\Pi})^b_\nu[\omega^c_\rho\omega^d_\sigma + \delta^d_\sigma\omega^c_\alpha\omega^\alpha_\rho]\right)\,, \end{aligned}$$

with

$$\omega_{ab} = \int_0^\infty {\rm d}u\epsilon^{-2u}e^{-u\hat{\Pi}_a^{a'}}G_{a'b'}e^{-u\hat{\Pi}_b^{b'}} \tag{8.64}$$

$$G_{ab} = \partial_a B_b - \partial_b B_a = \omega_{ac}\left(\delta + \hat{\Pi}\right)^c_{\ b} + (\delta + \hat{\Pi})^c_a \omega_{cb}\,, \tag{8.65}$$

where B_a is the helicity-1 mode. Since partially massless gravity (resp. bi-gravity) should only have 4 (resp. 7) propagating degrees of freedom—since the helicity-0 must be pure gauge—and since from the above action we see that the helicity-0 mode always interact with the helicity-1 modes in bi-gravity and in massive gravity, we can conclude that there is no partially massless theory of gravity or bi-gravity [32, 43].

8.3.2.3 Galileon Duality

There are actually two (completely equivalent) ways to introduce the Stückelberg fields. Rather than the procedure (8.59)

$$\begin{cases} \text{Dynamical Metric I}: & g_{\mu\nu}(x) \\ \text{Dynamical Metric II}: & F_{\mu\nu} = f_{AB}(\phi)\partial_\mu\phi^A\partial_\nu\phi^B \\ \text{Relations between Coordinates}: & \tilde{x}^A = \phi^A(x) = x^A + \partial^A\pi(x) \end{cases} \tag{8.66}$$

we can of course use instead

$$\begin{cases} \text{Dynamical Metric I}: & \tilde{G}_{AB}(\tilde{x}) = g_{\mu\nu}(Z)\partial_A Z^\mu \partial_B Z^\nu \\ \text{Dynamical Metric II}: & f_{AB}(\tilde{x}) \\ \text{Relations between Coordinates}: & x^\mu = Z^\mu(\tilde{x}) = \tilde{x}^\mu + \partial^\mu\rho(\tilde{x}) \end{cases} \tag{8.67}$$

This leads to a remarkable property. For every Galileon field $\pi(x)$ one can define a dual Galileon field via the implicit field-dependent coordinate transformation [34]

$$\tilde{x}^A = \phi^A(x) = x^A + \partial^A\pi(x) \tag{8.68}$$

$$x^\mu = Z^\mu(\tilde{x}) = \tilde{x}^\mu + \partial^\mu\rho(\tilde{x})\,. \tag{8.69}$$

Considers a Galileon operator in D dimensions

$$\mathcal{L}_n(\pi) = \pi\epsilon\epsilon(\hat{\Pi})^{n-1}\eta^{D-n+1}\,, \tag{8.70}$$

Then every Galileon field Lagrangian in D dimensions

$$\mathcal{L}(\pi) = \sum_{n=2}^{D+1} c_n \mathcal{L}_n(\pi) \tag{8.71}$$

admits a dual formulation as a Galileon

$$\mathscr{L}(\rho) = \sum_{n=2}^{D+1} p_n \mathscr{L}_n(\rho) \tag{8.72}$$

with distinct coefficients [34]

$$p_n = \frac{1}{n}\sum_{k=2}^{D+1}(-1)^k c_k \frac{k(d-k+1)!}{(n-k)!(d-n+1)!}\,. \tag{8.73}$$

The coupling to other matter fields transforms in a local way under this duality, [35].

This could have interesting consequences for understanding the features associated with the strong coupling and the Vainshtein mechanism in this types of theories.

8.3.2.4 Quasi-Dilaton Massive Gravity

To finish, let us present another extension of massive gravity known as quasi-dilaton. The same arguments found previously can be applied to generic cosmological solutions on quasi-dilaton massive gravity [12]

$$S_E = \int \mathrm{d}^4x\sqrt{-g}\left\{\frac{M_{\rm Pl}^2}{2}\left[R - \frac{\omega}{M_{\rm Pl}^2} g^{\mu\nu}\partial_\mu\sigma\partial_\nu\sigma - \frac{m^2}{4}\mathscr{U}[\tilde{K}]\right] + \mathscr{L}_M(g_{\mu\nu},\psi)\right\}, \tag{8.74}$$

with

$$\tilde{\mathscr{K}}^{\mu}_{\ \nu} = \delta^\mu_\nu - e^{\sigma/M_{\rm Pl}}\sqrt{g^{\mu\alpha}\partial_\alpha\phi^a\partial_\nu\phi^b\eta_{ab}}\,. \tag{8.75}$$

This model avoids the no-FRW argument formulated previously thanks to the quasi-dilaton field σ. Generically one finds a non-zero kinetic term for the helicity-0 mode, showing that the general cosmological solutions are healthy.

A generalized version of the quasi-dilaton was shown to provide stable self-accelerating solutions [14,16]. For the generalized quasi-dilaton, the action takes the same form as in (8.74) with the expression (8.75) for with the generalized expression for the tensor $\tilde{\mathscr{K}}^{\mu}_{\ \nu}$,

$$\tilde{\mathscr{K}}^{\mu}_{\ \nu} = \delta^\mu_\nu - e^{\sigma/M_{\rm Pl}}\sqrt{g^{\mu\alpha}\partial_\alpha\phi^a\partial_\nu\phi^b\tilde{\eta}_{ab}}\,, \tag{8.76}$$

and

$$\tilde{\eta}_{\mu\nu} = \eta_{\mu\nu} - \frac{\alpha_\sigma}{\Lambda_3^3}e^{-2\sigma/M_{\rm Pl}}\partial_\mu\sigma\partial_\nu\sigma\,. \tag{8.77}$$

Of course this theory can be generalized to arbitrary reference metrics $\eta_{\mu\nu} \rightarrow f_{\mu\nu}$, but it makes more physical sense to keep Minkowski as the reference metric. Effectively this corresponds to a theory of massive gravity with a dynamical mass and couplings and a dynamical reference metric governed by the quasi-dilaton scalar field. The self-accelerating solutions in that generalized theory were shown to be free of instabilities, making them particularly appealing [14, 16].

8.4 Summary

In these proceedings we have first established that massive gravity is a useful toy model to understand higher dimensional theories. They potentially exhibit both self-acceleration and self-tuning (degravitating) solutions. FRW solutions (fully homogeneous and isotropic) cannot directly emerge from massive gravity. Instead one can consider solutions which are inhomogeneous beyond the observable Universe, which is actually closer to natural concepts of modern cosmology. Inhomogeneous or anisotropic solutions (or both simultaneously) do exist in massive gravity. Not all such solutions are stable but some are.

We have shown how for Partially Massless gravity, the Higuchi bound was automatically satisfied for any choice of matter. Unfortunately the decoupling limit makes it easy to see the absence of partially massless (bi)gravity.

For massive gravity on a fixed FRW reference metric, the bound is in conflict with observations (it would effectively impose the mass to be much larger than the Hubble parameter at early time which would be ruled out observationally).

For bi-gravity on the other hand the Higuchi is almost always satisfied regardless of the choice of matter as long as $H \ll H_f$, where H is the Hubble parameter of the metric to which matter couples to.

Finally we have shown how to extend massive graviton to include a quasi-dilaton scalar field which admits stable self-accelerating solutions.

Needless to say this is still very early days for the cosmology in massive gravity and bi-gravity and the amount of different subclasses of models considered illustrates how rich and yet complex the cosmology of massive (bi-)gravity is.

Acknowledgements AJT is supported by Department of Energy Early Career Award DE-SC0010600. AJT wishes to thank the organizers of the seventh Aegean Summer School for an excellent meeting.

References

1. Y. Akrami, T.S. Koivisto, D.F. Mota, M. Sandstad, J. Cosmology Astroparticle Phys. **1310**, 046 (2013). [arXiv:1306.0004 [hep-th]]; Y. Akrami, T.S. Koivisto, M. Sandstad, arXiv:1302.5268 [astro-ph.CO]; Y. Akrami, T.S. Koivisto, M. Sandstad, J. High Energy Pract. **1303**, 099 (2013). [arXiv:1209.0457 [astro-ph.CO]]

2. Y. Akrami, T.S. Koivisto, M. Sandstad, J. High Energy Pract. **1303**, 099 (2013). [arXiv:1209.0457 [astro-ph.CO]]
3. M. Andrews, K. Hinterbichler, J. Stokes, M. Trodden, Class. Quant. Gravity **30**, 184006 (2013). [arXiv:1306.5743 [hep-th]]
4. N. Arkani-Hamed, S. Dimopoulos, G. Dvali, G. Gabadadze, arXiv:hep-th/0209227
5. M. Berg, I. Buchberger, J. Enander, E. Mortsell, S. Sjors, J. Cosmology Astroparticle Phys. **1212**, 021 (2012). [arXiv:1206.3496 [gr-qc]]
6. A.H. Chamseddine, V. Mukhanov, J. High Energy Pract. **1108**, 091 (2011). [arXiv:1106.5868 [hep-th]]
7. C. Charmousis, R. Gregory, N. Kaloper, A. Padilla, J. High Energy Pract. **0610**, 066 (2006). [hep-th/0604086]
8. D. Comelli, M. Crisostomi, F. Nesti, L. Pilo, J. High Energy Pract. **1203**, 067 (2012). [arXiv:1111.1983 [hep-th]]. [Erratum-J. High Energy Pract. **1206**, 020 (2012)]
9. D. Comelli, F. Nesti, L. Pilo, J. Cosmology Astroparticle Phys. **1405**, 036 (2014). arXiv:1307.8329 [hep-th]; D. Comelli, F. Nesti, L. Pilo, J. High Energy Pract. **1307**, 161 (2013). arXiv:1305.0236 [hep-th]
10. G. D'Amico, C. de Rham, S. Dubovsky, G. Gabadadze, D. Pirtskhalava, A.J. Tolley, Phys. Rev. D **84**, 124046 (2011). [arXiv:1108.5231 [hep-th]]
11. G. D'Amico, G. Gabadadze, L. Hui, D. Pirtskhalava, Phys. Rev. D **87**(6), 064037 (2013). [arXiv:1206.4253 [hep-th]]
12. G. D'Amico, G. Gabadadze, L. Hui, D. Pirtskhalava, Phys. Rev. D **87**(6), 064037 (2013). [arXiv:1206.4253 [hep-th]]
13. G. D'Amico, G. Gabadadze, L. Hui, D. Pirtskhalava, Class. Quant. Grav. **30**, 184005 (2013). [arXiv:1304.0723 [hep-th]]
14. A. De Felice, S. Mukohyama, Phys. Lett. B **728**, 622 (2014). [arXiv:1306.5502 [hep-th]]
15. A. De Felice, A.E. Gumrukcuoglu, S. Mukohyama, Phys. Rev. Lett. **109**, 171101 (2012). [arXiv:1206.2080 [hep-th]]
16. A. De Felice, A.E. Gumrukcuoglu, S. Mukohyama, Phys. Rev. D **88**, 124006 (2013). [arXiv:1309.3162 [hep-th]]
17. A. De Felice, A.E. Gmrkolu, C. Lin, S. Mukohyama, Class. Quant. Gravity **30**, 184004 (2013). [arXiv:1304.0484 [hep-th]]
18. A. De Felice, A.E. Gmrkolu, C. Lin, S. Mukohyama, J. Cosmology Astroparticle Phys. **1305**, 035 (2013). [arXiv:1303.4154 [hep-th]]
19. A. De Felice, A.E. Gmrkolu, C. Lin, S. Mukohyama, Class. Quant. Gravity **30**, 184004 (2013). [arXiv:1304.0484 [hep-th]]
20. C. de Rham, J. High Energy Pract. **0801**, 060 (2008). [arXiv:0707.0884 [hep-th]]
21. C. de Rham, Can. J. Phys. **87**, 201 (2009). [arXiv:0810.0269 [hep-th]]
22. C. de Rham, Living Rev. Rel. **17**, 7 (2014). arXiv:1401.4173 [hep-th]
23. C. de Rham, G. Gabadadze, Phys. Rev. D **82**, 044020 (2010). [arXiv:1007.0443 [hep-th]]
24. C. de Rham, A.J. Tolley, J. Cosmology Astroparticle Phys. **1005**, 015 (2010). [arXiv:1003.5917 [hep-th]]
25. C. de Rham, S. Renaux-Petel, J. Cosmology Astroparticle Phys. **1301**, 035 (2013). [arXiv:1206.3482 [hep-th]]
26. C. de Rham, S. Hofmann, J. Khoury, A.J. Tolley, J. Cosmology Astroparticle Phys. **0802**, 011 (2008). [arXiv:0712.2821 [hep-th]]
27. C. de Rham, G. Dvali, S. Hofmann, J. Khoury, O. Pujolas, M. Redi, A.J. Tolley, Phys. Rev. Lett. **100**, 251603 (2008). [arXiv:0711.2072 [hep-th]]
28. C. de Rham, J. Khoury, A.J. Tolley, Phys. Rev. Lett. **103**, 161601 (2009). [arXiv:0907.0473 [hep-th]]
29. C. de Rham, J. Khoury, A.J. Tolley, Phys. Rev. D **81**, 124027 (2010). [arXiv:1002.1075 [hep-th]]
30. C. de Rham, G. Gabadadze, A.J. Tolley, Phys. Rev. Lett. **106**, 231101 (2011). [arXiv:1011.1232 [hep-th]]

31. C. de Rham, G. Gabadadze, L. Heisenberg, D. Pirtskhalava, Phys. Rev. D **83**, 103516 (2011). [arXiv:1010.1780 [hep-th]]
32. C. de Rham, K. Hinterbichler, R.A. Rosen, A.J. Tolley, Phys. Rev. D **88**(2), 024003 (2013). [arXiv:1302.0025 [hep-th]]
33. C. de Rham, G. Gabadadze, L. Heisenberg, D. Pirtskhalava, Phys. Rev. D **87**(8), 085017 (2013). [arXiv:1212.4128]
34. C. de Rham, M. Fasiello, A. J. Tolley, Phys. Lett. B **733**(2), 46 (2014). [arXiv:1308.2702 [hep-th]]
35. C. de Rham, L. Keltner, A.J. Tolley, Phys. Rev. D **90**, 024050 (2014) arXiv:1403.3690 [hep-th]
36. C. Deffayet, Phys. Lett. B **502**, 199 (2001). [hep-th/0010186]
37. C. Deffayet, G. Esposito-Farese, A. Vikman, Phys. Rev. D **79**, 084003 (2009). [arXiv:0901.1314 [hep-th]]
38. G.R. Dvali, G. Gabadadze, M. Porrati, Phys. Lett. B **485**, 208 (2000). [hep-th/0005016]
39. G. Dvali, G. Gabadadze, M. Shifman, Published in Minneapolis 2002, Continuous advances in QCD, pp. 566–581. arXiv:hep-th/0208096
40. G. Dvali, G. Gabadadze, M. Shifman, Phys. Rev. D **67**, 044020 (2003). [arXiv:hep-th/0202174]
41. G. Dvali, S. Hofmann, J. Khoury, Phys. Rev. D **76**, 084006 (2007). [arXiv:hep-th/0703027]
42. M. Fasiello, A.J. Tolley, J. Cosmology Astroparticle Phys. **1211**, 035 (2012). [arXiv:1206.3852 [hep-th]]
43. M. Fasiello, A. J. Tolley, J. Cosmology Astroparticle Phys. **1312**, 002 (2013). [arXiv:1308.1647 [hep-th]]
44. R.P. Geroch, J.H. Traschen, Phys. Rev. D **36**, 1017 (1987) [Conf. Proc. C **861214**, 138 (1986)]
45. D. Gorbunov, K. Koyama, S. Sibiryakov, Phys. Rev. D **73**, 044016 (2006). [hep-th/0512097]
46. P. Gratia, W. Hu, M. Wyman, Phys. Rev. D **86**, 061504 (2012). [arXiv:1205.4241 [hep-th]]
47. P. Gratia, W. Hu, M. Wyman, Class. Quant. Gravity **30**, 184007 (2013). [arXiv:1305.2916 [hep-th]]
48. S. Groot Nibbelink, M. Peloso, M. Sexton, Eur. Phys. J. C **51**, 741 (2007). [hep-th/0610169]
49. A.E. Gumrukcuoglu, C. Lin, S. Mukohyama, J. Cosmology Astroparticle Phys. **1111**, 030 (2011). [arXiv:1109.3845 [hep-th]]
50. A.E. Gumrukcuoglu, C. Lin, S. Mukohyama, J. Cosmology Astroparticle Phys. **1203**, 006 (2012). [arXiv:1111.4107 [hep-th]]
51. A.E. Gumrukcuoglu, C. Lin, S. Mukohyama, Phys. Lett. B **717**, 295 (2012). [arXiv:1206.2723 [hep-th]]
52. S.F. Hassan, R.A. Rosen, J. High Energy Pract. **1107**, 009 (2011). [arXiv:1103.6055 [hep-th]]
53. S.F. Hassan, R.A. Rosen, J. High Energy Pract. **1202**, 126 (2012).[arXiv:1109.3515 [hep-th]]
54. S.F. Hassan, R.A. Rosen, A. Schmidt-May, J. High Energy Pract. **1202**, 026 (2012). [arXiv:1109.3230 [hep-th]]
55. A. Higuchi, Nucl. Phys. B **282**, 397 (1987)
56. K. Hinterbichler, R.A. Rosen, J. High Energy Pract. **1207**, 047 (2012). [arXiv:1203.5783 [hep-th]]
57. K. Hinterbichler, J. Stokes, M. Trodden, Phys. Lett. B **725**, 1 (2013). [arXiv:1301.4993 [astro-ph.CO]]
58. G.W. Horndeski, Int. J. Theory Phys **10**, 363 (1974)
59. M. Jaccard, M. Maggiore, E. Mitsou, Phys. Rev. D **88**(4), 044033 (2013). [arXiv:1305.3034 [hep-th]]
60. N. Khosravi, G. Niz, K. Koyama, G. Tasinato, J. Cosmology Astroparticle Phys. **1308**, 044 (2013). [arXiv:1305.4950 [hep-th]]
61. N. Khosravi, H.R. Sepangi, S. Shahidi, Phys. Rev. D **86**, 043517 (2012). [arXiv:1202.2767 [gr-qc]]
62. T. Kobayashi, M. Siino, M. Yamaguchi, D. Yoshida, Phys. Rev. D **86**, 061505 (2012). [arXiv:1205.4938 [hep-th]]
63. K. Koyama, Phys. Rev. D **72**, 123511 (2005). [hep-th/0503191]
64. G. Leon, J. Saavedra, E.N. Saridakis, Class. Quant. Gravity **30**, 135001 (2013). [arXiv:1301.7419 [astro-ph.CO]]

65. M.A. Luty, M. Porrati, R. Rattazzi, J. High Energy Pract. **0309**, 029 (2003). [hep-th/0303116]
66. L. Modesto, S. Tsujikawa, Phys. Lett. B **727**, 48 (2013). [arXiv:1307.6968 [hep-th]]
67. S. Mukohyama, Phys. Rev. D **90**, 024050 (2014). arXiv:1309.2146 [hep-th]
68. A. Nicolis, R. Rattazzi, J. High Energy Pract. **0406**, 059 (2004). [hep-th/0404159]
69. A. Nicolis, R. Rattazzi, E. Trincherini, Phys. Rev. D **79**, 064036 (2009). [arXiv:0811.2197 [hep-th]]
70. N.A. Ondo, A.J. Tolley, J. High Energy Pract. **1311**, 059 (2013). [arXiv:1307.4769 [hep-th]]
71. N. Tamanini, E.N. Saridakis, T.S. Koivisto, J. Cosmology Astroparticle Phys. **1402**, 015 (2014). [arXiv:1307.5984 [hep-th]]
72. G. Tasinato, K. Koyama, G. Niz, Class. Quant. Gravity **30**, 184002 (2013). [arXiv:1304.0601 [hep-th]]
73. A. Tolley, Helicity zero mode in cosmological solutions in massive gravity and bigravity contribution to the Workshop on Cosmic Acceleration, Carnegie Mellon University, Pittsburgh, PA August 24, 2012-August 26, 2012
74. A.I. Vainshtein, Phys. Lett. B **39**, 393 (1972)
75. B. Vakili, N. Khosravi, Phys. Rev. D **85**, 083529 (2012). [arXiv:1204.1456 [gr-qc]]
76. M.S. Volkov, Phys. Rev. D **86**, 061502 (2012). [arXiv:1205.5713 [hep-th]]
77. M.S. Volkov, Phys. Rev. D **86**, 104022 (2012). [arXiv:1207.3723 [hep-th]]
78. M.S. Volkov, J. High Energy Pract. **1201**, 035 (2012). [arXiv:1110.6153 [hep-th]]
79. M. von Strauss, A. Schmidt-May, J. Enander, E. Mortsell, S. F. Hassan, J. Cosmology Astroparticle Phys. **1203**, 042 (2012). [arXiv:1111.1655 [gr-qc]]
80. D.-J. Wu, Y.-S. Piao, Y.-F. Cai, Phys. Lett. B **721**, 7 (2013). [arXiv:1301.4326 [hep-th]]
81. M. Wyman, W. Hu, P. Gratia, Phys. Rev. D **87**(8), 084046 (2013). [arXiv:1211.4576 [hep-th]]

Part III
Further Modifications at Large Distances

Chapter 9
Higher-Spin Theory and Space-Time Metamorphoses

M.A. Vasiliev

Abstract Introductory lectures on higher-spin gauge theory given at seventh Aegean workshop on non-Einstein theories of gravity. The emphasis is on qualitative features of the higher-spin gauge theory and peculiarities of its space-time interpretation. In particular, it is explained that Riemannian geometry cannot play a fundamental role in the higher-spin gauge theory. The higher-spin symmetries are argued to occur at ultra high energy scales beyond the Planck scale. This suggests that the higher-spin gauge theory can help to understand Quantum Gravity. Various types of higher-spin dualities are briefly discussed.

9.1 Introduction

Higher-spin (HS) gauge theories form a class of theories exhibiting infinite-dimensional symmetries which go beyond conventional lower-spin symmetries. The primary goal of these lectures is to focus on qualitative aspects of HS gauge theories avoiding technical details as much as possible. The emphasis is on possible consequences of HS symmetries for our understanding of space-time. It will be explained in particular that in the setup of HS gauge theories the usual concepts of Riemannian geometry such as metric, local event and space-time dimension cannot play a fundamental role. The HS symmetries will be argued to occur at ultra high energy scales beyond the Planck scale. Having a potential to describe transPlanckian energies, HS gauge theory can shed light on the problem of Quantum Gravity. Various aspects of HS dualities including AdS/CFT and duality with quantum mechanics are briefly discussed.

M.A. Vasiliev (✉)
I.E. Tamm Department of Theoretical Physics, Lebedev Physical Institute, Leninsky Prospect 53, 119991 Moscow, Russia
e-mail: vasiliev@lpi.ru

E. Papantonopoulos (ed.), *Modifications of Einstein's Theory of Gravity at Large Distances*, Lecture Notes in Physics 892, DOI 10.1007/978-3-319-10070-8_9

9.2 Lower-Spin Global Symmetries

The fundamental example of a lower-spin symmetry is provided by the Poincaré symmetry which underlies relativistic theories. It acts on coordinates of Minkowski space-time as $\delta x^a = \epsilon^a + \epsilon^a{}_b x^b$ where ϵ^a and ϵ^{ab} are parameters of infinitesimal translations and Lorentz rotations, respectively. One can write

$$\delta x^a = [T, x^a], \qquad T = \epsilon^n P_a + \epsilon^{ab} M_{ab}, \tag{9.1}$$

where

$$P_a = \frac{\partial}{\partial x^a}, \qquad M_{ab} = x_a \frac{\partial}{\partial x^b} - x_b \frac{\partial}{\partial x^a}$$

are the generators of the Poincaré algebra $iso(d-1,1)$ obeying the commutation relations

$$[M_{ab}, P_c] = P_a \eta_{bc} - P_b \eta_{ac},$$

$$[M_{ab}, M_{cd}] = M_{ad}\eta_{bc} - M_{bd}\eta_{ac} - M_{ac}\eta_{bd} + M_{bc}\eta_{ad},$$

$$[P_a, P_b] = 0,$$

where η_{ab} is the Minkowski metric.

The Poincaré algebra admits the (anti-) de Sitter deformation l with

$$[P_a, P_b] = \Lambda M_{ab},$$

which describes symmetries of either anti-de Sitter space at $\Lambda < 0$ ($l = o(d-1,2)$) or de Sitter space at $\Lambda > 0$ ($l = o(d,1)$). At $\Lambda = 0$, $l = iso(d-1,1)$ describes the symmetries of Minkowski space.

Supersymmetry is the extension of the Poincaré symmetry by supergenerators Q_A obeying relations

$$\{Q_A, Q_B\} = \gamma^a_{AB} P_a,$$

$$[M_{ab}, Q_A] = \sigma_{abA}{}^B Q_B, \qquad \sigma_{ab} = \frac{1}{4}[\gamma_a, \gamma_b],$$

where $A, B = 1, 2, 3, 4$ are the Majorana spinor indices in four dimensions. Note that, being fermions, supergenerators obey anticommutation relations.

Internal symmetry generators T_i are space-time invariant

$$[T_i, P_a] = 0, \qquad [T^i, M_{ab}] = 0.$$

In particular, the symmetries of the Standard Model $T_i \in su(3) \times su(2) \times u(1)$ are of this type.

To complete the list of symmetries that play a role in conventional lower-spin theories it remains to mention conformal (super)symmetries. These will be discussed in some more detail below.

9.3 Local Symmetries

A useful viewpoint is that any global symmetry is the remnant of a local symmetry with parameters like $\varepsilon^a(x), \varepsilon^{ab}(x), \varepsilon^\alpha(x), \varepsilon^i(x)$ being arbitrary functions of space-time coordinates. Local symmetries are symmetries of the full theory. Global symmetries are symmetries of some its particular solution.

For example, the infinitesimal diffeomorphisms $\delta x^a = \varepsilon^a(x)$ are symmetries of GR while the global symmetries with $\varepsilon^a(x) = \epsilon^a + \epsilon^a{}_b x^b$ are symmetries of the Minkowski solution $g_{ab} = \eta_{ab}$ of the Einstein equations.

Let

$$S = \int_{M^d} L(\varphi(x), \partial_a \varphi(x), \ldots)$$

be invariant under a global symmetry g with parameters ϵ^n $(n = a, \alpha, i, \ldots)$. Letting the symmetry parameters be arbitrary functions of space-time coordinates, $\epsilon^n \to \varepsilon^n(x)$, we obtain that

$$\delta S = -\int_{M^d} J^a_n(\varphi) \partial_a \varepsilon^n(x)$$

since δS should be zero at $\partial_a \varepsilon^n(x) = 0$. $J^a_n(\varphi)$ are conserved currents since $\partial_a J^a_n(\varphi) = 0$ by virtue of the field equations $\delta S = 0$.

The local symmetry is achieved with the aid of gauge fields A^n_a that have the transformation law

$$\delta A^n_a = \partial_a \varepsilon^n + \ldots ,$$

where the ellipsis denotes possible field-dependent terms. The following modification of the action

$$S \longrightarrow S + \Delta S + \ldots , \qquad \Delta S = \int_{M^d} J^a_n(\varphi) A^n_a(x)$$

preserves local symmetry in the lowest order in interactions. The term ΔS describes the so-called Noether current interactions.

There is, however, a subtlety if $\varphi(x)$ were themselves gauge fields with gauge parameters ε'. In this case it may happen that $J^a_n(\varphi)$ is not invariant under the ε'

symmetry. Hence the Noether current interaction for several gauge fields may be obstructed by gauge symmetries.

Localization of various types of lower-spin symmetries leads to important classes of gauge field theories.

9.3.1 Yang–Mills Fields

The Yang–Mills theory is responsible for the localization of internal symmetries. For a Lie algebra l with generators T_i, Yang–Mills fields $A^i_a(x)$ and symmetry parameters ε^i are valued in l

$$A_a(x) = A^i_a(x)T_i\,, \qquad \varepsilon(x) = \varepsilon^i(x)T_i\,.$$

The Yang–Mills gauge transformation is

$$\delta A_a(x) = D_a\varepsilon(x)\,,$$

where

$$D_a\varepsilon(x) = \partial_a\varepsilon(x) + [A_a(x)\,,\varepsilon(x)]$$

is the covariant derivative. The commutator of the covariant derivatives gives the Yang–Mills curvature

$$[D_a\,, D_b] = R_{ab}\,, \qquad R_{ab} = \partial_a A_b - \partial_b A_a + [A_a, A_b]\,,$$

which has the transformation law

$$\delta R_{ab} = [R_{ab}, \varepsilon]\,.$$

Needless to say that the Yang–Mills fields play a prominent role in the modern theory of non-gravitational fundamental interactions, i.e. the Standard Model.

9.3.2 Einstein–Cartan Gravity and Supergravity

Localization of the Poincaré symmetry leads to the Cartan formulation of Einstein gravity. The Yang–Mills gauge fields $A^n_\nu = (e_\nu{}^a, \omega_\nu{}^{ab})$ associated with the Poincaré algebra include the frame field (vielbein) $e_\nu{}^a$ and the Lorentz connection $\omega_\nu{}^{ab}$. The frame field $e_\nu{}^a$ relates base indices ν with the fiber ones a. (In Minkowski space in Cartesian coordinates, where $e_\nu{}^a = \delta^a_\nu$, the two types of indices can be identified.) The gauge transformations have the form

$$\delta e_\nu{}^a(x) = \partial_\nu \varepsilon^a(x) + \omega_\nu{}^a{}_b(x)\varepsilon^b(x) - \varepsilon^a{}_b(x) e_\nu{}^b(x) + \Delta e_\nu{}^a\,,$$

$$\delta \omega_\nu{}^{ab}(x) = \partial_\nu \varepsilon^{ab}(x) + \omega_\nu{}^a{}_c(x)\varepsilon^{cb}(x) - \omega_\nu{}^b{}_c(x)\varepsilon^{ca}(x) + \Delta\omega_\nu{}^{ab}\,.$$

Here $\Delta e_\nu{}^a$ and $\Delta\omega_\nu{}^{ab}$ denote some corrections to the Yang–Mills transformation law, which are proportional to the curvatures

$$R_{\nu\mu}{}^a = \partial_\nu e_\mu{}^a + \omega_\nu{}^a{}_b e_\mu{}^b - (\nu \leftrightarrow \mu)\,, \qquad R_{\nu\mu}{}^{ab} = \partial_\nu \omega_\mu{}^{ab} + \omega_\nu{}^a{}_c \omega_\mu{}^{cb} - (\nu \leftrightarrow \mu)\,.$$

The zero-torsion constraint $R_{\nu\mu}{}^a = 0$ expresses the Lorentz connection in terms of the frame field and its derivatives: $\omega = \omega(e, \partial e)$. In this case $R_{\nu\mu}{}^{\rho\sigma}$ equals to the Riemann tensor. Recall that the relation of the metric with the frame field is $g_{\nu\mu} = e_\nu{}^a e_\mu{}^b \eta_{ab}$.

Localization of supersymmetry extends the gravitational fields by the spin-3/2 gauge field gravitino $\psi_{\nu\alpha}$ with the gauge transformation law

$$\delta\psi_{\nu\alpha} = D_\nu \varepsilon_\alpha + \dots .$$

Gauge theories of this type are called supergravities, constituting a very interesting class of extensions of the theory of gravity. (See e.g. [1] and references therein. Note that the construction of supergravity in terms of the gauge fields of the supersymmetry algebra was suggested in [2].)

9.3.3 *Spontaneous Symmetry Breaking*

Generally, one should distinguish between the symmetry G of some equations and a symmetry $\tilde{G}$ of some their particular solution. For example, for the case of the Higgs field $H^i(x) = H^i_0 + h^i(x)$, the unbroken part $\tilde{G} \subset G$ is a residual symmetry of H^i_0: $\tilde{G} = SU(3) \times U(1)$ in the Standard Model. For H^i_0 having a non-zero dimension $[H^i_0] = cm^{-1} \sim GeV$, spontaneous symmetry breaking is a low-energy effect. In other words, the symmetry restores at $E > H^i_0$.

In the unbroken regime, the gauge fields associated with the usual lower-spin symmetries describe massless particles of spin one $A_\nu{}^i$, spin $3/2$ $\psi_{\nu\alpha}$ and spin two $e_\nu{}^a$, $\omega_\nu{}^{ab}$.

9.4 General Properties of HS Theory

The key question is whether it is possible to go to larger HS symmetries. If yes, what are HS symmetries and HS counterparts of the lower-spin theories including GR? What are physical motivations for their study and possible outputs?

9.4.1 Fronsdal Fields

As shown by Fronsdal [3], all symmetric massless HS fields are gauge fields. They are described by rank-s symmetric tensors $\phi_{\nu_1\ldots\nu_s}$ obeying the double tracelessness condition $\phi^{\rho}{}_{\rho}{}^{\mu}{}_{\mu\nu_5\ldots\nu_s} = 0$. The gauge transformation is

$$\delta\phi_{\nu_1\ldots\nu_s}(x) = \partial_{(\nu_1}\varepsilon_{\nu_2\ldots\nu_s)}(x)\,, \tag{9.2}$$

where the gauge parameter is symmetric and traceless

$$\varepsilon^{\mu}{}_{\mu\nu_3\ldots\nu_{s-1}} = 0\,. \tag{9.3}$$

The field equations have the form

$$\mathscr{R}_{\nu_1\ldots\nu_s}(x) = 0\,,$$

where the Ricci-like tensor $\mathscr{R}_{\nu_1\ldots\nu_s}(x)$ is

$$\mathscr{R}_{\nu_1\ldots\nu_s}(x) = \Box\phi_{\nu_1\ldots\nu_s}(x) - s\,\partial_{(\nu_1}\partial^{\mu}\phi_{\nu_2\ldots\nu_s\mu)}(x) + \frac{s(s-1)}{2}\partial_{(\nu_1}\partial_{\nu_2}\phi^{\mu}{}_{\nu_3\ldots\nu_s\mu)}(x)\,.$$

The gauge invariant Fronsdal action is

$$S = \int_{M^d}\left(\frac{1}{2}\phi^{\nu_1\ldots\nu_s}\mathscr{R}_{\nu_1\ldots\nu_s}(\phi) - \frac{1}{8}s(s-1)\phi_{\mu}{}^{\mu\nu_3\ldots\nu_s}\mathscr{R}^{\rho}{}_{\rho\nu_3\ldots\nu_s}(\phi)\right).$$

9.4.2 No-Go and the Role of $(A)dS$

In the 1960s of the last century it was argued by Weinberg [4] and Coleman and Mandula [5] that HS symmetries cannot be realized in a nontrivial local field theory in Minkowski space. In the seventies it was shown by Aragone and Deser [6] that HS gauge symmetries are incompatible with GR within an expansion over Minkowski space. The general belief was that nontrivial interactions of massless HS fields cannot be introduced.

Nevertheless, in the 1980s, it was shown by light-cone [7, 8] and covariant methods [9, 10] that some non-gravitational HS interactions can be constructed at least at the cubic order. These results suggested that some consistent HS theory should exist.

The further progress resulted from the observation that the consistent formulation of the HS gauge theory requires a curved background instead of the flat Minkowski. The most symmetric curved cousins of the flat Minkowski space are de Sitter and anti-de Sitter spaces. That HS theories admit consistent interactions including the gravitational interaction in $(A)dS$ background was shown in [11, 12]. In agreement with the no-go statements, the limit of zero cosmological constant $\Lambda \to 0$ turns out to be singular so that, indeed, HS theories with unbroken HS symmetries do not exist in the Minkowski background.

9.4.3 HS Symmetries Versus Riemannian Geometry

The HS symmetries and the space-time symmetries do not commute simply because HS generators are higher-rank Lorentz tensors

$$[T^a, T^{HS}] = T^{HS}, \qquad [T^{ab}, T^{HS}] = T^{HS}.$$

However, the same commutation relations imply that HS generators transform the space-time generators to the HS generators. Since the gauge fields for space-time generators are the gravitational frame field and Lorentz connection, this implies that HS transformations map the gravitational fields (metric) to the HS fields.

This simple observation has the far-going consequence that the Riemannian geometry is not appropriate for the HS theory, implying in particular that the concept of local event may become illusive in the HS theory!

Though it is not appropriate to use the metric tensor in the HS theory, we do not want to give up the coordinate independence of GR. Fortunately, this can be achieved in the framework of the formalism of differential forms.

Differential forms are totally antisymmetric tensors. A p-form is a rank-p totally antisymmetric tensor $\omega(x) = \theta^{\nu_1} \dots \theta^{\nu_p} \omega_{\nu_1 \dots \nu_p}(x)$ where θ^ν are anticommuting symbols (variables)

$$\theta^\nu \theta^\mu = -\theta^\mu \theta^\nu$$

usually called differentials $\theta^\nu = dx^\nu$. The invariant differentiation is provided by the exterior (de Rham) derivative

$$d = \theta^\nu \frac{\partial}{\partial x^\nu}, \qquad d^2 = 0.$$

This formalism is covariant because, due to the total antisymmetrization of indices, symmetric Christoffel symbols drop out from the covariant derivatives. In this language, the connections $A = \theta^\nu A^i_\nu T_i$ are one-forms, while the curvatures $R = D^2$ with $D = d + A$ are two-forms.

Farther elaboration of this language in application to HS theory leads eventually to a deeper understanding of fundamental concepts of space-time including its dimension.

9.4.4 HS Gauge Theory, Quantum Gravity and String Theory

As explained in more detail below, the HS symmetry is in a certain sense maximal relativistic symmetry. Hence one can speculate that it cannot result from spontaneous breakdown of a larger symmetry. This implies that the HS symmetries are manifest at ultrahigh energies above any scale including the Planck scale. If this is true, the HS gauge theory should capture effects of Quantum Gravity. This opens a unique possibility for the theoretical study of the unreachable by experimental tests energy scale of Quantum Gravity by means of the highly restrictive HS symmetry.

On the other hand, since the lower-spin symmetries are subalgebras of the HS symmetries, it is natural to expect that the lower-spin theories can correspond to low-energy limits of the HS theory with spontaneously broken HS symmetries.

A related issue is a connection of HS theory with String Theory. A natural conjecture is that String Theory can be interpreted as a spontaneously broken theory of the HS type, where $s > 2$ fields acquire nonzero masses. An interesting recent conjecture [13] is that String Theory can be identified with the full quantum HS theory.

9.4.5 HS AdS/CFT Correspondence

That the HS gauge theories are most naturally formulated in the anti-de Sitter background makes them interesting from the perspective of AdS/CFT correspondence [14–16]. Various aspects of the HS holography were discussed by many authors starting from [17–19] (see also [20, 21]). However, the concrete proposal is due to Klebanov and Polyakov [22] who conjectured that the AdS_4 HS theory is dual to $3d$ the vectorial conformal models. This hypothesis was successfully checked by Giombi and Yin [23], that triggered a lot of interest to the HS holography. The conjecture of Klebanov and Polyakov was later extended to the fermionic boundary systems [24, 25] as well to the AdS_3/CFT_2 correspondence [26–28].

The HS holography has several features which give a hope that its analysis may help to uncover the origin of AdS/CFT. Indeed, as discussed in some more detail below, a progress in this direction has been achieved in [29]. It should be stressed that the HS holography does not rely on supersymmetry and is a weak–weak duality that therefore can be checked directly on the both sides. For more detail on the HS holography we refer the reader to [30, 31].

9.5 Global HS Symmetry: Idea of Construction

The simplest way to figure out what is a HS symmetry is via the AdS/CFT correspondence. Namely, the global HS symmetry of the most symmetric AdS_{d+1} solution can be identified with the maximal symmetry of the d-dimensional free conformal fields. In the most cases the latter are identified with the massless scalar and/or spinor.

Consider KG massless equation in d-dimensional Minkowski space

$$\Box C(x) = 0\,, \qquad \Box = \eta^{ab}\frac{\partial^2}{\partial x^a \partial x^b}\,.$$

The conformal HS symmetry is the symmetry of this equation. What is this symmetry? Its structure was first elaborated for $d = 3$ in [32] and soon after by Eastwood [33] for any d.

Of course, this symmetry contains the Poincaré transformations as well as the scale transformation (dilatation)

$$\delta C(x) = \varepsilon D C(x)\,, \qquad D = x^a \frac{\partial}{\partial x^a} + \frac{d}{2} - 1$$

and the special conformal transformations

$$\delta C(x) = \varepsilon_a K^a C(x)\,, \qquad K^a = (x^2\eta^{ab} - 2x^a x^b)\frac{\partial}{\partial x^b} + (2-d)x^a\,.$$

Altogether P_a, M_{ab}, K^a and D form the conformal Lie algebra $o(d,2)$.

To figure out the structure of the whole conformal HS algebra it is useful to consider an auxiliary problem.

9.5.1 Auxiliary Problem

Consider the equations

$$D\mathcal{C}_A(x) = 0\,, \tag{9.4}$$

where $\mathcal{C}_A(x)$ is a set of fields valued in some space V (the label A) and

$$D = d + \omega(x)\,, \qquad \omega_A{}^B(x) = \omega^\Omega(x) T_{\Omega A}{}^B$$

is a covariant derivative acting in the space V treated as a $gl(V)$-module. That is $\omega(x)$ is some $gl(V)$-connection. The covariant derivative D is demanded to be flat, i.e.

$$D^2 = 0\,. \tag{9.5}$$

Clearly, Eqs. (9.4) and (9.5) are invariant under the gauge transformation

$$\delta\mathcal{C}_A(x) = -\varepsilon_A{}^B(x)\mathcal{C}_B(x)\,, \qquad \varepsilon_A{}^B(x) = \varepsilon^\Omega(x) T_{\Omega A}{}^B(x)\,,$$

$$\delta\omega(x) = D\varepsilon(x) := d\varepsilon(x) + \omega(x)\varepsilon(x) - \varepsilon(x)\omega(x)\,,$$

where indices are implicit. The condition that the equations remain invariant for some fixed $\omega(x) = \omega_0(x)$ restricts the gauge parameters $\varepsilon^{\Omega}(x)$ to the parameters $\varepsilon^{\Omega}_{gl}(x)$ obeying the conditions

$$\delta\omega_0(x) = 0 \quad \longrightarrow \quad D_0\varepsilon^{\Omega}_{gl}(x) = 0\,, \qquad D_0 := d + \omega_0\,.$$

Since $D_0^2 = 0$, $\varepsilon^{\Omega}_{gl}(x)$ is reconstructed (locally) in terms of $\varepsilon^{\Omega}_{gl}(x_0)$ at any x_0. $\varepsilon^{\Omega}_{gl}(x_0)$ are the global symmetry parameters of the equation $D_0\mathscr{C}(x) = 0$.

Alternatively, one can write a solution in the pure gauge form

$$\omega_0(x) = g^{-1}(x)dg(x)\,, \qquad \mathscr{C}(x) = g^{-1}(x)\mathscr{C}\,, \qquad \varepsilon_{gl}(x) = g^{-1}(x)\epsilon g(x)\,.$$

For $g(x_0) = 1$ this gives $\mathscr{C} = \mathscr{C}(x_0)$ and $\epsilon = \varepsilon_{gl}(x_0)$.

9.5.2 Massless Scalar Field Unfolded

Minkowski space is described by a flat Poincaré-connection $\omega(x) = e^a(x)P_a + \frac{1}{2}\omega^{ab}(x)M_{ab}$. In Cartesian coordinates $e^a(x) = \theta^a$ and $\omega^{ab} = 0$.

Introduce an infinite set of zero-forms, which are traceless symmetric tensors

$$C_{a_1\ldots a_n}(x) = C_{(a_1\ldots a_n)}(x)\,, \quad \eta^{bc}C_{bca_3\ldots a_n}(x) = 0\,. \tag{9.6}$$

The unfolded system of equations equivalent to the Klein–Gordon equation has the form

$$dC_{a_1\ldots a_n}(x) = \theta^b C_{a_1\ldots a_n b}(x)\,. \tag{9.7}$$

Since the fields $C_{a_1\ldots a_n}(x)$ are symmetric while $\theta^b \wedge \theta^c = -\theta^c \wedge \theta^b$, the system (9.7) is formally consistent. (Equivalently, the covariant derivative associated with the Eq. (9.7) rewritten in the form (9.4) is flat.)

The first two equations imply

$$\partial_a C(x) = C_a(x)\,, \qquad \partial_a C_b(x) = C_{ab}(x) \longrightarrow C_{ab}(x) = \partial_a\partial_b C(x)\,.$$

Since $C_{ab}(x)$ is traceless this implies

$$\Box C(x) = 0\,. \tag{9.8}$$

All other equations express higher tensor components via higher derivatives of the scalar field

$$C_{a_1\ldots a_n}(x) = \partial_{a_1}\ldots\partial_{a_n}C(x)\,. \tag{9.9}$$

This formula explains the meaning of $C_{a_1...a_n}(x)$ as a basis of the space of all on-mass-shell nontrivial derivatives of $C(x)$. It should be noted that the space of $C_{a_1...a_n}(x)$ is analogous (in some sense dual) to the space of single-particle states. Via Eq. (9.7) the set of fields $C_{a_1...a_n}(x)$ at any given $x = x_0$ determines $C(x)$ in some neighborhood of x_0, thus providing a locally complete set of "initial data".

9.5.3 Any d

From the unfolded form of the massless scalar field equations it follows that the conformal HS algebra h in d dimensions is the algebra of linear transformations of the infinite-dimensional space V of various traceless symmetric tensors $C, C_a, C_{ab} \ldots$, i.e. $h = gl(V)$. Since the space V is infinite dimensional, such a definition is not fully satisfactory, requiring a more precise definition of the appropriate class of operators. In practice, the idea is that the basis operators of the conformal HS algebra h should reproduce the HS symmetry transformations represented by finite-order differential operators.

A careful definition of h was given by Eastwood in [33] by different methods. As shown in [32], the construction for $d = 3$ significantly simplifies in the framework of the spinorial formalism. Since this formulation is most relevant in the context of the AdS_4/CFT_3 HS holography we explain it in some more detail.

9.6 Conformal HS Algebra in $d = 3$

9.6.1 3d Multispinors

Convenience of the language of spinors in $3d$ theories is due to the following well-known isomorphisms of the $3d$ Lorentz algebra: $o(2,1) \sim sp(2,R) \sim sl_2(R)$. Three dimensional spinors in Minkowski signature are real

$$\chi^\dagger_\alpha = \chi_\alpha\,, \qquad \alpha = 1,2\,.$$

The $sp(2,R)$ invariant tensor $\epsilon^{\alpha\beta} = -\epsilon^{\beta\alpha}$ relates lower and upper indices

$$\chi^\alpha = \epsilon^{\alpha\beta}\chi_\beta\,, \qquad \chi_\alpha = \chi^\beta\epsilon_{\beta\alpha}\,.$$

Because a two-by-two antisymmetric matrix is unique up to a factor, the antisymmetrization of $3d$ spinor indices is equivalent to their contraction

$$A_{\alpha,\beta} - A_{\beta,\alpha} = \epsilon_{\alpha\beta}A_{\gamma,}{}^{\gamma}\,.$$

As a result, irreducible modules of the Lorentz algebra are represented by various totally symmetric multispinors $A_{\alpha_1...\alpha_n}$. As a consequence, rank-k traceless symmetric tensors in the tensor notations are equivalent to the rank-$2k$ totally symmetric multispinors

$$A_{a_1\dots a_m} \sim A_{\alpha_1\dots\alpha_{2m}}\,, \qquad A^b{}_{ba_3\dots a_m} = 0\,.$$

(The reader can compare the number of independent components of the both objects).

The explicit relation between the two formalisms is established with the help of the 2×2 real symmetric matrices $\sigma^n_{\alpha\beta}$

$$A_{\alpha\beta} = \sigma^n_{\alpha\beta} A_n\,, \qquad \sigma^n_{\alpha\beta} = \sigma^n_{\beta\alpha}\,.$$

9.6.2 Spinorial Form of 3d Massless Equations

In $d = 3$, the space V of all traceless symmetric tensors is equivalent to the space of even functions of the commuting spinor variable y^α

$$C(y|x) = \sum_{n=0}^{\infty} C^{\alpha_1\dots\alpha_{2n}}(x) y_{\alpha_1}\dots y_{\alpha_{2n}}\,.$$

In these terms, the unfolded equations for a massless scalar take the form

$$\theta^{\alpha\beta}\left(\frac{\partial}{\partial x^{\alpha\beta}} + \frac{\partial^2}{\partial y^\alpha \partial y^\beta}\right) C(y|x) = 0 \qquad (9.10)$$

with $C(-y|x) = C(y|x)$. The same equation with odd $C(-y|x) = -C(y|x)$ describes a $3d$ massless spinor field $C_\alpha(x) = \frac{\partial}{\partial y^\alpha} C(y|x)\Big|_{y=0}$ [32].

9.6.3 3d HS Symmetry

The $3d$ bosonic conformal HS algebra is the algebra of various differential operators $\epsilon(y, \frac{\partial}{\partial y})$ obeying

$$\epsilon(-y, -\frac{\partial}{\partial y}) = \epsilon(y, \frac{\partial}{\partial y})\,.$$

The transformation law is

$$\delta C(y|x) = \varepsilon_{gl}(y, \frac{\partial}{\partial y}|x) C(y|x)\,, \qquad (9.11)$$

where

$$\varepsilon_{gl}(y, \frac{\partial}{\partial y}|x) = \exp\left[-x^{\alpha\beta}\frac{\partial^2}{\partial y^\alpha \partial y^\beta}\right]\epsilon(y, \frac{\partial}{\partial y})\exp\left[x^{\alpha\beta}\frac{\partial^2}{\partial y^\alpha \partial y^\beta}\right]. \tag{9.12}$$

We leave it to the reader to check that this transformation indeed maps a solution of (9.10) to a solution. For any polynomial $\epsilon(y, \frac{\partial}{\partial y})$, $\epsilon_{gl}(y, \frac{\partial}{\partial y}|x)$ is polynomial as well. $\epsilon_{gl}(y, \frac{\partial}{\partial y})$ provides the generating function for parameters of the global HS transformations.

The $3d$ conformal algebra $sp(4) \sim o(3,2)$ is a subalgebra of the HS conformal algebra with the generators

$$P_{\alpha\beta} = \frac{\partial^2}{\partial y^\alpha \partial y^\beta}, \qquad K^{\alpha\beta} = y^\alpha y^\beta,$$
$$M_{\alpha\beta} = y_\alpha \frac{\partial}{\partial y^\beta} + y_\beta \frac{\partial}{\partial y^\alpha}, \qquad D = y^\alpha \frac{\partial}{\partial y^\alpha} + 1. \tag{9.13}$$

It is not difficult to check how formula (9.11) reproduces the standard conformal transformations for massless scalar and spinor in three dimensions.

9.6.4 Weyl Algebra and Star Product

The Weyl algebra A_n is the associative algebra of polynomials of oscillators $\hat{Y}_A$ obeying the commutation relations

$$[\hat{Y}_A, \hat{Y}_B] = 2iC_{AB}, \qquad A, B, \ldots = 1, \ldots 2n, \qquad C_{AB} = -C_{BA} \tag{9.14}$$

with a nondegenerate C_{AB}. Taking into account that

$$\hat{Y}_A = \begin{pmatrix} y^\alpha \\ i\frac{\partial}{\partial y^\beta} \end{pmatrix}$$

obey the Heisenberg commutation relations (9.14), we conclude that the $3d$ conformal HS algebra (to be identified with the AdS_4 HS algebra) is the Lie algebra associated with the even part of the Weyl algebra A_2.

In practice, it is convenient to replace any operator

$$\hat{f}(\hat{Y}) = \sum_{n=0}^{\infty} \frac{1}{n!} f^{A_1 \ldots A_n} \hat{Y}_{A_1} \ldots \hat{Y}_{A_n}$$

with symmetric $f^{A_1\dots A_n}$ by its Weyl symbol $f(Y)$ which is the function of commuting variables Y^A ($Y^AY^B = Y^BY^A$), that has the same power series expansion

$$f(Y) = \sum_{n=0}^{\infty} \frac{1}{n!} f^{A_1\dots A_n} Y_{A_1} \dots Y_{A_n} .$$

The Weyl star product is defined by the rule that $(f * g)(Y)$ is the symbol of $\hat{f}(\hat{Y})\hat{g}(\hat{Y})$. In particular, this implies

$$[Y_A, Y_B]_* = 2iC_{AB} , \qquad [a , b]_* = a * b - b * a .$$

One can also see that

$$\{Y_A , f(Y)\}_* = 2Y_A f(Y) , \qquad [Y_A , f(Y)]_* = 2i \frac{\partial}{\partial Y^A} f(Y) ,$$

where

$$Y^A = C^{AB} Y_B .$$

The star product is concisely described by the Weyl–Moyal formula

$$(f_1 * f_2)(Y) = f_1(Y) \exp [i \overleftarrow{\partial^A} \overrightarrow{\partial^B} C_{AB}] f_2(Y) , \quad \partial^A := \frac{\partial}{\partial Y_A} , \tag{9.15}$$

which can be proven using the Campbell–Hausdorff formula for $\exp J^A \hat{Y}_A$.

By its definition, the star product (9.15) is associative $(f * g) * h = f * (g * h)$ and regular in the sense that the star product of any two polynomials of Y is a polynomial.

The star product also admits the following useful integral representation

$$(f_1 * f_2)(Y) = \frac{1}{\pi^{2M}} \int dSdT \exp(-iS_A T_B C^{AB}) f_1(Y + S) f_2(Y + T) .$$

9.7 HS Symmetry in AdS_4

9.7.1 Spinor Language in Four Dimension

The HS theory in four dimensions is most naturally formulated in terms of two-component spinors which language is closely related to the twistor theory. Here the key fact is that $2\times 2 = 4$. Minkowski coordinates are represented by 2×2 Hermitian matrices

$$X^n \Rightarrow X^{\alpha\dot{\alpha}} = \sum_{n=0}^{3} X^n \sigma_n^{\alpha\dot{\alpha}}, \qquad \sigma_n^{\alpha\dot{\alpha}} = (I^{\alpha\dot{\alpha}}, \vec{\sigma}^{\alpha\dot{\alpha}}),$$

where $I^{\alpha\dot{\alpha}}$ is the unit matrix and $\vec{\sigma}^{\alpha\dot{\alpha}}$ are Pauli matrices. $\alpha, \beta, \ldots = 1, 2, \dot{\alpha}, \dot{\beta}, \ldots = 1, 2$ are two-component spinor indices.

In these terms

$$\det |X^{\alpha\dot{\alpha}}| = (X^0)^2 - (X^1)^2 - (X^2)^2 - (X^3)^2 .$$

This relation establishes the well-known isomorphism for the four-dimensional Lorentz algebra $sl(2, \mathbb{C}) \sim o(3, 1)$.

The dictionary between tensors and multispinors is provided by the σ-matrices

$$\sigma^a_{\alpha\dot{\alpha}}, \qquad \sigma^{ab}_{\alpha\beta} = \sigma^{[a}_{\alpha\dot{\alpha}} \sigma^{b]\dot{\alpha}}{}_{\beta}, \qquad \bar{\sigma}^{ab}_{\dot{\alpha}\dot{\beta}} = \sigma^{[a}_{\alpha\dot{\alpha}} \sigma^{b]\alpha}{}_{\dot{\beta}},$$

where the two-component indices are raised and lowered by the two-by-two antisymmetric form $\varepsilon_{\alpha\beta}$,

$$y^\alpha = \varepsilon^{\alpha\beta} y_\beta , \qquad y_\alpha = y^\beta \varepsilon_{\beta\alpha} , \qquad \varepsilon_{\alpha\gamma} \varepsilon^{\beta\gamma} = \delta^\beta_\alpha , \qquad \varepsilon_{12} = \varepsilon^{12} = 1 .$$

These relations show that a pair of dotted and undotted indices is equivalent to a vector index, while the pairs of symmetrized indices of the same type are equivalent to the second-rank antisymmetric tensors.

9.7.2 AdS_4 HS Algebra

The identification of the $3d$ conformal HS symmetry with the AdS_4 HS symmetry implies that the global symmetry of the most symmetric vacuum of the bosonic HS theory is represented by the Lie algebra associated with the even part of the Weyl algebra A_2. To have $4d$ Lorentz symmetry manifest, it is most convenient to realize A_2 as the algebra of mutually conjugate operators y_α and $\bar{y}_{\dot{\alpha}}$ that obey the star-product commutation relations

$$[y_\alpha , y_\beta]_* = 2i\,\varepsilon_{\alpha\beta} , \qquad [\bar{y}_{\dot{\alpha}} , \bar{y}_{\dot{\beta}}]_* = 2i\,\varepsilon_{\dot{\alpha}\dot{\beta}} .$$

Historically, the AdS_4 HS algebra was originally found in [34] by different methods from the analysis of the HS fields in AdS_4 while its relation to the Weyl algebra was found in [35].

This realization is convenient for the analysis of the properties of the HS algebra. Spin-s generators are represented by the homogeneous polynomials $T_s(y, \bar{y})$ of degree $2(s-1)$. The commutation relations have the following structure

$$[T_{s_1}\,,T_{s_2}] = T_{s_1+s_2-2} + T_{s_1+s_2-4} + \ldots + T_{|s_1-s_2|+2}\,.$$

Once a spin $s > 2$ appears, the HS algebra contains an infinite tower of higher spins. Indeed, since $[T_s, T_s]$ gives rise to T_{2s-2}, further commutators then lead to higher and higher spins. Note also that $[T_s, T_s]$ contains the generators T_2 of the AdS_4 algebra $o(3,2) \sim sp(4)$.

The HS gauge fields in four dimensions are the one-forms

$$\omega(Y|X) = \sum_{n,m=0}^{\infty} \frac{1}{2n!m!}\omega_{\alpha_1\ldots\alpha_n\,,\dot{\alpha}_1\ldots\dot{\alpha}_m}(X)y^{\alpha_1}\ldots y^{\alpha_n}\bar{y}^{\dot{\alpha}_1}\ldots\bar{y}^{\dot{\alpha}_m}\,,$$

where $Y_A = (y_\alpha, \bar{y}_{\dot{\alpha}})$ are commuting spinor variables and X are local coordinates of AdS_4. The HS curvatures and gauge transformations are

$$R(Y|X) = d\omega(Y|X) + \omega(Y|X) * \omega(Y|X)\,, \tag{9.16}$$

$$\delta\omega(Y|X) = D\epsilon(Y|X) = d\epsilon(Y|X) + [\omega(Y|X)\,,\epsilon(Y|X)]_*\,. \tag{9.17}$$

The symmetry algebra of a single boundary scalar field called $hu(1,0|4)$ contains every spin in one copy. Conventional symmetries are associated with spins $s \leq 2$, forming finite-dimensional subalgebras of the HS algebra. For example, the maximal finite-dimensional subalgebra of $hu(1,0|4)$ is $u(1) \oplus o(3,2)$ where $u(1)$ is associated with the unit element of the star-product algebra.

More generally, there are three series of $4d$ HS superalgebras, namely $hu(n,m|4)$, $ho(n,m|4)$ and $husp(2n,2m|4)$. Spin-one fields of the respective HS theories are the Yang–Mills fields of the Lie groups $G = U(n)\times U(m)$, $O(n)\times O(m)$ and $Usp(2n) \times Usp(2m)$, respectively. Fermions belong to the bifundamental modules of the two components of G. All odd spins are in the adjoint representation of G. Even spins carry the opposite symmetry second rank representation of G. Namely, in the $hu(n,m|4)$ HS theories they are still in the adjoint representation of $U(n) \times U(m)$, while in the $ho(n,m|4)$ and $husp(2n,2m|4)$ HS theories even spins carry rank-two symmetric representation of $O(n) \times O(m)$ and antisymmetric representation of $Usp(2n) \times Usp(2m)$, respectively. The $ho(1,0|4)$ HS theory is the minimal one only containing even spins $s = 0, 2, 4, 6, \ldots$.

The HS theories have the important feature that their particle spectrum always contains a colorless graviton and a colorless scalar which are both invariant under the spin-one Yang–Mills internal symmetries. It is interesting to note that the presence of the colorless scalar field in the spectrum, which is a standard ingredient of the modern cosmological models, is one of the predictions of the HS symmetry.

9.8 Free HS Fields in Four Dimension

9.8.1 Vacuum Solution

Whatever form they have, nonlinear HS field equations will be formulated in terms of the HS curvatures. Hence, any connection $\omega(Y|X)$ that has zero curvature solves the nonlinear HS equations of motion. Such solutions include in particular the AdS_4 connection because AdS_4 is described by the flat gravitational connections of $sp(4)$ which is a subalgebra of the HS algebra.

The AdS_4 vacuum solution solves the equations

$$R_0 = 0$$

for $\omega_0 \in sp(4) \sim o(3,2)$ that has the form

$$\omega_0(Y|X) = \frac{1}{4i}(w^{\alpha\beta}(X)y_\alpha y_\beta + \bar{w}^{\dot{\alpha}\dot{\beta}}(X)\bar{y}_{\dot{\alpha}}\bar{y}_{\dot{\beta}} + 2\lambda h^{\alpha\dot{\beta}}(X)y_\alpha \bar{y}_{\dot{\beta}}) .$$

We leave it to the reader to check that these equations indeed describe AdS_4.

Fluctuations describe small deviations of all massless fields from the vacuum

$$\omega = \omega_0 + \omega_1 , \qquad R_1 = D_0\omega_1 := d\omega_1 + [\omega_0 , \omega_1]_* .$$

Since we know free massless field equations, we anticipate them to result from the linearization of the full nonlinear system. The key question is in which form the free massless field equations will follow from the full nonlinear system? The appropriate form is provided by the Central on-shell theorem.

9.8.1.1 Central on-Shell Theorem

The full unfolded system for the free massless fields of all spins can be formulated in terms of the one-form $\omega(Y|X)$ and zero-form $C(Y|X)$ as follows [36]:

$$R_1(Y \mid X) = \overline{H}^{\dot{\alpha}\dot{\beta}} \frac{\partial^2}{\partial \bar{y}^{\dot{\alpha}} \partial \bar{y}^{\dot{\beta}}} C(0, \bar{y} \mid X) + H^{\alpha\beta} \frac{\partial^2}{\partial y^\alpha \partial y^\beta} C(y, 0 \mid X) , \qquad (9.18)$$

$$\tilde{D}_0 C(Y \mid X) = 0 , \qquad (9.19)$$

where

$$H^{\alpha\beta} = h^{\alpha}{}_{\dot{\alpha}} \wedge h^{\beta\dot{\alpha}} , \quad \overline{H}^{\dot{\alpha}\dot{\beta}} = h_{\alpha}{}^{\dot{\alpha}} \wedge h^{\alpha\dot{\beta}}$$

are the basis two-forms in four dimension,

$$R_1(Y \mid X) = D_0^{ad}\omega(Y \mid X)\,,$$

$$D_0^{ad} = D^L - \lambda h^{\alpha\dot\beta}\left(y_\alpha \frac{\partial}{\partial \bar y^{\dot\beta}} + \frac{\partial}{\partial y^\alpha}\bar y_{\dot\beta}\right), \qquad \tilde D_0 = D^L + \lambda h^{\alpha\dot\beta}\left(y_\alpha \bar y_{\dot\beta} + \frac{\partial^2}{\partial y^\alpha \partial \bar y^{\dot\beta}}\right),$$

and the Lorentz covariant derivative is

$$D^L A = d_X - \left(\omega^{\alpha\beta} y_\alpha \frac{\partial}{\partial y^\beta} + \bar\omega^{\dot\alpha\dot\beta}\bar y_{\dot\alpha}\frac{\partial}{\partial \bar y^{\dot\beta}}\right).$$

Since the system of equations (9.18) and (9.19) contains the exhaustive information about free massless fields, including all their dual formulations, it is called Central on-shell theorem.

The pattern of Eqs. (9.18) and (9.19) is as follows. The gauge fields of different spins are described by the homogeneous polynomials in Y

$$\omega^s(\nu y, \nu \bar y|X) = \nu^{2(s-1)}\omega(y, \bar y|X)\,.$$

The zero-forms associated with the spin s obey

$$C^s(\nu y, \nu^{-1}\bar y|X) = \nu^{\pm 2s} C(y, \bar y|X)\,.$$

This implies that a set of one-forms associated with a massless spin s contains a finite number of components while a set of zero-forms contains an infinite number of components. Altogether, these fields describe an infinite set of spins $s = 0, 1/2, 1, 3/2, 2, 5/2\ldots$

$$\omega^s_{\alpha_1\ldots\alpha_n,\dot\beta_1\ldots\dot\beta_m} : \quad n+m = 2(s-1)\,, \qquad C^s_{\alpha_1\ldots\alpha_n,\dot\beta_1\ldots\dot\beta_m} : \quad |n-m| = 2s\,. \tag{9.20}$$

The zero-forms $C(Y|X)$ encode the gauge invariant HS curvatures and spin-zero matter fields along with all their derivatives that remain non-zero on the dynamical field equations. Dynamical fields include the frame-like fields $\omega^s_{\alpha_1\ldots\alpha_{s-1},\dot\beta_1\ldots\dot\beta_{s-1}}$ and the scalar $C(0,0|x)$. The frame-like fields reduce to the Fronsdal fields upon gauge fixing of the Lorentz-like Stueckelberg gauge symmetries in the linearized HS gauge transformation (9.17).

All other fields from the list (9.20) are expressed by Eqs. (9.18) and (9.19) via higher derivatives of the dynamical fields. The derivatives come in the dimensionless combination

$$\lambda^{-1}\frac{\partial}{\partial x}\,, \qquad \lambda^2 = -\Lambda$$

with the inverse radius λ of the background AdS space. As a result, the HS interactions, that contain higher derivatives, turn out to be nonanalytic in the cosmological constant Λ of the background AdS space.

9.8.1.2 Examples

In the spin-zero sector, the Central on-shell theorem just reproduces the unfolded equations for a scalar field. Indeed, the set of all multispinors $C^0_{\alpha_1\ldots\alpha_n,\dot\beta_1\ldots\dot\beta_n}$ with the equal numbers of dotted and undotted spinor indices provides the spinorial realization of the set of all symmetric traceless tensors $C_{a_1\ldots a_n}$, $C^b{}_{ba_3\ldots a_n} = 0$ in four dimension.

Leaving the derivation of the Maxwell equations in the spin-one sector to the reader, we consider the case of spin two. Here the gauge fields include the Lorentz connection $\omega_{\alpha\beta}, \bar\omega_{\dot\alpha\dot\beta}$ and the frame field $\omega_{\alpha,\dot\beta}$. The zero-forms $C_{\alpha_1\alpha_2\alpha_3\alpha_4}(X)$ and $\bar C_{\dot\alpha_1\dot\alpha_2\dot\alpha_3\dot\alpha_4}(X)$ describe the Weyl tensor in terms of two-component spinors. Higher components $C^s_{\alpha_1\ldots\alpha_n,\dot\beta_1\ldots\dot\beta_m}$ with $|n-m| = 4$ describe all its non-trivial derivatives.

Consider first Eq. (9.18). The equation $R_{\alpha,\dot\beta} = 0$ is the usual zero-torsion condition that expresses the Lorentz connection via the vierbein. The other equations have the form

$$R_{\alpha\beta} = H^{\gamma\delta} C_{\alpha\beta\gamma\delta}\,, \qquad R_{\dot\alpha\dot\beta} = \bar H^{\dot\gamma\dot\delta} \bar C_{\dot\alpha\dot\beta\dot\gamma\dot\delta}\,. \tag{9.21}$$

These imply that a nonzero part of the Riemann tensor belongs to the Weyl tensor. This is equivalent to saying that the Ricci tensor is zero which, in turn, is equivalent to the Einstein equations in the vacuum.

In the tensorial language the same equations read as

$$R_{\nu\mu}{}^a = 0\,, \qquad R_{\nu\mu}{}^{ab} = e_\nu{}^c e_\mu{}^d C_{cd,}{}^{ab}\,, \qquad C_{ab,}{}^b{}_c = 0\,. \tag{9.22}$$

This implies the Einstein equations since $R_{\nu\mu} = R_{\nu\rho}{}^\rho{}_\mu = 0$. In addition, the system (9.22) implies that $C_{cd,}{}^{ab}$ coincides with the Weyl tensor.

Analogously, the Central on-shell theorem for higher spins imposes the Fronsdal equations $\mathscr{R}_{\nu_1\ldots\nu_s} = 0$ and expresses the generalized HS Weyl tensors in terms of derivatives of the Fronsdal fields.

9.9 Nonlinear Higher-Spin Theory

In this section we briefly summarize the construction of nonlinear HS equations. The reader not interested in technical details is advised to go directly to Sect. 9.9.4.

9.9.1 Idea of Construction

The idea is to look for nonlinear HS field equations in the form of a nonlinear deformation of the Central on-shell theorem. The first step is to replace the linearized HS curvatures and covariant derivatives by the full non-Abelian ones:

$$R(y,\overline{y} \mid X) = d\omega(y,\overline{y} \mid X) + \omega(y,\overline{y} \mid X) * \omega(y,\overline{y} \mid X)\,,$$

$$\tilde{D}C(y,\overline{y} \mid X) = dC(y,\overline{y} \mid X) + \omega(y,\overline{y} \mid X) * C(y,\overline{y} \mid X) - C(y,\overline{y} \mid X) * \omega(y,-\overline{y} \mid X)$$

trying to find a deformation of the form

$$R(Y \mid X) = \overline{H}^{\dot\alpha\dot\beta} \frac{\partial^2}{\partial\overline{y}^{\dot\alpha}\partial\overline{y}^{\dot\beta}} C(0,\overline{y} \mid X) + H^{\alpha\beta} \frac{\partial^2}{\partial y^{\alpha}\partial y^{\beta}} C(y,0 \mid X) + \dots\,,$$

$$\tilde{D}C(Y \mid X) + \dots = 0\,,$$

where further nonlinear corrections have to be determined from the formal consistency of the HS equations. Having the form of the generalized Bianchi identities, the consistency of the HS equations also guarantees their gauge invariance.

Field equations of such a form are called unfolded which means that all dynamical fields are differential forms and that the exterior derivative of any field is expressed via the exterior product of the fields themselves. As discussed in some more detail in Sect. 9.10, this form of dynamical equations is useful in many respects.

Being possible in a few first orders, the straightforward construction of the nonlinear deformation quickly gets complicated. The trick is to find a larger algebra g' such that an appropriate substitution

$$\omega \to W = \omega + \omega C + \omega C^2 + \dots$$

into $W \in g'$ reconstructs nonlinear equations via the flatness condition

$$dW + W \wedge W = 0\,.$$

The problem is to find appropriate restrictions on W that reconstruct the nonlinear HS equations in all orders.

This is achieved via the doubling of spinors

$$\omega(Y|X) \longrightarrow W(Z;Y|X)\,, \qquad C(Y|X) \longrightarrow B(Z;Y|X)$$

accompanied by the equations that determine the dependence on the additional spinorial variables Z^A in terms of the "initial data"

$$\omega(Y|X) = W(0;Y|X)\,, \qquad C(Y|X) = B(0;Y|X)\,,$$

where $\omega(Y|X)$ and $C(Y|X)$ are the HS fields of the Central on-shell theorem. To rewrite the evolution along the additional variables Z^A covariantly, it is useful to introduce a connection $S(Z,Y|X) = dZ^A S_A$ in the Z^A-space.

9.9.2 HS Star Product

Nonlinear HS field equations are formulated in terms of the specific star product

$$(f * g)(Z;Y) = \frac{1}{(2\pi)^4} \int d^4U\, d^4V \exp[iU^A V^B C_{AB}] \times f(Z+U;Y+U)g(Z-V;Y+V)\,, \tag{9.23}$$

where $C_{AB} = (\varepsilon_{\alpha\beta}, \bar{\varepsilon}_{\dot{\alpha}\dot{\beta}})$ is the $4d$ charge conjugation matrix and U^A, V^B are real integration variables. The normalization is such that 1 is a unit element of the star-product algebra, i.e. $f * 1 = 1 * f = f$. The star product (9.23) is associative and provides a particular realization of the Weyl algebra since

$$[Y_A, Y_B]_* = -[Z_A, Z_B]_* = 2iC_{AB} \qquad [Y_A, Z_B]_* = 0\,. \tag{9.24}$$

It results from the normal ordering with respect to the elements

$$b_A = \frac{1}{2i}(Y_A - Z_A)\,, \qquad a_A = \frac{1}{2}(Y_A + Z_A)\,,$$

which satisfy

$$[a_A, a_B]_* = [b_A, b_B]_* = 0\,, \quad [a_A, b_B]_* = C_{AB}$$

and can be interpreted as creation and annihilation operators. In fact, the star product (9.23) describes the normal ordering with respect to the oscillators a_A and b_A as is most evident from the following consequences of (9.23):

$$b_A * f(b,a) = b_A f(b,a)\,, \qquad f(b,a) * a_A = f(b,a) a_A\,.$$

An important property of the star product (9.23) is that it admits the inner Klein operator

$$\Upsilon = \exp i Z_A Y^A$$

which behaves as $(-1)^N$, where N is the spinor number operator. One can easily see that

$$\Upsilon * \Upsilon = 1,$$

$$\Upsilon * f(Z;Y) = f(-Z;-Y) * \Upsilon$$

and

$$(\Upsilon * f)(Z;Y) = \exp i Z_A Y^A\, f(Y;Z)\,.$$

With respect to the decomposition of Majorana spinors into two-component spinors, $Y_A = (y_\alpha, \bar{y}_{\dot{\alpha}})$, $\bar{y}_{\dot{\alpha}} = (y_\alpha)^\dagger$, the left and right inner Klein operators

$$\kappa = \exp i z_\alpha y^\alpha \,, \qquad \bar{\kappa} = \exp i \bar{z}_{\dot{\alpha}} \bar{y}^{\dot{\alpha}} \tag{9.25}$$

act analogously on the undotted and dotted spinors, respectively,

$$\kappa * f(z, \bar{z}; y, \bar{y}) = f(-z, \bar{z}; -y, \bar{y}) * \kappa \,, \quad \bar{\kappa} * f(z, \bar{z}; y, \bar{y}) = f(z, -\bar{z}; y, -\bar{y}) * \bar{\kappa} \,,$$

$$\kappa * \kappa = \bar{\kappa} * \bar{\kappa} = 1 \,, \qquad \kappa * \bar{\kappa} = \bar{\kappa} * \kappa \,.$$

9.9.3 The Full Nonlinear System

As shown in [37], the equations of motion of the four-dimensional HS theory can be formulated in terms of the three types of fields

$$W = dX^\nu W_\nu(Z, Y; K|X) \,, \qquad S = dZ^A S_A(Z, Y; K|X) \,, \qquad B(Z, Y; K|X) \,.$$

The fields W and S are, respectively, one-forms in the four-dimensional space-time with the coordinates X^ν and spinor space with the coordinates Z_A. The spinorial variables Z_A and Y_A are commuting while dZ_A are anticommuting differentials

$$Z_A Z_B = Z_B Z_A \,, \quad Y_A Y_B = Y_B Y_A \,, \quad Z_A Y_B = Y_B Z_A \,, \quad dZ_A dZ_B = -dZ_B dZ_A \,.$$

dZ_A commute with Z_B and Y_B but anticommute with the anticommuting space-time differentials dX^ν

$$dX_\nu dX_\mu = -dX_\mu dX_\nu \,, \qquad dZ_A dX_\nu = -dX_\nu dZ_A \,.$$

K denotes a pair of Klein operators $K = (k, \bar{k})$ that obey the relations

$$k^2 = \bar{k}^2 = 1 \,, \qquad k\bar{k} = \bar{k}k \,, \tag{9.26}$$

$$k w_\alpha = -w_\alpha k \,, \qquad \bar{k} w_\alpha = w_\alpha \bar{k} \,, \qquad k \bar{w}_{\dot{\alpha}} = \bar{w}_{\dot{\alpha}} k \,, \qquad \bar{k} \bar{w}_{\dot{\alpha}} = -\bar{w}_{\dot{\alpha}} \bar{k} \tag{9.27}$$

for $w_\alpha = (dz_\alpha, z_\alpha, y_\alpha)$, $\bar{w}_{\dot{\alpha}} = (d\bar{z}_{\dot{\alpha}}, \bar{z}_{\dot{\alpha}}, \bar{y}_{\dot{\alpha}})$. The important difference between $(k, \bar{k})$ and $(\kappa, \bar{\kappa})$ (9.25) is that the former anticommute with the respective differentials dz^α and $d\bar{z}^{\dot{\alpha}}$ while the latter commute.

The system of nonlinear HS equations in $d = 4$ reads as [37]

$$dW = W * W \,, \tag{9.28}$$

$$dB = W * B - B * W \,, \tag{9.29}$$

$$dS = W * S - S * W\,, \tag{9.30}$$

$$S * B = B * S\,, \tag{9.31}$$

$$S * S = -i(dZ^A dZ_A + dz^\alpha dz_\alpha F_*(B) * k\kappa + d\bar{z}^{\dot{\alpha}} d\bar{z}_{\dot{\alpha}} \bar{F}_*(B) * \bar{k}\bar{\kappa})\,. \tag{9.32}$$

$F_*(B)$ in Eq. (9.32) is some star-product function of the field B. The simplest case of the linear functions

$$F_*(B) = \eta B\,, \qquad \bar{F}_*(B) = \bar{\eta} B\,,$$

where η is some phase factor (its absolute value can be absorbed into a redefinition of B), leads to the class of pairwise nonequivalent nonlinear HS theories. The cases of $\eta = 1$ and $\eta = \exp\frac{i\pi}{2}$ are particularly interesting, corresponding to the so called A and B HS models. These two cases are distinguished by the property that they respect parity [25].

Expanding all fields in powers of k and $\bar{k}$ we obtain for $U = W, S, B$

$$U(Z, Y; K|X) = \sum_{i,j=0}^{1} k^i \bar{k}^j U_{ij}(Z, Y|X)\,.$$

Since the relations (9.26) and (9.27) are invariant under the reflections $k \to -k$ and $\bar{k} \to -\bar{k}$, and taking into account that the r.h.s. of Eq. (9.32) contains k and $\bar{k}$ explicitly, it follows that the system (9.28)–(9.32) is invariant under the following involutive map

$$\tau(W(Y,Z;K|X)) = \ W(Y,Z;-K|X)\,, \qquad \tau(S(Y,Z;K|X)) = S(Y,Z;-K|X)\,,$$
$$\tau(B(Y,Z;K|X)) = -\ B(Y,Z;-K|X)\,.$$

As a result, the full system of fields decomposes into τ-even and τ-odd fields. Clearly, the τ-even fields form a subsystem of the full system while the τ-odd fields can be consistently truncated away. This truncation is applied in most of applications. The dynamical role of the τ-even and τ-odd fields is different.

The τ-even fields we call dynamical since they describe massless fields of various spins. These are W_{ii}^{dyn}, S_{ii}^{dyn} and $B_{i\,1-i}^{dyn}$. Each of them appears in two copies because $i = 1, 2$. As shown in [38], this doubling is inevitable in presence of fermions.

Each member of the infinite set of the τ-odd fields describes at most a finite number of degrees of freedom. To stress that they carry no local degrees of freedom, they were called auxiliary in [37]. It is also appropriate to call them *moduli* fields since the finite number of degrees of freedom carried by each of these fields can be interpreted as a kind of coupling constants of the theory. In particular, this was demonstrated in [39] where it was shown that the mass parameter of the matter fields in the $3d$ HS theory results from a non-zero vacuum value of one of the moduli fields. The moduli fields include $W^{mod}_{i\,1-i}$, $S^{mod}_{i\,1-i}$ and B^{mod}_{ii}. Truncating away the moduli fields greatly reduces the moduli space of the theory. In particular, the moduli responsible for the massive boundary deformation can be argued to belong to this sector.

The perturbative analysis performed around the following vacuum solution

$$B_0 = 0\,, \qquad S_0 = dZ^A Z_A\,, \qquad W_0 = \frac{1}{2}\omega_0^{AB}(X) Y_\mu Y_\nu\,,$$

where W_0 obeys

$$dW_0 + W_0 \star W_0 = 0$$

so that $\omega_0^{AB}(X)$ describes the AdS_4, reproduces the Central on-shell theorem in the first-order [37]. This means that the nonlinear system (9.28)–(9.32) indeed provides a nonlinear deformation of the free equations of massless fields of all spins. Note that the specific form of the star product (9.23) is crucial for this analysis.

The HS equations exhibit manifest gauge invariance under the gauge transformations

$$\delta W = d\varepsilon + [W\,, \varepsilon]_*\,, \qquad \delta S = [S\,, \varepsilon]_*\,, \qquad \delta B = [B\,, \varepsilon]_*\,, \qquad \varepsilon = \varepsilon(Z; Y; K|X)\,.$$

The nonlinear HS equations are formally consistent and regular: perturbatively, there are no divergences due to star products of the non-polynomial elements resulting from the inner Klein operators κ and $\bar{\kappa}$ [40].

9.9.4 Properties of HS Interactions

Let us briefly discuss some of the most important properties of the nonlinear HS equations.

First of all, HS interactions contain higher derivatives. This property is closely related to nonanaliticity of the HS interactions in the cosmological constant $\Lambda = -\rho^{-2}$ which appears in the dimensionless combination $\rho\partial$ where ρ is the AdS radius while ∂ denotes the space-time derivative. This has the effect that background HS gauge fields contribute to the higher-derivative terms in the evolution equations. As a result, the evolution is determined mostly by the HS fields rather than by the metric. This provides the realization of the anticipated property that Riemannian geometry is not an appropriate tool in the HS theory.

In the HS theory, HS fields source lower-spin fields in particular via the $\omega * \omega$-like terms. Other way around, lower-spin fields source HS fields via the C^2 terms. In particular, gravity sources the HS fields and vice-versa. Among other things this implies that the Einstein gravity cannot be obtained as a consistent truncation of the HS theory.

A remarkable feature of the HS equations is that their nontrivial part is only represented by Eq. (9.32) which does not contain the space-time derivative d. This suggests that not only Riemannian geometry but even usual coordinates do not play a fundamental role in the system. In fact, this is a general property of unfolded dynamical equations the particular case of which is represented by the nonlinear equations (9.28)–(9.32).

9.10 Unfolded Dynamics

9.10.1 General Setup

The unfolded form of dynamical equations provides a covariant generalization of the first-order form of differential equations

$$\dot{q}^i(t) = \varphi^i(q(t)),$$

which is convenient in many respects. In particular, initial values can be given in terms of the values of variables $q^i(t_0)$ at any given point t_0. As a result, in the first-order formulation, the number of degrees of freedom equals to the number of dynamical variables.

Unfolded dynamics is a multidimensional generalization achieved via the replacement of the time derivative by the de Rham derivative

$$\frac{\partial}{\partial t} \to d = \theta^\nu \partial_\nu$$

and the dynamical variables q^i by a set of differential forms

$$q^i(t) \to W^\Omega(\theta, x) = \theta^{\nu_1} \dots \theta^{\nu_p} W^\Omega_{\nu_1 \dots \nu_p}(x)$$

to reformulate a system of partial differential equations in the first-order covariant form

$$dW^\Omega(\theta, x) = G^\Omega(W(\theta, x)). \tag{9.33}$$

Here $G^\Omega(W)$ are some functions of the "supercoordinates" W^Ω

$$G^\Omega(W) = \sum_n f^\Omega{}_{\Lambda_1 \dots \Lambda_n} W^{\Lambda_1} \dots W^{\Lambda_n}.$$

Since $d^2 = 0$, at $d > 1$ the functions $G^\Lambda(W)$ cannot be arbitrary but have to obey the compatibility conditions

$$G^\Lambda(W) \frac{\partial G^\Omega(W)}{\partial W^\Lambda} \equiv 0. \tag{9.34}$$

(Recall that all products of the differential forms $W(\theta, x)$ are the wedge products due to anticommutativity of θ^ν.) Let us stress that these are conditions on the functions $G^\Lambda(W)$ rather than on W.

The idea of the unfolded formulation was put forward in the paper [40] where it was realized that the full system of nonlinear equations can be searched in the form (9.33) as a deformation of the Central on-shell theorem.

As a consequence of the compatibility conditions (9.34) the system (9.33) is manifestly invariant under the gauge transformation

$$\delta W^\Omega = d\varepsilon^\Omega + \varepsilon^\Lambda \frac{\partial G^\Omega(W)}{\partial W^\Lambda},$$

where the gauge parameter $\varepsilon^\Omega(x)$ is a $(p_\Omega - 1)$-form if W^Ω is a p_Ω-form. Strictly speaking, this is true for the class of *universal* unfolded systems in which the compatibility conditions (9.34) hold independently of the dimension d of space-time, i.e. (9.34) should be true disregarding the fact that any $(d+1)$-form is zero. Let us stress that all unfolded systems, which appear in HS theories including those considered in these lectures, are universal.

The unfolded formulation can be applied to the description of invariant functionals of the system in question. Here it is useful to distinguish between the off-shell and on-shell unfolded dynamical systems.

As demonstrated in Sect. 9.5.2, most of the relations contained in unfolded equations impose constraints expressing some new fields in terms of derivatives of the old ones. In the *off-shell* case the unfolded equations just express all fields in terms of derivatives of some ground fields, imposing no differential restrictions on the latter. In the scalar-field example of Sect. 9.5.2, to make the system off-shell one should relax the tracelessness condition in (9.6). In this case, the pattern of the unfolded system (9.7) is given by the set of constraints (9.9) which express the higher tensors $C_{a_1 \ldots a_n}(x)$ via derivatives of the ground scalar field $C(x)$. The *on-shell* unfolded equations not only express all fields in terms of derivatives of the ground fields, but also impose differential restrictions on the latter. In the scalar-field example this is the Klein–Gordon equation (9.8).

As shown in [41], the variety of invariant functionals associated with the unfolded equations (9.33) is described by the cohomology of the operator

$$Q = G^\Omega \frac{\partial}{\partial W^\Omega}, \tag{9.35}$$

which obeys

$$Q^2 = 0$$

as a consequence of (9.34). By virtue of (9.33), Q-closed p-form functions $L_p(W)$ are d-closed, giving rise to the gauge invariant functionals

$$S = \int_{\Sigma^p} L_p .$$

In the off-shell case they can be used to construct invariant action functionals while in the on-shell case they describe conserved charges. (For more detail and examples see [41].) Also, in the on-shell case, S can play a róle of the Hamilton–Jacobi action which becomes a functional of boundary conditions in the context of holographic duality.[1]

9.10.2 *Properties*

The unfolded formulation of partial differential equations has a number of remarkable properties.

- First of all, it has general applicability: every system of partial differential equations can be reformulated in the unfolded form.
- Due to using the exterior algebra formalism, the system is invariant under diffeomorphisms, being coordinate independent.
- Interactions can be understood as nonlinear deformations of $G^{\Omega}(W)$.
- Degrees of freedom are represented by the subset of zero-forms $C^I(x_0) \in \{W^{\Omega}(x_0)\}$ at any $x = x_0$. This is analogous to the fact that $q^i(t_0)$ describe degrees of freedom in the first-order form of ordinary differential equations. The zero-forms $C^I(x_0)$ realize an infinite-dimensional module dual to the space of single-particle states of the system. In the HS theory it is realized as a space of functions of auxiliary variables like $C(y, \bar{y}|x_0)$. This space is an analogue of the phase space in the Hamiltonian approach.
- It is worth to mention that the same property of the unfolded dynamics provides a tool to control unitarity in presence of higher derivatives via the requirement that the space of zero-forms like $C(y, \bar{y})$ admits a positive-definite norm preserved by the unfolded equations in question.

The above list of remarkable properties of the unfolded formulation is far from being complete. In particular, the unfolded formulation admits a nice interpretation in terms of Lie algebra cohomology (for more detail see [42]), L_∞ algebra [43], Q-manifolds and many more (for more detail see e.g., [41, 44] and references therein). The most striking feature of this formulation is however that it makes it possible to describe one and the same dynamical system in space-times of different dimensions.

9.11 Space-Time Metamorphoses

Unfolded dynamics exhibits independence of the "world-volume" space-time with coordinates x. Instead, geometry is encoded by the functions $G^{\Omega}(W)$ in the "target space" of fields W^{Ω}. Indeed, the universal unfolded equations make sense in any

[1] I am grateful to Ioannis Papadimitriou for the stimulating discussion of this point.

space-time independently of a particular realization of the de Rham derivative d. For instance one can extend space time by adding additional coordinates z

$$dW^{\Omega}(x) = G^{\Omega}(W(x)), \quad x \to X = (x,z), \quad d_x \to d_X = d_x + d_z, \quad d_z = dz^u \frac{\partial}{\partial z^u}.$$

The unfolded equations reconstruct the X-dependence in terms of values of the fields $W^{\Omega}(X_0) = W^{\Omega}(x_0, z_0)$ at any X_0. Clearly, to take $W^{\Omega}(x_0, z_0)$ in space M_X with coordinates X_0 is the same as to take $W^{\Omega}(x_0)$ in the space $M_x \subset M_X$ with coordinates x.

The problem becomes most interesting provided that there is a nontrivial vacuum connection along the additional coordinates z. This is in particular the case of AdS/CFT correspondence where the conformal flat connection at the boundary is extended to the flat AdS connection in the bulk with z being a radial coordinate of the Poincaré type.

Generally, the unfolding can be interpreted as some sort of a covariant twistor transform

$$\begin{array}{ccc} & C(Y|x) & \\ \eta \swarrow & & \searrow \nu \\ M(x) & & T(Y). \end{array}$$

Here $W^{\Omega}(Y|x)$ are functions on the "correspondence space" C with local coordinates Y, x. The space-time M has local coordinates x. The twistor space T has local coordinates Y.

The unfolded equations reconstruct the dependence of $W^{\Omega}(Y|x)$ on x in terms of the function $W^{\Omega}(Y|x_0)$ on T at some fixed x_0. The restriction of $W^{\Omega}(Y|x)$ or some its Y-derivatives to $Y = 0$ gives dynamical fields $\omega(x)$ in M which, in the on-shell case, solve their dynamical field equations. Hence, similarly to the Penrose transform (see [45] and references therein), unfolded equations map functions on T to solutions of the dynamical field equations in M.

In these terms, the holographic duality can be interpreted as the duality between different space-times M that can be associated with the same twistor space. This phenomenon has a number of interesting realizations.

9.11.1 AdS_4/CFT_3 HS Holography

The AdS_4/CFT_3 HS holography [22] relates the HS gauge theory in AdS_4 to the quantum theory of conformal currents in three dimensions. To see how it works, let us first discuss the unfolded equations for free massless fields and currents on the $3d$ boundary.

The unfolded equations for conformal massless fields in three dimensions are [32,46]

$$(\frac{\partial}{\partial x^{\alpha\beta}} \pm i \frac{\partial^2}{\partial y^\alpha \partial y^\beta})C_j^\pm(y|x) = 0\,, \qquad \alpha,\beta = 1,2\,, \quad j = 1,\dots \mathcal{N}\,.$$

The equations for $3d$ conformal conserved currents have the form of rank-two equations [47]

$$\left\{ \frac{\partial}{\partial x^{\alpha\beta}} - \frac{\partial^2}{\partial y^{(\alpha} \partial u^{\beta)}} \right\} J(u,\ y|x) = 0\,. \tag{9.36}$$

$J(u,\ y|x)$ contains all $3d$ HS currents along with their derivatives.

Elementary $3d$ conformal currents, which are conformal primaries, contain currents of all spins

$$J(u,0|x) = \sum_{2s=0}^{\infty} u^{\alpha_1} \dots u^{\alpha_{2s}} J_{\alpha_1\dots\alpha_{2s}}(x)\,, \quad \tilde{J}(0,y|x) = \sum_{2s=0}^{\infty} y^{\alpha_1} \dots y^{\alpha_{2s}} \tilde{J}_{\alpha_1\dots\alpha_{2s}}(x)$$

along with the additional scalar current

$$J^{asym}(u,y|x) = u_\alpha y^\alpha J^{asym}(x)\,.$$

Their conformal dimensions are

$$\Delta J_{\alpha_1\dots\alpha_{2s}}(x) = \Delta \tilde{J}_{\alpha_1\dots\alpha_{2s}}(x) = s+1 \qquad \Delta J^{asym}(x) = 2\,.$$

The unfolded equations express all other components of $J(u,y|x)$ in terms of derivatives of the primaries, also imposing the differential equations on the primaries, which are just the conservation conditions

$$\frac{\partial}{\partial x^{\alpha\beta}} \frac{\partial^2}{\partial u_\alpha \partial u_\beta} J(u,0|x) = 0\,, \qquad \frac{\partial}{\partial x^{\alpha\beta}} \frac{\partial^2}{\partial y_\alpha \partial y_\beta} \tilde{J}(0,y|x) = 0$$

for all currents except for the scalar ones that do not obey any differential equations.

The rank-two equation is obeyed by

$$J(u,\ y\,|x) = \sum_{i=1}^{\mathcal{N}} C_i^-(u+y|x)\, C_i^+(y-u|x)\,.$$

This simple formula gives the explicit realization of the HS conformal conserved currents in terms of bilinear combinations of derivatives of free massless fields in three dimensions.

Generally, the rank-two fields and, hence conserved currents, can be interpreted as bi-local fields in the twistor space. In this respect they are somewhat analogous to space-time bi-local fields also used for the description of currents (see e.g [48, 49] and references therein).

To relate $3d$ currents to $4d$ massless fields it remains to extend the $3d$ current equation to the $4d$ massless equations. This is easy to achieve in the unfolded dynamics via the extension of the $3d$ coordinates $x^{\alpha\beta}$ to the $4d$ coordinates $X^{\alpha\dot{\beta}}$, extending the $3d$ equations to

$$\left(\frac{\partial}{\partial X^{\alpha\dot{\alpha}}} + \frac{\partial^2}{\partial y^\alpha \partial \bar{y}^{\dot{\beta}}}\right) C(y, \bar{y}|X) = 0 \,. \tag{9.37}$$

These are just the free unfolded equations for $4d$ massless fields of all spins in Minkowski space, i.e. at $\Lambda = 0$.

The analysis in AdS_4, which is also simple, is performed analogously. In this case, $x^{\alpha\beta} = \frac{1}{2}(X^{\alpha\beta} + X^{\beta\alpha})$ are boundary coordinates, while $z^{-1} = X^{\alpha\beta}\epsilon_{\alpha\beta}$ is the radial coordinate. (For more detail see [29].) At the non-linear level, the full HS theory in AdS_4 turns out to be equivalent to the theory of $3d$ currents of all spins interacting through conformal HS gauge fields [29].

9.11.2 *sp*(8) *Invariant Setup*

Another example of the application of unfolded dynamics is related to the $sp(8)$ extension of conformal symmetry in the theory of massless fields in four dimensions. As was shown by Fronsdal [50], the tower of all $4d$ massless fields is $sp(8)$ symmetric. The $sp(8)$ symmetry extends conformal symmetry $su(2,2) \subset sp(8)$ that acts on every massless field. The generators in $sp(8)/su(2,2)$ mix fields of different spins in the tower of massless fields of all spins $0 \leq s < \infty$.

Indeed, Eq. (9.11), that describe gauge invariant combinations of massless fields, are covariant constancy conditions for 0-forms $C(y, \bar{y}|x)$ valued in the space of functions of spinor variables y_α and $\bar{y}_{\dot{\alpha}}$. Hence, symmetries of these equations contain $sp(8)$ realized by bilinears (9.13) with indices α taking four values.

Fronsdal has shown that the space-time $\mathscr{M}_4$ appropriate for geometric realization of $Sp(8)$ is ten-dimensional with local coordinates $X^{AB} = X^{BA}$, where $A = (\alpha, \dot{\alpha}) = 1, 2, 3, 4$. Applying the construction of Sect. 9.10, it is easy to derive the equations for massless fields in $\mathscr{M}_4$.

9.11.2.1 From Four to Ten

Indeed, unfolded $4d$ massless equations can be easily uplifted to $\mathscr{M}_4$ as follows [46]:

$$dX^{AB}\left(\frac{\partial}{\partial X^{AB}}+\frac{\partial^2}{\partial Y^A\partial Y^B}\right)C(Y|X)=0\,,\qquad A,B=1,\dots 4\,, \tag{9.38}$$

where, for the sake of simplicity, the massless equations are presented in the Cartesian-like coordinates. Note that to obtain the proper $\lambda \to 0$ limit from Eq. (9.11), it is necessary to rescale the spinor variables

$$y_\alpha \to \lambda^{\frac{1}{2}} y_\alpha\,,\qquad \bar{y}_{\dot\alpha} \to \lambda^{\frac{1}{2}} \bar{y}_{\dot\alpha} \tag{9.39}$$

before taking the limit.

If the indices A, B take just two values, Eq. (9.38) describe $3d$ massless fields invariant under $Sp(4)$ which is the $3d$ conformal group [32,46].

By the general argument in the beginning of Sect. 9.11, Eq. (9.38) describe the same dynamics as the original massless field equations in $4d$ Minkowski space because they consist of the usual $4d$ equations (9.37) for the coordinates $X^{\alpha\dot\beta}$ supplemented with the equations describing the evolution along the additional spinning coordinates $X^{\alpha\beta}$ and $X^{\dot\alpha\dot\beta}$. The key question is what are independent dynamical variables in $\mathscr{M}_4$? From (9.38) it is clear that these are the fields $C(0|X)$ and $Y^A C_A(0|X)$. Indeed, all other components of $C(Y|X)$ are expressed by Eq. (9.38) via X-derivatives of $C(0|X)$ and $Y^A C_A(0|X)$. It turns out that $C(0|X)$ describes all $4d$ massless fields of integer spins while $C_A(0|X)$ describes all $4d$ massless fields of half-integer spins. So, $C(0|X)$ and $C_A(0|X)$ serve as certain hyperfields for the HS multiplets.

The nontrivial field equations in $\mathscr{M}_4$ are [46]

$$\left(\frac{\partial^2}{\partial X^{AB}\partial X^{CD}}-\frac{\partial^2}{\partial X^{CB}\partial X^{AD}}\right)C(X)=0 \tag{9.40}$$

for bosons and

$$\left(\frac{\partial}{\partial X^{AB}}C_C(X)-\frac{\partial}{\partial X^{CB}}C_A(X)\right)=0 \tag{9.41}$$

for fermions. These equations are interesting in many respects. First of all, they are overdetermined. This is what makes it possible to describe the four-dimensional massless fields by virtue of differential equations in the ten-dimensional space $\mathscr{M}_4$. Another interesting feature is that Eqs. (9.40) and (9.41) contain no index contraction and hence no metric tensor.

9.11.2.2 From Ten to Four

It is instructive to see how the usual space-time picture re-appears from the ten-dimensional one. Remarkably, in this setup, the conventional four-dimensional picture results from the identification of a concept of local event simultaneously with the metric tensor. Referring for more detail of the derivation to the original paper [51], we just summarize the final results.

Time in $\mathcal{M}_M$ is a parameter t along a time-like direction in $\mathcal{M}_4$ represented by any positive-definite matrix T^{AB}

$$X^{AB} = T^{AB} t \,.$$

Usual space in $\mathcal{M}_M$ is identified with the space of local events at a given time. Coordinates of the space of local events x^n are required to have the property that the differential equations in question admit "initial data" localized at any point of space-time, i.e. represented by the δ-functions $\delta(x^n - x^n_0)$ with various x^n_0. Since the system of equations in question is overdetermined, the analysis of this issue is not quite trivial. The final result is [51] that, for Eqs. (9.40),(9.41), the space of local events in $\mathcal{M}_4$ is represented by a Clifford algebra with

$$X^{AB} = x^n \gamma_n^A{}_C T^{BC}$$

formed by matrices $\gamma_n^A{}_B$ that obey

$$\{\gamma_n \,, \gamma_m\} = 2 g_{nm} \,, \tag{9.42}$$

where g_{nm} is the spatial metric tensor of R^3.

Thus, the three-dimensional space of the $4d$ Minkowski space appears as the space R^3 of local events. In this analysis, the metric tensor appears just after the identification of coordinates that parametrize local events with the generators of the Clifford algebra. In a certain sense, this construction is opposite to the original Dirac's construction where the γ-matrices were introduced as a square root of the metric tensor. Here, the metric tensor appears from the definition of the γ-matrices that represent local events.

Analogous analysis can be performed in some other dimensions. In particular in [51–53] it was shown that Eq. (9.38) at $M = 2, 4, 8, 16$ describe free conformal fields of all spins in $d = 3, 4, 6, 10$.

It should be noted that different $sp(2M)$-symmetric field equations in the same space $\mathcal{M}_M$, like e.g. the higher-rank equations of [47], have spaces of local events of different dimensions. The resulting picture is somewhat analogous to the brane picture in String Theory allowing the co-existence of objects of different dimensions in the same space. The difference is however that the "HS branes" in the $sp(2M)$ setup are not localized as a particular surface embedded into $\mathcal{M}_M$. Instead, different choices of a representative surface is a matter of the gauge choice. This example gives another manifestation of the general property that higher symmetries may affect such fundamental concepts as local event and space-time dimension.

9.12 HS Theory and Quantum Mechanics

Classical HS theory has several interesting links with quantum mechanics.

One is that unfolded dynamics in the spinor (twistor) formulation distinguishes between positive and negative frequencies

$$\Big(\frac{\partial}{\partial X^{AB}} \pm i \frac{\partial^2}{\partial Y^A \partial Y^B}\Big) C^{\pm}(Y|X) = 0\,. \tag{9.43}$$

Indeed, since the time parameter $t = \frac{1}{M} X^{AB} T_{AB}$ is associated with any positive-definite T_{AB}, the sign in the exponential

$$C(X) = C^+(X) + C^-(X)\,, \qquad C^{\pm}(X) = \int d^M \xi c^{\pm}(\xi) \exp \pm i \xi_A \xi_B X^{AB}$$

is associated with the positive and negative frequencies. Hence, the unfolded equations for massless fields in $\mathscr{M}_M$ effectively quantize the model.

Another is the holographic duality between relativistic HS theory and nonrelativistic quantum mechanics. To this end, consider the reduction of Eq. (9.43) to the time arrow setting $X^{AB} = \delta^{AB} t$. The pullback of Eq. (9.43) to the time axis gives

$$i \frac{\partial}{\partial t} C^{\pm}(Y|t) = \pm \frac{\partial^2}{\partial Y^A \partial Y^B} \delta^{AB} C^{\pm}(Y|t)\,. \tag{9.44}$$

We observe that this equation has the form of the non-relativistic Schrodinger equation for a free particle in the space with coordinates Y^A. Indeed, its right-hand side acquires the form of Laplacian in the variables Y^A while $C^{\pm}$ play a role of ψ and $\bar{\psi}$.

By the general argument of the beginning of this section, the two systems are equivalent, i.e. the relativistic HS theory in the X-space is equivalent to the nonrelativistic theory in the twistor space. In particular, this equivalence manifests itself in the equivalence of their symmetry algebra. As demonstrated in [54,55], the symmetry algebra of the Schrodinger equation is just the HS algebra of Sect. 9.6.

The Schrodinger equation (9.44) has zero potential. An interesting question is what are dual HS theories for one or another nonzero potential. In the case of harmonic potential the answer is known [29]. The HS equations in *AdS* and *dS* space-times are dual to the quantum-mechanical models with the proper and upside down harmonic potentials, respectively. (Not surprisingly, the *dS* geometry corresponds to the unstable quantum mechanics.)

Since the HS theory has a potential to unify gravity with quantum mechanics, one can speculate that it may be able to shed light on the both ingredients. Since full HS theory is nonlinear, its identification with quantum mechanics at the linearized level may suggest that, at ultrahigh energies, the HS theory may affect the fundamentals of quantum mechanics itself, making it nonlinear with the gravitationally small coupling constant!

9.13 To String Theory via Multiparticle Symmetry

Properties of the HS theory are to large extent determined by the properties of the HS algebra. It has been long anticipated that the HS theory should be related somehow to String Theory. To materialize this idea it is most important to find a HS algebra rich enough to underly the full fledged String Theory. Recently it was conjectured [56] that such a symmetry can be associated with a multiparticle symmetry that acts on all multiparticle states of the HS theory.

Mathematically, this symmetry algebra can be defined as the Lie algebra associated with the universal enveloping algebra of the HS algebra of Sect. 9.6. It has a number of features that make it promising as a candidate for a string-like extension of the HS theory. In particular, it contains the original HS algebra as a subalgebra. Acting on all multiparticle states of HS theory it has enough room for mixed symmetry fields which appear in String Theory.

If this idea will indeed work, it will allow to interpret String Theory as a theory of bound states of the HS theory in striking analogy with the conjecture of [13].

Summary and Conclusion

The HS gauge theories contain gravity along with infinite towers of other fields with various spins including ordinary matter fields. An interesting feature of any HS model is that it always contains a scalar field associated with graviton, which carries no internal indices. It is tempting to speculate that this scalar may play a role in cosmology and, specifically, for inflation.

The HS theory contains non-minimal higher-derivative interactions that make it a kind of a nonlocal theory with unusual properties. In particular, many of the standard tools of GR based on Riemannian geometry may not be applicable to the HS theory as a consequence of the fact that the HS symmetry transforms a spin-two field to HS fields. In practice, this implies that in HS theories one has to be careful with the conventional interpretation of physical phenomena in terms of the metric tensor. In particular, this should be taken into account in the analysis of black hole physics in the framework of HS theory.

The HS gauge theories exist in any dimension [57]. However, the HS theories available so far are analogues of pure supergravity with no matter multiplets included. This makes it difficult to analyze the important issue of spontaneous breakdown of the HS symmetry which is necessary to introduce a mass scale analogous to the string tension. In fact, it can be argued that, to achieve a spontaneous breakdown of the HS symmetry, the string-like extension of the HS theory is needed. It was recently conjectured [56] that such an extension can be provided by a multiparticle theory to be identified with the quantum HS theory and String Theory.

(continued)

Another exciting feature of the HS theory is that it exhibits a remarkable interplay between classical and quantum physics. This suggests that the further analysis may shed some light on both gravity and quantum mechanics at transplanckian energies which is the regime to be described by the HS theory.

HS theories not only have interesting holographic duals but also, being formulated in terms of unfolded dynamics approach, can shed light on the very origin of holographic duality. It can be argued [29] that the holographic duality links such models in space-times of different dimensions, that have equivalent form of their unfolded equations.

There are many important directions of the research of HS theories I had no chance to touch in these lectures.

One of the most interesting is the construction of exact solutions of HS equations. Most of exact solutions available so far, one way or another result from the solution of $3d$ HS theory obtained in [39]. There are two main types of exact black-hole type solutions of the nonlinear HS equations available in the literature. The first one is represented by the flat connections associated with the BTZ-like black holes in the $3d$ HS theory (see [58] and references therein; the interpretation of the usual BTZ black hole [59] as a solution of the HS equations was given in [60]). The second type includes black hole solutions in AdS_4 with the nonzero curvature tensor [61, 62]. Some other solutions were considered e.g. in [63–65]. Analysis of their properties in the context of the HS holographic duality and beyond is an important direction of the current research.

Among other activities we should mention analysis of the action principle in HS theory at the cubic level (see, e.g., [66–72]) and beyond [73, 74] as well as the further progress in understanding HS holography (see e.g., [75–83]) including holographic RG flows [84–86] and conformal correlators of HS currents [31, 87–92].

Since in this short review it is hard even to list all important research directions in the HS theory, we refer the reader to other reviews [30, 31, 44, 58, 93–97] as well as to the contribution of Ricardo Troncoso to this workshop [98], where more detail and references on various aspects of the HS theory can be found.

Acknowledgements The author is grateful to Olga Gelfond for useful comments on the manuscript and to the organizers of the seventh Aegean workshop on non-Einstein theories of gravity on Paros for the warm atmosphere and hospitality. This research was supported in part by RFBR Grant No 14-02-01172.

References

1. D.Z. Freedman, A. Van Proeyen, Supergravity. (Cambridge University Press, Cambridge, 2012), 607 p
2. A.H. Chamseddine, P.C. West, Nucl. Phys. B **129**, 39 (1977)
3. C. Fronsdal, Phys. Rev. **D18**, 3624 (1978)
4. S. Weinberg, Phys. Rev. **138**, B988–B1002 (1965)
5. S.R. Coleman, J. Mandula, Phys. Rev. **159**, 1251–1256 (1967)
6. C. Aragone, S. Deser, Phys. Lett. **B86**, 161 (1979)
7. A.K.H. Bengtsson, I. Bengtsson, L. Brink, Nucl. Phys. **B227**, 31 (1983)
8. A.K.H. Bengtsson, I. Bengtsson, L. Brink, Nucl. Phys. **B227**, 41 (1983)
9. F.A. Berends, G.J.H. Burgers, H. Van Dam, Z. Phys. **C24**, 247–254 (1984)
10. F.A. Berends, G.J.H. Burgers, H. van Dam, Nucl. Phys. **B260**, 295 (1985)
11. E.S. Fradkin, M.A. Vasiliev, Phys. Lett. **B189**, 89–95 (1987)
12. E.S. Fradkin, M.A. Vasiliev, Nucl. Phys. **B291**, 141 (1987)
13. C.-M. Chang, S. Minwalla, T. Sharma, X. Yin, J. Phys. A **46**, 214009 (2013). [arXiv:1207.4485 [hep-th]]
14. J.M. Maldacena, Adv. Theory Math. Phys. **2**, 231 (1998). [arXiv:hep-th/9711200]. [Int. J. Theory Phys. **38** 1113, (1999)]
15. S.S. Gubser, I.R. Klebanov, A.M. Polyakov, Phys. Lett. B **428**, 105 (1998). [arXiv: hep-th/9802109]
16. E. Witten, Adv. Theory Math. Phys. **2**, 253 (1998). [arXiv:hep-th/9802150]
17. S.E. Konstein, M.A. Vasiliev, V.N. Zaikin, J. High Energy Pract. **0012**, 018 (2000). [hep-th/0010239]
18. B. Sundborg, Nucl. Phys. Proc. Suppl. **102**, 113 (2001). [arXiv:hep-th/0103247]
19. E. Witten, Talk at the John Schwarz 60-th birthday symposium. (2011). http://theory.caltech.edu/jhs60/witten/1.html
20. A. Mikhailov, (2002). arXiv:hep-th/0201019
21. E. Sezgin, P. Sundell, Nucl. Phys. B **644**, 303 (2002). [arXiv:hep-th/0205131] . [Erratum-Nucl. Phys. B **660**, (2003) 403]
22. I.R. Klebanov, A.M. Polyakov, Phys. Lett. B **550**, 213 (2002). [arXiv:hep-th/0210114]
23. S. Giombi, X. Yin, J. High Energy Pract. **1009**, 115 (2010). [arXiv:0912.3462 [hep-th]]
24. R.G. Leigh, A.C. Petkou, J. High Energy Pract. **0306**, 011 (2003). [hep-th/0304217]
25. E. Sezgin, P. Sundell, J. High Energy Pract. **0507**, 044 (2005). [hep-th/0305040]
26. M. Henneaux, S.-J. Rey, J. High Energy Pract. **1012**, 007 (2010). [arXiv:1008.4579 [hep-th]]
27. A. Campoleoni, S. Fredenhagen, S. Pfenninger, S. Theisen, J. High Energy Pract. **1011**, 007 (2010). [arXiv:1008.4744 [hep-th]]
28. M.R. Gaberdiel, R. Gopakumar, Phys. Rev. D **83**, 066007 (2011). [arXiv:1011.2986 [hep-th]]
29. M.A. Vasiliev, J. Phys. A **46**, 214013 (2013). [arXiv:1203.5554 [hep-th]]
30. M.R. Gaberdiel, R. Gopakumar, J. Phys. A **46**, 214002 (2013). [arXiv:1207.6697 [hep-th]]
31. S. Giombi, X. Yin, J. Phys. A **46**, 214003 (2013). [arXiv:1208.4036 [hep-th]]
32. O.V. Shaynkman, M.A. Vasiliev, Theor. Math. Phys. **128**, 1155 (2001). [hep-th/0103208]. [Teor. Mat. Fiz. **128**, 378 (2001)]
33. M.G. Eastwood, Ann. Math. **161** 1645, (2005). [hep-th/0206233]
34. E.S. Fradkin, M.A. Vasiliev, Ann. Phys. **177**, 63 (1987)
35. M.A. Vasiliev, Fortsch. Phys. **36**, 33 (1988)
36. M.A. Vasiliev, Ann. Phys. **190**, 59 (1989)
37. M.A. Vasiliev, Phys. Lett. B **285**, 225 (1992)
38. S.E. Konshtein, M.A. Vasiliev, Nucl. Phys. B **312**, 402 (1989)
39. S.F. Prokushkin, M.A. Vasiliev, Nucl. Phys. B **545**, 385 (1999). [hep-th/9806236]
40. M.A. Vasiliev, Class. Quantum Gravity **8**, 1387 (1991)
41. M.A. Vasiliev, Int. J. Geom. Methods Mod. Phys. **3**, 37 (2006). [hep-th/0504090]
42. M.A. Vasiliev, Nucl. Phys. B **793**, 469 (2008). [arXiv:0707.1085 [hep-th]]

43. T. Lada, J. Stasheff, Int. J. Theory Phys. **32**, 1087 (1993). [hep-th/9209099]
44. X. Bekaert, S. Cnockaert, C. Iazeolla, M.A. Vasiliev, (2005). hep-th/0503128
45. R.J. Baston, M.G. Eastwood, Oxford mathematical monographs. (Clarendon, Oxford, 1989), 213 p.
46. M.A. Vasiliev, Phys. Rev. D **66**, 066006 (2002). [hep-th/0106149].
47. O.A. Gelfond, M.A. Vasiliev, Theory Math. Phys. **145**, 1400 (2005). [hep-th/0304020]. [Teor. Mat. Fiz. **145**, 35 (2005)]
48. A. Jevicki, K. Jin, Q. Ye, J. Phys. A **46**, 214005 (2013). [arXiv:1212.5215 [hep-th]]
49. N.M. Nikolov, Y.S. Stanev, I.T. Todorov, J. Phys. A **35**, 2985 (2002). [hep-th/0110230]
50. C. Fronsdal, Massless particles, ortosymplectic symmetry and another type of Kaluza-Klein theory. In: *Essays on Supersymmetry*. Mathematical Physics Studies, vol. 8 (Reidel, Dordrecht, 1986). Preprint UCLA/85/TEP/10
51. M.A. Vasiliev, In: *Multiple Facets of Quantization and Supersymmetry*, ed. by M. Olshanetsky, et al. (2001), pp. 826–872. [hep-th/0111119]
52. I.A. Bandos, J. Lukierski, D.P. Sorokin, Phys. Rev. D **61**, 045002 (2000). [arXiv: hep-th/9904109].
53. I. Bandos, X. Bekaert, J.A. de Azcarraga, D. Sorokin, M. Tsulaia, J. High Energy Pract. **0505**, 031 (2005). [hep-th/0501113]
54. M. Valenzuela, (2009). arXiv:0912.0789 [hep-th]
55. X. Bekaert, E. Meunier, S. Moroz, J. High Energy Pract. **1202**, 113 (2012). [arXiv:1111.3656 [hep-th]]
56. M.A. Vasiliev, Class. Quant. Gravity **30**, 104006 (2013). [arXiv:1212.6071 [hep-th]]
57. M.A. Vasiliev, Phys. Lett. B **567**, 139 (2003). [hep-th/0304049]
58. M. Ammon, M. Gutperle, P. Kraus, E. Perlmutter, J. Phys. A **46**, 214001 (2013).
59. M. Banados, C. Teitelboim, J. Zanelli, Phys. Rev. Lett. **69**, 1849 (1992). [hep-th/9204099]
60. V.E. Didenko, A.S. Matveev, M.A. Vasiliev, Theory Math. Phys. **153**, 1487 (2007). [hep-th/0612161]. [Teor. Mat. Fiz. **153** 158, (2007)]
61. V.E. Didenko, M.A. Vasiliev, Phys. Lett. B **682**, 305 (2009). [arXiv:0906.3898 [hep-th]]. [Erratum-Phys. Lett. B **722**, 389 (2013)]
62. C. Iazeolla, P. Sundell, J. High Energy Pract. **1112**, 084 (2011). [arXiv:1107.1217 [hep-th]]
63. E. Sezgin, P. Sundell, Nucl. Phys. B **762**, 1 (2007). [hep-th/0508158]
64. E. Sezgin, P. Sundell, (2005). hep-th/0511296
65. C. Iazeolla, E. Sezgin, P. Sundell, Nucl. Phys. B **791**, 231 (2008). [arXiv:0706.2983 [hep-th]]
66. R.R. Metsaev, Nucl. Phys. B **759**, 147 (2006). [hep-th/0512342]
67. K. Alkalaev, J. High Energy Pract. **1103**, 031 (2011). [arXiv:1011.6109 [hep-th]]
68. M.A. Vasiliev, Nucl. Phys. B **862**, 341 (2012). [arXiv:1108.5921 [hep-th]]
69. I.L. Buchbinder, T.V. Snegirev, Y..M. Zinoviev, Nucl. Phys. B **864**, 694 (2012). [arXiv:1204.2341 [hep-th]]
70. R.R. Metsaev, Phys. Lett. B **720**, 237 (2013). [arXiv:1205.3131 [hep-th]]
71. E. Joung, L. Lopez, M. Taronna, J. High Energy Pract. **1301**, 168 (2013). [arXiv:1211.5912 [hep-th]].
72. N. Boulanger, D. Ponomarev, E.D. Skvortsov, J. High Energy Pract. **1305**, 008 (2013). [arXiv:1211.6979 [hep-th]]
73. N. Boulanger, P. Sundell, J. Phys. A **44**, 495402 (2011). [arXiv:1102.2219 [hep-th]]
74. N. Boulanger, N. Colombo, P. Sundell, J. High Energy Pract. **1210** 043, (2012). [arXiv:1205.3339 [hep-th]]
75. J. Maldacena, A. Zhiboedov, J. Phys. A **46**, 214011 (2013). [arXiv:1112.1016 [hep-th]]
76. J. Maldacena, A. Zhiboedov, Class. Quant. Gravity **30**, 104003 (2013). [arXiv:1204.3882 [hep-th]]
77. V.E. Didenko, E.D. Skvortsov, J. Phys. A **46**, 214010 (2013). [arXiv:1207.6786 [hep-th]]
78. H. Afshar, A. Bagchi, R. Fareghbal, D. Grumiller, J. Rosseel, Phys. Rev. Lett. **111**, 121603 (2013). [arXiv:1307.4768 [hep-th]]
79. S. Giombi, I.R. Klebanov, J. High Energy Pract. **1312**, 068 (2013). [arXiv:1308.2337 [hep-th]]
80. A.A. Tseytlin, Nucl. Phys. B **877**, 598 (2013). [arXiv:1309.0785 [hep-th]]

81. R.R. Metsaev, (2013). arXiv:1311.7350 [hep-th]
82. S. Giombi, I.R. Klebanov, B.R. Safdi, (2014). arXiv:1401.0825 [hep-th]
83. S. Giombi, I.R. Klebanov, A.A. Tseytlin, (2014). arXiv:1402.5396 [hep-th]
84. M.R. Douglas, L. Mazzucato, S.S. Razamat, Phys. Rev. D **83**, 071701 (2011). [arXiv:1011.4926 [hep-th]]
85. I. Sachs, (2013). arXiv:1306.6654 [hep-th]
86. R.G. Leigh, O. Parrikar, A.B. Weiss, (2014). arXiv:1402.1430 [hep-th]
87. A. Zhiboedov, (2012). arXiv:1206.6370 [hep-th]
88. N. Colombo, P. Sundell, (2012). arXiv:1208.3880 [hep-th]
89. V.E. Didenko, E.D. Skvortsov, J. High Energy Pract. **1304**, 158 (2013). [arXiv:1210.7963 [hep-th]]
90. O.A. Gelfond, M.A. Vasiliev, Nucl. Phys. B **876**, 871 (2013). [arXiv:1301.3123 [hep-th]]
91. Y.S. Stanev, Nucl. Phys. B **876**, 651 (2013). [arXiv:1307.5209 [hep-th]]
92. I. Florakis, D. Sorokin, M. Tsulaia, (2014). arXiv:1401.1645 [hep-th]
93. M.A. Vasiliev, Int. J. Mod. Phys. D **5**, 763 (1996). [hep-th/9611024]
94. M.A. Vasiliev, In: *The Many Faces of the Superworld*, ed. by M.A. Shifman, (1999), pp. 533–610. [hep-th/9910096]
95. X. Bekaert, N. Boulanger, P. Sundell, Rev. Mod. Phys. **84**, 987 (2012). [arXiv:1007.0435 [hep-th]]
96. A. Sagnotti, J. Phys. A **46**. 214006 (2013). [arXiv:1112.4285 [hep-th]]
97. V.E. Didenko, E.D. Skvortsov, (2014). arXiv:1401.2975 [hep-th]
98. A. Perez, D. Tempo, R. Troncoso, (2014). arXiv:1402.1465 [hep-th]

Chapter 10
Higher Spin Black Holes

Alfredo Pérez, David Tempo, and Ricardo Troncoso

Abstract We review some relevant results in the context of higher spin black holes in three-dimensional spacetimes, focusing on their asymptotic behaviour and thermodynamic properties. For simplicity, we mainly discuss the case of gravity nonminimally coupled to spin-three fields, being nonperturbatively described by a Chern–Simons theory of two independent $sl(3, \mathbb{R})$ gauge fields. Since the analysis is particularly transparent in the Hamiltonian formalism, we provide a concise discussion of their basic aspects in this context; and as a warming up exercise, we briefly analyze the asymptotic behaviour of pure gravity, as well as the BTZ black hole and its thermodynamics, exclusively in terms of gauge fields. The discussion is then extended to the case of black holes endowed with higher spin fields, briefly signaling the agreements and discrepancies found through different approaches. We conclude explaining how the puzzles become resolved once the fall off of the fields is precisely specified and extended to include chemical potentials, in a way that it is compatible with the asymptotic symmetries. Hence, the global charges become completely identified in an unambiguous way, so that different sets of asymptotic conditions turn out to contain inequivalent classes of black hole solutions being characterized by a different set of global charges.

10.1 Introduction

Fundamental particles of spin greater than two are hitherto unknown, which from a purely theoretical point of view, appears to agree with the widespread belief that massless fields of spin $s > 2$ are doomed to suffer from inconsistencies. Indeed, the lore is reflected through a well-known claim in the context of supergravity (see e.g.,

---

A. Pérez • D. Tempo
Centro de Estudios Científicos (CECs), Casilla 1469, Valdivia, Chile
e-mail: aperez@cecs.cl; tempo@cecs.cl

R. Troncoso (✉)
Centro de Estudios Científicos (CECs), Casilla 1469, Valdivia, Chile

Universidad Andrés Bello, Av. Repblica 440, Santiago, Chile
e-mail: troncoso@cecs.cl

E. Papantonopoulos (ed.), *Modifications of Einstein's Theory of Gravity at Large Distances*, Lecture Notes in Physics 892, DOI 10.1007/978-3-319-10070-8_10

[1]), which asserts that the maximum number of local supersymmetries is bounded by eight; otherwise, since the supersymmetry generators act as raising or lowering operators for spin, a supermultiplet would contain fields of spin greater than two. In turn, through the Kaluza–Klein mechanism, this also sets an upper bound on the spacetime dimension to be at most eleven. The supposed inconsistency of higher spin fields relies on solid no-go theorems (see [2] for a good review about this subject). In particular, it is worth mentioning the result of Aragone and Deser [3], which states that the higher spin gauge symmetries of the free theory around flat spacetime, cannot be preserved once the field is minimally coupled to gravity.

A consistent way to circumvent the incompatibility of higher spin gauge symmetries with interactions was pioneered by Vasiliev [4, 5], who was able to formulate the field equations for a whole tower of nonminimally coupled fields of spin $s = 0, 1, 2, \ldots, \infty$, in presence of a cosmological constant (For recent reviews see e.g., [6, 7]). It is worth pointing out that, since the hypotheses of the Coleman–Mandula theorem are not fulfilled by Vasiliev theory, spacetime and gauge symmetries become inherently mixed in an unaccustomed form [8]. It then goes without saying that the very existence of Vasiliev theory, naturally suggests a possible reformulation of supergravity theories from scratch, which would may in turn elucidate new alternative approaches to strings and M-theory. Indeed, in eleven dimensions and in presence of a negative cosmological constant, a supergravity theory that shares some of these features, as the mixing of spacetime and gauge symmetries, is known to exist [9].

In order to gain some insights about this counterintuitive subject, one may instead follow the less ambitious approach of finding a simpler set up that still captures some of the relevant features that characterize the dynamics of higher spin fields. In this sense, the three-dimensional case turns out to be particularly appealing, since the dynamics is described through a standard field theory with a Chern–Simons action [10–12]. The generic theory can be further simplified, since it admits a consistent truncation to the case of a finite number of nonpropagating fields with spin $s = 2, 3, \ldots, N$. Hence the simplest case with the desired properties corresponds to $N = 3$, so that the theory describes gravity with negative cosmological constant, nonminimally coupled to an interacting spin-three field. The remarkable simplification of the theory then allows the possibility of finding different classes of exact black hole solutions endowed with a nontrivial spin-three field, as the ones in [13, 14], and [15], respectively. However, despite the simplicity of these solutions, the subject has not been free of controversy, mainly due to the puzzling discrepancies that have been found in the characterization of their global charges and their entropy.

The purpose of this brief review, is overviewing some of the relevant results about this ongoing subject, as well as explaining how the apparent tension between different approaches is fully resolved once the chemical potentials are suitably identified along the lines of [15, 16], so that the asymptotic symmetries, and hence the global charges, are completely characterized in an unambiguous way.

It is worth highlighting that the action principle in terms of the metric and the spin-three field is currently known as a weak field expansion of the spin-three field up to quadratic order [17]. Thus, in order to deal with the full nonperturbative treatment of the higher spin black hole solutions, it turns out to be useful to describe them only in terms of gauge fields and the topology of the manifold, without making any reference neither to the metric nor to the spin-three field.

Since the analysis becomes particularly transparent in the Hamiltonian formalism, in the next section we concisely discuss some of their basic aspects in the context of Chern–Simons theories in three dimensions. As a useful warming up exercise, in Sect. 10.3, the asymptotic behaviour of pure gravity with negative cosmological constant [18], as well as the BTZ black hole [19, 20] and its thermodynamics, are briefly analyzed exclusively in terms of gauge fields. Section 10.4 is devoted to the case of gravity coupled to spin-three fields, including the asymptotic behaviour described in [21, 22], the higher spin black hole solution of [13, 23], and its thermodynamics [24, 25], briefly signaling the agreements and discrepancies found through different approaches. We conclude with Sect. 10.5, where it is explained how these puzzling differences become fully resolved once the fall off of the fields is precisely specified, so that different sets of asymptotic conditions turn out to contain inequivalent classes of black hole solutions [15, 16] being characterized by a different set of global charges.

10.2 Basic Aspects and Hamiltonian Formulation of Chern–Simons Theories in Three Dimensions

In three-dimensional spacetimes, gauge theories described by a Chern–Simons action are much simpler than their corresponding Yang–Mills analogues, in the sense that less structure is required in order to formulate them. Indeed, the manifold M, locally described by a set of coordinates x^μ, is only endowed with a gauge field $A = A^I_\mu T_I dx^\mu$, where T_I stand for the generators of a Lie algebra $\mathfrak{g}$, which is assumed to admit an invariant nondegenerate bilinear form $g_{IJ} = \langle T_I, T_J \rangle$. These ingredients are enough to construct the action, given by

$$I_{CS}[A] = \frac{k}{4\pi} \int_M \left\langle AdA + \frac{2}{3} A^3 \right\rangle , \tag{10.1}$$

where k is a constant, and wedge product between forms has been assumed. Consequently, the action does not require the existence of a spacetime metric, but it is sensitive to the topology of M. The field equations imply the vanishing of curvature, i.e., $F = dA + A^2 = 0$, so that the connection becomes locally flat on shell, and then the theory is devoid of local propagating degrees of freedom. Note that the action (10.1) is already in Hamiltonian form. Indeed, if the topology of M is of the form $M = \Sigma \times \mathbb{R}$, where Σ stands for the spacelike section, the connection splits as $A = A_i dx^i + A_t dt$, and hence the action (10.1) reduces to

$$I_H = -\frac{k}{4\pi}\int_M dtd^2x\varepsilon^{ij}\left\langle A_i\dot{A}_j - A_tF_{ij}\right\rangle \ , \tag{10.2}$$

up to a boundary term. It is then apparent that A_i correspond to the dynamical fields, whose Poisson brackets are given by $\left\{A_i^I(x), A_j^J(x')\right\} = \frac{2\pi}{k}g^{IJ}\varepsilon_{ij}\delta(x-x')$, while A_t become Lagrange multipliers associated to the constraints $G = \frac{k}{4\pi}\varepsilon^{ij}F_{ij}$. Then, the smeared generator of the gauge transformations reads

$$G(\Lambda) = \int_\Sigma d^2x\,\langle\Lambda G\rangle \ ,$$

so that $\delta A_i = \{A_i, G(\Lambda)\} = \partial_i\Lambda + [A_i, \Lambda]$ (see, e.g., [26–28]). However, when Σ has a boundary, according to the Regge–Teitelboim approach [29], the generator of the gauge transformations has to be improved by a boundary term $Q(\Lambda)$, i.e.,

$$\tilde{G}(\Lambda) = G(\Lambda) + Q(\Lambda) \ , \tag{10.3}$$

being such that its functional variation is well-defined everywhere. This implies that the variation of the conserved charge associated to an asymptotic gauge symmetry, generated by a Lie algebra valued parameter Λ, is determined by the dynamical fields at a fixed time slice at the boundary, which reads

$$\delta Q(\Lambda) = -\frac{k}{2\pi}\int_{\partial\Sigma}\langle\Lambda\delta A_\theta\rangle\,d\theta \ , \tag{10.4}$$

where $\partial\Sigma$ stands for the boundary of the spacelike section Σ.

The transformation law of the Lagrange multipliers, $\delta A_t = \partial_t\Lambda + [A_t, \Lambda]$, is then recovered requiring the improved action to be invariant under gauge transformations. Note that on-shell, by virtue of the identity $\mathscr{L}_\xi A_\mu = \nabla_\mu(\xi^\nu A_\nu) + \xi^\nu F_{\nu\mu}$, diffeomorphisms $\delta_\xi A_\mu = -\mathscr{L}_\xi A_\mu$ are equivalent to gauge transformations with parameter $\Lambda = -\xi^\mu A_\mu$, and hence, the variation of the generator of an asymptotic symmetry spanned by an asymptotic killing vector ξ^μ, reads

$$\delta Q(\xi) = \frac{k}{2\pi}\int_{\partial\Sigma}\xi^\mu\left\langle A_\mu\delta A_\theta\right\rangle d\theta \ . \tag{10.5}$$

This means that the variation of the total energy of the system, which takes into account the contribution of all the constraints, is given by

$$\delta E = \delta Q(\partial_t) = \frac{k}{2\pi}\int_{\partial\Sigma}\langle A_t\delta A_\theta\rangle\,d\theta \ . \tag{10.6}$$

It should be stressed that the whole canonical structure only makes sense provided the variation of the canonical generators can be integrated. This can be generically done once a precise set of asymptotic conditions is specified, which in turn determines the asymptotic symmetries. This will be explicitly discussed in the next section for the case of pure gravity with negative cosmological constant, as well as in Sect. 10.4, and further elaborated in Sect. 10.5 in the case of gravity coupled to a spin-three field.

10.3 General Relativity with Negative Cosmological Constant in Three Dimensions

As it was shown in [30, 31] General Relativity in vacuum can be described in terms of a Chern–Simons action. In the case of negative cosmological constant the corresponding Lie algebra is of the form $\mathfrak{g} = \mathfrak{g}_+ + \mathfrak{g}_-$, where $\mathfrak{g}_\pm$ stand for two independent copies of $sl(2,\mathbb{R})$, which will be assumed to be described by the same set of matrices L_i, with $i = -1, 0, 1$, given by

$$L_{-1} = \begin{pmatrix} 0 & 0 \\ 1 & 0 \end{pmatrix} \quad ; \quad L_0 = \begin{pmatrix} -\frac{1}{2} & 0 \\ 0 & \frac{1}{2} \end{pmatrix} \quad ; \quad L_1 = \begin{pmatrix} 0 & -1 \\ 0 & 0 \end{pmatrix} , \tag{10.7}$$

so that the $sl(2,\mathbb{R})$ algebra reads

$$\left[L_i, L_j\right] = (i-j)\, L_{i+j} \ . \tag{10.8}$$

The connection then splits in two independent $sl(2,\mathbb{R})$-valued gauge fields, according to $A = A^+ + A^-$, while the invariant nondegenerate bilinear form is chosen such that the action (10.1) reduces to

$$I = I_{CS}\left[A^+\right] - I_{CS}\left[A^-\right] , \tag{10.9}$$

so that the bracket now corresponds to just the trace, i.e., in the representation of (10.7), $\langle\cdots\rangle = \mathrm{tr}(\cdots)$, and the level is fixed by the AdS radius and the Newton constant as $k = \frac{l}{4G}$. The link between the gauge fields and spacetime geometry is made through

$$A^\pm = \omega \pm \frac{e}{l} , \tag{10.10}$$

where ω and e correspond to the spin connection and the dreibein, respectively. The field equations, $F^\pm = 0$, then imply that the spacetime curvature is constant and the torsion vanishes, while the metric is recovered from

$$g_{\mu\nu} = 2\mathrm{tr}\left(e_\mu e_\nu\right) , \tag{10.11}$$

which is manifestly invariant under the diagonal subgroup of $SL(2,\mathbb{R}) \times SL(2,\mathbb{R})$, that corresponds to the local Lorentz transformations. Note that diffeomorphisms can always be expressed in terms of the remaining gauge symmetries.

10.3.1 Brown–Henneaux Boundary Conditions

As explained in [32], the asymptotic behaviour of gravity with negative cosmological constant, as originally described by Brown and Henneaux [18], can be readily formulated in terms of the gauge fields $A^{\pm}$. The gauge can be chosen such that the radial dependence is entirely captured by the group elements

$$g_{\pm} = e^{\pm\rho L_0} , \tag{10.12}$$

so that the asymptotic form of the connections is given by

$$A^{\pm} = g_{\pm}^{-1} a^{\pm} g_{\pm} + g_{\pm}^{-1} dg_{\pm} , \tag{10.13}$$

where $a^{\pm} = a_{\theta}^{\pm} d\theta + a_{t}^{\pm} dt$, read

$$a^{\pm} = \pm \left(L_{\pm 1} - \frac{2\pi}{k} \mathscr{L}_{\pm} L_{\mp 1} \right) dx^{\pm} , \tag{10.14}$$

with $x^{\pm} = \frac{t}{l} \pm \theta$, and the functions $\mathscr{L}_{\pm}$ depend only on time and the angular coordinate.

The asymptotic form of the dynamical fields $a_{\theta}^{\pm}$ is preserved under gauge transformations, $\delta a_{\theta}^{\pm} = \partial_{\theta} \Lambda^{\pm} + \left[a_{\theta}^{\pm}, \Lambda^{\pm} \right]$, generated by

$$\Lambda^{\pm} \left(\varepsilon_{\pm} \right) = \varepsilon_{\pm} L_{\pm 1} \mp \varepsilon_{\pm}' L_0 + \frac{1}{2} \left(\varepsilon_{\pm}'' - \frac{4\pi}{k} \varepsilon_{\pm} \mathscr{L}_{\pm} \right) L_{\mp 1} , \tag{10.15}$$

where $\varepsilon_{\pm}$ are arbitrary functions of t, θ, provided the functions $\mathscr{L}_{\pm}$ transform as

$$\delta \mathscr{L}_{\pm} = \varepsilon_{\pm} \mathscr{L}_{\pm}' + 2 \mathscr{L}_{\pm} \varepsilon_{\pm}' - \frac{k}{4\pi} \varepsilon_{\pm}''' . \tag{10.16}$$

Hereafter, prime denotes the derivative with respect to θ. Furthermore, requiring the components of the gauge fields along time, $a_{t}^{\pm}$, to be mapped into themselves under the same gauge transformations, together with the transformation laws in (10.16), implies that the functions $\mathscr{L}_{\pm}$ and the parameters $\varepsilon_{\pm}$ are chiral, i.e.,

$$\partial_{\mp} \mathscr{L}_{\pm} = 0 \ , \quad \partial_{\mp} \varepsilon_{\pm} = 0 \, . \tag{10.17}$$

Note that the first condition in (10.17) means that the field equations have to be fulfilled in the asymptotic region.

Consequently, according to (10.4), the variation of the canonical generators associated to the asymptotic gauge symmetries generated by $\Lambda = \Lambda^{+} + \Lambda^{-}$, in this case reduces to

$$\delta Q(\Lambda) = \delta Q_+ \left(\Lambda^+\right) - \delta Q_- \left(\Lambda^-\right) , \tag{10.18}$$

with

$$\delta Q_\pm \left(\Lambda^\pm\right) = -\frac{k}{2\pi} \int \left\langle \Lambda^\pm \delta a_\theta^\pm \right\rangle d\theta = -\int \varepsilon_\pm \delta \mathscr{L}_\pm d\theta , \tag{10.19}$$

which can be readily integrated as

$$Q_\pm \left(\Lambda^\pm\right) = -\int \varepsilon_\pm \mathscr{L}_\pm d\theta . \tag{10.20}$$

Therefore, since the canonical generators fulfill $\delta_{\Lambda_1} Q[\Lambda_2] = \{Q[\Lambda_2], Q[\Lambda_1]\}$, their algebra can be directly obtained by virtue of (10.16), which reduces to two copies of the Virasoro algebra with the same central extension $c = \frac{3l}{2G}$ [18]. Expanding in Fourier modes, according to $\mathscr{L} = \frac{1}{2\pi}\sum_m \mathscr{L}_m e^{im\theta}$, the algebra explicitly reads

$$i\left\{\mathscr{L}_m, \mathscr{L}_n\right\} = (m-n)\mathscr{L}_{m+n} + \frac{k}{2} m^3 \delta_{m+n,0} , \tag{10.21}$$

for both copies.

10.3.2 BTZ Black Hole and Its Thermodynamics

The asymptotic conditions described above, manifestly contain the BTZ black hole solution [19,20], being described by

$$a^\pm = \pm\left(L_{\pm 1} - \frac{2\pi}{k}\mathscr{L}_\pm L_{\mp 1}\right) dx^\pm , \tag{10.22}$$

when $\mathscr{L}_\pm$ are nonnegative constants. Indeed, by virtue of Eqs. (10.10) and (10.11), the spacetime metric is recovered in normal coordinates:

$$ds^2 = l^2 \left[d\rho^2 + \frac{2\pi}{k}\left(\mathscr{L}_+ \left(dx^+\right)^2 + \mathscr{L}_- \left(dx^-\right)^2\right) \right.$$
$$\left. - \left(e^{2\rho} + \frac{4\pi^2}{k^2}\mathscr{L}_+\mathscr{L}_- e^{-2\rho}\right) dx^+ dx^- \right] . \tag{10.23}$$

As shown in [33] (see also [34]), the topology of the Euclidean black hole corresponds to $\mathbb{R}^2 \times S^1$, where $\mathbb{R}^2$ stands for the one of the $\rho - \tau$ plane, and $\tau = -it$ is the Euclidean time, fulfilling $0 \leq \tau < \beta$, where $\beta = T^{-1}$ is the inverse of the Hawking temperature. Since $\mathbb{R}^2$ can be mapped into a disk through a conformal compactification, the black hole topology is then equivalent to the one of a solid torus.

As explained in the introduction, and for later purposes, afterwards we will perform the remaining analysis exclusively in terms of the gauge fields (10.22) and the topology of the manifold, without making any reference to the spacetime metric.

The simplest gauge covariant object that is sensitive to the global properties of the manifold turns out to be the holonomy of the gauge field around a closed cycle $\mathscr{C}$, defined as

$$\mathscr{H}_{\mathscr{C}} = P\exp\left(\int_{\mathscr{C}} A_\mu dx^\mu\right) , \tag{10.24}$$

which is an element of the gauge group. Hence, since in this case the gauge group corresponds to $SL(2,\mathbb{R}) \times SL(2,\mathbb{R})$, the holonomy around $\mathscr{C}$ is of the form $\mathscr{H}_{\mathscr{C}} = \mathscr{H}_{\mathscr{C}}^+ \otimes \mathscr{H}_{\mathscr{C}}^-$, with

$$\mathscr{H}_{\mathscr{C}}^\pm = P\exp\left(\int_{\mathscr{C}} A_\mu^\pm dx^\mu\right) . \tag{10.25}$$

As the topology of the manifold is the one of a solid torus, there are two inequivalent classes of cycles: (I) the ones that wind around the handle, and (II) those that do not. This means that the former ones are noncontractible, while the latter can be continuously shrunk to a point. Then, the holonomies along contractible cycles are trivial, i.e.,

$$\mathscr{H}_{\mathscr{C}_{II}}^\pm = -1 , \tag{10.26}$$

where the negative sign is due to the fact that, according to (10.35), we are dealing with the fundamental (spinorial) representation of $SL(2,\mathbb{R})$; while the holonomies along noncontractible cycles $\mathscr{H}_{\mathscr{C}_I}^\pm$ are necessarily nontrivial. Indeed, it is easy to verify that this is the case for the gauge fields that describe the BTZ black hole (10.22). For simplicity, we explicitly carry out the computation in the static case, i.e., for $\mathscr{L} := \mathscr{L}_\pm$, since the inclusion of rotation is straightforward.

A simple noncontractible cycle in this case is parameterized by $\rho = \rho_0$, and $\tau = \tau_0$, with ρ_0, τ_0 constants, so that the corresponding holonomies around it read

$$\mathscr{H}_\theta^\pm = e^{2\pi a_\theta^\pm} . \tag{10.27}$$

These holonomies are then fully characterized, up to conjugacy by elements of $SL(2,\mathbb{R})$, by the eigenvalues of $2\pi a_\theta^\pm$, given by

$$\lambda_\pm^2 = 2\pi^2 \mathrm{tr}\left[\left(a_\theta^\pm\right)^2\right] = \frac{8\pi^3}{k}\mathscr{L} , \tag{10.28}$$

and hence, since $\mathscr{L}$ is nonnegative, they are manifestly nontrivial.

Analogously, a simple contractible cycle is parameterized by $\rho = \rho_0$, and $\theta = \theta_0$. Since the holonomies around this cycle are trivial, the conditions in (10.26) reduce to

$$\mathcal{H}_\tau^\pm = e^{\beta a_\tau^\pm} = e^{i\beta a_t^\pm} = -1 \,, \tag{10.29}$$

and since the cycle winds once, the eigenvalues of $i\beta a_t$ are given by $\pm i\pi$, which equivalently implies that

$$\beta^2 \mathrm{tr}\left[\left(a_t^\pm\right)^2\right] = 2\pi^2 \,. \tag{10.30}$$

Therefore, the triviality of the holonomies around this cycle amounts to fix the Euclidean time period as

$$\beta = l\sqrt{\frac{\pi k}{2\mathscr{L}}} \,, \tag{10.31}$$

in full agreement with the Hawking temperature.

Note that the variation of the total energy (10.6) in this case reads

$$\delta E = \frac{k}{2\pi}\int_{\partial\Sigma} \left(\langle a_t^+ \delta a_\theta^+\rangle - \langle a_t^- \delta a_\theta^-\rangle\right) d\theta = \frac{4\pi}{l}\delta\mathscr{L} \,, \tag{10.32}$$

from which, by virtue of (10.31) and the first law, implies that

$$\delta S = \beta\delta E = \delta\left(4\pi\sqrt{2\pi k\mathscr{L}}\right) \,, \tag{10.33}$$

which means that the entropy can be expressed in terms of the global charges (10.20), as

$$S = 4\pi\sqrt{2\pi k\mathscr{L}} \,. \tag{10.34}$$

The black hole entropy found in this way agrees with the standard result obtained in the metric formalism. Indeed, according to (10.23), in the static case the event horizon is located at $e^{2\rho_+} = \frac{2\pi}{k}\mathscr{L}$, so that its area is given by $A = 4\pi l\sqrt{\frac{2\pi}{k}\mathscr{L}}$, and hence (10.34) is equivalent to the Bekenstein–Hawking formula $S = \frac{A}{4G}$.

10.4 Higher Spin Gravity in 3D

As explained in the introduction, gravity with negative cosmological constant, nonminimally coupled to an interacting spin-three field can be described in terms of a Chern–Simons theory [10–12]. The action is then of the form (10.1), and as in the case of pure gravity, the corresponding Lie algebra is of the form $\mathfrak{g} = \mathfrak{g}_+ + \mathfrak{g}_-$, but where now $\mathfrak{g}_\pm$ are enlarged to two independent copies of $sl(3,\mathbb{R})$. Both copies of the algebra will be assumed to be spanned by the same set of matrices L_i, W_m, with $i = -1, 0, 1$, and $m = -2, -1, 0, 1, 2$, given by (see e.g., [22])

$$L_{-1} = \begin{pmatrix} 0 & -2 & 0 \\ 0 & 0 & -2 \\ 0 & 0 & 0 \end{pmatrix} \quad ; \quad L_0 = \begin{pmatrix} 1 & 0 & 0 \\ 0 & 0 & 0 \\ 0 & 0 & -1 \end{pmatrix} \quad ; \quad L_1 = \begin{pmatrix} 0 & 0 & 0 \\ 1 & 0 & 0 \\ 0 & 1 & 0 \end{pmatrix} ,$$

$$W_{-2} = \begin{pmatrix} 0 & 0 & 8 \\ 0 & 0 & 0 \\ 0 & 0 & 0 \end{pmatrix} \quad ; \quad W_{-1} = \begin{pmatrix} 0 & -2 & 0 \\ 0 & 0 & 2 \\ 0 & 0 & 0 \end{pmatrix} \quad ; \quad W_0 = \frac{2}{3}\begin{pmatrix} 1 & 0 & 0 \\ 0 & -2 & 0 \\ 0 & 0 & 1 \end{pmatrix} , \tag{10.35}$$

$$W_1 = \begin{pmatrix} 0 & 0 & 0 \\ 1 & 0 & 0 \\ 0 & -1 & 0 \end{pmatrix} \quad ; \quad W_2 = \begin{pmatrix} 0 & 0 & 0 \\ 0 & 0 & 0 \\ 2 & 0 & 0 \end{pmatrix} ,$$

whose commutation relations read

$$\begin{aligned} \left[L_i, L_j\right] &= (i-j)\, L_{i+j} , \\ \left[L_i, W_m\right] &= (2i-m)\, W_{i+m} , \\ \left[W_m, W_n\right] &= -\frac{1}{3}(m-n)\left(2m^2+2n^2-mn-8\right) L_{m+n} , \end{aligned} \tag{10.36}$$

so that the subset of generators L_i span the algebra $sl(2,\mathbb{R})$ in the so-called principal embedding.

The invariant nondegenerate bilinear form can also be chosen so that the action (10.1) reads

$$I = I_{CS}\left[A^+\right] - I_{CS}\left[A^-\right] , \tag{10.37}$$

where $A^\pm$ stand for the gauge fields that correspond to both copies of $sl(3,\mathbb{R})$, and now the bracket is given by a quarter of the trace in the representation of (10.35), i.e., $\langle\cdots\rangle = \frac{1}{4}\mathrm{tr}(\cdots)$. As in the case of pure gravity, the level is also chosen as $k = \frac{l}{4G}$.

It is useful to introduce a generalization of the dreibein and the spin connection, which relate with the gauge fields according to

$$A^\pm = \omega \pm \frac{e}{l} , \tag{10.38}$$

so that the spacetime metric and the spin-three field can be recovered as

$$g_{\mu\nu} = \frac{1}{2}\text{tr}\left(e_\mu e_\nu\right) \; ; \; \varphi_{\mu\nu\rho} = \frac{1}{3!}\text{tr}\left(e_{(\mu} e_\nu e_{\rho)}\right) \,, \tag{10.39}$$

being manifestly invariant under the diagonal subgroup of $SL(3,\mathbb{R}) \times SL(3,\mathbb{R})$, which corresponds to an extension of the local Lorentz group. The remaining gauge symmetries are then not only related to diffeomorphisms, but also with the higher spin gauge transformations. It is worth pointing out that, since the metric transforms in a nontrivial way under the action of the higher spin gauge symmetries, some standard geometric and physical notions turn out to be ambiguous, since they are no longer invariant. This last observation can be regarded as an additional motivation to explore the physical properties of the theory directly in terms of its original variables, given by the gauge fields $A^\pm$.

10.4.1 Asymptotic Conditions with W_3 Symmetries

A consistent set of asymptotic conditions for the theory described above was found in [21, 22]. Using the gauge choice as in [32], the radial dependence can be completely absorbed by $SL(3,\mathbb{R})$ group elements of the form (10.12), so that the asymptotic behaviour of the gauge fields can be written as in Eq. (10.13), where $a^\pm$ are now given by

$$a^\pm = \pm\left(L_{\pm1} - \frac{2\pi}{k}\mathscr{L}_\pm L_{\mp1} - \frac{\pi}{2k}\mathscr{W}_\pm W_{\mp2}\right) dx^\pm \,, \tag{10.40}$$

and $\mathscr{L}_\pm$, $\mathscr{W}_\pm$ stand for arbitrary functions of t, θ. The asymptotic symmetries can then be readily found following the same steps as in the case of pure gravity, previously discussed in Sect. 10.3.1.

The asymptotic form of the fields $a_\theta^\pm$ is maintained under gauge transformations generated by

$$\begin{aligned}
\Lambda^\pm\left(\varepsilon_\pm, \chi_\pm\right) = {} & \varepsilon_\pm L_{\pm1} + \chi_\pm W_{\pm2} \mp \varepsilon'_\pm L_0 \mp \chi'_\pm W_{\pm1} + \frac{1}{2}\left(\chi''_\pm - \frac{8\pi}{k}\mathscr{L}_\pm \chi_\pm\right) W_0 \\
& + \frac{1}{2}\left(\varepsilon''_\pm - \frac{4\pi}{k}\varepsilon_\pm \mathscr{L}_\pm + \frac{8\pi}{k}\mathscr{W}_\pm \chi_\pm\right) L_{\mp1} - \left(\frac{\pi}{2k}\mathscr{W}_\pm \varepsilon_\pm + \frac{7\pi}{6k}\mathscr{L}'_\pm \chi'_\pm\right. \\
& \left. + \frac{\pi}{3k}\chi_\pm \mathscr{L}''_\pm + \frac{4\pi}{3k}\mathscr{L}_\pm \chi''_\pm - \frac{4\pi^2}{k^2}\mathscr{L}_\pm^2 \chi_\pm - \frac{1}{24}\chi''''_\pm\right) W_{\mp2} \\
& \mp \frac{1}{6}\left(\chi'''_\pm - \frac{8\pi}{k}\chi_\pm \mathscr{L}'_\pm - \frac{20\pi}{k}\mathscr{L}_\pm \chi'_\pm\right) W_{\mp1} \,,
\end{aligned} \tag{10.41}$$

which depend on two arbitrary parameters per copy, $\varepsilon_\pm$, $\chi_\pm$, being functions of t and θ, provided the transformation law of the fields $\mathscr{L}_\pm$, $\mathscr{W}_\pm$ reads

$$\delta\mathscr{L}_\pm = \varepsilon_\pm\mathscr{L}'_\pm + 2\mathscr{L}_\pm\varepsilon'_\pm - \frac{k}{4\pi}\varepsilon'''_\pm - 2\chi_\pm\mathscr{W}'_\pm - 3\mathscr{W}_\pm\chi'_\pm\,, \tag{10.42}$$

$$\begin{aligned}\delta\mathscr{W}_\pm &= \varepsilon_\pm\mathscr{W}'_\pm + 3\mathscr{W}_\pm\varepsilon'_\pm - \frac{64\pi}{3k}\mathscr{L}_\pm^2\chi'_\pm + 3\chi'_\pm\mathscr{L}''_\pm + 5\mathscr{L}'_\pm\chi''_\pm + \frac{2}{3}\chi_\pm\mathscr{L}'''_\pm\\ &\quad - \frac{k}{12\pi}\chi'''''_\pm - \frac{64\pi}{3k}\left(\chi_\pm\mathscr{L}'_\pm - \frac{5k}{32\pi}\chi'''_\pm\right)\mathscr{L}_\pm\,.\end{aligned} \tag{10.43}$$

Then, the time component of the gauge fields $a_t^\pm$, is preserved under the gauge transformations generated by (10.41), with the transformation rules in (10.42), (10.43), provided the fields and the parameters are chiral:

$$\partial_\mp\mathscr{L}_\pm = \partial_\mp\mathscr{W}_\pm = 0\,, \tag{10.44}$$

$$\partial_\mp\varepsilon_\pm = \partial_\mp\chi_\pm = 0\,. \tag{10.45}$$

As in the case of pure gravity, the chirality of the fields in Eq. (10.44) reflects the fact that the field equations in the asymptotic region are satisfied.

The variation of the canonical generators that correspond to the asymptotic symmetries spanned by (10.41) now reads

$$\delta Q_\pm\left(\Lambda^\pm\right) = -\frac{k}{2\pi}\int\left\langle\Lambda^\pm\delta a_\theta^\pm\right\rangle d\theta = -\int\left(\varepsilon_\pm\delta\mathscr{L}_\pm - \chi_\pm\delta\mathscr{W}_\pm\right)d\theta\,, \tag{10.46}$$

and then integrates as

$$Q_\pm\left(\Lambda^\pm\right) = -\int\left(\varepsilon_\pm\mathscr{L}_\pm - \chi_\pm\mathscr{W}_\pm\right)d\theta\,. \tag{10.47}$$

This means that generic gauge fields that fulfill the asymptotic conditions described here, do not only carry spin-two charges associated to $\mathscr{L}_\pm$, whose zero modes are related to the energy and the angular momentum, but they also possess spin-three charges corresponding to $\mathscr{W}_\pm$.

The algebra of the canonical generators can be straightforwardly recovered from the transformation law of the fields in (10.42), (10.43) and it is found to be given by two copies of the W_3 algebra with the same central extension as in pure gravity, i.e., $c = \frac{3l}{2G}$. Once the fields are expanded in modes, the Poisson bracket algebra is such that both copies fulfill

$$\begin{aligned}i\left\{\mathscr{L}_m,\mathscr{L}_n\right\} &= (m-n)\,\mathscr{L}_{m+n} + \frac{k}{2}m^3\delta_{m+n,0}\,,\\ i\left\{\mathscr{L}_m,\mathscr{W}_n\right\} &= (2m-n)\,\mathscr{W}_{m+n}\,,\\ i\left\{\mathscr{W}_m,\mathscr{W}_n\right\} &= \frac{1}{3}(m-n)\left(2m^2-mn+2n^2\right)\mathscr{L}_{m+n} + \frac{16}{3k}(m-n)\,\Lambda_{m+n}\\ &\quad + \frac{k}{6}m^5\delta_{m+n,0}\,,\end{aligned} \tag{10.48}$$

where

$$\Lambda_n = \sum_m \mathscr{L}_{n-m}\mathscr{L}_m \, , \tag{10.49}$$

so that the algebra is manifestly nonlinear.

It has also been shown that once the asymptotic conditions (10.40) are expressed in a suitable "decoupling" gauge choice, they admit a consistent vanishing cosmological constant limit, so that the asymptotic symmetries are spanned by a higher spin extension of the BMS_3 algebra with an appropriate central extension [35] (see also [36]). Related results along these lines, including Hamiltonian reduction [37], unitarity [38], and the analysis of cosmologies endowed with higher spin fields have been discussed in [39–42].

10.4.2 Higher Spin Black Hole Proposal and Its Thermodynamics

It is simple to verify that, for the case of constant functions $\mathscr{L}_\pm$ and $\mathscr{W}_\pm$, the asymptotic conditions described in the previous subsection do not accommodate black holes carrying nontrivial spin-three charges. This is because once the holonomies along a thermal cycle are required to be trivial, the spin-three charges $\mathscr{W}_\pm$ are forced to vanish. Thus, with the aim of finding black holes solutions which could in principle be endowed with spin-three charges, a different set of asymptotic conditions was proposed in [13] (see Sect. 10.5) and further analyzed in [43, 44]. Indeed, this set includes interesting new black holes solutions, which in the static case are described by three constants, and the gauge fields are of the form (10.13), with

$$\begin{aligned} a^\pm &= \pm\left(L_{\pm1} - \frac{2\pi}{k}\tilde{\mathscr{L}} L_{\mp1} \mp \frac{\pi}{2k}\tilde{\mathscr{W}} W_{\mp2}\right) dx^\pm \\ &+ \tilde{\mu}\left(W_{\pm2} - \frac{4\pi}{k}\tilde{\mathscr{L}} W_0 + \frac{4\pi^2}{k^2}\tilde{\mathscr{L}}^2 W_{\mp2} \pm \frac{4\pi}{k}\tilde{\mathscr{W}} L_{\mp1}\right) dx^\mp \, . \end{aligned} \tag{10.50}$$

The precise form of the $SL(3,\mathbb{R})$ group elements $g_\pm = g_\pm(\rho)$, which was further specified in [23], would be needed in order to reconstruct the metric and the spin-three field according to Eq. (10.39). In the case of $sl(3,\mathbb{R})$ gauge fields, the conditions that guarantee the triviality of their holonomies around the thermal circle, since the representation in (10.35) is vectorial, now read

$$\mathscr{H}_\tau^\pm = e^{i\beta a_t^\pm} = 1 \, , \tag{10.51}$$

which turn out to be equivalent to

$$\mathrm{tr}\left[\left(a_t^{\pm}\right)^3\right] = 0 \ ; \ \beta^2 \mathrm{tr}\left[\left(a_t^{\pm}\right)^2\right] = 8\pi^2 \,. \tag{10.52}$$

For the gauge fields (10.50), conditions (10.52) reduce to

$$64\pi\tilde{\mathscr{L}}^2\tilde{\mu}\left(32\pi\tilde{\mathscr{L}}\tilde{\mu}^2 - 9k\right) + 27k\tilde{\mathscr{W}}\left(32\pi\tilde{\mathscr{L}}\tilde{\mu}^2 + k\right) - 864\pi k\tilde{\mathscr{W}}^2\tilde{\mu}^3 = 0 \,, \tag{10.53}$$

$$\frac{l^2\pi k}{2}\left(\tilde{\mathscr{L}} - 3\tilde{\mu}\tilde{\mathscr{W}} + \frac{32\pi}{3k}\tilde{\mu}^2\tilde{\mathscr{L}}^2\right)^{-1} = \beta^2 \,, \tag{10.54}$$

respectively, which for the branch that is connected to the BTZ black hole, being such that $\tilde{\mu} \to 0$ when $\tilde{\mathscr{W}} \to 0$, can be solved for β and $\tilde{\mu}$ in terms of $\tilde{\mathscr{L}}$ and $\tilde{\mathscr{W}}$, according to

$$\beta = \frac{l}{2}\sqrt{\frac{\pi k}{2\tilde{\mathscr{L}}}}\frac{2C-3}{C-3}\left(1 - \frac{3}{4C}\right)^{-1/2} , \tag{10.55}$$

$$\tilde{\mu} = \frac{3}{4}\sqrt{\frac{kC}{2\pi\tilde{\mathscr{L}}}}\frac{1}{2C-3} , \tag{10.56}$$

where the constant C is defined through

$$\frac{C-1}{C^{3/2}} = \sqrt{\frac{k}{32\pi\tilde{\mathscr{L}}^3}}\tilde{\mathscr{W}} \,. \tag{10.57}$$

A proposal to deal with the global charges and the thermodynamics of this black hole solution, being based on a holographic approach, was put forward in [13, 23]. The bulk field equations were identified with the Ward identities for the stress tensor and the spin-three current of an underlying dual CFT in two dimensions, so that the integration constant $\tilde{\mathscr{L}}$ was interpreted as the stress tensor, while $\tilde{\mathscr{W}}$ and $\tilde{\mu}$ were associated to the spin-three current and its source, respectively. According to this prescription, the first law of thermodynamics implies that the variation of the entropy should be given by

$$\delta\tilde{S} = \frac{4\pi}{l}\beta\left(\delta\tilde{\mathscr{L}} - \tilde{\mu}\delta\tilde{\mathscr{W}}\right) , \tag{10.58}$$

which by virtue of (10.55), (10.56) integrates as

$$\tilde{S} = 4\pi\sqrt{2\pi k\tilde{\mathscr{L}}}\sqrt{1 - \frac{3}{4C}} , \tag{10.59}$$

so that the trivial holonomy conditions around the thermal circle agree with the integrability conditions of thermodynamics.

It is worth mentioning that the black hole entropy formula (10.59) remarkably agrees with the result found in [45], which was obtained from a completely different approach. Indeed, the computation of the free energy was carried out directly in the dual CFT with extended conformal symmetry in two dimensions, exploiting the properties of the partition function under the S-modular transformation, making then no reference to the holonomies in the bulk.

These approaches have been reviewed in [46–48], and further results about black hole thermodynamics along these lines have been found in [49–58].

However, it should be stressed that identifying the integration constants $\tilde{\mathscr{L}}$ and $\tilde{\mathscr{W}}$ with global charges, appears to be very counterintuitive from the point of view of the canonical formalism. This is because, in spite of the fact that the components of the gauge fields along $dx^{\pm}$ for the black hole solution (10.50) agree with the ones of the asymptotic fall-off in (10.40), once a nonvanishing constant $\tilde{\mu}$ is included, the additional terms along $dx^{\mp}$ amount to a severe modification of the asymptotic form of the dynamical fields $a_\theta^{\pm}$, so that the expression for the global charges in Eq. (10.47) no longer applies for this class of black hole solutions. Hence, as shown in [24], in full analogy with what occurs in the case of three-dimensional General Relativity coupled to scalar fields with slow fall-off at infinity [59, 60], the effect of modifying the asymptotic behaviour is such that the total energy acquires additional nonlinear contributions in the deviation of the fields with respect to the reference background. Indeed, the variation of the total energy can be obtained directly from (10.6), which for the case of the black hole solution (10.50), reads

$$\begin{aligned} \delta E &= \frac{k}{2\pi}\int \left(\langle a_t^{+}\delta a_\theta^{+}\rangle - \langle a_t^{-}\delta a_\theta^{-}\rangle\right) d\theta \;, \\ &= \frac{4\pi}{l}\left[\delta\tilde{\mathscr{L}} - \frac{32\pi}{3k}\delta(\tilde{\mathscr{L}}^2\mu^2) + \tilde{\mu}\delta\tilde{\mathscr{W}} + 3\tilde{\mathscr{W}}\delta\tilde{\mu}\right] . \end{aligned} \tag{10.60}$$

Note that (10.60) is not an exact differential. This is natural because the variation of the total energy not only includes the variation of the mass, but also the contribution coming from all the constraints. Therefore, in order to suitably disentangle the mass (internal energy) from the work terms, one should provide a consistent set of asymptotic conditions that allows the precise identification of the global charges as well as the chemical potentials. This is discussed in Sect. 10.5. Nonetheless, the expression (10.60) provides a nice shortcut to compute the black hole entropy, circumventing the explicit computation of higher spin charges and their chemical potentials [24, 25]. This is because, by virtue of the first law, the inverse temperature β acts as an integrating factor, being such that the product $\beta\delta E$ becomes an exact differential that corresponds to the variation of the entropy, i.e.,

$$\delta S = \beta\delta E = \delta\left[4\pi\sqrt{2\pi k\tilde{\mathscr{L}}}\left(1-\frac{3}{2C}\right)^{-1}\sqrt{1-\frac{3}{4C}}\right] , \tag{10.61}$$

so that the black hole entropy is given by

$$S = 4\pi\sqrt{2\pi k\tilde{\mathscr{L}}}\left(1-\frac{3}{2C}\right)^{-1}\sqrt{1-\frac{3}{4C}}\,. \tag{10.62}$$

As explained in [25], the entropy (10.62) can be recovered from a suitable generalization of the Bekenstein–Hawking formula, given by

$$S = \frac{A}{4G}\cos\left[\frac{1}{3}\arcsin\left(3^{3/2}\frac{\varphi_+}{A^3}\right)\right], \tag{10.63}$$

which depends on the reparameterization invariant integrals of the pullback of the metric and the spin-3 field at the spacelike section of the horizon, i.e., on the horizon area A and its spin-3 analogue:

$$\varphi_+^{1/3} := \int_{\partial\Sigma_+}\left(\varphi_{\mu\nu\rho}\frac{dx^\mu}{d\sigma}\frac{dx^\nu}{d\sigma}\frac{dx^\rho}{d\sigma}\right)^{1/3} d\sigma\,. \tag{10.64}$$

It is worth highlighting that, for the static case, and in the weak spin-three field limit, our expression for the entropy (10.63) reduces to

$$S = \frac{A}{4G}\left(1-\frac{3}{2}\left(g^{\theta\theta}\right)^3\varphi_{\theta\theta\theta}^2+\mathscr{O}\left(\varphi^4\right)\right)\Big|_{\rho_+}, \tag{10.65}$$

in full agreement with the result found in [17], which was obtained from a completely different approach. Indeed, in [17] the action was written in terms of the metric and the perturbative expansion of the spin-three field up to quadratic order, so that the correction to the area law in (10.65) was found by means of Wald's formula [61].

Further results about black hole thermodynamics and along these lines have been found in [53, 54, 62–65], and the variation of the total energy (10.60) has also been recovered through different methods in [43, 44].

Since the entropy is expected to be an intrinsic property of the black hole, the fact that the nonperturbative expression for the entropy S in Eq. (10.62) differs from $\tilde{S}$ in (10.59) by a factor that characterizes the presence of the spin-three field, i.e., $S = \tilde{S}\left(1-\frac{3}{2C}\right)^{-1}$, is certainly disturbing. Indeed, curiously, a variety of different approaches either lead to $\tilde{S}$ or S, in [13, 45, 52, 58], and [25, 62–64], respectively, or even to both results [53, 54] for the black hole entropy.

As explained in [24,25], the discrepancy of these results stems from the mismatch in the definition of global charges aforementioned, which turns out to be inherited by the entropy once computed through the first law, even in the weak spin-three field limit.

Nonetheless, some puzzles still remain to be clarified, as it is the question about how the entropy (10.62) fulfills the first law of thermodynamics in the grand canonical ensemble, which is related to whether the black hole solution (10.50) actually carries or not a global a spin-three charge. This is discussed in the next Sect. 10.5.

10.5 Solving the Puzzles: Asymptotic Conditions Revisited and Different Classes of Black Holes

As explained in [15, 16], the puzzles mentioned above become resolved once the asymptotic conditions are extended so as to admit a generic choice of chemical potentials associated to the higher spin charges, so that the original asymptotic W_3 symmetries are manifestly preserved by construction. In this way, any possible ambiguity is removed. This can be seen as follows. At a slice of fixed time, according to (10.40), the asymptotic behaviour of the dynamical fields is of the form

$$a_\theta^\pm = \left(L_{\pm 1} - \frac{2\pi}{k} \mathscr{L}_\pm L_{\mp 1} - \frac{\pi}{2k} \mathscr{W}_\pm W_{\mp 2} \right) d\theta \ , \tag{10.66}$$

which is maintained under the gauge transformations $\Lambda^\pm$, defined through (10.41), with (10.42) and (10.43). In order to determine the asymptotic form of the gauge fields along time evolution, note that the field equations $F_{ti} = 0$ read

$$\dot{A}_i = \partial_i A_t + [A_i, A_t] \ ,$$

which implies that the time evolution of the dynamical fields corresponds to a gauge transformation parameterized by A_t. Hence, in order to preserve the asymptotic symmetries along the evolution in time, the Lagrange multipliers must be of the allowed form (10.41), i.e., $a_t^\pm = \Lambda^\pm$. Thus, following [15], the chemical potentials are included in the time component of the gauge fields only, so that the asymptotic form of the gauge fields is given by

$$a^\pm = \pm \left(L_{\pm 1} - \frac{2\pi}{k} \mathscr{L}_\pm L_{\mp 1} - \frac{\pi}{2k} \mathscr{W}_\pm W_{\mp 2} \right) dx^\pm \pm \frac{1}{l} \Lambda^\pm(\nu_\pm, \mu_\pm) dt \ , \tag{10.67}$$

where $\nu_\pm$, $\mu_\pm$ stand for arbitrary fixed functions of t, θ without variation ($\delta\nu_\pm = \delta\mu_\pm = 0$), that correspond to the chemical potentials. Note that, since the asymptotic form of the dynamical fields (10.66) is unchanged as compared with (10.40), the expression for the global charges remains the same, i.e., at a fixed t slice, the global charges are again given by (10.47), so that the asymptotic symmetries are still generated by two copies of the W_3 algebra.

Consistency then requires that the asymptotic form of $a_t^\pm$, should also be preserved under the asymptotic symmetries, which implies that the field equations have to be fulfilled in the asymptotic region, and the parameters of the asymptotic symmetries satisfy "deformed chirality conditions", which read

$$\begin{aligned} l\dot{\mathscr{L}}_{\pm} &= \pm\left(1+\nu_{\pm}\right)\mathscr{L}'_{\pm} \mp 2\mu_{\pm}\mathscr{W}'_{\pm}\,, \\ l\dot{\mathscr{W}}_{\pm} &= \pm\left(1+\nu_{\pm}\right)\mathscr{W}'_{\pm} \pm \frac{2}{3}\mu_{\pm}\left(\mathscr{L}'''_{\pm} - \frac{16\pi}{k}\left(\mathscr{L}^{2}_{\pm}\right)'\right)\,, \end{aligned} \tag{10.68}$$

and

$$\begin{aligned} l\dot{\chi}_{\pm} &= \pm\left(1+\nu_{\pm}\right)\chi'_{\pm} \pm 2\mu_{\pm}\varepsilon'_{\pm}\,, \\ l\dot{\varepsilon}_{\pm} &= \pm\left(1+\nu_{\pm}\right)\varepsilon'_{\pm} \mp \frac{2}{3}\mu_{\pm}\left(\chi'''_{\pm} - \frac{32\pi}{k}\chi'_{\pm}\mathscr{L}_{\pm}\right)\,, \end{aligned} \tag{10.69}$$

respectively, where for simplicity, in Eqs. (10.68), (10.69), the chemical potentials associated to the spin-two and spin-three charges, given by $\nu_{\pm}$ and $\mu_{\pm}$, were assumed to be constants.

Therefore, by construction, the functions $\mathscr{L}_{\pm}$, $\mathscr{W}_{\pm}$ are really what they mean, since their Poisson brackets fulfill the W_3 algebra with the same central extension. Note that this is so regardless the choice of chemical potentials, because the canonical generators do no depend on the Lagrange multipliers.

The asymptotic conditions given by (10.67) then provide the required extension of the ones in [21, 22], since the latter are recovered when the chemical potentials are switched off, i.e., for $\nu_{\pm} = 0$, $\mu_{\pm} = 0$. In this case, Eqs. (10.68) and (10.69) reduce to (10.44) and (10.45), respectively, expressing the fact that the fields and the parameters become chiral.

From a different perspective, the case of $\nu_{\pm} = -1$, $\mu_{\pm} = 1$ has also been discussed in [66].

It is worth emphasizing that since the Lagrange multipliers appear in the improved action through the improved generators (10.3), the interpretation of $\nu_{\pm}$, $\mu_{\pm}$ as chemical potentials, is also guaranteed by construction. Note that this corresponds to the standard procedure one follows in the case of Reissner–Nordstrm black holes, where the chemical potential associated to the electric charge corresponds to the time component of the electromagnetic field, being the Lagrange multiplier of the $U(1)$ constraint.

The extended asymptotic conditions (10.67), in the case of constant functions $\mathscr{L}_{\pm}$, $\mathscr{W}_{\pm}$ and chemical potentials $\nu_{\pm}$, $\mu_{\pm}$, then accommodate a new class of black hole solutions, endowed not only with mass and angular momentum, but also with nontrivial well-defined spin-three charges [15]. Their asymptotic and thermodynamical properties are further discussed in [16], where it is explicitly shown that for this solution, there is no tension between the different approaches mentioned above.

Note that in the standard approach for black hole thermodynamics, the temperature and the chemical potential for the angular momentum do not explicitly appear in the fields. Instead, they enter through the identifications involving the Euclidean time and the angle, so that the range of the coordinates is not fixed and depends on the solution. The presence of nonvanishing chemical potentials $\nu_{\pm}$ associated to the spin-two charges, then allows performing the description keeping the range of the coordinates fixed once and for all, i.e., $0 \leq \theta < 2\pi$ and $0 \leq \tau < 2\pi l$, which amounts to introduce a non trivial lapse and shift in the metric formalism. Both approaches are indeed equivalent, but in the case of higher spin black holes, since the chemical potentials that correspond to the spin-three charges cannot be absorbed into the modular parameter of the torus, it becomes conceptually safer to follow the latter approach, since all the chemical potentials become introduced and treated unambiguously in the same footing.

Otherwise, for instance, if the chemical potentials were not introduced along the thermal circles, but instead along additional non-vanishing components of the gauge fields along the conjugate null directions, as in the case of [13], the asymptotic form of the gauge fields would be given by

$$a^{\pm} = \pm \left(L_{\pm 1} - \frac{2\pi}{k} \tilde{\mathscr{L}}_{\pm} L_{\mp 1} - \frac{\pi}{2k} \tilde{\mathscr{W}}_{\pm} W_{\mp 2} \right) dx^{\pm} \pm \Lambda^{\pm} \left(\tilde{\nu}_{\pm}, \tilde{\mu}_{\pm} \right) dx^{\mp} , \tag{10.70}$$

which severely modifies the components of the dynamical fields $a_{\theta}^{\pm}$, in a way that is incompatible with the asymptotic W_3 symmetry. This is because at a fixed t slice, the terms proportional to $\tilde{\mu}_{\pm}$ contribute to $a_{\theta}^{\pm}$ with additional terms of the form

$$\begin{aligned} a_{\theta}^{\pm} &= \left(L_{\pm 1} - \frac{2\pi}{k} \tilde{\mathscr{L}}_{\pm} L_{\mp 1} - \frac{\pi}{2k} \tilde{\mathscr{W}}_{\pm} W_{\mp 2} \right) + \left(\tilde{\nu}_{\pm} L_{\pm 1} + \tilde{\mu}_{\pm} W_{\pm 2} \right) \\ &\quad + \left[\frac{1}{2} \left(-\frac{4\pi}{k} \tilde{\nu}_{\pm} \tilde{\mathscr{L}}_{\pm} + \frac{8\pi}{k} \tilde{\mathscr{W}}_{\pm} \tilde{\mu}_{\pm} \right) L_{\mp 1} - \left(\frac{\pi}{2k} \tilde{\mathscr{W}}_{\pm} \tilde{\nu}_{\pm} - \frac{4\pi^2}{k^2} \tilde{\mathscr{L}}_{\pm}^2 \tilde{\mu}_{\pm} \right) W_{\mp 2} \right] \\ &\quad - \frac{4\pi}{k} \tilde{\mathscr{L}}_{\pm} \tilde{\mu}_{\pm} W_0 , \end{aligned} \tag{10.71}$$

that are not of highest (or lowest) weight, and hence incompatible with the asymptotic conditions (10.67) that implement the Hamiltonian reduction of the current algebra associated to $sl(3,\mathbb{R})$ to the W_3 algebra. Indeed, in this case, the asymptotic symmetries that preserve the asymptotic form of a_{θ} are shown to be spanned by two copies of the Bershardsky–Polyakov algebra W_3^2 [67,68], corresponding to the other non trivial (so-called diagonal) embedding of $sl\,(2,\mathbb{R})$ into $sl\,(3,\mathbb{R})$ [16]. Therefore, in spite of dealing with the same action, the effect of this drastic modification of the boundary conditions amounts to deal with a completely different theory, being characterized by a different field content, and hence with an inequivalent spectrum, so that their corresponding black hole solutions, as the one in (10.50), are characterized by another set of global charges of lower spin.

It is worth pointing out that our procedure to incorporate chemical potentials can be straightforwardly extended to the case of $\mathfrak{g}_{\pm} = sl\,(N,\mathbb{R})$, regardless the way in which $sl\,(2,\mathbb{R})$ is embedded, as well as to the case of infinite-dimensional higher spin algebras.

Some closing remarks are in order. It should be mentioned that the case of three-dimensional gravity nonminimally coupled with spin-three fields, also appears to be consistently formulated in the second-order formalism by introducing a suitable set of auxiliary fields [69]. Besides, in the case of spin-three and higher, consistent sets of asymptotic conditions have also been proposed in [21, 70, 71], while exact solutions and their properties have been explored in [72–76]. In the context of higher spin supergravity in three dimensions, the asymptotic structure was analyzed in [77], and exact solutions have also been found in [78–80]. Moreover, along the lines of holography and the corresponding dual CFT theory with extended conformal symmetry at the boundary [81–83], further interesting results can also be found in [84–91].

Acknowledgements We thank G. Barnich, X. Bekaert, E. Bergshoeff, C. Bunster, A. Campoleoni, R. Canto, D. Grumiller, M. Henneaux, J. Jottar, C. Martínez, J. Matulich, J. Ovalle, R. Rahman, S-J Rey, J. Rosseel, C. Troessaert and M. Vasiliev for stimulating discussions. R.T. also thanks E. Papantonopoulos and the organizers of the Seventh Aegean Summer School, "Beyond Einstein's theory of gravity", for the opportunity to give this lecture in a wonderful atmosphere. This work is partially funded by the Fondecyt grants N° 1130658, 1121031, 11130260, 11130262. The Centro de Estudios Científicos (CECs) is funded by the Chilean Government through the Centers of Excellence Base Financing Program of Conicyt.

References

1. P. Van Nieuwenhuizen, "Supergravity." Phys. Rept. **68**, 189 (1981)
2. X. Bekaert, N. Boulanger, P. Sundell, How higher-spin gravity surpasses the spin two barrier: no-go theorems versus yes-go examples. Rev. Mod. Phys. **84**, 987 (2012).[arXiv:1007.0435 [hep-th]]
3. C. Aragone, S. Deser, Consistency problems of hypergravity. Phys. Lett. B **86**, 161 (1979)
4. M.A. Vasiliev, "Consistent equation for interacting gauge fields of all spins in (3 + 1)-dimensions."Phys. Lett. B **243**, 378 (1990)
5. M.A. Vasiliev, "Nonlinear equations for symmetric massless higher spin fields in (A)dS(d)."Phys. Lett. B **567**, 139 (2003). [hep-th/0304049]
6. X. Bekaert, S. Cnockaert, C. Iazeolla, M.A. Vasiliev, "Nonlinear higher spin theories in various dimensions." (2005). [hep-th/0503128]
7. A. Sagnotti, Notes on strings and higher spins. J. Phys. A **46**, 214006 (2013). [arXiv:1112.4285 [hep-th]]
8. M.A. Vasiliev, "Higher-Spin Theory and Space-Time Metamorphoses." arXiv:1404.1948 [hep-th]
9. R. Troncoso, J. Zanelli, New gauge supergravity in seven-dimensions and eleven-dimensions. Phys. Rev. D **58**, 10170 (1998). [hep-th/9710180]
10. M.P. Blencowe, "A consistent interacting massless higher spin field theory in D = (2 + 1)." Class. Quant. Gravity **6**, 443 (1989)
11. E. Bergshoeff, M.P. Blencowe, K.S. Stelle, "Area preserving diffeomorphisms and higher spin algebra." Commun. Math. Phys. **128**, 213 (1990)
12. M.A. Vasiliev, "Higher spin gauge theories in four-dimensions, three-dimensions, and two-dimensions." Int. J. Mod. Phys. D **5**, 763 (1996). [hep-th/9611024]
13. M. Gutperle, P. Kraus, Higher spin black holes. J. High Energy Phys. **1105**, 022 (2011). [arXiv:1103.4304 [hep-th]]

14. A. Castro, E. Hijano, A. Lepage-Jutier, A. Maloney, Black holes and singularity resolution in higher spin gravity. J. High Energy Phys. **1201**, 031 (2012). [arXiv:1110.4117 [hep-th]]
15. M. Henneaux, A. Prez, D. Tempo, R. Troncoso, "Chemical potentials in three-dimensional higher spin anti-de Sitter gravity." J. High Energy Phys. **1312**, 048 (2013). [arXiv:1309.4362 [hep-th]]
16. C. Bunster, M. Henneaux, A. Prez, D. Tempo, R. Troncoso, "Generalized black holes in three-dimensional spacetime." JHEP **1405**, 031 (2014). [arXiv:1404.3305 [hep-th]]
17. A. Campoleoni, S. Fredenhagen, S. Pfenninger, S. Theisen, Towards metric-like higher-spin gauge theories in three dimensions. J. Phys. A **46**, 214017 (2013). [arXiv:1208.1851 [hep-th]]
18. J.D. Brown, M. Henneaux, "Central charges in the canonical realization of asymptotic symmetries: an example from three-dimensional gravity."Commun. Math. Phys. **104**, 207 (1986)
19. M. Banados, C. Teitelboim, J. Zanelli, The black hole in three-dimensional space-time. Phys. Rev. Lett. **69**, 1849 (1992). [hep-th/9204099]
20. M. Banados, M. Henneaux, C. Teitelboim, J. Zanelli, Geometry of the (2+1) black hole. Phys. Rev. D **48**, 1506 (1993. [gr-qc/9302012]
21. M. Henneaux, S.-J. Rey, "Nonlinear $W_{infinity}$ as asymptotic symmetry of three-dimensional higher spin anti-de sitter gravity." J. High Energy Phys. **1012**, 007 (2010). [arXiv:1008.4579 [hep-th]]
22. A. Campoleoni, S. Fredenhagen, S. Pfenninger, S. Theisen, "Asymptotic symmetries of three-dimensional gravity coupled to higher-spin fields." J. High Energy Phys. **1011**, 007 (2010). [arXiv:1008.4744 [hep-th]]
23. M. Ammon, M. Gutperle, P. Kraus, E. Perlmutter, Spacetime geometry in higher spin gravity. J. High Energy Phys. **1110**, 053 (2011). [arXiv:1106.4788 [hep-th]]
24. A. Prez, D. Tempo, R. Troncoso, "Higher spin gravity in 3D: black holes, global charges and thermodynamics." Phys. Lett. B **726**, 444 (2013). [arXiv:1207.2844 [hep-th]]
25. A. Pérez, D. Tempo, R. Troncoso, "Higher spin black hole entropy in three dimensions." J. High Energy Phys. **1304**, 143 (2013). [arXiv:1301.0847 [hep-th]]
26. A.P. Balachandran, G. Bimonte, K.S. Gupta, A. Stern, "Conformal edge currents in Chern-Simons theories." Int. J. Mod. Phys. A **7**, 4655 (1992). [hep-th/9110072]
27. M. Banados, "Global charges in Chern-Simons field theory and the (2+1) black hole."Phys. Rev. D **52**, 5816 (1996). [hep-th/9405171]
28. S. Carlip, "Conformal field theory, $(2+1)$-dimensional gravity, and the BTZ black hole."Class. Quant. Gravity **22**, R85 (2005). [gr-qc/0503022]
29. T. Regge, C. Teitelboim, "Role of surface integrals in the Hamiltonian formulation Of general relativity." Ann. Phys. **88**, 286 (1974)
30. A. Achucarro, P.K. Townsend, "A Chern-Simons action for three-dimensional anti-De Sitter supergravity theories." Phys. Lett. B **180**, 89 (1986)
31. E. Witten, "$(2 + 1)$-dimensional gravity as an exactly soluble system." Nucl. Phys. B **311**, 46 (1988)
32. O. Coussaert, M. Henneaux, P. van Driel, The asymptotic dynamics of three-dimensional Einstein gravity with a negative cosmological constant. Class. Quant. Gravity **12**, 2961 (1995). [gr-qc/9506019]
33. S. Carlip, C. Teitelboim, Aspects of black hole quantum mechanics and thermodynamics in (2+1)-dimensions. Phys. Rev. D **51**, 622 (1995). [gr-qc/9405070]
34. J.M. Maldacena, A. Strominger, AdS(3) black holes and a stringy exclusion principle. J. High Energy Phys. **9812**, 005 (1998). [hep-th/9804085]
35. H.A. Gonzalez, J. Matulich, M. Pino, R. Troncoso, Asymptotically flat spacetimes in three-dimensional higher spin gravity. J. High Energy Phys. **1309**, 016 (2013). [arXiv:1307.5651 [hep-th]]
36. H. Afshar, A. Bagchi, R. Fareghbal, D. Grumiller, J. Rosseel, "Higher spin theory in 3-dimensional flat space." Phys. Rev. Lett. **111**, 121603 (2013). [arXiv:1307.4768 [hep-th]]
37. H. A. González, M. Pino, "Boundary dynamics of asymptotically flat 3D gravity coupled to higher spin fields." arXiv:1403.4898 [hep-th]

38. D. Grumiller, M. Riegler, J. Rosseel, "Unitarity in three-dimensional flat space higher spin theories." arXiv:1403.5297 [hep-th]
39. C. Krishnan, S. Roy, "Higher Spin Resolution of a Toy Big Bang." Phys. Rev. D **88**, 044049 (2013) [arXiv:1305.1277 [hep-th]]
40. B. Burrington, L.A. Pando Zayas, N. Rombes, "On resolutions of cosmological singularities in higher-spin gravity. " (2013). [arXiv:1309.1087 [hep-th]]
41. C. Krishnan, S. Roy, Desingularization of the Milne Universe (2013). [arXiv:1311.7315 [hep-th]]
42. C. Krishnan, A. Raju, S. Roy, "A Grassmann path from AdS_3 to flat space. " (2013). arXiv:1312.2941 [hep-th]
43. G. Compre, W. Song, $\mathscr{W}$ symmetry and integrability of higher spin black holes. J. High Energy Phys. **1309**, 144 (2013). [arXiv:1306.0014 [hep-th]]
44. G. Compre, J.I. Jottar, W. Song, Observables and microscopic entropy of higher spin black holes. J. High Energy Phys. **1311**, 054 (2013). [arXiv:1308.2175 [hep-th]]
45. M.R. Gaberdiel, T. Hartman, K. Jin, "Higher spin black holes from CFT." J. High Energy Phys. **1204**, 103 (2012). [arXiv:1203.0015 [hep-th]]
46. M.R. Gaberdiel, R. Gopakumar, Minimal model holography. J. Phys. A **46**, 214002 (2013). [arXiv:1207.6697 [hep-th]]
47. M. Ammon, M. Gutperle, P. Kraus, E. Perlmutter, Black holes in three dimensional higher spin gravity: a review. J. Phys. A **46**, 214001 (2013). [arXiv:1208.5182 [hep-th]]
48. K. Jin, Higher spin gravity and exact holography. PoS Corfu **2012**, 086 (2013). [arXiv:1304.0258 [hep-th]]
49. B. Chen, J. Long, Y.-n. Wang, Black holes in truncated higher spin AdS_3 gravity. J. High Energy Phys. **1212**, 052 (2012). [arXiv:1209.6185 [hep-th]]
50. J.R. David, M. Ferlaino, S.P. Kumar, Thermodynamics of higher spin black holes in 3D. J. High Energy Phys. **1211**, 135 (2012). [arXiv:1210.0284 [hep-th]]
51. B. Chen, J. Long, Y. -N. Wang, "Phase Structure of Higher Spin Black Hole." JHEP **1303**, 017 (2013). [arXiv:1212.6593]
52. P. Kraus, T. Ugajin, An entropy formula for higher spin black holes via conical singularities. J. High Energy Phys. **1305**, 160 (2013). [arXiv:1302.1583 [hep-th]]
53. J. de Boer, J.I. Jottar, Thermodynamics of higher spin black holes in AdS_3. J. High Energy Phys. **1401**, 023 (2014). [arXiv:1302.0816 [hep-th]]
54. J. de Boer, J.I. Jottar, "Entanglement entropy and higher spin holography in AdS_3." (2013). arXiv:1306.4347 [hep-th]
55. M. Beccaria, G. Macorini, On the partition functions of higher spin black holes. J. High Energy Phys. **1312**, 027 (2013). [arXiv:1310.4410 [hep-th]]
56. M. Beccaria, G. Macorini, "Analysis of higher spin black holes with spin-4 chemical potential." (2013). arXiv:1312.5599 [hep-th]
57. B. Chen, J. Long, J. -J. Zhang, "Holographic Rényi entropy for CFT with W symmetry." arXiv:1312.5510 [hep-th]
58. S. Datta, J.R. David, M. Ferlaino, S.P. Kumar, Higher spin entanglement entropy from CFT. (2014). arXiv:1402.0007 [hep-th]
59. M. Henneaux, C. Martínez, R. Troncoso, J. Zanelli, "Black holes and asymptotics of $2+1$ gravity coupled to a scalar field." Phys. Rev. D **65**, 104007 (2002). [arXiv:hep-th/0201170]
60. M. Henneaux, C. Martínez, R. Troncoso, J. Zanelli, "Asymptotically anti-de Sitter spacetimes and scalar fields with a logarithmic branch. " Phys. Rev. D **70**, 044034 (2004). [hep-th/0404236]
61. R.M. Wald, "Black hole entropy is the noether charge." Phys. Rev. D **48**, 3427 (1993). [gr-qc/9307038]
62. M. Ammon, A. Castro, N. Iqbal, Wilson lines and entanglement entropy in higher spin gravity. J. High Energy Phys. **1310**, 110 (2013). [arXiv:1306.4338 [hep-th]]
63. L. McGough, H. Verlinde, Bekenstein-Hawking entropy as topological entanglement entropy. J. High Energy Phys. **1311**, 208 (2013). [arXiv:1308.2342 [hep-th]]

64. W. Li, F.-L. Lin, C.-W. Wang, Modular properties of 3D higher spin theory. J. High Energy Phys. **1312**, 094 (2013). [arXiv:1308.2959 [hep-th]]
65. A. Chowdhury, A. Saha, Phase structure of higher spin black holes (2013). arXiv:1312.7017 [hep-th]
66. M. Gutperle, E. Hijano, J. Samani, Lifshitz black holes in higher spin gravity (2013). arXiv:1310.0837 [hep-th]
67. A.M. Polyakov, Gauge transformations and diffeomorphisms. Int. J. Mod. Phys. A **5**, 833 (1990).
68. M. Bershadsky, "Conformal field theories via Hamiltonian reduction. "Commun. Math. Phys. **139**, 71 (1991)
69. I. Fujisawa, R. Nakayama, "Second-order formalism for 3D spin-3 gravity."Class. Quant. Gravity **30**, 035003 (2013). [arXiv:1209.0894 [hep-th]]
70. A. Campoleoni, S. Fredenhagen, S. Pfenninger, Asymptotic W-symmetries in three-dimensional higher-spin gauge theories. " J. High Energy Phys. **1109**, 113 (2011). [arXiv:1107.0290 [hep-th]]
71. M.R. Gaberdiel, T. Hartman, Symmetries of holographic minimal models. J. High Energy Phys. **1105**, 031 (2011). [arXiv:1101.2910 [hep-th]]
72. P. Kraus, E. Perlmutter, "Partition functions of higher spin black holes and their CFT duals. "J. High Energy Phys. **1111**, 061 (2011). [arXiv:1108.2567 [hep-th]]
73. H.-S. Tan, "Aspects of three-dimensional spin-4 gravity." J. High Energy Phys. **1202**, 035 (2012). [arXiv:1111.2834 [hep-th]]
74. M. Gary, D. Grumiller, R. Rashkov, "Towards non-AdS holography in 3-dimensional higher spin gravity." J. High Energy Phys. **1203**, 022 (2012). [arXiv:1201.0013 [hep-th]]
75. M. Banados, R. Canto, S. Theisen, "The action for higher spin black holes in three dimensions."J. High Energy Phys. **1207**, 147 (2012). [arXiv:1204.5105 [hep-th]]
76. M. Ferlaino, T. Hollowood, S.P. Kumar, "Asymptotic symmetries and thermodynamics of higher spin black holes in AdS3," Phys. Rev. D **88**, 066010 (2013). [arXiv:1305.2011 [hep-th]]
77. M. Henneaux, G. Lucena Gómez, J. Park, S.-J. Rey, "Super- W(infinity) asymptotic symmetry of higher-spin AdS_3 supergravity." J. High Energy Phys. **1206**, 037 (2012). [arXiv:1203.5152 [hep-th]]
78. B. Chen, J. Long, Y. -N. Wang, "Conical Defects, Black Holes and Higher Spin (Super-) Symmetry." JHEP **1306**, 025 (2013). [arXiv:1303.0109 [hep-th]]
79. S. Datta, J.R. David, Black holes in higher spin supergravity. J. High Energy Phys. **1307**, 110 (2013). [arXiv:1303.1946 [hep-th]]
80. H.S. Tan, "Exploring three-dimensional higher-spin supergravity based on sl(N |N - 1) Chern-Simons theories." J. High Energy Phys. **1211**, 063 (2012). [arXiv:1208.2277 [hep-th]]
81. M.R. Gaberdiel, R. Gopakumar, A. Saha, "Quantum W-symmetry in AdS_3," J. High Energy Phys. **1102**, 004 (2011). [arXiv:1009.6087 [hep-th]]
82. M.R. Gaberdiel, R. Gopakumar, "An AdS_3 dual for minimal model CFTs." Phys. Rev. D **83**, 066007 (2011). [arXiv:1011.2986 [hep-th]]
83. M.R. Gaberdiel, R. Gopakumar, T. Hartman, S. Raju, "Partition functions of holographic minimal models." J. High Energy Phys. **1108**, 077 (2011). [arXiv:1106.1897 [hep-th]]
84. A. Castro, R. Gopakumar, M. Gutperle, J. Raeymaekers, Conical defects in higher spin theories. J. High Energy Phys. **1202**, 096 (2012). [arXiv:1111.3381 [hep-th]]
85. M. Ammon, P. Kraus, E. Perlmutter, "Scalar fields and three-point functions in $D = 3$ higher spin gravity. " J. High Energy Phys. **1207**, 113 (2012). [arXiv:1111.3926 [hep-th]]
86. M.R. Gaberdiel, P. Suchanek, "Limits of minimal models and continuous orbifolds. " J. High Energy Phys. **1203**, 104 (2012). [arXiv:1112.1708 [hep-th]]
87. P. Kraus, E. Perlmutter, Probing higher spin black holes. J. High Energy Phys. **1302**, 096 (2013). [arXiv:1209.4937 [hep-th]]
88. S. Banerjee, A. Castro, S. Hellerman, E. Hijano, A. Lepage-Jutier, A. Maloney, S. Shenker, "Smoothed transitions in higher spin AdS gravity."Class. Quant. Grav. **30**, 104001 (2013). [arXiv:1209.5396 [hep-th]]

89. A. Campoleoni, T. Prochazka, J. Raeymaekers, A note on conical solutions in 3D Vasiliev theory. J. High Energy Phys. **1305**, 052 (2013). [arXiv:1303.0880 [hep-th]]
90. M.R. Gaberdiel, K. Jin, E. Perlmutter, "Probing higher spin black holes from CFT. " J. High Energy Phys. **1310**, 045 (2013). [arXiv:1307.2221 [hep-th]]
91. A. Campoleoni, S. Fredenhagen, On the higher-spin charges of conical defects. Phys. Lett. B **726**, 387 (2013). [arXiv:1307.3745]

Chapter 11
Chern–Simons Forms and Gravitation Theory

Jorge Zanelli

Abstract The Chern–Simons (**CS**) form started as a curious obstruction in mathematics 40 years ago, to become a central object in theoretical physics. CS terms are central features in high temperature superconductivity and in recently discovered topological insulators. In classical physics, the minimal coupling in electromagnetism and the action for a mechanical system in Hamiltonian form are examples of CS functionals. CS forms are also the natural generalization of the minimal coupling between the electromagnetic field and an even-dimensional membrane. Here, a cursory review of the role of CS forms in gravitation theories is presented at an introductory level.

11.1 Introduction

Chern–Simons forms are mathematical structures related to integral topological invariants known as characteristic classes, that describe the mapping between a manifold and a gauge group. The idea is that a gauge connection $\mathbf{A}(x)$ can be viewed as a mapping between the spacetime manifold $\mathscr{M}$ and a Lie algebra $\mathbb{L}$. A characteristic class is an integer that counts how many times the gauge group is covered as x takes values on $\mathscr{M}$.

In 1972, S.-S. Chern and J. Simons were looking for a combinatorial formula to express a characteristic class known as the Pontryagin character. However, in their own words, …*This process got stuck by the emergence of a boundary term which did not yield to a simple combinatorial analysis. The boundary term seemed interesting in its own right and it and its generalization are the subject of this paper* [1]. They turned their failed attempt into an important mathematical discovery, because they had the intuition that the existence of the annoying boundary term should have a profound meaning.

J. Zanelli (✉)
Centro de Estudios Científicos (CECs), Casilla 1469, Valdivia, Chile

Universidad Andrés Bello, República 440, Santiago, Chile
e-mail: z@cecs.cl

E. Papantonopoulos (ed.), *Modifications of Einstein's Theory of Gravity at Large Distances*, Lecture Notes in Physics 892, DOI 10.1007/978-3-319-10070-8_11

In the 40 years since their discovery, CS forms have opened new areas of study in mathematics and several excellent books aimed at their many applications in physics have been written [2, 3], for which these notes are no substitute. Our purpose here is to merely collect a few useful observations that could help to understand the role of CS forms in gravitation. There is an abundant literature describing solutions and other applications of CS gravities, like [4–6]. Here, the emphasis is on the construction of the action principles, and in the geometric features that make CS forms particularly interesting for gravity. More extended discussions can be found in other reviews by this author such as [7–9].

11.2 Chern–Simons Forms in Physics

CS forms appear in gauge theories, however their usefulness is not because they are gauge-invariant, but **quasi-invariant**: under a gauge transformation they transform by the addition of a total derivative, just like an abelian connection [10]. Consider a Yang-Mills connection (vector potential) $\mathbf{A}$ for a nonabelian gauge field theory. Under a gauge transformation, $\mathbf{A}$ changes as

$$\mathbf{A}_\mu(x) \to \mathbf{A}'_\mu(x) = g^{-1}(x)[\mathbf{A}_\mu(x) + \partial_\mu]g(x), \tag{11.1}$$

where $x \in M$, $g(x) \in \mathbb{G}$ defines a gauge transformation that can be continuously connected to the identity. Then a CS form $\mathscr{C}$, constructed with the connection $\mathbf{A}$ transforms as

$$\mathscr{C}(\mathbf{A}') = \mathscr{C}(\mathbf{A}) + d\Omega, \tag{11.2}$$

precisely as an abelian connection. Clearly, an abelian connection 1-form is a particular case of a CS form, but in general $\mathscr{C}(\mathbf{A})$ are $(2n+1)$-forms with integer $n \geq 0$.

There are two instances in classical physics where a function changes by a total derivative giving rise to nontrivial effects: the Lagrangian change under a symmetry transformation, $L(q, \dot{q})dt \to L(q', \dot{q}')dt' = L(q, \dot{q})dt + d\Omega(q, t)$, and the minimal coupling between the electromagnetic potential A_μ and an external current, $\int A_\mu(x) j^\mu(x) d^n x$.

These are not completely distinct situations. According to Noether's theorem, a symmetry transformation that changes the action by a boundary term gives rise to a conserved current. This current in turn couples to the dynamical variables as a source for the classical equations. In classical mechanics, the conserved charges are constants of the Hamiltonian flow in phase space, like the energy-momentum or the angular momentum of an isolated system. In electrodynamics, requiring gauge invariance of the coupling between the gauge potential and the external current implies the conservation law for the electric current (electric charge conservation).

It is reassuring that neither the minimal coupling, $A_\mu(x)j^\mu$, nor the conservation law $\partial_\mu j^\mu = 0$ require a metric, (here ∂ is the ordinary derivative, and j is a contravariant vector density). This makes the coupling and the conservation equation valid in any coordinate basis and for any background geometry. This ultimately means that, if the current is produced by a point charge, the coupling is insensitive to shape of the particle's worldline, and independent of the geometry of the ambient spacetime. Thus, regardless of how the particle twists and turns in its evolution, or the metric properties of spacetime where the interaction takes place, the coupling remains consistently gauge invariant and the current is absolutely conserved.

A $(2n+1)$ CS form makes a good candidate Lagrangian for a gauge-invariant, background metric-independent theory in $2n+1$ dimensions, a possibility that has been extensively explored in theoretical physics over the past 30 years. It is less obvious but equally true that the CS forms define gauge-invariant couplings between a non-abelian connection $\mathbf{A}$ and a charged $2p$-brane in analogy to the minimal coupling [10, 11]. This can be seen as the generalization of the minimal coupling between the electromagnetic connection A (a $(0+1)$-CS form) and the current generated by a charged point particle (0-brane).

11.2.1 Construction of CS Forms

The fundamental object in a gauge theory is the **connection**, a generalization of the abelian vector potential. Typically, the connection is a matrix-valued one-form[1] field,

$$\mathbf{A} = \mathbf{A}_\mu dx^\mu = A^a_\mu \mathbf{K}_a dx^\mu, \tag{11.3}$$

where $\{\mathbf{K}_a; a = 1, 2, \ldots, N\}$ is a basis of the Lie algebra $\mathbb{L}$ associated to $\mathbb{G}$. Under the action of $\mathbb{G}$, the connection transforms as in (11.1), while the curvature two-form (field strength) $\mathbf{F} = d\mathbf{A} + \mathbf{A} \wedge \mathbf{A}$, transforms homogeneously,

$$\mathbf{F} \xrightarrow{g} \mathbf{F}' = g^{-1}\mathbf{F}g. \tag{11.4}$$

In the electromagnetic case the gauge group is abelian and therefore the curvature, defined by the electric and magnetic fields, is gauge-invariant. All gauge-invariant quantities have directly observable local features and this is why $\vec{\mathbf{E}}$ and $\vec{\mathbf{B}}$ can be directly measured by the forces they produce. In the non-abelian case, the $2k$-form

$$P_{2k} = \langle \mathbf{F}^k \rangle, \tag{11.5}$$

[1] In what follows, differential forms will be used throughout unless otherwise indicated. We follow the notation and conventions of [12].

is also invariant under $\mathbb{G}$ by construction and is therefore observable. Here $\langle\cdots\rangle$ stands for a symmetric, multilinear operation in the Lie algebra (a generalized trace). The integrals of an invariant of this kind (or more generally, the trace of any polynomial in **F**), like the Euler or the Pontryagin forms are the **characteristic classes**. They capture the topological nature of the mapping between the spacetime manifold and the Lie algebra $\mathbb{L}$ in which the connection one form $\mathbf{A}: M \mapsto \mathbb{L}$ takes its values.

A **CS form** is defined in association with a characteristic class. To fix ideas, consider $\langle\cdots\rangle$ to be the ordinary trace in a particular representation.[2] Then, P_{2k} is a homogeneous polynomial in the curvature **F** associated to a gauge connection **A** such that [2]:

i. It is invariant under gauge transformations (11.4), which is expressed as $\delta_{gauge} P_{2k} = 0$.
ii. It is closed, $dP_{2k} = 0$. Hence, it is locally exact, $P_{2k} = d\mathscr{C}_{2k-1}$, where $\mathscr{C}$ is some $(2k-1)$-form,
iii. Its integral over a $2k$-dimensional compact manifold, orientable and without boundary, is a topological invariant,

$$\int_M P_{2k} = c_{2k} z(M)\,, z \in \mathbf{Z}. \tag{11.6}$$

Condition (**i**) is satisfied by virtue of the cyclic property of the trace of products of 2-forms. Condition (**ii**) is a consequence of the Bianchi identity, $D\mathbf{F} = d\mathbf{F} + [\mathbf{A}, \mathbf{F}] \equiv 0$. Finally, (**iii**) means that, although P_{2k} is an exact form in a local chart, globally it may not be. It is precisely the fact that $\mathscr{C}$ could not always be globally defined, what caught the attention of Chern and Simons and led to the identification of CS forms in [1]. These CS forms can be expressed as the trace of a polynomial in **A** and $d\mathbf{A}$ that *cannot* be written as a local function involving only the curvature **F**. This makes the CS forms rather cumbersome to write, but its exact expression is not needed in order to establish its most important property: *Under a gauge transformation* (11.1), $\mathscr{C}_{2n-1}$ *changes by a locally exact form (a total derivative in a coordinate patch).*

The proof is elementary: Since the homogeneous polynomial P_{2k} is invariant under gauge transformations, performing a gauge transformation on it gives $\delta_{gauge} P_{2k} = \delta_{gauge} d(\mathscr{C}_{2k-1}) = d(\delta_{gauge}\mathscr{C}_{2k-1})$. Since P_{2k} is invariant, one concludes that $d(\delta_{gauge}\mathscr{C}_{2k-1}) = 0$, and by Poincaré's lemma, this last equation implies that locally $\delta_{gauge}\mathscr{C}_{2k-1} = d\Omega$.

This is a nontrivial result: although the nonabelian connection **A** transforms inhomogeneously, as in (11.1), the CS form transforms in the same way as an abelian connection. This is sufficient to ensure that a CS $(2n-1)$-form defines a gauge invariant action in a $(2n-1)$-dimensional manifold,

[2] Some Lie algebras have more than one way to define a trace. The algebra of rotations, for example, has two.

$$\delta_{gauge} I[A] = \int_{M^{2n-1}} \delta_{gauge} \mathscr{C}_{2n-1} = \int_{M^{2n-1}} d\Omega, \tag{11.7}$$

which vanishes for appropriate boundary conditions.

CS actions are exceptional because, unlike most physical actions, such as a free particle, Maxwell or Yang-Mills, the CS form does not require a metric. The gauge invariance of the action does not depend on the shape of the manifold M^{2n-1}; a metric structure may not even be defined on it. This is a welcome feature in a gravitation theory in which the geometry is dynamical. A consequence of this is that in CS gravity theories, the metric is a derived (composite) object and not a fundamental field to be quantized. This in turn implies that concepts such as the energy-momentum tensor and the inertial mass must be regarded as phenomenological constructs of classical or semi-classical nature, as emerging phenomena.

11.2.2 Gravitation and Diffeomorphism Invariance

It is commonly said that the fundamental symmetry of gravity is the group of general coordinate transformations[3] These transformations do form a group whose action is certainly local, but this is not a useful symmetry and much less a unique feature of gravity. Indeed, physics does not depend on the choice of coordinates, and therefore any action, for whatever physical system, must be coordinate-invariant; otherwise one has made a mistake somewhere. Hence, all meaningful statements derived from an action principle must be coordinate-invariant as well.

Any well defined physical theory must be invariant under general changes of coordinates: coordinates are labels introduced by humans in order to describe where, when and how events occur, and to communicate with other humans. Coordinates, together with the units for measuring space, time, temperature, pressure, tension, etc., are conventional, and objective situations cannot depend on the coordinates physicists employ to describe them.

General coordinate invariance is explicitly recognized in Lagrangian mechanics, where the choice of coordinates is left completely arbitrary. In other words, general coordinate transformations are not a distinctive symmetry of gravity, it is the invariance of the laws of Nature under changes in the form humans choose to describe it. We experience this every time we write or Maxwell's or Schrödinger's equations in spherical coordinates, in order to render more transparent the presence of boundaries or sources with spherical symmetry. In such cases, the coordinates are adapted to the symmetry of the physical situation, but that does not imply that coordinates could not be chosen otherwise.

[3]Those transformations are often called the diffeomorphisms, although this means "coordinate diffeos", not to be confused with the diffeomorphisms of the spacetime manifold.

The confusion seems to start from the assertion that diffeomorphisms are local translations,

$$x^{\mu} \to x'^{\mu} = x^{\mu} + \xi^{\mu}(x). \tag{11.8}$$

These are gauge-like transformations in the sense that ξ^{μ} is an arbitrary function of x, but here the analogy with gauge transformations stops. Under coordinate transformations, a vector transforms as

$$v'^{\mu}(x') = L^{\mu}_{\nu}(x)v^{\nu}(x), \text{ where } L^{\mu}_{\nu}(x) = \frac{\partial x'^{\mu}}{\partial x^{\nu}} = \delta^{\mu}_{\nu} + \frac{\partial \xi^{\mu}}{\partial x^{\nu}}. \tag{11.9}$$

In gauge transformations, the new field has the same argument as the old one, whereas here the argument of v'^{μ} in (11.9) is x' and not x, which is a different point on the manifold. One could try to write (11.9) in a form similar to a gauge transformation,

$$\begin{aligned} v'^{\mu}(x) &= L^{\mu}_{\nu}(x)v^{\nu}(x) - \xi^{\lambda}(x)\partial_{\lambda}v^{\mu}(x) \\ &= v^{\mu}(x) + \frac{\partial \xi^{\mu}}{\partial x^{\nu}}v^{\nu}(x) - \xi^{\lambda}(x)\partial_{\lambda}v^{\mu}(x) \end{aligned} \tag{11.10}$$

While the first two terms on the right of (11.10) correspond to the way a vector transforms under the gauge group, the last term, represents a drift, produced by the fact that the translation actually shifts the point in the manifold. This type of term is not present in a gauge transformation of the type (11.1), which means that the diffeomorphism group does not act as a local symmetry as in a standard gauge theory, like Yang-Mills.

The best way to describe gauge transformations is in the language of fibre bundles. A fibre bundle is locally a direct product of a base manifold and a group, each fibre being a copy of the orbit of the group. In the case of the diffeomorphism group, the "fibres" would lie along the base. Therefore this structure is not locally a product and the group of coordinate transformations on a manifold do not define a standard fibre bundle structure. Basically, the problem is that the translation group does not take a field at a given point into a different field at the same point, but changes the arguments of the fields, something gauge transformations never do.

Apart from the obvious fact that the translations form a rather trivial group whose gauging could hardly describe the richness of gravity, it is apparent that the translation symmetry is violated by the spacetime curvature, unless spacetime happens to be exceptionally symmetric, like Minkowski or (anti-) de Sitter. In a genuine gauge theory the gauge invariance is respected everywhere, by all solutions of classical equations, and by all conceivable off-shell fields in the quantum theory. This is a key feature that makes gauge symmetries extremely useful in the quantum description: the invariance is an inherent property of the fields in the action and is not spoiled by dynamics, be it classical or quantum.

11.2.3 Lorentz Transformations

Einstein's starting point of General Relativity was the observation that gravity can be neutralized by free fall. In a small freely falling laboratory, the effect of gravity can be eliminated so that the laws of physics there are indistinguishable from those observed in an inertial laboratory, in Minkowski space. This trick is a local one: the lab has to be small enough and the time span of the experiments must be short enough. Under these conditions, the experiments will be indistinguishable from those performed in absence of gravity. In other words, in a local neighbourhood, spacetime is Lorentz invariant. In order to make this invariance manifest, it is necessary to perform an appropriate coordinate transformation to a particular reference system, viz., a freely-falling one. Conversely, Einstein argued, in the absence of gravity the gravitational field could be mocked by applying an acceleration to the laboratory.

This idea is known as the principle of equivalence meaning that, in a small spacetime region, gravitation and acceleration are equivalent effects. A freely falling observer defines a local inertial system. For a small enough region, freely falling projectiles trace straight lines, and the discrepancies with Euclidean geometry become negligible. Particle collisions mediated by short range forces, such as those between billiard balls, molecules or subnuclear particles, satisfy the conservation laws of energy and momentum valid in special relativity.

Since physical phenomena in a small neighbourhood of any spacetime should be Lorentz invariant, and since Lorentz transformations can be performed independently at every point, gravity must be endowed with local Lorentz symmetry. Hence, Einstein's observation makes gravitation a gauge theory for the group $SO(3,1)$, the first nonabelian gauge theory ever proposed [13, 14].

Note that while the Lorentz group can act independently at each spacetime point, the translations are a symmetry only in maximally symmetric spacetimes. The invariance of gravitation theory under $SO(3,1)$ is a minimal requirement, the complete group of invariance could be larger, $\mathbb{G} \supseteq SO(3,1)$. Natural options are the de Sitter ($SO(4,1)$), anti-de Sitter ($SO(3,2)$), conformal ($SO(4,2)$) and Poincaré ($ISO(3,1)$) groups, or some of their supersymmetric extensions.

11.3 First Order Gravity

In order to make the gauge symmetry of gravitation manifest it is best to use fields that correspond to some nontrivial representation of the Lorentz group, $SO(D-1,1)$, where D is the spacetime dimension. This is most effectively done if the metric and affine features of the geometry are represented by two one-form fields, the vielbein $e^a(x) = e^a_\mu dx^\mu$ and the Lorentz connection $\omega^a_{\ b}(x) = \omega^a_{\ b\mu} dx^\mu$, treated independently. This construction embraces the Principle of Equivalence, fully exploiting it to describe the geometry [15, 16].

Spacetime is postulated to be a smooth D-dimensional manifold M, of Lorentzian signature $(-1, 1, 1, \cdots, 1)$. At every point on $x \in M$ there is a D-dimensional tangent space T_x, which is a good approximation of the manifold M in the neighborhood of x. This tangent space corresponds to the reference frame of a freely falling observer mentioned in the Equivalence Principle. Every tangent space is a replica of Minkowski space, invariant under the action of the Lorentz group. This endows the spacetime manifold with a *collection of vector spaces parametrized by the manifold*, $\{T_x, x \in M\}$, something that mathematicians call a fibre bundle. In this case this is called the *tangent bundle*, where the basis is the spacetime and the fibres are the tangent spaces on which the Lorentz group acts locally. The essential point is that the manifold M, labelled by the coordinates x^μ is the spacetime where we live, and the collection of tangent spaces over it is where the symmetry group acts.

11.3.1 The Vielbein

The fact that any measurement carried out in spacetime can be translated to one in a freely falling frame, means that there is an isomorphism between tensors on M and tensors on T_x, represented by means of a linear mapping, also called *"soldering form"* or **vielbein**. It is sufficient to define this mapping on a complete set of vectors such as the coordinate separation dx^μ between two infinitesimally close points on M. The corresponding separation in T_x is defined to be

$$dz^a = e^a_\mu(x)dx^\mu, \tag{11.11}$$

where z^a represent an orthonormal coordinate basis in the tangent space. For this reason the vielbein is also viewed as a local orthonormal frame. Since T_x is a standard Minkowski space, it has a natural metric (η_{ab}) which induces a metric on M through the isomorphism e^a_μ. In fact,

$$\begin{aligned} ds^2 = \eta_{ab} dz^a dz^b &= \eta_{ab}\, e^a_\mu(x)dx^\mu\, e^b_\nu(x)dx^\nu \\ &= g_{\mu\nu}(x)dx^\mu dx^\nu, \end{aligned} \tag{11.12}$$

where $g_{\mu\nu}(x) \equiv \eta_{ab}\, e^a_\mu(x)\, e^b_\nu(x)$ is the induced metric on M from the tangent space metric. This relation can be read as the vielbein being "the square root" of the metric. Given $e^a_\mu(x)$ one can find the metric and therefore, all the metric properties of spacetime are contained in the vielbein. The converse, however, is not true: given the metric, there exist infinitely many choices of vielbein related by $O(D-1, 1)$ transformations that give the same metric.

By the definition (11.11) the vielbein one-forms transform as the components of (contravariant) vector under local Lorentz rotations of T_x, $SO(D-1, 1)$, as

$$e^a(x) \xrightarrow{\Lambda} e'^a(x) = \Lambda^a{}_b(x)e^b(x), \tag{11.13}$$

where the matrix $\Lambda(x)$ leaves the metric in the tangent space unchanged,

$$\Lambda^a_c(x)\Lambda^b_d(x)\eta_{ab} = \eta_{cd}\,, \tag{11.14}$$

and the metric $g_{\mu\nu}(x)$ is also invariant under Lorentz transformations.

11.3.2 The Lorentz Connection

A connection is required in order to ensure that the differential structure remains invariant under local Lorentz transformations $\Lambda(x)$. The role of the connection is to compensate the rotation experienced by a vector relative to the local Lorentz frames when parallel transported between tangent spaces on neighboring points, T_x and T_{x+dx}. Consider a field $\phi^a(x)$ that transforms as a vector under Lorentz rotations defined on the tangent T_x. The covariant derivative, compares the components of $\phi^a(x)$ and those of $\phi^a_{||}(x)$, the field obtained by parallel transport of $\phi^a(x+dx)$ from $x+dx$ to x. The final expression is [17, 18],

$$dx^\mu[\partial_\mu\phi^a(x) + \omega^a_{\ b\mu}(x)\phi^b(x)] = dx^\mu D_\mu\phi^a(x)$$

or, more compactly,

$$D\phi^a(x) = d\phi^a + \omega^a_b\phi^a. \tag{11.15}$$

where $\omega^a_{\ b} = \omega^a_{\ b\mu}dx^\mu$ is the **connection one-form.**

This new expression is also a vector under Lorentz transformations at x provided ω transforms as a connection, namely

$$\omega^a{}_b(x) \xrightarrow{\Lambda} \omega'^a{}_b(x) = \Lambda^a{}_c(x)\omega^c{}_d(x)\Lambda^d{}_b(x) + \Lambda^a{}_c(x)d\Lambda^c{}_b(x). \tag{11.16}$$

11.3.3 Lorentz Invariant Tensors

The group $SO(D-1,1)$ has two invariant tensors, the Minkowski metric, η_{ab}, already mentioned, and the totally antisymmetric Levi-Civita tensor, $\epsilon_{a_1a_2\cdots a_D}$. They are the same in every tangent space, they are constant; moreover, they are also Lorentz-invariant. These two conditions imply

$$d\eta_{ab} = D\eta_{ab} = 0, \tag{11.17}$$

$$d\epsilon_{a_1a_2\cdots s_D} = D\epsilon_{a_1a_2\cdots a_D} = 0. \tag{11.18}$$

The first condition, the requirement that the Lorentz connection be compatible with the metric structure of the tangent space, implies that the Lorentz connection must be antisymmetric, $\eta_{ac}\omega^c{}_b = -\eta_{bc}\omega^c{}_a$. The second, implies a more obscure identity, $\epsilon_{b_1 a_2 \cdots a_D}\omega^{b_1}{}_{a_1} + \epsilon_{a_1 b_2 \cdots a_D}\omega^{b_2}{}_{a_2} + \cdots + \epsilon_{a_1 a_2 \cdots b_D}\omega^{b_D}{}_{a_D} = 0$. This second relation does not impose further restrictions on the components of the Lorentz connection. Thus, the number of independent components of $\omega^a_{b\mu}$ is $D^2(D-1)/2$, which is less than the number of independent components of the Christoffel symbol ($D^2(D+1)/2$).

11.3.4 Curvature

One of the fundamental properties of exterior calculus is that the second exterior derivative applied to a continuously differentiable p-form, vanishes identically, $d(d\alpha_p) = d^2\alpha_p = 0$. A consequence of this is that the square of the covariant derivative operator is not a differential operator, but an algebraic operator, the curvature two-form,

$$\begin{aligned} D^2\phi^a &= d[d\phi^a + \omega^a{}_b\phi^b] + \omega^a{}_b[d\phi^b + \omega^b{}_c\phi^c] \\ &= [d\omega^a{}_b + \omega^a{}_c\omega^c{}_b]\phi^b. \end{aligned} \tag{11.19}$$

The two-form within brackets in this last expression is a second rank Lorentz tensor known as the **curvature two-form** (see, e.g., [17, 18] for a formal definition of R^a_b),

$$R^a{}_b = d\omega^a{}_b + \omega^a{}_c \wedge \omega^c{}_b = \frac{1}{2}R^a{}_{b\mu\nu}dx^\mu \wedge dx^\nu. \tag{11.20}$$

The fact that $\omega^a_b(x)$ and the gauge potential in Yang-Mills theory, $A^a_b = A^a_{b\mu}dx^\mu$, are both 1-forms and their transformation laws have the same form, reflects the fact that they are both connections of a gauge group. The curvature $R^a{}_b$ is completely analogous to the Yang-Mills curvature (field strength), $\mathbf{F} = d\mathbf{A} + \mathbf{A}\wedge\mathbf{A}$.

11.3.5 Torsion

The two independent geometrical ingredients, ω and e, play different roles as underscored by their different transformation properties under the Lorentz group, and define two independent tensors involving their derivatives, one is the curvature (11.20), and the other is the covariant derivative of the vielbein, also known the **torsion 2-form**,

$$T^a = de^a + \omega^a_b \wedge e^b, \tag{11.21}$$

In contrast with T^a, the curvature $R^a_{\ b}$ depends only on ω and is not the covariant derivative of something else. In a manifold with torsion, one can split the connection into a torsion-free part $\bar{\omega}$, and the so-called contorsion κ,

$$de^a + \bar{\omega}^a_{\ b} \wedge e^b \equiv 0, \quad \text{and} \quad T^a = \kappa^a_{\ b} \wedge e^b .$$

In this case, the curvature two-form reads

$$R^a_{\ b} = \bar{R}^a_{\ b} + \bar{D}\kappa^a_{\ b} + \kappa^a_{\ c}\kappa^c_{\ b}, \tag{11.22}$$

where $\bar{R}^a_{\ b}$ and $\bar{D}$ are the curvature and the covariant derivative constructed out of the torsion-free connection. It is the purely metric part of the curvature two-form, $\bar{R}^a{}_b$, that relates to the Riemann curvature, $R^{\alpha\beta}{}_{\mu\nu}$, through

$$\bar{R}^a{}_b = \frac{1}{2} e^a_{\ \alpha} e^b_{\ \beta} R^{\alpha\beta}{}_{\mu\nu} dx^\mu dx^\nu . \tag{11.23}$$

11.3.6 Bianchi Identities

Acting with the covariant derivative on the curvature yields an important property,

$$DR^a_{\ b} = dR^a_{\ b} + \omega^a_{\ c} \wedge R^c_{\ b} - \omega^c_{\ b} \wedge R^a_{\ c} \equiv 0 . \tag{11.24}$$

This is known as Bianchi identity, because it is not a set of equations but it is satisfied for *any* well defined connection 1-form, and therefore it does not restrict the form of $\omega^a_{\ b\mu}$ in any way. The Bianchi identity can be checked explicitly by substituting (11.22) in the second term of (11.24), and it implies that the curvature R^{ab} is "transparent" to the exterior covariant derivative, $D(R^a{}_b\phi^b) = R^a{}_b \wedge D\phi^b$.

Another direct consequence of this identity is that by taking successive exterior derivatives of e^a and ω^{ab} no new independent tensors are generated, in particular,

$$DT^a = R^a_{\ b} \wedge e^b . \tag{11.25}$$

The physical implication is that if no other fields are introduced, there is a very limited number of possible Lagrangians that can be constructed out of these fields in any given dimension [19].

11.4 Gravity Actions

All geometric features of the spacetime manifold M are captured by the two fundamental fields e^a and $\omega^a{}_b$. Hence, the action principle for a purely gravitational system could be expressed by a functional $I[e, \omega]$, with these two fields independently varied. In this scheme, the metric is a derived expression and is not a fundamental field to be varied in the action .

We postulate the action to be a local functional of the one-forms e^a, $\omega^a_{\ b}$, their exterior derivatives and exterior products of them. In addition, the two invariant tensors of the Lorentz group, η_{ab}, and $\epsilon_{a_1 \cdots a_D}$ can be used to raise, lower and contract indices. Invariance under general coordinate transformations is automatically guaranteed as exterior forms are coordinate-independent by construction.

The use of only exterior products of forms excludes the metric, its inverse and the Hodge $*$-dual (see [15] and [16] for more on this). This postulate also excludes tensors like the Ricci tensor[4] $R_{\mu\nu} = E^\lambda_a \eta_{bc} e^c_\mu R^{ab}_{\ \ \lambda\nu}$, or $R_{\alpha\beta} R_{\mu\nu} R^{\alpha\mu\beta\nu}$, except in very special combinations like the Gauss-Bonnet form, that can be expressed as exterior products of forms.

The action principle cannot depend on the choice of basis in the tangent space and hence Lorentz invariance should be ensured. A sufficient condition to have Lorentz invariant field equations is to demand the Lagrangian itself to be Lorentz invariant, but this is not really necessary. If the Lagrangian is *quasi-invariant* so that it changes by a total derivative—and the action changes by a boundary term—, still gives rise to covariant field equations in the bulk, provided the fields satisfy appropriate boundary conditions.

11.4.1 Lorentz-Invariant Lagrangians

Let us consider first Lorentz invariant Lagrangians. They must scalar D-forms consisting of linear combinations of products of e^a, $R^a_{\ b}$, T^a, contracted with η_{ab} and $\epsilon_{a_1 \cdots a_D}$ (no ω). Such invariants are listed in the following table [19]:

Invariant	Form type
$\mathscr{P}_{2k} =: R^{a_1}_{\ \ a_2} R^{a_2}_{\ \ a_3} \cdots R^{a_k}_{\ \ a_1}$	2k-form
$\Upsilon_k =: e_{a_1} R^{a_1}_{\ \ a_2} R^{a_2}_{\ \ a_3} \cdots R^{a_k}_{\ \ b} e^b$, odd k	(2k+2)-form
$\tau_k =: T_{a_1} R^{a_1}_{\ \ a_2} R^{a_2}_{\ \ a_3} \cdots R^{a_k}_{\ \ b} T^b$, even k	(2k+4)-form
$\zeta_k =: e_{a_1} R^{a_1}_{\ \ a_2} R^{a_2}_{\ \ a_3} \cdots R^{a_k}_{\ \ b} T^b$	(2k+3)-form
$\mathscr{E}_n =: \epsilon_{a_1 a_2 \cdots a_D} R^{a_1 a_2} R^{a_3 a_4} \ldots R^{a_{n-1} a_n}$, even n	2n-form
$L_p =: \epsilon_{a_1 a_2 \cdots a_D} R^{a_1 a_2} R^{a_3 a_4} \ldots R^{a_{2p-1} a_{2p}} e^{a_{2p+1}} \cdots e^{a_N}$	N-form

Among these local Lorentz invariants, the Pontryagin forms $\mathscr{P}_{2k}$ and the Euler forms $\mathscr{E}_{2n}$ define **topological invariants** in $4k$ and $2n$ dimensions, respectively: Their integrals on compact manifolds without boundary have integral spectra,

$$\Omega_n \int_{M^{2n}} \mathscr{E}_{2n} \in \mathbb{Z}, \qquad \tilde{\Omega}_k \int_{M^{4k}} \mathscr{P}_{4k} \in \mathbb{Z}, \tag{11.26}$$

[4]Here E^λ_a is the inverse vielbein, $E^\lambda_a e^b_{\ \lambda} = \delta^a_b$.

where Ω_n and $\tilde{\Omega}_k$ are normalization coefficients determined uniquely by the dimension of the manifold. Thus, in every even dimension $D = 2n$ there is a topological invariant of the Euler family. If the dimension is a multiple of four, there are invariants of the Pontryagin family as well, of the form

$$\mathscr{P}_D = \mathscr{P}_{4k_1}\mathscr{P}_{4k_2}\cdots\mathscr{P}_{4k_r}, \tag{11.27}$$

where $4(k_1 + k_2 + \cdots + k_r) = D$, so the number of Pontryagin invariants grows with the number of different ways to express the number $D/4$ as a sum of integers. As shown in [19], for large D this number grows approximately as by the Hardy-Ramanujan formula, $\sim \frac{1}{\sqrt{3}D}\exp[\pi\sqrt{D/6}]$.

11.4.2 Lovelock Theories

If torsion is set to zero the first and second order theories coincide. Moreover, τ_k, ζ_k and Υ_k vanish, and we are led to the following theorem [15, 20]: In the absence of torsion, the most general action for gravity $I[e,\omega]$, invariant under Lorentz transformations, takes the form

$$I_D[e,\omega] = \kappa\int_M \sum_{p=0}^{[D/2]} \alpha_p L_p^D \tag{11.28}$$

where α_p are arbitrary constants, and L_p^D is given by

$$L_p^D = \epsilon_{a_1\cdots a_D} R^{a_1a_2}\cdots R^{a_{2p-1}a_{2p}} e^{a_{2p+1}}\cdots e^{a_D}. \tag{11.29}$$

The Lovelock series is an arbitrary linear combination where each term L_p^D is the continuation to dimension D of all the lower-dimensional Euler forms. In even dimensions, the last term in the series is the Euler form of the corresponding dimension, $L_{D/2}^D = \mathscr{E}_D$. Let us examine a few examples.

• **D = 2**: The Lovelock Lagrangian reduces to 2 terms, the two-dimensional Euler form and the spacetime volume (area),

$$\begin{aligned} I_2 &= \kappa\int_M \epsilon_{ab}[\alpha_1 R^{ab} + \alpha_0 e^a e^b] \\ &= \kappa\int_M \sqrt{|g|}\,(\alpha_1 R + 2\alpha_0)\, d^2x \\ &= \kappa\alpha_1\cdot \mathbf{E}_2 + 2\kappa\alpha_0\cdot V_2. \end{aligned} \tag{11.30}$$

This action has as a local extremum for $V_2 = 0$ and does not produce a very interesting dynamical theory for the geometry unless matter is included. If the manifold M has Euclidean metric and a prescribed boundary, the first term becomes a boundary term and the action is extremized by a minimal surface, like a soap bubble, the famous Plateau problem [21].

• **D** $=$ **3** and **D** $=$ **4**: The action reduces to the Einstein-Hilbert and cosmological constant terms. In four dimensions, the action also admits the Euler invariant $\mathscr{E}_4$,

$$\begin{aligned} I_4 &= \kappa \int_M \epsilon_{abcd} \left[\alpha_2 R^{ab} R^{cd} + \alpha_1 R^{ab} e^c e^d + \alpha_0 e^a e^b e^c e^d\right] \\ &= -\kappa \int_M \sqrt{|g|} \left[\alpha_2 \left(R^{\alpha\beta\gamma\delta} R_{\alpha\beta\gamma\delta} - 4 R^{\alpha\beta} R_{\alpha\beta} + R^2\right) + 2\alpha_1 R + 24\alpha_0\right] d^4x \\ &= -\kappa\alpha_2 \cdot \mathscr{E}_4 - 2\alpha_1 \int_M \sqrt{|g|} R d^4x - 24\kappa\alpha_0 \cdot V_4. \end{aligned} \tag{11.31}$$

• **D** $=$ **5**: The Euler form $\mathscr{E}_4$, also known as the Gauss-Bonnet density, provides the first nontrivial generalization of Einstein gravity occurring in five dimensions,

$$\epsilon_{abcde} R^{ab} R^{cd} e^e = \sqrt{|g|} \left[R^{\alpha\beta\gamma\delta} R_{\alpha\beta\gamma\delta} - 4 R^{\alpha\beta} R_{\alpha\beta} + R^2\right] d^5x. \tag{11.32}$$

The fact that this term could be added to the Einstein-Hilbert action in five dimensions seems to have been known for many years, and is commonly attributed to Lanczos [22].

11.4.3 Torsional Series

Lovelock's theorem assumes $T^a = 0$ to be an identity, which contradicts the assumption that e^a and $\omega^a_{\ b}$ are two independent features of the geometry, to be treated on equal footing. For $D \leq 4$ and in the absence of fermionics or non-minimally coupled fields, this is a consequence of the equation obtained by varying with respect to $\omega^a{}_b$, so that imposing the torsion-free constraint may be seen as an unnecessary but harmless restriction. In fact, for 3 and 4 dimensions, the Lorentz connection can be algebraically solved from its own field equation and, by the implicit function theorem, the first order and the second order actions have the same extrema and define equivalent theories, $I[\omega, e] = I[\omega(e, \partial e), e]$.

In general, even if the torsion-free condition is often a solution of the field equations, it does not generically follow from them. Thus, it is reasonable to consider generalizations of the Lovelock action in which torsion *is not* assumed to vanish identically, adding all possible Lorentz invariants involving torsion that would vanish if $T^a = 0$ [19]. This means allowing for combinations of $\mathscr{P}_{2k}$, Υ_k, τ_k and ζ_k. For example, a possible contribution to the Lagrangian in fifteen dimensions could be $\mathscr{P}_2 \Upsilon_1 \tau_0 \zeta_0$. Other examples are: $\zeta_0 = e^a T_a$ in $D = 3$, and $\Upsilon_1 = e^a e^b R_{ab}$, $\tau_0 = T^a T_a$, in $D = 4$. Some linear combinations of these may yield a torsional topological invariants. For example,

$$\mathscr{K}_4 = \tau_0 - \Upsilon_1 = T^a T_a - e^a e^b R_{ab}, \tag{11.33}$$

is the well-known Nieh-Yan form, whose integral yields a linear combination of Pontryagin numbers associated to $SO(n, 5-n$ and $SO(m, 4-m)$ [23] (for details and extensive discussions, see [19]).

11.4.4 Chern–Simons Series

For every characteristic class in a given dimension $2n$, there is a $(2n-1)$ CS form can be associated. Thus, for the Euler forms $\mathscr{E}$, the Pontryagin forms $\mathscr{P}$, and for every torsional topological form $\mathscr{K}$, there is a corresponding CS form such that

$$d\mathscr{C}^E = \mathscr{E}, \qquad d\mathscr{C}^P = \mathscr{P}, \qquad d\mathscr{C}^K = \mathscr{K}. \tag{11.34}$$

The CS forms associated to the Euler invariant are less obvious to obtain and, as we will see next, they correspond to particular combinations of Lovelock Lagrangians L_p.

These CS forms are ready to be integrated in one dimension less than the dimension for which the characteristic classes are defined. For example, the Pontryagin and the Nieh-Yan forms in four dimensions, $\mathscr{P}_4$ and $\mathscr{N}_4$, respectively have their corresponding CS three-forms,

$$\mathscr{C}_3^P = \omega^a{}_b d\omega^b{}_a + \frac{2}{3}\omega^a{}_b\omega^b{}_c\omega^c{}_a, \quad d\mathscr{C}_3^P = R^{ab}R_{ab} = \mathscr{P}_4, \tag{11.35}$$

$$\mathscr{C}_3^K = e^a T_a, \qquad d\mathscr{C}_3^K = T^a T_a - e^a e^b R_{ab} = \mathscr{K}_4. \tag{11.36}$$

The most general gravity Lagrangian in D dimensions is a linear combination of D-form Lorentz invariants and quasi-invariants of the families described above. While the Lovelock series has a simple systematic rule for any dimension (11.28), there is no simple recipe for the torsional Lagrangians and there is not even a systematic rule to tell how many terms appear in a given dimension, and moreover the number of torsional invariants grows exponentially with D. This proliferation is not just an aesthetic problem, but it involves a huge number of indeterminate dimensionful parameters in the theory, making it hopelessly unmanageable.

11.4.5 Dynamical Content of Lovelock Theory

The Lovelock theory is the natural generalization of GR in spacetimes of dimension greater than four. In the absence of torsion these theories generically describe the same number of degrees of freedom as the Einstein-Hilbert theory, $D(D-3)/2$ [24].

The action (11.28) describes the only ghost-free effective theory for a spin-2 field obtained from string theory at low energy [15, 25]. The unexpected and nontrivial lack of ghosts seems reflects the fact that for vanishing torsion the Lovelock action yields second order field equations for the metric, so that the propagators behave as k^{-2} instead of $k^{-2} + k^{-4}$, as would be the case in a generic theory involving arbitrary scalar combinations of the curvature tensor, like in $f(R)$ theories.

Extremizing the action (11.28) with respect to e^a and ω^{ab}, yields

$$\delta I_D = \int [\delta e^a \mathscr{E}_a + \delta\omega^{ab}\mathscr{E}_{ab}] = 0, \tag{11.37}$$

up to surface terms. The condition for $\delta I_D = 0$ under arbitrary infinitesimal variations is that $\mathscr{E}_a$ and $\mathscr{E}_{ab}$ vanish,

$$\mathscr{E}_a = \sum_{p=0}^{[\frac{D-1}{2}]} \alpha_p (D-2p)\mathscr{E}_a^{(p)} = 0\,, \qquad \mathscr{E}_{ab} = \sum_{p=1}^{[\frac{D-1}{2}]} \alpha_p p(D-2p)\mathscr{E}_{ab}^{(p)} = 0, \tag{11.38}$$

where we have defined

$$\mathscr{E}_a^{(p)} := \epsilon_{ab_2\cdots b_D} R^{b_2b_3}\cdots R^{b_{2p}b_{2p+1}} e^{b_{2p+2}}\cdots e^{b_D}, \tag{11.39}$$

$$\mathscr{E}_{ab}^{(p)} := \epsilon_{aba_3\cdots a_D} R^{a_3a_4}\cdots R^{a_{2p-1}a_{2p}} T^{a_{2p+1}} e^{a_{2p+2}}\cdots e^{a_D}. \tag{11.40}$$

These equations contain first derivatives of e^a and $\omega^a_{\ b}$. If one furthermore assumes—as is usually done—that the torsion vanishes identically,

$$T^a = de^a + \omega^a{}_b e^b = 0, \tag{11.41}$$

then the second equation (11.38) is automatically satisfied. Moreover, the torsion-free condition can be solved for ω as a function of the inverse vielbein (E_a^μ) and its derivatives,

$$\omega^a{}_{b\mu} = -E_b^\nu(\partial_\mu e_\nu^a - \Gamma^\lambda_{\mu\nu} e_\lambda^a), \tag{11.42}$$

where $\Gamma^\lambda_{\mu\nu}$ is symmetric in $\mu\nu$ and can be identified as the Christoffel symbol (torsion-free affine connection). Substituting this expression for the Lorentz connection back into (11.39) yields second order field equations for the metric, identical to the ones obtained from varying the Lovelock action written in terms of the Riemann tensor and the metric,

$$I_D[g] = \int_M d^D x\sqrt{g}\left[\alpha_0' + \alpha_1' R + \alpha_2'(R^{\alpha\beta\gamma\delta}R_{\alpha\beta\gamma\delta} - 4R^{\alpha\beta}R_{\alpha\beta} + R^2) + \cdots\right].$$

The remarkable feature of GR that the field equations for the metric are second order and not fourth order, in spite of the Lagrangian involving second derivatives of $g_{\mu\nu}$, is a consequence of the fact that the action can also be written using only wedge products and exterior derivatives of form fields, without using the $*$-Hodge dual. In $f(R)$ theories, for example, this condition is not respected and they generically give rise to fourth order field equations. The standard, purely metric theory is also called "second order formalism" because it involves up to second derivatives of the metric. In the presence of fermionic matter or non-minimal couplings, however, torsion does not necessarily vanish and therefore a purely metric formulation would not be equivalent to the first order one. Therefore, the first and second order Lagrangians are much more than two different *formalisms*, they correspond to inequivalent theories.

One important feature that makes the behaviour of Lovelock theories very different for $D \leq 4$ and for $D > 4$ is that in the former case the field equations (11.38) are linear in the curvature tensor, while in the latter case the equations are generically nonlinear in R^{ab}. For $D \leq 4$ the Eq. (11.40) imply the vanishing of torsion, which is no longer true for $D > 4$. In fact, the field equations evaluated in some configurations may leave some components of the curvature and torsion tensors completely undetermined. For example, Eq. (11.38) has the form of a polynomial in R^{ab} times T^a, and it often happens that the polynomial vanishes identically, leaving the torsion tensor completely undetermined.

However, the configurations for which the equations leave some components of R^{ab} or T^a undetermined form sets of measure zero in the space of solutions. In a generic case, outside of these degenerate configurations, the Lovelock theory has the same number of degrees of freedom as ordinary gravity [24]. The problem of degeneracy, however, is a major issue in determining the time evolution of certain dynamical systems, usually associated with the splitting of the phase space into causally disconnected regions and irreversible loss of degrees of freedom [26]. These features might be associated to a dynamical dimensional reduction in gravitation theories [27], and has been shown to survive even at the quantum level [28].

11.4.6 Euler-CS Forms and the Extension of Lorentz Symmetry

The proliferation issue is a serious weakness of the theory. The coefficients in front of each term in the Lagrangian are not only arbitrary but dimensionful. This problem already occurs in 4 dimensions, where Newton's constant and the cosmological constant are dimensionful, and it only gets worse at higher dimensions and leaves little room for optimism in a quantum version of the theory. Dimensionful parameters in the action are potentially dangerous because they are likely to acquire uncontrolled quantum corrections. This is what makes ordinary gravity nonrenormalizable in perturbation theory: In 4 dimensions, Newton's constant has dimensions of [mass]$^{-2}$ in natural units. This means that as the order in perturbation series increases, more powers of momentum will occur in the Feynman graphs, making the ultraviolet divergences increasingly worse. Concurrently, the radiative corrections to these bare parameters require the introduction of infinitely many counterterms into the action to render them finite [29], and an illness that requires infinite amount of medication is incurable.

One solution for this problem would be if those coefficients were somehow fixed by a symmetry principle. But a symmetry that can protect the theory in the quantum regime should be unbreakable by quantum corrections, and not just an approximate feature of the effective classical descendant. A symmetry that is only present in the classical limit but is not a feature of the quantum theory is said to be anomalous. An anomalous symmetry is an mirage of the classical limit that does not correspond to a true symmetry of the microscopic world.

A good way to make a symmetry the parameters would be if they could be absorbed in a rescaling of the fields. In odd dimensions there is a unique choice of coefficients in the Lovelock action that yields a theory in which the local Lorentz symmetry is enlarged to de Sitter, anti-de Sitter or Poincar'e gauge groups. The resulting action has no dimensionful parameters and can be seen to depend on a unique (dimensionless) constant. This coefficient can be further shown to be quantized, following an argument similar to the one that yields Dirac's quantization of the product of magnetic and electric charge [30]. All these miraculous properties can be traced back to the fact that the particular choice of coefficients in that Lagrangian turns the Lovelock Lagrangian into a CS form for an enhanced gauge symmetry.

The coefficients α_p in the Lovelock action (11.28) have dimensions l^{D-2p}. This is because the canonical dimension of the vielbein is $[e^a] = l$, while the Lorentz connection has dimensions $[\omega^{ab}] = l^0$, as a true gauge field. This reflects the fact that gravity is naturally a gauge theory for the Lorentz group, where e^a plays the role of a matter field, *not* a connection field but a vector under Lorentz transformations.

Consider combining the two fundamental fields e^a, and ω^{ab} into a bigger 1-form W^{AB} as

$$W^{AB} = \begin{bmatrix} \omega^{ab} & e^a \ell^{-1} \\ -e^b \ell^{-1} & 0 \end{bmatrix} \tag{11.43}$$

where $a, b, .. = 1, 2, ..D$, while $A, B, \ldots = 1, 2, .., D+1$. This one-form defines a new connection antisymmetric in $A-B$ with $D(D+1)/2$ independent components, that can therefore accommodate the generators for the de Sitter or anti-de Sitter group, depending on how the Minkowski metric is extended to $D+1$ dimensions, $\eta_{ab} \rightarrow \Upsilon_{AB}$. The curvature $F^{AB} = dW^{AB} + W^A_{\ C} W^{CB}$, is easily shown to be a combination of the Lorentz curvature, the vielbein and torsion,

$$F^{AB} = \begin{bmatrix} R^{ab} \pm \ell^{-2} e^a e^b & \ell^{-1} T^a \\ -\ell^{-1} T^b & 0 \end{bmatrix}. \tag{11.44}$$

In this way, the D-dimensional Lorentz group is embedded into the de-Sitter or anti-de Sitter groups,

$$SO(D-1,1) \hookrightarrow \begin{cases} SO(D,1), & \Upsilon_{AB} = \mathrm{diag}(\eta_{ab}, +1) \\ SO(D-1,2), & \Upsilon_{AB} = \mathrm{diag}(\eta_{ab}, -1) \end{cases}, \tag{11.45}$$

or into the Poincaré group $ISO(D-1,1)$ in the limit $\ell \to \infty$.

With the new curvature F one can define the Euler invariant,

$$\mathscr{E}_{D+1} = \epsilon_{A_1 A_2 \cdots A_{D+1}} F^{A_1 A_2} \cdots F^{A_D A_{D+1}}, \tag{11.46}$$

which is nontrivial provided the field W^{AB} is viewed as a connection in $D+1$ dimensions, which requires that the D-dimensional spacetime be considered as a submanifold embedded in $D+1$ dimensional manifold $\bar{M}$ ($M \subset \bar{M}$), or as its boundary, ($M = \partial\bar{M}$). Obviously, since F^{AB} is a two-form, this also requires that $D+1 = 2n$.

Substituting the expression (11.44) in (11.46) produces a polynomial of the form

$$\epsilon_{a_1 b_1 \cdots a_n b_n c_{D+1}} (R^{a_1 b_1} \pm \ell^{-2} e^{a_1} e^{b_1}) \cdots (R^{a_n b_n} \pm \ell^{-2} e^{a_n} e^{b_n}) T^{c_{D+1}}. \tag{11.47}$$

It is simple algebra to observe that since $T^a = De^a$, and $DR^{ab} \equiv 0$, this last expression can be written as the exterior derivative of a polynomial in R^{ab} and e^a,

$$\mathscr{E}_{D+1} = d\mathscr{C}^E_{2n-1}, \tag{11.48}$$

which yields the expression for the Euler-CS form in $2n-1$ dimensions. For example, for $D = 3$ this yields

$$L^E_3 = \kappa\epsilon_{abc}(R^{ab}e^c \pm \frac{1}{3l^2} e^a e^b e^c), \tag{11.49}$$

which can be recognized as the Einstein-Hilbert Lagrangian with cosmological constant in three dimensions. The important feature is that since the Euler form $\mathscr{E}_{D+1}$ is invariant under an extension of the three-dimensional Lorentz group, $G \supseteq SO(2,1)$, where $G = SO(2,2)$ or $SO(3,1)$, the three-dimensional theory described by (11.49) inherits this enlarged symmetry. The same construction directly yields the $(2n-1)$-dimensional (A)dS-invariant Lagrangian as

$$L^E_{2n-1} = \kappa \sum_{p=0}^{n-1} \frac{(\pm 1)^{p+1} l^{2p-D}}{(D-2p)} \binom{n-1}{p} \epsilon[R]^p [e]^{2n-1-2p}, \tag{11.50}$$

where $\epsilon[R]^p[e]^{2n-1-2p}$ stands for L^{2n-1}_p in (11.29).

Clearly (11.50) is a particular case of a Lovelock Lagrangian in which all the coefficients $\bar{\alpha}_p$ have been fixed to so that the symmetry is embedded in a larger group as in (11.45). Here κ is an arbitrary dimensionless constant, and ℓ, the only dimensionful parameter of the Lagrangian can be absorbed in a rescaling of the vielbein, so that the resulting CS Lagrangian not only has enlarged gauge symmetry, it is also scale invariant.

Finally, it can also be observed that a similar CS construction can be done for the torsional topological invariants, leading to CS forms that also have enlarged gauge symmetry and scale invariance, but since there is no simple formula for the torsional Lorentz invariants and torsional topological densities, the corresponding CS forms have to be defined in each dimension separately. For more details, the reader is urged to consult a more extensive review like [7, 31].

11.5 Summary

The relevance of gauge symmetry in physics cannot be overemphasized. Establishing that all interactions in nature originate on this invariance principle is one of the great achievements of physics in the twentieth century. Gauge symmetry explains the subnuclear interactions, the functioning of stars, the chemistry of life and the geometry of the universe at large. Einstein's discovery that the equivalence principle means that gravity is a gauge theory for the Lorentz group. The spacetime geometry is described by two independent notions, metricity and affinity, each represented by a fundamental field that transforms under the local Lorentz symmetry in a definite representation.

It is an even more remarkable feature that in odd dimensions these two fundamental objects can combine to become a connection for an enlarged gauge symmetry, the de Sitter, anti-de Sitter or Poincaré groups. The result is a CS theory that has no arbitrary free parameters, no dimensionful couplings, and whose gauge invariance is independent of the spacetime geometry. None of these features seems random. One cannot help feeling that something profound and beautiful lies in these structures. Whether the CS theories of gravity, or some more ambitious underlying extension turns out to be the way to understand the connection between gravitation and quantum mechanics, remains to be seen. However, the fact that CS forms are singled out in gravity, the fact that they play such an important role in the couplings between gauge fields and sources, their deep relation with quantum mechanics, strongly suggest that there is some meaning to it. This does not look like a contingent result of natural chaos.

Acknowledgements I have benefited from hundreds interesting discussions with many colleagues and students at CECs and elsewhere over the past decade. They have taught me almost everything that is contained here I could not name them all and I don't want to be unfair to anyone. Needless to say, any errors are my own responsibility. Last but not least, thanks are due to the organizers of the Aegean School in Paros for the stimulating atmosphere, and the editors of these proceedings for their patience with this author. This work has been partially supported by FONDECYT grant 1140155. The Centro de Estudios Científicos (CECs) is funded by the Chilean Government through the *Centers of Excellence Base Financing Program* of CONICYT.

References

1. S.S. Chern, J. Simons, Characteristic forms and geometric invariants. Ann. Math. **99**, 48 (1974)
2. M. Nakahara, *Geometry, Topology and Physics*, Graduate Student Series in Physics, 2nd edn. (Taylor & Francis, Boca Raton, 2003)
3. J. Baez, J.P. Munian, *Gauge Fields, Knots and Gravity* (World Scientific, Singapore, 1994)
4. J. Crisóstomo, R. Troncoso, J. Zanelli, Black hole scan. Phys. Rev. **D62**, 084013 (2000); [hep-th/0003271]
5. C. Charmousis, Higher order gravity theories and their black hole solutions. Lect. Notes Phys. **769**, 299 (2009). [arXiv:0805.0568/gr-qc]
6. C. Garraffo, G. Giribet, The lovelock black holes. Mod. Phys. Lett. A **23**, 1801 (2008); [arXiv:0805.3575/gr-qc]
7. J. Zanelli, *Lecture Notes on Chern-Simons (Super-)gravities*, 2nd edn (February 2008), hep-th/0502193
8. J. Zanelli, Chern-Simons forms in gravitation theories. Class. Quant. Grav. **29**, 133001 (2012) [arXiv:1208.3353 [hep-th]]
9. J. Zanelli, *Gravitation Theory and Chern-Simons Forms*, in "Geometric, Algebraic and Topological Methods for Quantum Field Theory", Proceedings of the 2011 Villa de Leyva Summer School, Villa de Leyva, Colombia, 4–22 July 2011, ed. by A. Cardona, C. Neira-Jimanez, H. Ocampo, S. Paycha, A.F. Reyes-Lega
10. J. Zanelli, *Uses of Chern-Simons actions*, in "Ten Years of AdS/CFT: A Workshop Celebrating the Tenth Anniversary of the Maldacena Conjecture", AIP Conf. Proc. vol. 1031, 115 (2008) [arXiv:0805.1778 [hep-th]]
11. J.D. Edelstein, J. Zanelli, *Sources for Chern-Simons theories*, in "Quantum Mechanics of Fundamental Systems: The Quest for Beauty and Simplicity" [arXiv:0807.4217 [hep-th]]
12. B. Schutz, *Geometrical Methods of Mathematical physics* (Cambridge University Press, Cambridge, 1980)
13. L. O'Raifeartaigh, N. Straumann, Gauge theory: Historical origins and some modern developments. Rev. Mod. Phys. **72**, 1 (2000)
14. J.D. Jackson, L.B. Okun, Historical roots of gauge invariance. Rev. Mod. Phys. **73**, 663 (2001) [hep-ph/0012061]
15. B. Zumino, Gravity theories in more than four dimensions. Phys. Rep. **137**, 109 (1986)
16. T. Regge, On broken symmetries and gravity. Phys. Rep. **137**, 31 (1986)
17. T. Eguchi, P. Gilkey, A.J. Hanson, Gravitation, gauge theories and differential geometry. Phys. Rept. **66**, 213 (1980)
18. M. Goeckeler, T. Schuecker, *Differential Geometry, Gauge Theories and Gravity* (Cambridge University Press, Cambridge, 1987)
19. A. Mardones, J. Zanelli, Lovelock-Cartan theory of gravity. Class. Quant. Grav. **8**, 1545 (1991)
20. D. Lovelock, The Einstein tensor and its generalizations. J. Math. Phys. **12**, 498 (1971)
21. J. Douglas, Solution of the problem of Plateau. Trans. Am. Math. Soc. **33**, 263 (1931)
22. C. Lanczos, A remarkable property of the Riemann-Christoffel tensor in four dimensions. Ann. Math. **39**, 842 (1938)
23. O. Chandía and J. Zanelli, *Topological invariants, instantons and chiral anomaly on spaces with torsion*, Phys. Rev. **D55**, 7580–7585 (1997), [hep-th/9702025]; *Supersymmetric particle in a spacetime with torsion and the index theorem*, **D58**, 045014 (1998); [hep-th/9803034].
24. C. Teitelboim, J. Zanelli, Dimensionally continued topological gravitation theory in hamiltonian form. Class. Quant. Grav. **4**, L125 (1987); and in *Constraint Theory and Relativistic Dynamics*, G. Longhi, L. Lussana, (eds.) (World Scientific, Singapore, 1987)
25. B. Zwiebach, Curvature squared terms and string theories. Phys. Lett. **B156**, 315 (1985)
26. J. Saavedra, R. Troncoso, J. Zanelli, Degenerate dynamical systems. J. Math. Phys **42**, 4383 (2001); [hep-th/0011231]
27. M. Hassaïne, R. Troncoso, J. Zanelli, Poincaré-invariant gravity with local supersymmetry as a gauge theory for the M-algebra. Phys. Lett. **B 596**, 132 (2004); [hep-th/0306258]

28. F. de Micheli, J. Zanelli, Quantum degenerate systems. J. Math. Phys. **53**, 102112 (2012). [arXiv:1203.0022 [hep-th]]
29. G. 't Hooft, M. Veltman, One loop divergences in the theory of gravitation. Ann. Inst. H. Poincaré (Phys. Theor.) **A20**, 69–94 (1974). G. 't Hooft, Quantum gravity: A fundamental problem and some radical ideas, 1978 Cargese Summer School, and also in NATO Adv. Study Inst. Ser. B Phys.**44**, 323 (1979)
30. J. Zanelli, Quantization of the gravitational constant in odd dimensions. Phys. Rev. **D51**, 490 (1995); [hep-th/9406202]
31. L. Huerta, J. Zanelli, Introduction to Chern-Simons theories. PoS ICFI **2010**, 004 (2010)

Chapter 12
Holographic Chern–Simons Theories

H. Afshar, A. Bagchi, S. Detournay, D. Grumiller, S. Prohazka, and M. Riegler

Abstract Chern–Simons theories in three dimensions are topological field theories that may have a holographic interpretation for suitable chosen gauge groups and boundary conditions on the fields. Conformal Chern–Simons gravity is a topological model of three-dimensional gravity that exhibits Weyl invariance and allows various holographic descriptions, including Anti-de Sitter, Lobachevsky and flat space holography. The same model also allows to address some aspects that arise in higher spin gravity in a considerably simplified setup, since both types of models have gauge symmetries other than diffeomorphisms. In these lectures we summarize briefly recent results.

12.1 Introduction

Chern–Simons theories in three dimensions have a wide range of applications in mathematics and physics (see [1–7] for various reviews). The bulk action

$$S_{\text{CS}}[A] = \frac{k_{\text{CS}}}{4\pi} \int_{\mathscr{M}} \text{tr}\big(A \wedge \text{d}A + \tfrac{2}{3}\, A \wedge A \wedge A\big) \tag{12.1}$$

H. Afshar
Centre for Theoretical Physics, University of Groningen, Nijenborgh 4, 9747 AG Groningen, The Netherlands
e-mail: h.r.afshar@rug.nl

A. Bagchi
Indian Institute of Science Education and Research, Pune 411008, India
e-mail: a.bagchi@iiserpune.ac.in

S. Detournay
Physique Théorique et Mathématique, Université Libre de Bruxelles and International Solvay Institutes, Campus Plaine C.P. 231, B-1050 Bruxelles, Belgium
e-mail: sdetourn@ulb.ac.be

D. Grumiller (✉) • S. Prohazka • M. Riegler
Institute for Theoretical Physics, Vienna University of Technology, Wiedner Hauptstrasse 8-10/136, A-1040 Vienna, Austria
e-mail: grumil@hep.itp.tuwien.ac.at; prohazka@hep.itp.tuwien.ac.at; rieglerm@hep.itp.tuwien.ac.at

E. Papantonopoulos (ed.), *Modifications of Einstein's Theory of Gravity at Large Distances*, Lecture Notes in Physics 892, DOI 10.1007/978-3-319-10070-8_12

depends on a dimensionless coupling constant, the Chern–Simons level k_{CS}, a Lie-algebra valued connection 1-form A and a manifold $\mathscr{M}$ that often has some boundary $\partial\mathscr{M}$. In these lectures we always assume that $\mathscr{M}$ topologically is either a filled cylinder or a filled torus.

While the Lagrange-3-form in the action (12.1) is not gauge invariant, the equations of motion are gauge invariant,

$$F = \mathrm{d}A + A \wedge A = 0\,, \tag{12.2}$$

and show that locally all solutions are pure gauge. The theory is topological in the sense that its action does not depend on the metric, and also topological in the sense that the theory has no local physical degrees of freedom (see [8] for a review on topological field theories).

Thus, all physical excitations are of global nature, and if $\mathscr{M}$ has a boundary one can picture the excitations as edge states localized at the boundary, much like in the Anti-de Sitter/conformal field theory (AdS/CFT) correspondence.

The precise boundary conditions imposed on the connection A are a crucial input in the specification of the model, and the same bulk action can describe completely different physical systems, depending on the specific choice of boundary data.

Prominent examples of Chern–Simons theories with special boundary conditions are Einstein gravity with negative cosmological constant [9, 10] and higher spin theories [11, 12], some aspects of which are reviewed below.

In these lectures we focus mostly on a specific theory of gravity, conformal Chern–Simons gravity (CSG) [13–15]. Its bulk action is similar to the Chern–Simons action (12.1), but depends on a connection that is not a fundamental field, namely on the Christoffel connection.

$$S_{\text{CSG}} = \frac{k}{4\pi}\int_{\mathscr{M}} \mathrm{d}^3x\, \epsilon^{\alpha\beta\gamma}\, \Gamma^{\mu}{}_{\alpha\nu}\left(\partial_\beta \Gamma^{\nu}{}_{\gamma\mu} + \tfrac{2}{3}\,\Gamma^{\nu}{}_{\beta\sigma}\Gamma^{\sigma}{}_{\gamma\mu}\right) \tag{12.3}$$

Consequently, the equations of motion obtained by varying the action (12.1) with respect to the metric do not imply flatness of the geometry, but only conformal flatness.

$$C_{\mu\nu} = \tfrac{1}{2}\,\varepsilon_{\mu}{}^{\alpha\beta}\nabla_\alpha R_{\beta\nu} + (\mu \leftrightarrow \nu) = 0 \tag{12.4}$$

The quantity $C_{\mu\nu}$ is the Cotton tensor, which vanishes in three dimensions if and only if spacetime is conformally flat (see for instance [16]).

Thus, as opposed to three-dimensional Einstein gravity with negative cosmological constant, which allows only locally AdS solutions and thus only AdS holography, CSG has also some non-AdS solutions and is thus a simple model that allows to study non-AdS holography. Moreover, CSG has an additional gauge symmetry, namely Weyl symmetry

$$g_{\mu\nu} \to e^{2\Omega}\, g_{\mu\nu}\,, \tag{12.5}$$

so that metrics that are not diffeomorphic to each other can nevertheless be gauge equivalent. All these properties are shared by higher spin gravity, which is why CSG can be regarded as a simple toy model for higher spin gravity and non-AdS holography (see [17, 18] for the higher spin perspective and [19, 20] for the CSG perspective).

We address now which boundary conditions are possible in CSG. In principle, any conformally flat metric is an allowed background. However, for practical applications it usually makes sense to consider backgrounds that have at least one Killing vector, e.g., associated with asymptotic time translations. In that case, a Kaluza–Klein reduction to two dimensions reduces CSG to a specific non-linear Maxwell–Einstein theory [21]. This theory in turn can be mapped to a specific Dilaton–Maxwell–Einstein theory, whose classical solutions can be found globally [22]. It turns out that all such solutions have additional Killing vectors: they are either maximally symmetric, i.e., have six Killing vectors, or they have four Killing vectors.

The first option allows to study AdS holography, flat space holography and de Sitter holography. The second option allows to study Lobachevsky holography. In the rest of these lectures we review some of these holographic setups and recent results. In Sect. 12.2 we review AdS holography. In Sect. 12.3 we address Lobachevsky holography. In Sect. 12.4 we focus on flat space holography, in particular in the context of quantum gravity toy models.

12.2 Anti-de Sitter Holography

Holography provides a map between quantum gravity in $d+1$ dimensions and quantum field theories in d dimensions. While holographic correspondences exist that involve specific types of non-unitary theories—see [23, 24] and references therein—for many purposes one would like to insist on unitarity.

As we shall review in Sects. 12.2.1 and 12.4.2, in three-dimensional gravity unitarity prefers spacetimes with AdS asymptotics for quantization of parity even theories and asymptotically flat spacetimes for quantization of parity odd theories. There are two pure gravity models without local degrees of freedom in three dimensions, parity even *Einstein-Hilbert gravity* (EHG) and parity odd *conformal Chern–Simons gravity* (CSG). These models can be written as Chern–Simons topological gauge theories of level k_{CS} for SO(2,2) AdS [10, 25] and SO(3,2) conformal [26] groups respectively, with a *proper* non-degenerate bilinear form. The AdS algebra

$$[J_a, J_b] = \epsilon_{abc} J^c, \qquad [J_a, P_b] = \epsilon_{abc} P^c, \qquad [P_a, P_b] = \Lambda \epsilon_{abc} J^c\,, \tag{12.6}$$

admits two different non-degenerate bilinear forms. In case of EHG this would be [10],

$$\mathrm{tr}(J_a, P_b) = \tfrac{1}{2}\eta_{ab}\,. \tag{12.7}$$

The Chern–Simons theory based on this algebra and this bilinear form can be decomposed as the sum of two Chern–Simons actions of SO(2,1) gauge group with opposite levels. The conformal algebra on the other hand has a unique bilinear form.

In this formalism, the dreibein e^a, and the (dualized) spin connection ω^a, are gauge fields in the translation P_a and the rotation J_a generators and the gauge transformations $A_\mu \to A_\mu + D_\mu \varepsilon$ generate diffeomorphisms on-shell [10] when the gauge parameter ϵ depends linearly on fields, $\varepsilon^a = A^a{}_\mu \xi^\mu$,

$$\delta_\xi A^a{}_\mu = \partial_\mu \xi \cdot A^a + \xi \cdot \partial A^a{}_\mu + \xi^\nu F^a{}_{\mu\nu}\,, \tag{12.8}$$

The asymptotic analysis for EHG on AdS was first done by Brown and Henneaux in [27] where they recognized that under suitable boundary conditions the asymptotic symmetries of this theory are given by two copies of the Virasoro algebra with the same central charge. A detailed analysis for CSG with AdS boundary conditions was done in [19, 20, 28]. In the following subsection we address the main aspects of these results.

12.2.1 Conformal Chern–Simons Gravity

Before discussing the first order formulation of CSG as a CS gauge theory of SO(3,2), we review the asymptotic analysis of (12.3) in the metric formulation in which the metric g is the dynamical field [19, 20]. In Gaussian normal coordinates, consistent asymptotically locally AdS boundary conditions on the metric are,

$$\mathrm{d}s^2 = g_{\mu\nu}\mathrm{d}x^\mu \mathrm{d}x^\nu = e^{2\phi}\Big[\mathrm{d}\rho^2 + \overbrace{\Big(\gamma^{(0)}_{\alpha\beta}e^{2\rho} + \gamma^{(1)}_{\alpha\beta}e^{\rho} + \gamma^{(2)}_{\alpha\beta} + \cdots\Big)}^{\gamma_{\alpha\beta}}\mathrm{d}x^\alpha \mathrm{d}x^\beta\Big]\,, \tag{12.9}$$

where ρ is the "radial" coordinate and x^α the "boundary coordinates" (for instance, light-cone coordinates $x^\pm$). The equations of motion (12.4) for the choices $\gamma^{(0)}_{+-} = \frac{1}{2}$, $\gamma^{(0)}_{\pm\pm} = 0$ and $\gamma^{(1)}_{--} = 0$ impose the restrictions

$$\gamma^{(2)}_{++} = \mathscr{L}(x^+)\,, \qquad \gamma^{(2)}_{--} = \bar{\mathscr{L}}(x^-) \quad \text{and} \quad \partial_-^2\gamma^{(1)}_{++} = \gamma^{(1)}_{++}\gamma^{(2)}_{--}\,. \tag{12.10}$$

The most general variation of the line-element that we permit is

$$\delta\left(\mathrm{d}s^2\right) = e^{2\phi}\left(2\delta\phi\,\mathrm{d}\rho^2 + [2\gamma_{\alpha\beta}\,\delta\phi + \delta\gamma_{\alpha\beta}]\,\mathrm{d}x^\alpha \mathrm{d}x^\beta\right)\,. \tag{12.11}$$

which indicates different scenarios in deforming the boundary metric, namely the *trivial*, *fixed* and *free* Weyl factor, $\phi = 0$, $\delta\phi = 0$ and $\delta\phi \neq 0$, respectively. Here we consider the last case with $\phi = f(x^+, x^-)$ (for possible radial dependence see [20]). After adding a suitable boundary term for having a well-defined variational principle, the full on-shell variation of the action reads

$$\delta S_{\mathrm{CSG}}\big|_{\mathrm{EOM}} = \frac{1}{2}\int_{\partial M} \mathrm{d}^2x\sqrt{-\gamma^{(0)}}\left(T^{\alpha\beta}\,\delta\gamma^{(0)}_{\alpha\beta} + J^{\alpha\beta}\,\delta\gamma^{(1)}_{\alpha\beta}\right). \tag{12.12}$$

The response functions $T^{\alpha\beta}$ and $J^{\alpha\beta}$ are Brown–York stress tensor and partially massless response with conformal weights $\Delta = 2$ and $\Delta = 1$, respectively, whose non-zero components are given by

$$T_{\pm\pm} = \mp\frac{k}{\pi}\left(\gamma^{(2)}_{\pm\pm} + \tfrac{1}{2}\partial^2_\pm f\right) \tag{12.13}$$

$$J_{++} = \frac{k}{2\pi}\,\gamma^{(1)}_{++} \quad \text{with} \quad \left(\partial^2_- - \tfrac{\pi}{k}T_{--}\right)J_{++} = 0\,. \tag{12.14}$$

For the BTZ black hole [29] we obtain

$$M_{\mathrm{BTZ}} = 2kr_+r_-\,, \qquad\qquad J_{\mathrm{BTZ}} = k(r_+^2 + r_-^2)\,, \tag{12.15}$$

where $|r_+| \geq |r_-|$ are the inner and outer horizon radii, respectively (with the usual definitions of mass, $M = -\int \mathrm{d}\varphi T^t_t$, and angular momentum, $J = -\int \mathrm{d}\varphi T^t_\varphi$ where $x^\pm = t \pm \varphi$). As compared to EHG the role of mass and angular momentum is exchanged: for real $r_\pm$ the angular momentum J_{BTZ} is non-negative, whereas the mass M_{BTZ} can have either sign, exactly like in "exotic" gravity theories [30].

The asymptotic Weyl factor $\phi = f$ gives in general a contribution to the asymptotic charges, since CSG is only invariant under diffeomorphism and Weyl rescaling up to a boundary term. Conservation of the corresponding charges in turn requires cancellation of these anomalies by imposing the following conditions on the Weyl factor and its variation,

$$\partial_+\partial_- f = 0\,, \qquad\qquad \partial_t(f\,\partial_\varphi\delta f) = \text{total }\ \varphi\text{-derivative}\,. \tag{12.16}$$

Particularly simple choices are $f = f(x^+)$ or $f = f(x^-)$. The non-vanishing 2-point functions are given by ($z = \varphi + it$):

$$\langle J_{++}(z,\bar z)J_{++}(0,0)\rangle = \frac{2k\,\bar z}{z^3} \tag{12.17}$$

$$\langle T_{++}(z)T_{++}(0)\rangle = \frac{6k}{z^4} = -\langle T_{--}(\bar z)T_{--}(0)\rangle \tag{12.18}$$

These results show that one of the conformal weights of the partially massless mode is negative, $\bar h = -1/2$. This is precisely the conformal weight required for a semi-classical null state at level 2 [20], which is indeed reproduced on the gravity side through a 1-loop ghost determinant [31]. We can also read off the central charges of the dual CFT,

$$c = -\bar{c} = 12k\,. \tag{12.19}$$

In order to be explicit about the derivation of the asymptotic symmetry algebra, we now move to the first order formulation where CSG can be written in terms of three Lorentz valued variables (note that $k_{\text{CS}} = 2k$ here), e, ω and λ.

$$S^{(1)}_{\text{CSG}} = \frac{k_{\text{CS}}}{4\pi}\int_{\mathcal{M}} \text{tr}\left(\omega \wedge \left(d\omega + \tfrac{2}{3}\omega \wedge \omega\right) - 2\lambda \wedge T\right) \tag{12.20}$$

The spin-connection is solved in terms of the dreibein $\omega = \omega(e)$ by the torsion constraint, $T = de + e \wedge \omega = 0$, variation with respect to ω solves the Lagrange multiplier as $\lambda = S(e)$, where S is the Schouten one-form, and variation with respect to e gives the same field equation as in the metric formulation, $C(e) = 0$ where C is the Cotton one-form. It has been shown by Horne and Witten [26] that considering these variables (e, ω and λ) as gauge fields along translation, rotation and special conformal transformation generators and adding a Stückelberg field ϕ along the dilatation,

$$A_\mu = e^a{}_\mu P_a + \omega^a{}_\mu J_a + \lambda^a{}_\mu K_a + \phi_\mu D\,, \tag{12.21}$$

this action can be written as a Chern–Simons theory based on the SO(3,2) gauge group.

We exploit now the Chern–Simons formulation for canonically and asymptotically analyzing CSG. The fact that SO(3,2) contains SO(2,2) as a subgroup, suggests that we can study AdS boundary conditions in this setup.[1] Introducing the following state dependent one forms,

$$\begin{aligned} t^0 &= T_1 \text{d}t - T_2 \text{d}\varphi\,, \qquad t^1 = T_1 \text{d}\varphi - T_2 \text{d}t \qquad \text{and} \\ p^0 &= P_2 \text{d}t - P_1 \text{d}\varphi\,, \qquad p^1 = P_1 \text{d}t - P_2 \text{d}\varphi\,, \qquad p^2 = P_3(\text{d}t + \text{d}\varphi)\,, \end{aligned} \tag{12.22}$$

we present the AdS boundary conditions as follows [28],

$$\begin{aligned} e^0 &= -\ell e^f \left(e^\rho \text{d}t - p^0 + t^0 e^{-\rho}\right), \qquad e^1 = -\ell e^f \left(e^\rho \text{d}\varphi - p^1 - t^1 e^{-\rho}\right), \\ e^2 &= -\ell e^f \left(\text{d}\rho - p^2 e^{-\rho}\right), \\ \lambda^0 &= \tfrac{1}{2\ell} e^{-f} \left(e^\rho \text{d}t + p^0 + t^0 e^{-\rho}\right), \qquad \lambda^1 = \tfrac{1}{2\ell} e^{-f} \left(e^\rho \text{d}\varphi + p^1 - t^1 e^{-\rho}\right), \end{aligned}$$

[1] The same statement holds for SO(3,1), ISO(2,1) and SO(2,1)×$\mathbb{R}$ as subgroups of SO(3,2) corresponding to de Sitter, Flat and Lobachevsky boundary conditions [28].

$$\lambda^2 = \tfrac{1}{2\ell} e^{-f}\left(\mathrm{d}\rho + p^2 e^{-\rho}\right),$$
$$\omega^0 = e^\rho\,\mathrm{d}\varphi + t^1 e^{-\rho}, \qquad \omega^1 = e^\rho\,\mathrm{d}t - t^0 e^{-\rho},$$
$$\omega^2 = 0, \qquad \phi = \mathrm{d}f(t,\varphi) - p^2 e^{-\rho}.$$

Solving the flatness conditions (12.2) we find, $(\partial := \partial_+, \bar\partial := \partial_-)$

$$T_1 = -\tfrac{1}{2}(\mathcal{L}(x^+) - \bar{\mathcal{L}}(x^-)), \qquad T_2 = \tfrac{1}{2}(\mathcal{L}(x^+) + \bar{\mathcal{L}}(x^-)),$$
$$P_1 = -P_2 = \mathcal{P}(t,\varphi), \qquad P_3 = \bar\partial\mathcal{P}, \qquad \left(\bar{\mathcal{L}} - \bar\partial^2\right)\mathcal{P} = 0. \tag{12.23}$$

These are the analogue of (12.10). A general Lie algebra-valued generator of gauge transformations is

$$\varepsilon = \rho^a P_a + \tau^a J_a + \sigma^a K_a + \gamma D. \tag{12.24}$$

The boundary conditions given in (12.23) are preserved by gauge transformations (12.24) when,

$$\rho^0 = \ell e^f\left(a_2 e^\rho + (a_1+a_2)\mathcal{P} + a_4 e^{-\rho}\right), \quad \sigma^0 = -\tfrac{1}{2\ell} e^{-f}\left(a_2 e^\rho - (a_1+a_2)\mathcal{P} + a_4 e^{-\rho}\right),$$
$$\rho^1 = \ell e^f\left(a_1 e^\rho - (a_1+a_2)\mathcal{P} + a_3 e^{-\rho}\right), \quad \sigma^1 = -\tfrac{1}{2\ell} e^{-f}\left(a_1 e^\rho + (a_1+a_2)\mathcal{P} + a_3 e^{-\rho}\right),$$
$$\rho^2 = -\ell e^f\left(\partial_\varphi a_1 + d_1 e^{-\rho}\right), \qquad \sigma^2 = \tfrac{1}{2\ell} e^{-f}\left(\partial_\varphi a_1 - d_1 e^{-\rho}\right),$$
$$\tau^0 = -a_1 e^\rho + a_3 e^{-\rho}, \quad \tau^1 = -a_2 e^\rho + a_4 e^{-\rho}, \quad \tau^2 = \partial_\varphi a_2, \quad \gamma = \Omega + d_1 e^{-\rho}.$$

where the following relations should hold,

$$a_2 = -\tfrac{1}{2}\left(\epsilon(x^+) + \bar\epsilon(x^-)\right), \quad a_1 = -\tfrac{1}{2}\left(\epsilon(x^+) - \bar\epsilon(x^-)\right), \quad d_1 = -\bar\partial\mathcal{P}\epsilon(x^+)$$
$$a_3 = T_2 a_2 - T_1 a_1 - \tfrac{1}{2}\partial_\varphi^2 a_1, \qquad a_4 = T_1 a_2 - T_2 a_1 + \tfrac{1}{2}\partial_\varphi^2 a_2. \tag{12.25}$$

The variation of the state dependent functions in (12.23) with respect to these parameters are,

$$\delta\mathcal{L} = \partial\mathcal{L}\epsilon + 2\mathcal{L}\partial\epsilon - \tfrac{1}{2}\partial^3\epsilon, \qquad \delta\bar{\mathcal{L}} = \bar\partial\bar{\mathcal{L}}\bar\epsilon + 2\bar{\mathcal{L}}\bar\partial\bar\epsilon + \tfrac{1}{2}\bar\partial^3\bar\epsilon,$$
$$\delta\mathcal{P} = \partial\mathcal{P}\epsilon + \tfrac{3}{2}\mathcal{P}\partial\epsilon + \bar\partial\mathcal{P}\bar\epsilon - \tfrac{1}{2}\mathcal{P}\bar\partial\bar\epsilon, \qquad \delta_\Omega f = \Omega, \tag{12.26}$$

which are the analogue of (12.17). The conserved charges associated to these variations are given by,

$$Q = \frac{k_{\text{cs}}}{2\pi}\int \mathrm{d}\varphi\left[\epsilon(x^+)\mathcal{L}(x^+) + \bar\epsilon(x^-)\bar{\mathcal{L}}(x^-) + \Omega(x^+)\partial_\varphi f(x^+)\right]. \tag{12.27}$$

Defining the generators of these global symmetries as,

$$L_n = \tilde{G}[\epsilon = e^{inx^+}], \qquad \bar{L}_n = \tilde{G}[\bar{\epsilon} = e^{inx^-}] \quad \text{and} \quad J_n = \tilde{G}[\Omega = e^{inx^+}], \tag{12.28}$$

we compute the Poisson brackets and convert Poisson brackets into commutators by the prescription $i\{q, p\} = [\hat{q}, \hat{p}]$. The resulting algebra is $\text{Vir} \oplus \overline{\text{Vir}} \oplus \hat{u}(1)_k$. Finally, we Sugawara-shift the quantum L generator

$$L_m \to L_m + \frac{1}{4k} \sum_{n \in \mathbb{Z}} : J_n J_{m-n} : \tag{12.29}$$

In conclusion, the asymptotic symmetry algebra has the following non-zero commutators:

$$\begin{aligned}
[L_n, L_m] &= (n-m)\, L_{n+m} + \frac{c+1}{12}\,(n^3 - n)\,\delta_{n+m,0} \\
[\bar{L}_n, \bar{L}_m] &= (n-m)\, \bar{L}_{n+m} + \frac{\bar{c}}{12}\,(n^3 - n)\,\delta_{n+m,0} \\
[L_n, J_m] &= -m\, J_{n+m} \\
[J_n, J_m] &= 2k\, n\, \delta_{n+m,0}
\end{aligned} \tag{12.30}$$

The central charges are given by $c = -\bar{c} = 12k$ with $k = k_{\text{CS}}/2$. Note the quantum shift by one in the central charge of one copy of the Virasoro algebra. This is due to the normal ordering of J's introduced in (12.29). The relative sign of two central charges is a sign of non-unitarity. This is consistent with the parity odd nature of this theory; as mentioned before, flat boundary conditions seem more suitable for unitarity in the asymptotic analysis of parity odd models. For a detailed asymptotically flat analysis of CSG as a Chern–Simons gauge theory of SO(3,2) see [28] and in the metric formulation see Sect. 12.4 and [32].

12.2.2 Higher Spin Theories

In the introduction we alluded to some similarities between CSG and higher spin theories. In this subsection we make this statement more concrete and summarize some important properties of such theories.

Even though it is easy to write down the (Fronsdal-)equations [33] for free massless higher spin fields, the coupling of the fields for spins greater than two to gravity is severely constrained by various no-go theorems (for a review see [34]). Fradkin and Vasiliev [35] showed that consistent interacting higher spin gauge theories involving gravity need to be defined on a curved background and involve an infinite tower of massless higher spin fields [36], see e.g. [37, 38] for reviews.

One interesting aspect of higher spin gauge fields is that they might be connected to string theory in the tensionless limit in which the massive excitations of string theory become massless. It is conjectured that string theory is a broken phase of a higher spin gauge theory. For more details see [39] and references therein.

Another interesting aspect is that holographic correspondences between higher spin theories and field theories can be formulated, such as the conjectured duality in the large N limit of the critical three-dimensional $O(N)$ model and the minimal bosonic higher spin theory in AdS_4 [40–42] (for a review of various impressive checks of this conjecture see [43]).

We focus now on $2+1$ dimensions where the situation simplifies significantly. An action is known [44], namely the sum of two Chern–Simons actions (12.1) with opposite levels with the gauge group $SL(N)$ which is a natural generalization of EHG and corresponds to fields of spin $s = 3, 4, \dots, N$ coupled to gravity. This consistent truncation to a finite number of higher spin fields is not possible in higher dimensions [45]. Moreover, the dual field theories are two-dimensional, which allows a high degree of analytic control.

The Brown–Henneaux type of analysis reviewed in the previous subsection generalizes to higher spin fields for asymptotic AdS_3 [11, 12, 46, 47] and leads to asymptotic $\mathscr{W}_N \times \mathscr{W}_N$ [48, 49] symmetry algebras. Using the infinite dimensional higher spin algebras $hs[\lambda] \oplus hs[\lambda]$ as gauge algebra we get gravity coupled to massless fields with spins $s = 3, 4, \dots, \infty$ and, again for AdS_3, asymptotic symmetries of the form $\mathscr{W}_\infty[\lambda] \times \mathscr{W}_\infty[\lambda]$.

Gaberdiel and Gopakumar proposed [50] that the $hs[\lambda]$ theory coupled to an additional complex scalar field on AdS_3 is dual to a specific large-N limit of $\mathscr{W}_N$ minimal models on the CFT side. The duality is reviewed in [51].

Since the BTZ black hole can also be generalized to higher spin theories, new questions arise concerning gauge invariant characterizations of observables—like in CSG there are gauge symmetries that act on the metric but are not diffeomorphisms—and black hole thermodynamics (for a review of the proposed answers see [52, 53]).

An interesting possibility that we will exhibit in the next section—first for CSG and then for higher spin theories—is to realize higher spin holography for backgrounds other than AdS_3 [17], see [18, 54–56] for explicit constructions.

12.3 Lobachevsky Holography

Lobachevsky holography refers to asymptotic expansions of the line-element of the form

$$\mathrm{d}s^2 = \pm \mathrm{d}t^2 + \mathrm{d}\rho^2 + \sinh^2 \rho \mathrm{d}\varphi^2 + \dots \tag{12.31}$$

where the ellipsis refers to suitable expressions subleading as $\rho \to \infty$. Without subleading terms the line-element (12.31) describes a direct product manifold of the two-dimensional Lobachevsky plane $\mathbb{H}_2$ (famously depicted by M.C. Escher in his paintings "Circle Limits") and a line or S^1 corresponding to the time-direction (with upper sign: Euclidean time). Which subleading expressions are "suitable" depends on the specific theory.

In [57] boundary conditions suitable for CSG were formulated and their consistency was checked. Performing the Brown–Henneaux type of analysis reviewed in Sect. 12.2.1 then leads to the asymptotic symmetry algebra

$$\begin{aligned}
[L_n, L_m] &= (n-m)\, L_{n+m} + \frac{c}{12}\,(n^3-n)\,\delta_{n+m,0} \\
[L_n, J_m] &= -m\, J_{n+m} \\
[J_n, J_m] &= 2k\, n\, \delta_{n+m,0}\,.
\end{aligned} \tag{12.32}$$

The value of the central charge, $c = 24k$, is compatible with the limiting case of warped AdS holography [58]. The algebra above is similar to the AdS asymptotic symmetry algebra (12.30), with the following differences: there is no second copy of the Virasoro algebra and no quantum shift by one in the central charge. The appearance of a single Virasoro algebra and a $\hat{u}(1)$ current algebra suggests that the dual field theory, if it exists, is a warped CFT [59]. Some checks and aspects of this proposal—consistency of canonical charges, one-loop partition function, identification of non-perturbative states, aspects of the Lobachevsky $\leftrightarrow$ field theory map—are discussed in [57], but many open issues remain (some of which are also mentioned in that paper).

Amusingly, the higher spin side of the Lobachevsky story seems more straightforward, so let us switch now to higher spin theories. The first explicit example of non-AdS holography was worked out in [18] for spin-3 gravity (for more details see [54]). In this example one considers a bulk metric that is asymptotically $\mathbb{H}_2 \times \mathbb{R}$. In order to succeed it is crucial that the embedding of $\mathfrak{sl}(2)$ into $\mathfrak{sl}(3)$ yields at least one singlet under the $\mathfrak{sl}(2)$. Otherwise it turns out that one cannot reproduce the correct $\mathrm{d}t^2$ term in the line-element (12.31). The unique viable choice for spin-3 gravity is then the non-principal embedding of $\mathfrak{sl}(2)$ into $\mathfrak{sl}(3)$ (also called diagonal embedding). In this way we reproduce (12.31) (up to subleading terms) in the limit $\rho \to \infty$.

Besides the $\mathfrak{sl}(2)$ part given by the generators L_i with $i = 0, \pm 1$ this embedding contains the singlet S and "colored" doublets $\psi_j^{\pm}$ with $j = \pm\frac{1}{2}$. We write the connections as

$$a_\mu = \hat{a}_\mu^{(0)} + a_\mu^{(0)} + a_\mu^{(1)} \quad \text{and} \quad \bar{a}_\mu = \hat{\bar{a}}_\mu^{(0)} + \bar{a}_\mu^{(0)} + \bar{a}_\mu^{(1)}\,. \tag{12.33}$$

One set of connections reproducing (12.31) in the large ρ limit is given by

$$\hat{a}_\rho^{(0)} = L_0, \quad \hat{a}_\varphi^{(0)} = -\frac{1}{4} L_1, \quad \hat{\bar{a}}_\rho^{(0)} = -L_0, \quad \hat{\bar{a}}_\varphi^{(0)} = -L_{-1}, \quad \hat{\bar{a}}_t^{(0)} = \sqrt{3} S \tag{12.34a}$$

$$a_{\varphi}^{(0)} = \frac{2\pi}{k}\left(\frac{3}{2}\mathscr{W}_0(\varphi)S + \mathscr{W}_{\frac{1}{2}}^{+}(\varphi)\psi_{-\frac{1}{2}}^{+} - \mathscr{W}_{\frac{1}{2}}^{-}(\varphi)\psi_{-\frac{1}{2}}^{-} - \mathscr{L}(\varphi)L_{-1}\right), \tag{12.34b}$$

$$\bar{a}_{\varphi}^{(0)} = \frac{2\pi}{k}\left(\frac{3}{2}\bar{\mathscr{W}}_0(\varphi)S + \bar{\mathscr{W}}_{\frac{1}{2}}^{+}(\varphi)\psi_{-\frac{1}{2}}^{+} + \bar{\mathscr{W}}_{\frac{1}{2}}^{-}(\varphi)\psi_{-\frac{1}{2}}^{-} + \bar{\mathscr{L}}(\varphi)L_{-1}\right), \tag{12.34c}$$

$$\hat{a}_{t}^{(0)} = a_{\rho}^{(0)} = a_{t}^{(0)} = \bar{a}_{\rho}^{(0)} = \bar{a}_{t}^{(0)} = 0, \tag{12.34d}$$

$$a_{\mu}^{(1)} = \mathscr{O}(e^{-2\rho}) = \bar{a}_{\mu}^{(1)}, \tag{12.34e}$$

where the $\hat{a}_{\mu}^{(0)}$ ($\hat{\bar{a}}_{\mu}^{(0)}$) describe the part of the connection that reproduces the background, $a_{\mu}^{(0)}$ ($\bar{a}_{\mu}^{(0)}$) state dependent fluctuations that are of leading order for large ρ and $a_{\mu}^{(1)}$ ($\bar{a}_{\mu}^{(1)}$) are subleading terms.

As in the example in Sect. 12.2.1, in order to check whether or not the boundary conditions lead to interesting physics one has to find gauge transformations that preserve these boundary conditions and check that the resulting canonical boundary charge is finite at the boundary, nontrivial and conserved in time. After having determined a canonical boundary charge which satisfies these conditions one can determine the asymptotic symmetry algebra on the level of Poisson brackets. One can then replace $i\{\cdot,\cdot\} \to [\cdot,\cdot]$ and expand the fields appearing in (12.34) in terms of their Fourier modes in order to obtain the (semi-classical) symmetry algebra which determines essential properties of the dual quantum field theory.

In the case of the boundary conditions (12.34) the asymptotic symmetry algebra obtained this way consists of one copy of the semi-classical (large values of k_{CS}) $\mathscr{W}_3^{(2)}$ algebra, also known as Polyakov-Bershadsky Algebra [60, 61] and one copy of an affine $\hat{\mathfrak{u}}(1)$ algebra. This is the anticipated spin-3 generalization of the CSG result (12.32).

Since the $\mathscr{W}_3^{(2)}$ algebra is an infinite dimensional, non-linear, centrally extended algebra one has to introduce normal ordering prescription for the non-linear terms if we are interested in the regime where k_{CS} is of order one, i.e., in the quantum regime. The structure constants of the $\mathscr{W}_3^{(2)}$ algebra are functions of k_{CS}. Hence one has to check whether or not the algebra still satisfies the Jacobi identities after introducing normal ordering. And indeed, in order to be compatible with the Jacobi identities, some of the structure constants and the central charges obtain $\mathscr{O}(1)$ corrections in the quantum regime. The final result for the asymptotic symmetry algebra for connections obeying (12.34) is $\mathscr{W}_3^{(2)} \oplus \hat{\mathfrak{u}}(1)$.

After having found the quantum asymptotic symmetry algebra of spacetimes that are asymptotically $\mathbb{H}_2 \times \mathbb{R}$ one can also ask whether or not there are unitary representations of this algebra. In the case of Lobachevsky holography it is surprisingly easy to answer this question. There is only one value of the Chern Simons level k_{CS} where it is possible to obtain nontrivial unitary representations [18, 54]. The reason why this question is so easy to answer in this case is because the states that correspond to descendants of the "colored" doublet have to be absent, otherwise those states would always have norms with opposite signs spoiling unitarity. This leaves only two possible values of the level k_{CS} with only one of them leading to a nontrivial theory, which can be interpreted as the theory of a free boson with a coupling constant fixed by an additional gauge symmetry. The generalization of the unitarity discussion to the full $\mathscr{W}_N^{(2)}$ family is more involved, particularly for even N [62].

12.4 Flat Space Holography

The constructions reviewed above are all similar at a technical level. This has two reasons. First, we were always dealing with some Chern–Simons theory (12.1) supplemented by suitable boundary conditions (finding the latter was the main non-trivial task). Second, we were almost exclusively concerned with asymptotic symmetry algebras and did not specify in detail the precise field theory that is supposed to be dual to a given gravitational or higher spin theory, other than that it has to fall into representations of the corresponding asymptotic symmetry algebra (given that all these symmetry algebras are infinite dimensional and have specific values of the central charges predicted from the gravity calculation this puts already a lot of constraints on the dual two-dimensional field theory). In addition, all the constructions above referred to some curved asymptotic background.

In this section we go beyond this basic scenario, by allowing for non-topological theories like topologically massive gravity, by attempting to establish a more precise holographic correspondence to specific field theories, and by studying backgrounds that are locally and asymptotically flat. In Sect. 12.4.1 we review attempts to establish precise holographic correspondences between AdS quantum gravity and specific CFTs, before addressing the flat case in Sect. 12.4.2, where we shall come back to our starting point, CSG.

12.4.1 Introduction to Three-Dimensional Quantum Gravity in AdS

Quantum gravity is a notoriously difficult subject. As such, one strategy to tackle it is to consider toy models capturing some of its salient features. EHG in (2+1)-dimensions has emerged over the years as an archetypical model for quantum gravity in general, and AdS/CFT in particular. It differs in important respects from its (3+1)-dimensional counterpart: it has no bulk propagating degrees of freedom, and any solution to the equations of motion has constant curvature (i.e. is flat for vanishing cosmological constant $\Lambda := -1/\ell^2$; for reviews, see e.g. [63–65], and [66] p. 29 for a chronological list of references). Despite the remarkable observation that three-dimensional gravity could itself be formulated as a Chern–Simons theory of the form (12.1) [9, 10, 67] with a gauge group depending on Λ, it appeared at first sight too simple to be able to address the conundrums of quantum gravity. The situation changed dramatically through a series of seminal contributions in the negatively curved case $\Lambda < 0$ of which we cite three hereafter.

First, even though there are no *bulk* degrees of freedom, the presence of an asymptotic boundary in AdS_3 induces *boundary* degrees of freedom [64]. In particular, the phase space of AdS_3 gravity admits a non-trivial action of the two-dimensional conformal group with two sets of non-trivial Virasoro charges $L_n^{\pm}$ and non-vanishing central charge given by $c^{\pm} = \frac{3\ell}{2G}$. This appeared as the first hint of a deep connection between a gravity theory in AdS space and a conformal field theory in one dimension less.

Second, the AdS_3 phase space happens to contain black hole solutions, the BTZ black holes [68, 69] with the exciting prospect of addressing questions related to black hole physics in a simplified setting.

Third, assuming the existence of a dual CFT_2 of which BTZ black holes are particular thermal states, the BTZ Bekenstein-Hawking entropy could be reproduced by a counting of states using the Cardy formula[70].

Despite these striking and suggestive results, the precise nature of the corresponding dual CFT_2 (in pure gravity) remained elusive for another 10 years. In 2007, Witten revisited the subject and made a concrete proposal for the partition function of the CFT dual to pure three-dimensional gravity [71]. Assuming holomorphic factorization (motivated partially by the relation to Chern–Simons theory), he argued from the BTZ spectrum in AdS_3 gravity that the holomorphic part of the partition function should take the form (with $k = c/24$ quantized to integers)

$$Z(q) = \sum_{r=0}^{k} a_r J(q)^r, \quad J(q) = \frac{1}{q} + 196884q + \cdots \tag{12.35}$$

where $J(q)$ is the unique modular-invariant function on the upper half plane, which is holomorphic away from a single pole at the cusp. Therefore, the requirement that the partition function be of the form

$$Z(q) = Z_0(q) + O(q), \quad Z_0(q) = q^{-k} \prod_{n=2}^{\infty} \frac{1}{1-q^n}, \tag{12.36}$$

where $Z_0(q)$ captures the vacuum descendants and the "$O(q)$" piece the BTZ black holes (having $L_0 > 0$), uniquely fixes the form of the partition function. CFTs with partition functions (12.35) are called *extremal*, roughly because they have as few low-lying primaries as possible compatible with modular invariance, and display remarkable group- and number-theoretic properties.

It happens that AdS_3 gravity is simple enough that the quantum gravity partition function can be explicitly calculated as a sum over geometries. Maloney and Witten performed this computation [72] and found out that the result could *not* be interpreted as a CFT partition function, i.e., as a trace over some CFT Hilbert space. They concluded that either pure gravity in 2+1 dimensions simply did not exist quantum mechanically, or that additional contributions should be included. At any rate, the quantity they computed did not holomorphically factorize, thereby violating one of the assumptions of [71].

An alternative emerged few months later under the name *chiral gravity* [73, 74]. The idea was to modify pure gravity by supplementing if with the gravitational Chern–Simons term (12.3). The resulting theory is called *Topologically Massive Gravity* (TMG) [13, 75] with action

$$S_{\text{TMG}} = \frac{1}{16\pi}\int d^3x\sqrt{-g}\,\big(R+\frac{2}{\ell^2}\big) - \frac{1}{8k\mu}\,S_{\text{CSG}}\,. \tag{12.37}$$

One effect of the additional term (12.3) is to shift the values of the (asymptotically) conserved charges as compared to EHG. For Brown–Henneaux boundary conditions [76]

$$\Delta g_{rr} = \frac{f_{rr}}{r^4} + \mathcal{O}(\frac{1}{r^5}) \qquad \Delta g_{r\pm} = \frac{f_{r\pm}}{r^3} + \mathcal{O}(\frac{1}{r^4}) \qquad \Delta g_{\pm\pm} = f_{\pm\pm} + \mathcal{O}(\frac{1}{r}) \tag{12.38}$$

the corresponding Virasoro charges are given by

$$L_n^{\pm} = \frac{2}{\ell}\left(1 \pm \frac{1}{\mu\ell}\right)\int e^{inx^{\pm}}\, f_{\pm\pm}d\phi \tag{12.39}$$

with the corresponding central extensions [77]

$$c^{\pm} = \left(1 \pm \frac{1}{\mu\ell}\right)\frac{3\ell}{2G}\,. \tag{12.40}$$

Therefore, at the critical point $\mu\ell = 1$, one copy of the Virasoro algebra has vanishing central charge. If the theory is unitary then it must be chiral and one is left with a single copy of the Virasoro algebra. Alternatively, if the theory is non-unitary one encounters the structure of a specific type of logarithmic CFT where one chiral part of the stress tensor acquires a logarithmic partner [23, 24]. In the former case, holomorphic factorization would be explicitly implemented in the resulting theory, dubbed "chiral gravity" [78] (see also [79]). Chiral gravity (which could exist as a unitary truncation of the non-unitary logarithmic CFT that is dual to TMG at the critical point $\mu\ell = 1$) therefore appears as a candidate for the simplest and potentially solvable model including quantum black holes.

12.4.2 Flat Space Chiral Gravity

The above considerations regarded gravity theories with a negative cosmological constant. Could a similar logic be used to argue that flat space could be dual to a field theory of some kind? And if yes, what could it be?

It is tempting to use as guiding principle the ingredients that led to the first glimpses of AdS/CFT: asymptotic symmetries. The first caveat is that the asymptotic structure of flat space is more involved than that of AdS spaces (see e.g. [80]). However, the structure of its various asymptotic symmetry groups has been studied over the years, starting with [81]. For the case that will interest us in the following, the asymptotic symmetries of (2+1)-dimensional gravity at null infinity form the so-called BMS_3 algebra [82], with commutation relations

$$[L_m, L_n] = (m-n)\, L_{n+m} + \frac{c_1}{12}(n^3-n)\,\delta_{n+m,0} \tag{12.41a}$$

$$[L_m, M_n] = (m-n)\, M_{n+m} + \frac{c_2}{12}(n^3-n)\,\delta_{n+m,0} \tag{12.41b}$$

$$[M_m, M_n] = 0 \tag{12.41c}$$

It is generated by Virasoro generators L_n and supertranslations M_n. The latter are the modes of diffeomorphisms preserving the following boundary conditions at null infinity [32]:

$$g_{uu} = h_{uu} + O(\tfrac{1}{r}) \qquad g_{ur} = -1 + h_{ur}/r + O(\tfrac{1}{r^2}) \tag{12.42a}$$

$$g_{u\theta} = h_{u\theta} + O(\tfrac{1}{r}) \qquad g_{rr} = h_{rr}/r^2 + O(\tfrac{1}{r^3}) \tag{12.42b}$$

$$g_{r\theta} = h_1(\theta) + h_{r\theta}/r + O(\tfrac{1}{r^2}) \tag{12.42c}$$

$$g_{\theta\theta} = r^2 + (h_2(\theta) + uh_3(\theta))r + O(1) \tag{12.42d}$$

The flat counterpart of (12.39) is then given by

$$M_n = \frac{1}{16\pi G}\int \mathrm{d}\theta\, e^{in\theta}\,\big(h_{uu} + h_3\big) \tag{12.43a}$$

$$L_n = \frac{1}{16\pi G\mu}\int \mathrm{d}\theta\, e^{in\theta}\,\big(h_{uu} + h_3\big) + \frac{1}{16\pi G}\int \mathrm{d}\theta\, e^{in\theta} \times\big(inuh_{uu} + inh_{ur} + 2h_{u\theta} + \partial_u h_{r\theta} - h_3 h_1 - in\partial_\theta h_1\big) \tag{12.43b}$$

and the central extensions in (12.41) are computed as[32][2]

$$c_1 = \frac{3}{\mu G}, \qquad c_2 = \frac{3}{G}. \tag{12.44}$$

The phase space defined by the boundary conditions (12.42) contains an interesting two-parameter family of solutions recognized some time ago as the shift-boost orbifold of flat space [85]:

[2] It should be straightforward to generalize these results to other massive gravity theories like "new" massive gravity [83, 84].

$$ds^2 = 8m\,du^2 - 2dr du + 8j\,d\theta du + r^2 d\theta^2. \tag{12.45}$$

They represent cosmological solutions (here expressed in Eddington–Finkelstein coordinates)—in particular, they have a cosmological horizon, an associated Bekenstein–Hawking entropy and a Hawking temperature [86, 87]. We therefore have a classical phase space endowed with an action of an infinite-dimensional BMS_3 symmetry, and by analogy with the AdS_3 situation, one could expect that upon quantization states will form representation of that algebra, i.e. quantum gravity in flat space would be related to a BMS_3-invariant field theory. Although some hints in this direction have been given, it is fair to say these types of field theories remain relatively unexplored. Some aspects of the representation theory have been discussed in [88–92]. What is lacking as opposed to the exhaustive study of two-dimensional CFTs is the presence of concrete examples of such field theories. We review now a first concrete example of holography in flat spacetimes.

To this end, there is a limit that make our lives easier. Consider

$$\mu \to 0\,, \quad G \to \infty \quad \text{keeping} \quad \mu G := \frac{1}{8k} \quad \text{finite.} \tag{12.46}$$

In that limit, the M_n charges become trivial, the central term c_2 vanishes and the BMS_3 algebra reduces to a single copy of a Virasoro algebra! This can be further checked by looking at null vectors in the field theory and observing that in the above limit, there is indeed a consistent truncation of the representations of the algebra (12.41) to simply the Virasoro module [32]. On the bulk side, the Bekenstein-Hawking entropy of the above solutions (taking into account the Chern–Simons contribution [77, 93–95]) is

$$S = 8\pi k\sqrt{2m} = 2\pi\sqrt{\frac{c_1 L_0}{6}} \tag{12.47}$$

i.e., precisely a chiral half of the Cardy formula. This provides a check on the correctness of flat space holography.

One can go further. The vacuum flat space solution lies in (12.45) for $m = -\frac{1}{8}$ and $j = 0$, i.e., for $L_0 = -k = -\frac{c}{24}$, while the cosmological solutions have $L_0 > 0$. The spectrum therefore share strong similarities with that of AdS_3 gravity, as there is a gap between the vacuum and the first primary state. One can then follow the same reasoning as Witten, arguing that modular invariance uniquely fixes the partition function to be of the form (12.35). As a consequence, we can proceed with a comparison analogous to the one done in [71] for BTZ black holes. Consider a cosmological solution with $L_0 = 1$, at $k = 1$. Its (semi-classical) entropy is $S_{\text{BH}} = 4\pi \sim 12.57$. On the other hand, in the expansion (12.35), 196884 is the total number of states with $L_0 = 1$, representing one descendant of the vacuum state and 198883 primaries creating the corresponding cosmological solution. The entropy is thus $S_{\text{CFT}} = \ln 196883 \sim 12.19$, which matches with the geometrical entropy within a few percents (perfect agreement was not be expected since the semi-classical entropy is valid for large k and we used $k = 1$; the agreement gets better as k increases). This leads us to conjecture that CSG with the above boundary conditions—a theory which we call flat space chiral gravity—is dual to a chiral CFT with $c = 24k$.

This conjecture can be sharpened by further arguments, which we now present. The presence of the finite sized gap leads to the expectation that the dual CFT is an extremal CFT with $c = 24k$. An important caveat is that such CFTs need not exist for arbitrary values of k [96, 97], but at least for $k = 1$ the extremal CFT that could serve as a gravity dual has been previously identified by Witten [71] as the Monster CFT [98]. So we can sharpen our conjecture to the following [32]:
Flat space chiral gravity at Chern–Simons level $k = 1$ is dual to the Monster CFT.

Acknowledgements HA was supported by the Dutch stichting voor Fundamenteel Onderzoek der Materie (FOM). AB was supported by an INSPIRE award of the Department of Science and Technology, India. SD is a Research Associate of the Fonds de la Recherche Scientique F.R.S.-FNRS (Belgium). DG and SP were supported by the START project Y 435-N16 of the Austrian Science Fund (FWF) and the FWF projects I 952-N16 and I 1030-N27. MR was supported by the Austrian Science Fund (FWF) and the FWF project I 1030-N27. DG dedicates these proceedings contributions to the memory of his grandmother Gerda.

References

1. G.V. Dunne, Les Houches Summer School in Theoretical Physics. (1998)
2. M. Marino, Rev. Mod. Phys. **77**, 675 (2005). DOI 10.1103/RevModPhys.77.675
3. J. Zanelli, Proceedings of the 7th Mexican Workshop on Particles and Fields. (1999)
4. J. Zanelli, AIP Conf. Proc. **1031**, 115 (2008). DOI 10.1063/1.2971999
5. M. Marino, J. Phys. **A44**, 463001 (2011). DOI 10.1088/1751-8113/44/46/463001
6. J. Zanelli, Class. Quant. Grav. **29**, 133001 (2012). DOI 10.1088/0264-9381/29/13/133001
7. E.A. Bergshoeff, O. Hohm, W. Merbis, A.J. Routh, P.K. Townsend, (2014)
8. D. Birmingham, M. Blau, M. Rakowski, G. Thompson, Phys. Rept. **209**, 129 (1991)
9. A. Achucarro, P.K. Townsend, Phys. Lett. **B180**, 89 (1986)
10. E. Witten, Nucl. Phys. **B311**, 46 (1988)
11. M. Henneaux, S.J. Rey, JHEP **1012**, 007 (2010). DOI 10.1007/JHEP12(2010)007
12. A. Campoleoni, S. Fredenhagen, S. Pfenninger, S. Theisen, JHEP **1011**, 007 (2010). DOI 10.1007/JHEP11(2010)007
13. S. Deser, R. Jackiw, S. Templeton, Phys. Rev. Lett. **48**, 975 (1982)
14. S. Deser, R. Jackiw, S. Templeton, Ann. Phys. **140**, 372 (1982)
15. S. Deser, R. Jackiw, S. Templeton, Erratum-ibid. **185**, 406 (1988)
16. A. Garcia, F.W. Hehl, C. Heinicke, A. Macias, Class. Quant. Grav. **21**, 1099 (2004)
17. M. Gary, D. Grumiller, R. Rashkov, JHEP **1203**, 022 (2012). DOI 10.1007/JHEP03(2012)022
18. H. Afshar, M. Gary, D. Grumiller, R. Rashkov, M. Riegler, JHEP **1211**, 099 (2012). DOI 10.1007/JHEP11(2012)099
19. H. Afshar, B. Cvetkovic, S. Ertl, D. Grumiller, N. Johansson, Phys. Rev. **D84**, 041502(R) (2011). DOI 10.1103/PhysRevD.84.041502
20. H. Afshar, B. Cvetkovic, S. Ertl, D. Grumiller, N. Johansson, Phys. Rev. **D85**, 064033 (2012)
21. G. Guralnik, A. Iorio, R. Jackiw, S.Y. Pi, Ann. Phys. **308**, 222 (2003)
22. D. Grumiller, W. Kummer, Ann. Phys. **308**, 211 (2003)
23. D. Grumiller, N. Johansson, JHEP **07**, 134 (2008). DOI 10.1088/1126-6708/2008/07/134

24. D. Grumiller, W. Riedler, J. Rosseel, T. Zojer, J. Phys. **A46**, 494002 (2013). DOI 10.1088/1751-8113/46/49/494002
25. A. Achucarro, P.K. Townsend, Phys. Lett. **B180**, 89 (1986)
26. J.H. Horne, E. Witten, Phys. Rev. Lett. **62**, 501 (1989)
27. J.D. Brown, M. Henneaux, Commun. Math. Phys. **104**, 207 (1986)
28. H.R. Afshar, JHEP **1310**, 027 (2013). DOI 10.1007/JHEP10(2013)027
29. M. Bañados, C. Teitelboim, J. Zanelli, Phys. Rev. Lett. **69**, 1849 (1992)
30. P.K. Townsend, B. Zhang, Phys. Rev. Lett. **110**(24), 241302 (2013). DOI 10.1103/PhysRevLett.110.241302
31. M. Bertin, D. Grumiller, D. Vassilevich, T. Zojer, JHEP **1106**, 111 (2011). DOI 10.1007/JHEP06(2011)111
32. A. Bagchi, S. Detournay, D. Grumiller, Phys. Rev. Lett. **109**, 151301 (2012). DOI 10.1103/PhysRevLett.109.151301
33. C. Fronsdal, Phys. Rev. **D18**, 3624 (1978). DOI 10.1103/PhysRevD.18.3624
34. X. Bekaert, N. Boulanger, P. Sundell, Rev. Mod. Phys. **84**, 987 (2012). DOI 10.1103/RevModPhys.84.987
35. E. Fradkin, M.A. Vasiliev, Phys. Lett. **B189**, 89 (1987). DOI 10.1016/0370-2693(87)91275-5
36. M.A. Vasiliev, Phys. Lett. **B285**, 225 (1992). DOI 10.1016/0370-2693(92)91457-K
37. X. Bekaert, S. Cnockaert, C. Iazeolla, M. Vasiliev, 1st Solvay Workshop on Higher Spin Gauge Theories (2004)
38. V. Didenko, E. Skvortsov, (2014). (unpublished)
39. A. Sagnotti, J. Phys. **A46**, 214006 (2013). DOI 10.1088/1751-8113/46/21/214006
40. I. Klebanov, A. Polyakov, Phys. Lett. **B550**, 213 (2002). DOI 10.1016/S0370-2693(02)02980-5
41. E. Sezgin, P. Sundell, Nucl. Phys. **B644**, 303 (2002). DOI 10.1016/S0550-3213(02)00739-3, 10.1016/S0550-3213(02)00739-3
42. E. Sezgin, P. Sundell, JHEP **0507**, 044 (2005). DOI 10.1088/1126-6708/2005/07/044
43. S. Giombi, X. Yin, JHEP **1009**, 115 (2010). DOI 10.1007/JHEP09(2010)115
44. M. Blencowe, Class. Quant. Grav. **6**, 443 (1989). DOI 10.1088/0264-9381/6/4/005
45. C. Aragone, S. Deser, Phys. Lett. **B86**, 161 (1979). DOI 10.1016/0370-2693(79)90808-6
46. M.R. Gaberdiel, T. Hartman, JHEP **1105**, 031 (2011). DOI 10.1007/JHEP05(2011)031
47. A. Campoleoni, S. Fredenhagen, S. Pfenninger, JHEP **1109**, 113 (2011). DOI 10.1007/JHEP09(2011)113
48. A. Zamolodchikov, Theor. Math. Phys. **65**, 1205 (1985). DOI 10.1007/BF01036128
49. P. Bouwknegt, K. Schoutens, Phys. Rep. **223**, 183 (1993). DOI 10.1016/0370-1573(93)90111-P
50. M.R. Gaberdiel, R. Gopakumar, Phys. Rev. **D83**, 066007 (2011). DOI 10.1103/PhysRevD.83.066007
51. M.R. Gaberdiel, R. Gopakumar, J. Phys. **A46**, 214002 (2013). DOI 10.1088/1751-8113/46/21/214002
52. M. Ammon, M. Gutperle, P. Kraus, E. Perlmutter, J. Phys. **A46**, 214001 (2013). DOI 10.1088/1751-8113/46/21/214001
53. A. Perez, D. Tempo, R. Troncoso, (2014)
54. M. Riegler, Asymptotic Symmetry Algebras in Non-Anti-de-Sitter Higher-Spin Gauge Theories. Master's thesis, Vienna University of Technology (2012)
55. H. Afshar, A. Bagchi, R. Fareghbal, D. Grumiller, J. Rosseel, Phys. Rev. Lett. **111**, 121603 (2013). DOI 10.1103/PhysRevLett.111.121603
56. H.A. Gonzalez, J. Matulich, M. Pino, R. Troncoso, JHEP **1309**, 016 (2013). DOI 10.1007/JHEP09(2013)016
57. M. Bertin, S. Ertl, H. Ghorbani, D. Grumiller, N. Johansson, et al., JHEP **1306**, 015 (2013). DOI 10.1007/JHEP06(2013)015
58. D. Anninos, W. Li, M. Padi, W. Song, A. Strominger, JHEP **03**, 130 (2009). DOI 10.1088/1126-6708/2009/03/130

59. S. Detournay, T. Hartman, D.M. Hofman, Phys. Rev. **D86**, 124018 (2012). DOI 10.1103/PhysRevD.86.124018
60. M. Bershadsky, Commun. Math. Phys. **139**, 71 (1991). DOI 10.1007/BF02102729
61. A.M. Polyakov, Int.J.Mod.Phys. **A5**, 833 (1990). DOI 10.1142/S0217751X90000386
62. H. Afshar, T. Creutzig, D. Grumiller, Y. Hikida, P.B. Ronne, JHEP **1406**, 063 (2014). DOI 10.1007/JHEP06(2014)063
63. S. Carlip, J. Korean Phys. Soc. **28**, S447 (1995)
64. S. Carlip, Class. Quant. Grav. **22**, R85 (2005)
65. S. Carlip, *Quantum Gravity in 2+1 Dimensions* (Cambridge University Press, Cambridge, 1998)
66. O.J. Campos Dias, PhD Thesis, Universidade Tecnica de Lisboa, 2009
67. E. Witten, Nucl. Phys. **B323**, 113 (1989)
68. M. Banados, C. Teitelboim, J. Zanelli, Phys. Rev. Lett. **69**, 1849 (1992)
69. M. Banados, M. Henneaux, C. Teitelboim, J. Zanelli, Phys. Rev. **D48**, 1506 (1993)
70. A. Strominger, JHEP **02**, 009 (1998)
71. E. Witten, (2007). (unpublished)
72. A. Maloney, E. Witten, JHEP **1002**, 029 (2010). DOI 10.1007/JHEP02(2010)029
73. W. Li, W. Song, A. Strominger, JHEP **0804**, 082 (2008). DOI 10.1088/1126-6708/2008/04/082
74. A. Maloney, W. Song, A. Strominger, Chiral gravity, Log gravity and Extremal CFT (2009). Draft of paper sent to the authors
75. S. Deser, R. Jackiw, S. Templeton, Ann. Phys. **140**, 372 (1982). DOI 10.1016/0003-4916(82)90164-6
76. M. Henneaux, C. Martinez, R. Troncoso, Phys. Rev. **D79**, 081502 (2009). DOI 10.1103/PhysRevD.79.081502
77. P. Kraus, F. Larsen, JHEP **01**, 022 (2006)
78. W. Li, W. Song, A. Strominger, JHEP **04**, 082 (2008). DOI 10.1088/1126-6708/2008/04/082
79. A. Maloney, W. Song, A. Strominger, Phys. Rev. **D81**, 064007 (2010). DOI 10.1103/PhysRevD.81.064007
80. R.M. Wald, *General Relativity* (The University of Chicago Press, Chicago, 1984)
81. H. Bondi, M. van der Burg, A. Metzner, Proc. Roy. Soc. Lond. **A269**, 21 (1962)
82. G. Barnich, G. Compere, Class. Quant. Grav. **24**, F15 (2007). DOI 10.1088/0264-9381/24/5/F01,10.1088/0264-9381/24/11/C01
83. E.A. Bergshoeff, O. Hohm, P.K. Townsend, Phys. Rev. Lett. **102**, 201301 (2009). DOI 10.1103/PhysRevLett.102.201301
84. E.A. Bergshoeff, O. Hohm, P.K. Townsend, Phys. Rev. **D79**, 124042 (2009). DOI 10.1103/PhysRevD.79.124042
85. L. Cornalba, M.S. Costa, Phys. Rev. **D66**, 066001 (2002). DOI 10.1103/PhysRevD.66.066001
86. A. Bagchi, S. Detournay, R. Fareghbal, J. Simon, Phys. Rev. Lett. **110**, 141302 (2013). DOI 10.1103/PhysRevLett.110.141302
87. G. Barnich, JHEP **1210**, 095 (2012). DOI 10.1007/JHEP10(2012)095
88. A. Bagchi, R. Gopakumar, JHEP **0907**, 037 (2009). DOI 10.1088/1126-6708/2009/07/037
89. A. Bagchi, Phys. Rev. Lett. **105**, 171601 (2010). DOI 10.1103/PhysRevLett.105.171601
90. G. Barnich, A. Gomberoff, H.A. Gonzalez, Phys. Rev. **D87**(12), 124032 (2013). DOI 10.1103/PhysRevD.87.124032
91. G. Barnich, B. Oblak, Class. Quant. Grav. **31**, 152001 (2014). DOI 10.1088/0264-9381/31/15/152001
92. G. Barnich, B. Oblak, JHEP **1406**, 129 (2014). DOI 10.1007/JHEP06(2014)129
93. S.N. Solodukhin, Phys. Rev. **D74**, 024015 (2006). DOI 10.1103/PhysRevD.74.024015
94. A. Bouchareb, G. Clement, Class. Quant. Grav. **24**, 5581 (2007)
95. Y. Tachikawa, Class. Quant. Grav. **24**, 737 (2007). DOI 10.1088/0264-9381/24/3/014
96. M.R. Gaberdiel, JHEP **11**, 087 (2007). DOI 10.1088/1126-6708/2007/11/087
97. M.R. Gaberdiel, C.A. Keller, JHEP **09**, 079 (2008). DOI 10.1088/1126-6708/2008/09/079
98. J. Lepowsky, I.B. Frenkel, A. Meurman, in *Unified String Theories* (Santa Barbara, 1985), pp. 533–546

Chapter 13
Gravitational Duality, Topologically Massive Gravity and Holographic Fluids

P. Marios Petropoulos

Abstract Self-duality in Euclidean gravitational set ups is a tool for finding remarkable four-dimensional geometries. From a holographic perspective, self-duality sets a relationship between two a priori independent boundary data: the boundary energy–momentum tensor and the boundary Cotton tensor. This relationship, which can be viewed as resulting from a topological mass term for gravity boundary dynamics, survives under the Lorentzian signature and provides a tool for generating exact bulk Einstein spaces carrying, among others, nut charge. In turn, the holographic analysis exhibits perfect-fluid-like equilibrium states and the presence of non-trivial vorticity allows to show that infinite number of transport coefficients vanish.

13.1 Introduction

Gravitational duality is known to map the curvature form of a connection onto a dual curvature form. It allows for constructing self-dual, four-dimensional, Euclidean-signature geometries, which are in particular Ricci-flat. Many exact solutions to Einstein's vacuum equations have been obtained in this manner, such as Taub–NUT [1], Eguchi–Hanson [2, 3], or Atiyah–Hitchin [4] gravitational instantons.

The remarkable integrability properties underlying the above constructions have created the lore that in one way or another, integrability is related with self-duality,[1]

---

[1]Quoting Ward (1985, [5]):

> ... many (and perhaps all?) of the ordinary or partial differential equations that are regarded as being integrable or solvable may be obtained from the self-duality equations (or its generalizations) by reduction.

P.M. Petropoulos (✉)
Centre de Physique Théorique, Ecole Polytechnique, CNRS, Route de Saclay,
F-91128 Palaiseau, France
e-mail: marios@cpht.polytechnique.fr

E. Papantonopoulos (ed.), *Modifications of Einstein's Theory of Gravity at Large Distances*, Lecture Notes in Physics 892, DOI 10.1007/978-3-319-10070-8_13

in a general and somewhat loose sense. In particular, this statement applies to conformal self-duality conditions, either for Kähler or for Einstein spaces, which have delivered many exact geometries (LeBrun, Fubini–Study, Calderbank–Pedersen, Przanowski–Tod, Tod–Hitchin, ... [6–19]).

Conformally self-dual spaces can be Einstein—called then quaternionic. They can be asymptotically anti-de Sitter and analyzed from a (Euclidean) holographic perspective. Hence, it is legitimate to ask (i) how self-duality reveals holographically *i.e.* on the boundary data, (ii) whether its underlying integrability properties extend to Lorentzian three-dimensional boundaries and allow to obtain exact bulk Einstein spaces, and (iii) what the physical content is for a boundary fluid emerging from such exact bulk solutions.

The aim of these lecture notes is to provide a tentative answer to the above questions. They exhibit our present understanding of the subject, as it emerges from our works [20–23]. The exact reconstruction of the bulk Einstein geometry or, equivalently, the resummability of the Fefferman–Graham expansion are achieved assuming a specific relationship among the two *a priori* independent boundary data, which are the boundary metric $g_{\mu\nu}$ and the boundary momentum $F_{\mu\nu}$ interpreted as the boundary field theory energy–momentum tensor expectation value $T_{\mu\nu}$[2]:

$$wT_{\mu\nu} + C_{\mu\nu} = 0 \, . \tag{13.1}$$

Here $C_{\mu\nu}$ is the Cotton–York tensor of the boundary geometry. In the Euclidean case, (anti-)self-duality corresponds precisely to the choice $w = \pm 3k^3/\kappa$ (k is related to the cosmological constant, $\Lambda = -3k^2$, and κ to Newton's constant, $\kappa = 3k/8\pi G_{\rm N}$). Equation (13.1) appears as the natural extension of this duality—and integrability, in the spirit of the above discussion—requirement, irrespective of the signature of the metric, with arbitrary real w. This answers questions (i) and (ii). Furthermore, the boundary condition (13.1) can be recast as

$$\frac{\delta S}{\delta g_{\mu\nu}} = 0 \tag{13.2}$$

with

$$S = S_{\rm matter} + \frac{1}{w} \int \omega_3(\gamma) \, , \tag{13.3}$$

where $S_{\rm matter}$ is the action of the holographic boundary matter and $\omega_3(\gamma)$ the Chern–Simons density (γ is the boundary connection one-form). The reader will have recognized the dynamics of matter coupled to a topological mass term for gravity [24]. Exact bulk Einstein spaces satisfying this boundary dynamics turn out to provide laboratories for probing transport properties of three-dimensional

[2]The relation is $T_{\mu\nu} = \kappa F_{\mu\nu}$, given in Eq. (13.22).

holographic fluids, and this is an important spin-off of the present analysis that will answer question (iii).

13.2 The Ancestor of Holography

We will here review some basic facts about gravitational duality and their application to the filling-in problem, which can be considered as the ancestor of holography. All this will be illustrated in the example of asymptotically AdS Schwarzschild Taub–NUT geometry.

13.2.1 Curvature Decomposition and Self-Duality

The Cahen–Debever–Defrise decomposition, more commonly known as Atiyah–Hitchin–Singer [25, 26],[3] is a convenient taming of the 20 independent components of the Riemann tensor. In Cartan's formalism, these are captured by a set of curvature two-forms $(a, b, \ldots = 0, \ldots, 3)$

$$\mathscr{R}^a{}_b = \mathrm{d}\omega^a{}_b + \omega^a{}_c \wedge \omega^c{}_b = \frac{1}{2} R^a{}_{bcd} \theta^c \wedge \theta^d \ , \tag{13.4}$$

where $\{\theta^a\}$ are a basis of the cotangent space and $\omega^a{}_b = \Gamma^a{}_{bc}\theta^c$ the set of connection one-forms. We will assume the basis $\{\theta^a\}$ to be orthonormal with respect to the metric

$$\mathrm{d}s^2 = \delta_{ab}\theta^a\theta^b \ , \tag{13.5}$$

and the connection to be torsionless and metric—this latter statement is equivalent to $\omega_{ab} = -\omega_{ba}$, where the connection satisfies

$$\mathrm{d}\theta^a + \omega^a{}_b \wedge \theta^b = 0 \ . \tag{13.6}$$

The general holonomy group in four dimensions is $SO(4)$, and (13.5) is invariant under local transformations $\Lambda(x)$ such that

$$\theta^{a\prime} = \Lambda^{-1\,a}{}_b \theta^b \ ,$$

[3] See also [27] for a review.

under which the curvature two-form transform as[4]

$$\mathscr{R}^{a\prime}_{\ \ b} = \Lambda^{-1a}_{\ \ \ \ c}\mathscr{R}^{c}_{\ d}\Lambda^{d}_{\ b} \, .$$

Both ω_{ab} and $\mathscr{R}_{ab}$ are antisymmetric-matrix-valued forms, belonging to the representation **6** of $SO(4)$.

Four dimensions is a special case as $SO(4)$ is factorized into $SO(3) \times SO(3)$. Both connection and curvature forms are therefore reduced with respect to each $SO(3)$ factor as $\mathbf{3} \times \mathbf{1} + \mathbf{1} \times \mathbf{3}$, where **3** and **1** are respectively the vector and singlet representations. The connection and curvature decomposition leads to $(\lambda, \mu, \nu, \ldots = 1, 2, 3$ and $\epsilon_{123} = 1)$:

$$\Sigma_\lambda = \frac{1}{2}\left(\omega_{0\lambda} + \frac{1}{2}\epsilon_{\lambda\mu\nu}\omega^{\mu\nu}\right), \qquad A_\lambda = \frac{1}{2}\left(\omega_{0\lambda} - \frac{1}{2}\epsilon_{\lambda\mu\nu}\omega^{\mu\nu}\right), \tag{13.7}$$

$$\mathscr{S}_\lambda = \frac{1}{2}\left(\mathscr{R}_{0\lambda} + \frac{1}{2}\epsilon_{\lambda\mu\nu}\mathscr{R}^{\mu\nu}\right), \qquad \mathscr{A}_\lambda = \frac{1}{2}\left(\mathscr{R}_{0\lambda} - \frac{1}{2}\epsilon_{\lambda\mu\nu}\mathscr{R}^{\mu\nu}\right). \tag{13.8}$$

Using this decomposition, (13.4) reads:

$$\mathscr{S}_\lambda = \mathrm{d}\Sigma_\lambda - \epsilon_{\lambda\mu\nu}\Sigma^\mu \wedge \Sigma^\nu \, , \qquad \mathscr{A}_\lambda = \mathrm{d}A_\lambda + \epsilon_{\lambda\mu\nu}A^\mu \wedge A^\nu \, . \tag{13.9}$$

Usually $\mathscr{S}$ and $\mathscr{A}$ are referred to as self-dual and anti-self-dual components of the Riemann curvature. This follows from the definition of the dual forms (supported by the fully antisymmetric symbol[5] ϵ_{abcd})

$$\tilde{\mathscr{R}}^{a}_{\ b} = \frac{1}{2}\epsilon^{a\ \ d}_{\ bc}\mathscr{R}^{c}_{\ d} \, ,$$

borrowed from Yang–Mills. Under this involutive operation, $\mathscr{S}$ remains unaltered whereas $\mathscr{A}$ changes sign. Similar relations hold for the components (Σ, A) of the connection.

Following the previous reduction pattern, the basis of 6 independent two-forms can be decomposed in terms of two sets of singlets/vectors with respect to the two $SO(3)$ factors:

[4]Note the transformation of the connection: $\omega^{a\prime}_{\ \ b} = \Lambda^{-1a}_{\ \ \ \ c}\omega^{c}_{\ d}\Lambda^{d}_{\ b} + \Lambda^{-1a}_{\ \ \ \ c}\mathrm{d}\Lambda^{c}_{\ b}$.

[5]A remark is in order here for $D = 7$ and 8. The octonionic structure constants $\psi_{\alpha\beta\gamma}$ $\alpha, \beta, \gamma \in \{1, \ldots, 7\}$ and the dual G_2-invariant antisymmetric symbol $\psi^{\alpha\beta\gamma\delta}$ allow to define a duality relation in 7 and 8 dimensions with respect to an $SO(7) \supset G_2$, and an $SO(8) \supset \mathrm{Spin}_7$ respectively. Note, however, that neither $SO(7)$ nor $SO(8)$ is factorized, as opposed to $SO(4)$.

$$\phi^\lambda = \theta^0 \wedge \theta^\lambda + \frac{1}{2}\epsilon^\lambda{}_{\mu\nu}\theta^\mu \wedge \theta^\nu ,$$

$$\chi^\lambda = \theta^0 \wedge \theta^\lambda - \frac{1}{2}\epsilon^\lambda{}_{\mu\nu}\theta^\mu \wedge \theta^\nu .$$

In this basis, the 6 curvature two-forms $\mathscr{S}$ and $\mathscr{A}$ are decomposed as

$$\begin{pmatrix} \mathscr{S} \\ \mathscr{A} \end{pmatrix} = \frac{r}{2}\begin{pmatrix} \phi \\ \chi \end{pmatrix} ,$$

where the 6×6 matrix r reads:

$$r = \begin{pmatrix} A & C^+ \\ C^- & B \end{pmatrix} = \begin{pmatrix} W^+ & C^+ \\ C^- & W^- \end{pmatrix} + \frac{s}{6}\mathbf{I}_6 . \qquad (13.10)$$

The 20 independent components of the Riemann tensor are stored inside the symmetric matrix r as follows:

- $s = \operatorname{Tr} r = 2\operatorname{Tr} A = 2\operatorname{Tr} B = R/2$ is the scalar curvature.
- The 9 components of the traceless part of the Ricci tensor $S_{ab} = R_{ab} - \frac{R}{4}g_{ab}$ ($R_{ab} = R^c{}_{acb}$) are given in $C^+ = (C^-)^{\mathrm{t}}$ as

$$S_{00} = \operatorname{Tr} C^+ , \quad S_{0\lambda} = \epsilon_\lambda{}^{\mu\nu} C^-_{\mu\nu} , \quad S_{\lambda\mu} = C^+_{\lambda\mu} + C^-_{\lambda\mu} - \operatorname{Tr} C^+ \delta_{\lambda\mu} .$$

- The 5 entries of the symmetric and traceless W^+ are the components of the self-dual Weyl tensor, while W^- provides the corresponding 5 anti-self-dual ones.

In summary,

$$\mathscr{S}_\lambda = \mathscr{W}^+_\lambda + \frac{1}{12}s\phi_\lambda + \frac{1}{2}C^+_{\lambda\mu}\chi^\mu , \qquad (13.11)$$

$$\mathscr{A}_\lambda = \mathscr{W}^-_\lambda + \frac{1}{12}s\chi_\lambda + \frac{1}{2}C^-_{\lambda\mu}\phi^\mu , \qquad (13.12)$$

where

$$\mathscr{W}^+_\lambda = \frac{1}{2}W^+_{\lambda\mu}\phi^\mu, \quad \mathscr{W}^-_\lambda = \frac{1}{2}W^-_{\lambda\mu}\chi^\mu$$

are the self-dual and anti-self-dual Weyl two-forms respectively.

Given the above decomposition, the following nomenclature is used (see e.g. [27] for details):

Einstein $C^\pm = 0$ ($\Leftrightarrow R_{ab} = \frac{R}{4}g_{ab}$)
Ricci flat $C^\pm = 0, \quad s = 0$
Self-dual $\mathscr{A} = 0 \Leftrightarrow \{W^- = 0,\ C^\pm = 0,\ s = 0\}$

Anti-self-dual $\mathscr{S} = 0 \Leftrightarrow \{W^+ = 0,\ C^\pm = 0,\ s = 0\}$
Conformally self-dual $W^- = 0$
Conformally anti-self-dual $W^+ = 0$
Conformally flat $W^+ = W^- = 0$

Quaternionic spaces are Einstein and conformally self-dual (or anti-self-dual). Conformal self-duality can also be combined with Kähler structure. In either case, remarkable integrable structures emerge.

Quaternionic conditions can be elegantly implemented by Weyl tensor introducing the *on-shell Weyl tensor*, defined as the antisymmetric-matrix-valued two-from:

$$\hat{\mathscr{W}}^{ab} = \mathscr{R}^{ab} + k^2\theta^a \wedge \theta^b \,. \tag{13.13}$$

Decomposing the latter *à la* Atiyah–Hitchin–Singer, we obtain:

$$\hat{\mathscr{W}}_\lambda^+ = \mathscr{S}_\lambda + \frac{k^2}{2}\phi_\lambda = \mathscr{W}_\lambda^+ + \frac{1}{12}\left(s + 6k^2\right)\phi_\lambda + \frac{1}{2}C^+_{\lambda\mu}\chi^\mu \,, \tag{13.14}$$

$$\hat{\mathscr{W}}_\lambda^- = \mathscr{A}_\lambda + \frac{k^2}{2}\chi_\lambda = \mathscr{W}_\lambda^- + \frac{1}{12}\left(s + 6k^2\right)\chi_\lambda + \frac{1}{2}C^-_{\lambda\mu}\phi^\mu \,. \tag{13.15}$$

A quaternionic space is such that either $\hat{\mathscr{W}}^+$ or $\hat{\mathscr{W}}^-$ vanish.

13.2.2 The Filling-In Problem

A round three-sphere is a positive-curvature, maximally symmetric Einstein space with $SU(2) \times SU(2)$ isometry. Its metric can be expressed using the Maurer–Cartan forms of $SU(2)$:

$$\mathrm{d}\Omega_3^2 = \left(\sigma^1\right)^2 + \left(\sigma^2\right)^2 + \left(\sigma^3\right)^2 \tag{13.16}$$

with

$$\begin{cases} \sigma^1 = \sin\vartheta \sin\psi \,\mathrm{d}\varphi + \cos\psi \,\mathrm{d}\vartheta \\ \sigma^2 = \sin\vartheta \cos\psi \,\mathrm{d}\varphi - \sin\psi \,\mathrm{d}\vartheta \\ \sigma^3 = \cos\vartheta \,\mathrm{d}\varphi + \mathrm{d}\psi \,; \end{cases}$$

$0 \le \vartheta \le \pi, 0 \le \varphi \le 2\pi, 0 \le \psi \le 4\pi$ are the Euler angles.

A hyperbolic four-space H_4 is a negative-curvature, maximally symmetric Einstein space. It is a foliation over three-spheres and its metric reads:

$$\mathrm{d}s^2_{H_4} = \frac{\mathrm{d}r^2}{1 + k^2r^2} + k^2r^2\mathrm{d}\Omega_3^2 \,.$$

(we assumed $R_{ab} = -3k^2 g_{ab}$ for H_4). The conformal boundary of H_4 is reached at $r \to \infty$ as

$$\mathrm{d}s^2_{H_4} \underset{r\to\infty}{\longrightarrow} k^2 r^2 \mathrm{d}\Omega^2_3 \, .$$

In this sense, the round three-sphere is filled-in with H_4, the latter being the only regular metric filling-in this three-dimensional space.

The natural question to ask in view of the above is how to fill-in the more general Berger sphere S^3, which is a homogeneous but non-isotropic deformation of (13.16):

$$\mathrm{d}\Omega^2_{S^3} = \left(\sigma^1\right)^2 + \left(\sigma^2\right)^2 + 4n^2k^2\left(\sigma^3\right)^2 \qquad (13.17)$$

with nk constant. This metric is invariant under $SU(2) \times U(1)$, respectively generated by the Killings

$$\begin{cases} \xi_1 = -\sin\varphi \cot\vartheta \, \partial_\varphi + \cos\varphi \, \partial_\vartheta + \frac{\sin\varphi}{\sin\vartheta} \, \partial_\psi \\ \xi_2 = \cos\varphi \cot\vartheta \, \partial_\varphi + \sin\varphi \, \partial_\vartheta - \frac{\cos\varphi}{\sin\vartheta} \, \partial_\psi \\ \xi_3 = \partial_\varphi \, , \end{cases}$$

and ∂_ψ.

LeBrun studied the filling-in problem in general terms [8] and showed that an analytic three-metric can be regularly filled-in by a four-dimensional Einstein space that has self-dual (or anti-self-dual) Weyl tensor, *i.e.* by a *quaternionic* space. In modern holographic words, LeBrun's result states that requiring regularity makes the boundary metric a sufficient piece of data for reconstructing the bulk. Regularity translates into conformal self-duality, which effectively reduces by half the independent Cauchy data of the problem, as we will see in Sect. 13.3.2.

13.2.3 A Concrete Example

LeBrun's analysis is very general. We can illustrate it in the specific example of the Berger sphere S^3. We search therefore a four-dimensional foliation over S^3, which is Einstein. This leads to the Bianchi IX Euclidean Schwarzschild–Taub–NUT family on hyperbolic space (*i.e.* with $\Lambda = -3k^2$):

$$\mathrm{d}s^2 = \frac{\mathrm{d}r^2}{V(r)} + \left(r^2 - n^2\right)\left(\left(\sigma^1\right)^2 + \left(\sigma^2\right)^2\right) + 4n^2 V(r) \left(\sigma^3\right)^2 \qquad (13.18)$$

with

$$V(r) = \frac{1}{r^2 - n^2}\left[r^2 + n^2 - 2Mr + k^2\left(r^4 - 6n^2r^2 - 3n^4\right)\right] , \qquad (13.19)$$

where M and n are the mass and nut charge. Clearly the metric fulfills the boundary requirement since

$$\mathrm{d}s^2 \underset{r\to\infty}{\longrightarrow} r^2 \mathrm{d}\Omega^2_{S^3} ,$$

where $\mathrm{d}\Omega^2_{S^3}$ is given in (13.17).

The family of solutions at hand depends on 2 parameters, M and n, of which only the second remains visible on the conformal boundary. In that sense, the bulk is not fully determined by the boundary metric. However, regularity is not always guaranteed either, as $\mathrm{d}s^2$ is potentially singular at $r = +n$ or $r = -n$ (depending on whether the range for r is chosen positive or negative). Actually, this locus coincides with the fixed points of the Killing vector ∂_ψ, generating the extra $U(1)$.[6] In the present case, these are nuts and they are removable provided the space surrounding them is locally flat.

In order to make the above argument clear, let us focus for concreteness on $r = n$ (assuming thus $r > 0$), write $r = n + \epsilon$ and expand the metric using momentarily ϵ as radial coordinate:

$$\begin{aligned}\mathrm{d}s^2 \approx & \frac{\mathrm{d}\epsilon^2}{V(n) + \epsilon V'(n)} + 2n\epsilon \left(\mathrm{d}\vartheta^2 + \sin^2\vartheta \mathrm{d}\varphi^2\right) \\ & +4n^2 \left(V(n) + \epsilon V'(n)\right) \left(\mathrm{d}\psi + \cos\vartheta \mathrm{d}\varphi\right)^2 . \end{aligned} \tag{13.20}$$

Clearly to reconstruct locally flat space we must impose $V(n) = 0$ and $V'(n) = 1/2n$. The first of these requirements is equivalent to

$$M = n\left(1 - 4k^2n^2\right) , \tag{13.21}$$

and makes the second automatically satisfied. Under (13.21) and with $\tau = 2\sqrt{2n\epsilon}$ (proper time), Eq. (13.20) reads:

$$\mathrm{d}s^2 \approx \mathrm{d}\tau^2 + \frac{\tau^2}{4}\left(\mathrm{d}\psi^2 + \mathrm{d}\varphi^2 + \mathrm{d}\vartheta^2 + 2\cos\vartheta\, \mathrm{d}\psi\, \mathrm{d}\varphi\right) ,$$

which is indeed $\mathbb{R}^4$.

We can similarly analyze the behavior around $r = -n$. We then reach the same conclusion, with an overall change of sign in condition (13.21). These conditions

[6]In four dimensions, the fixed locus of an isometry is either a zero-dimensional or a two-dimensional space. The first case corresponds to a nut, the second to a bolt, and both can be removable singularities under appropriate conditions (see [28] for a complete presentation).

are nothing but conformal (anti-)self-duality requirements, as we see by computing the Weyl components of the curvature, $W^{\pm}$, in the decomposition (13.10):

$$W^{\pm} = \frac{M \mp n(1-4k^2n^2)}{(r\mp n)^3}\begin{pmatrix}-1 & 0 & 0\\ 0 & -1 & 0\\ 0 & 0 & 2\end{pmatrix}.$$

The regularity requirement for the family of Einstein spaces (13.18) is thus equivalent to demand the space be quaternionic. In that case, the boundary metric contains enough information for determining the bulk and solving thereby the filling-in problem for the Berger sphere.

For the quaternionic Schwarzschild–Taub–NUT geometries (13.18) with (13.21), the function $V(r)$ in (13.19) reads:

$$V(r) = \frac{r-n}{r+n}\left[1+k^2(r-n)(r+3n)\right].$$

These geometries belong to the general class of Calderbank–Pedersen [19], which is the family of quaternionic spaces with at least two commuting Killing fields.[7] They belong to a wide web of structures, and are in particular conformal to a family of spaces, which are Kähler and Weyl-anti-self-dual with vanishing scalar curvature, known as LeBrun geometries [29]. The limit $n \to \infty$ deserves a particular attention, as it corresponds to the pseudo-Fubini–Study[8] metric on $\widetilde{\mathbb{CP}}_2 = \frac{SU(2,1)}{U(2)}$. Further holographic properties of these geometries can be found in [30, 31].

13.3 Weyl Self-Duality from the Boundary

The filling-in problem was presented as the ancestor of holography in the sense that (i) it poses the problem of reconstructing the bulk out of the boundary and (ii) it raises the issue of regularity as a mean to relate *a priori* independent boundary data. The bonus is that in the present Euclidean approach, regularity condition appears as conformal self-duality requirement, which in turn makes Einstein's equations integrable and the bulk an exact solution.

The natural question to ask at this stage is how the bulk Weyl self-duality gets manifest on the boundary. In order to answer, we must perform a clear analysis of the independent boundary data following Fefferman–Graham approach and recast in these data the self-duality requirement.

[7]The metrics at hand are sometimes called *spherical Calderbank–Pedersen*, because they possess in total four Killings, of which three form an $SU(2)$ algebra.

[8]This is the non-compact Fubini–Study. The ordinary Fubini–Study corresponds to the compact $\mathbb{CP}_2 = \frac{SU(3)}{U(2)}$ and has positive cosmological constant.

13.3.1 The Fefferman–Graham Expansion

The work of LeBrun [8], quoted previously in the framework of the filling-in problem, led Fefferman and Graham to set up a systematic expansion for Einstein metrics in powers of a radial coordinate [32, 33]. The infinite set of coefficients are data of the boundary, expressed in terms of two independent ones: $g_{\mu\nu}$ and $F_{\mu\nu}$. From a Hamiltonian perspective, with the radial coordinate as evolution parameter, $g_{\mu\nu}$ and $F_{\mu\nu}$ are Cauchy data of "coordinate" and "momentum" type. The former is of geometric nature, the latter is not. In the holographic language, $g_{\mu\nu}$ corresponds to a non-normalizable mode and is the boundary metric, whereas $F_{\mu\nu}$ is related to a normalizable operator and carries information on the energy–momentum-tensor expectation value of the boundary field theory:

$$T_{\mu\nu} = \frac{3k}{8\pi G_{\rm N}} F_{\mu\nu} \, , \tag{13.22}$$

where $G_{\rm N}$ is four-dimensional Newton's constant.

The method of Fefferman–Graham is well suited for holography and has led to important developments (see e.g. [34–36]). It nicely fits the gravito-electric/gravito-magnetic split Hamiltonian formalism of four-dimensional gravity [37, 38]. In the Euclidean, this formalism is basically adapted to the self-dual/anti-self-dual splitting of the gravitational degrees of freedom presented in Sect. 13.2.1.

Let us summarize here the basic facts, leaving aside the rigorous and complete exhibition that can be found in the above references. In Palatini formulation, the four-dimensional (bulk) Einstein–Hilbert action reads:

$$I_{\rm EH} = -\frac{1}{32\pi G_{\rm N}} \int_{\mathscr{M}} \epsilon_{abcd} \left(\mathscr{R}^{ab} + \frac{k^2}{2} \theta^a \wedge \theta^b \right) \wedge \theta^c \wedge \theta^d \, .$$

As we already mentioned, $\theta^a, a = r, \lambda$ are basis elements of a coframe, orthonormal with respect to the signature $(+\eta + +)$. The first direction r is the holographic one, and $\mathrm{x} \equiv (t, x^1, x^2)$ are the remaining coordinates, surviving on the conformal boundary—with $t \equiv x^3$ in the Euclidean instance ($\eta = +$).

The most general form for the coframe is

$$\theta^r = N \frac{\mathrm{d}r}{kr} \, , \quad \theta^\lambda = N^\lambda \mathrm{d}r + \tilde{\theta}^\lambda \, ,$$

whereas the Levi–Civita connection generally reads:

$$\omega^{r\lambda} = q^{r\lambda} \mathrm{d}r + \mathscr{K}^\lambda \, , \quad \omega^{\mu\nu} = -\epsilon^{\mu\nu\lambda} \left(Q_\lambda \frac{\mathrm{d}r}{kr} + \mathscr{B}_\lambda \right) \, .$$

Without loss of generality, we can make the following gauge choice:

$$N = 1 \, , \quad N^\mu = q^{r\mu} = Q_\rho = 0 \, ,$$

leading to the Fefferman–Graham form for the bulk metric:

$$\mathrm{d}s^2 = \frac{\mathrm{d}r^2}{k^2r^2} + \eta_{\mu\nu}\tilde{\theta}^\mu\tilde{\theta}^\nu \,. \tag{13.23}$$

The connection is encapsulated in $\mathscr{K}^\mu$ and $\mathscr{B}_\lambda$. In Euclidean signature ($\eta = +$), these are vector-valued (with respect to the holonomy $SO(3)$ subgroups) connection one-forms, related to the (anti-)self-dual ones introduced in (13.7):

$$\mathscr{K}_\lambda = A_\lambda + \Sigma_\lambda \,, \quad \mathscr{B}_\lambda = A_\lambda - \Sigma_\lambda \,. \tag{13.24}$$

The zero-torsion condition (13.6) translates in this language into

$$\begin{cases} \mathscr{K}_\lambda \wedge \tilde{\theta}^\lambda = 0 \\ \mathrm{d}\tilde{\theta}^\lambda = \frac{1}{kr}\mathscr{K}^\lambda \wedge \mathrm{d}r - \epsilon^{\lambda\mu\nu}\mathscr{B}_\mu \wedge \tilde{\theta}_\nu \,. \end{cases} \tag{13.25}$$

With the present choice of gauge, all relevant information on the bulk geometry is stored inside $\left\{\tilde{\theta}^\lambda, \mathscr{K}^\lambda, \mathscr{B}^\lambda\right\}$. Assuming the metric be Einstein, leads to a very specific r-expansion of these vector-valued one-forms, in terms of the boundary data. This is the Fefferman–Graham expansion:

$$\tilde{\theta}^\lambda(r, \mathrm{x}) = kr\, E^\lambda(\mathrm{x}) + \sum_{\ell=0}^{\infty} \frac{1}{(kr)^{\ell+1}} F^\lambda_{[\ell+2]}(\mathrm{x}) \,, \tag{13.26}$$

$$\mathscr{K}^\lambda(r, \mathrm{x}) = -k^2 r\, E^\lambda(\mathrm{x}) + k\sum_{\ell=0}^{\infty} \frac{\ell+1}{(kr)^{\ell+1}} F^\lambda_{[\ell+2]}(\mathrm{x}) \,, \tag{13.27}$$

$$\mathscr{B}^\lambda(r, \mathrm{x}) = B^\lambda(\mathrm{x}) + \sum_{\ell=0}^{\infty} \frac{1}{(kr)^{\ell+2}} B^\lambda_{[\ell+2]}(\mathrm{x}) \,. \tag{13.28}$$

The boundary data are vector-valued one-forms. They are not all independent, and higher orders are derivatives of lower orders (we will meet an example of this "horizontal" relationship in a short while). Furthermore, due to the zero-torsion condition (13.25), further "vertical" relations exist order by order amongst the three sets. This is manifest when comparing (13.26) and (13.27), where the relations are algebraic. The forms in (13.26) and (13.28) are also related, in a differential manner, though.

The form E^λ is the boundary coframe. It is the first independent coefficient and it allows to reconstruct the three-dimensional boundary metric:

$$\mathrm{d}s^2_{\text{bry.}} = \lim_{r\to\infty} \frac{\mathrm{d}s^2}{k^2r^2} = \eta_{\mu\nu}E^\mu E^\nu \,.$$

The one-form B^{μ} appearing in the expansion of the magnetic component of the bulk connection, Eq. (13.28), is the boundary Levi–Civita connection, differentially related to the coframe (boundary zero-torsion condition):

$$\mathrm{d}E^{\lambda} = \epsilon^{\lambda\mu\nu} B_{\nu} \wedge E_{\mu} \,.$$

Other forms such as $F^{\mu}_{[2]} = F^{\mu}_{[2]\nu} E^{\nu}$ or $B^{\mu}_{[2]} = B^{\mu}_{[2]\nu} E^{\nu}$ are also geometric, respectively related to the Schouten and Cotton–York tensors[9]:

$$S^{\mu\nu} = -2k^2 F^{\mu\nu}_{[2]} \,, \quad C^{\mu\nu} = 2k^2 B^{\mu\nu}_{[2]} \,. \tag{13.29}$$

There is again a differential relationship among the two, following basically from the bulk zero-torsion condition (13.25), since by definition

$$C^{\mu\nu} = \eta^{\mu\rho\sigma} \nabla_{\rho} S^{\nu}{}_{\sigma} \tag{13.30}$$

($\eta^{\mu\rho\sigma} = \epsilon^{\mu\rho\sigma}/\sqrt{|g|}$).

Other curvature tensors of arbitrary order appear in the Fefferman–Graham expansion, all differentially related to the ones already described above. These tensors do *not* exhaust, however, all coefficients of the series (13.26), (13.27) and (13.28), as some infinite sequences of those are not of geometric nature, *i.e.* are not determined by the boundary metric itself (or by the coframe E^{μ}). Instead, they follow differentially from the second independent piece of data, $F^{\mu} \equiv F^{\mu}_{[3]}$, related to the energy–momentum expectation value according to (13.22). The interested reader will find a more complete exhibition of the Fefferman–Graham expansion in the literature, and particularly in [37, 38] for the gravito-electric/gravito-magnetic split formalism.

13.3.2 Self-Duality and Its Lorentzian Extension

Riemann Delf-Duality A word on Riemann self-duality is in order at this stage, before exploring the more subtle issue of Weyl self-duality.

Demanding the Riemann tensor be (anti-)self-dual (see end of Sect. 13.2.1) guarantees Ricci flatness and Weyl (anti-)self-duality. Such a requirement on the curvature is easily transported to the connection, using Eq. (13.9): the anti-self-dual connection $\mathscr{K}_{\mu} + \mathscr{B}_{\mu}$ (see Eq. (13.24)) of a self-dual Riemann is either vanishing

[9] In three dimensions, the Schouten tensor is defined as $S^{\mu\nu} = R^{\mu\nu} - \frac{R}{4} g^{\mu\nu}$, whereas the Cotton–York tensor is the Hodge-dual of the Cotton tensor, defined in Eq. (13.30). The latter replaces the always vanishing three-dimensional Weyl tensor. In particular, conformally flat boundaries have zero Cotton tensor and vice versa.

or a pure gauge (flat). This basically removes the corresponding degrees of freedom and gives an easy way to handle the problem via first-order differential equations.

The case of Bianchi foliations along the radial (holographic) direction, as the example we described in Sect. 13.2.3, has been largely analyzed in the literature (see [27] for a general discussion, [39] for Bianchi IX, or [40–42] for a more recent general and exhaustive Bianchi analysis). The requirement of (anti-)self-dual Riemann leads to the following equation:

$$\mathscr{K}_\mu \pm \mathscr{B}_\mu = \lambda_{\mu\nu}\sigma^\nu\,, \tag{13.31}$$

where σ^ν are the Maurer–Cartan forms of the Bianchi group, and $\lambda_{\mu\nu}$ a constant matrix parameterizing the homomorphisms mapping $SO(3)$ onto the Bianchi group. Expressing $\mathscr{K}_\mu$ and $\mathscr{B}_\mu$ in terms of the metric, (13.31) provides a set of first-order differential equations that have usually remarkable integrability properties. For concreteness, in the case of Bianchi IX ($SO(3)$) foliations, $\lambda_{\mu\nu} = 0$ or $\delta_{\mu\nu}$. The former case leads to the Lagrange equations, whereas the latter to the Darboux–Halphen system. Both systems are integrable, with celebrated solutions such as Eguchi–Hanson or BGPP for the first [2, 3, 43], and Taub–NUT or Atiyah–Hitchin for the second [1, 4].

Weyl Self-Duality Demanding Weyl (anti-)self-duality is not sufficient for setting $\mathscr{K}_\mu \pm \mathscr{B}_\mu$ as a pure gauge (flat connection). In the case of Bianchi foliations e.g. Eq. (13.31) is still valid but $\lambda_{\mu\nu}$ is a function of the radial coordinate r, and satisfies a first-order differential equation. The general structure of this equation (independently of any ansatz such as a Bianchi foliation) imposes a certain behavior and this is how Weyl (anti-)self-duality affects boundary conditions in a way that becomes transparent in the Fefferman–Graham large-r expansion.

We are specifically interested in quaternionic spaces, which are Einstein and conformally (anti-)self-dual. Thanks to the on-shell Weyl tensor (13.13), these requirements are simply either $\hat{\mathscr{W}}_\lambda^+ = 0$ (anti-self-dual) or $\hat{\mathscr{W}}_\lambda^- = 0$ (self-dual). Expressions (13.14) and (13.15), combined with (13.9) and (13.24)–(13.28), allow to establish the effect of Weyl (anti-)self-duality on the boundary one-forms. This appears as a hierarchy of algebraic equations[10]

$$k\left[(\ell+2)^2-1\right]F^\lambda_{[\ell+3]} \pm (\ell+2)B^\lambda_{[\ell+2]} = 0\,, \quad \forall \ell \geq 0$$

[10]When dealing with the Fefferman–Graham expansion together with Einstein dynamics, attention should be payed to the underlying variational principle. This sometimes requires Gibbons–Hawking boundary terms to be well posed. In the Hamiltonian language, these terms are generators of canonical transformations and in AdS/CFT their effect is known as holographic renormalization. These subtleties are discussed in [37, 38, 44–47], together with the specific role of the Chern–Simons boundary term, which produces the boundary Cotton tensor, and in conjunction with Dirichlet vs. Neumann boundary conditions. One should also quote the related works [48, 49], in the linearized version of gravitational duality though.

(the upper + sign corresponds to the self-dual case), of which only the first is independent:

$$3kF^{\lambda}_{[3]} \pm 2B^{\lambda}_{[2]} = 0\,. \tag{13.32}$$

The others follow from the already existing horizontal differential relationships. This *algebraic* equation between *a priori* independent boundary data is at the heart of conformal self-duality. In terms of the boundary energy–momentum and Cotton tensors (see (13.22) and (13.29)), Eq. (13.32) reads:

$$8\pi G_{\mathrm{N}} k^2 T^{\mu\nu} \pm C^{\mu\nu} = 0\,. \tag{13.33}$$

Several important comments are in order at this stage. Firstly, referring to the original problem of Sect. 13.2.2, Eq. (13.32) provides the filling-in boundary condition for some *a priori* given boundary metric (not necessarily a three-sphere as originally studied in [8]). This condition tunes algebraically the Cauchy data ("initial position" and "initial momentum"), in such a way that any boundary metric can be filled-in *regularly*. Following the intuition developed in the example of Sect. 13.2.3, we may slightly relax this condition and trade it for

$$wT^{\mu\nu} + C^{\mu\nu} = 0\,, \tag{13.34}$$

where we now allow for any real w and not solely $w = \pm 8\pi G_{\mathrm{N}} k^2$. The filling-in is still expected to occur, without guaranty for the regularity though.

Secondly, as discussed in the introduction, duality is underlying integrability. This statement is clear in the case of Riemann self-duality, where the key is the reduction of the differential order of the equations. For conformal self-duality it operates via an appropriate tuning of the boundary conditions, the effect of which would be better qualified as exactness rather than integrability: the equations of motion are not simplified, but the initial conditions select a specific corner of the phase space, which enables for exact solutions to emerge, *i.e.* for the Fefferman–Graham series to be resummable. Furthermore, even though self-duality (Riemann or Weyl) does not apply to the Lorentzian frame,[11] condition (13.34) remains consistent for a Lorentzian boundary, and is expected, following our heuristic arguments, to guarantee the resummability of the Fefferman–Graham expansion and lead to exact solutions. This is not a theorem, much like everything regarding the relationship between integrability and self-duality in general, but the idea seems to work, as we will see in Sect. 13.4.2.

[11] In four-dimensional metrics with Lorentzian signature, self-duality leads either to complex solutions, or to Minkowski and AdS_4, which are both self-dual *and* anti-self-dual (they have vanishing Riemann and vanishing Weyl, respectively).

Our last comment concerns the potential developments around Eq. (13.34), already announced, and discussed, in the introduction (Eq. (13.1)). This equation is a boundary condition, which, however, can follow from a three-dimensional variational principle. In order to enforce it via this principle, we must equip the boundary field theory with specific dynamics that incorporates three-dimensional gravity, in the form of a topological massive term, as suggested by Eqs. (13.2) and (13.3). The first term in S is the phenomenological holographic matter action, whereas the second is the Chern–Simons term with ω_3 the Lagrangian density given in terms of the boundary connection one-form γ:

$$\omega_3(\gamma) = \frac{1}{2}\mathrm{Tr}\left(\gamma \wedge \mathrm{d}\gamma + \frac{2}{3}\gamma \wedge \gamma \wedge \gamma\right) .$$

Conceptually, this is a non-trivial step as holography is not supposed *a priori* to endow the boundary theory with gravitational dynamics. It raises three questions:

1. What are the allowed boundary geometries, given certain assumptions on the energy–momentum tensor?
2. What are the bulk geometries that reproduce holographically the boundary data? Are those exact Einstein spaces, *i.e.* is the corresponding Fefferman–Graham expansion resummable in accordance with the above discussion?
3. Are there situations where gravitational degrees of freedom emerge?

We will answer questions 1 and 2, at least in some specific framework, leaving open interesting extensions. As we will see, in some situations, the boundary geometry is really a topologically massive gravity vacuum—as if the three-dimensional Einstein–Hilbert term were effectively present in (13.3). We will not delve into question 3, because this is a definitely different direction of investigation. The interested reader may find [46] useful and inspiring regarding that issue.

13.4 Application to Holographic Fluids

The purpose of the present part is to answer questions 1 and 2 raised in Sect. 13.3.2. Solving Eq. (13.34) is possible, provided some assumptions are made both on the energy–momentum tensor, and on the boundary metric. These assumptions are motivated by our goal to probe transport coefficients for holographic fluids, without performing linear-response analysis. For that we must study equilibrium configurations of the fluid in various exact non-trivial backgrounds and design accordingly the boundary data. These satisfy Eq. (13.34) and are integrable *i.e.* the corresponding Fefferman–Graham expansion is resummable.

13.4.1 *Fluids at Equilibrium in Papapetrou–Randers Backgrounds*

Hydrodynamic Description A given bulk configuration (geometry possibly supplemented with other fields) provides a boundary geometry, and a finite-temperature and finite-density state of the—generally unknown—microscopic boundary theory. It has expectation value $T_{\mu\nu}$ for the energy–momentum tensor, satisfying

$$\nabla_{\mu} T^{\mu\nu} = 0 , \tag{13.35}$$

and possibly other conserved currents. This state may be close to a hydrodynamic configuration and is potentially described within the hydrodynamic approximation. This assumes, among others, local thermodynamic equilibrium. For this description to hold, it is necessary that the scale of variation of the diverse quantities describing the fluid be large compared to any microscopic scale (such as the mean free path). We will work in this framework and furthermore suppose the fluid neutral, as the only bulk degrees of freedom are gravitational in our case.

The relativistic fluid is described in terms of a velocity field u(x), as well as of local thermodynamic quantities like $T(\mathrm{x}), p(\mathrm{x}), \varepsilon(\mathrm{x}), s(\mathrm{x})$, obeying an equation of state and thermodynamic identities

$$\begin{cases} sT &= \varepsilon + p \\ \mathrm{d}\varepsilon &= T\mathrm{d}s \, . \end{cases}$$

All these enter the energy–momentum tensor. The energy–momentum tensor of a neutral hydrodynamic system can be expanded in derivatives of the hydrodynamic variables, namely

$$T^{\mu\nu} = T^{\mu\nu}_{(0)} + T^{\mu\nu}_{(1)} + T^{\mu\nu}_{(2)} + \cdots , \tag{13.36}$$

where the subscript denotes the number of covariant derivatives. The validity of this derivative expansion is subject to the above assumptions regarding the scale of variation. The zeroth order energy–momentum tensor is the so called perfect-fluid energy–momentum tensor:

$$T^{\mu\nu}_{(0)} = \varepsilon u^{\mu} u^{\nu} + p \, \Delta^{\mu\nu} , \tag{13.37}$$

where $\Delta^{\mu\nu} = u^{\mu} u^{\nu} + g^{\mu\nu}$ is the projector onto the space orthogonal to u. This corresponds to a fluid being locally in equilibrium, in its proper frame.[12] The

[12]Defining the local proper frame, *i.e.* the velocity field u, is somewhat ambiguous in relativistic fluids. A possible choice is the *Landau frame*, where the non-transverse part of the energy–momentum tensor vanishes when the pressure is zero. This will be our choice.

conservation of the perfect-fluid energy–momentum tensor leads to the relativistic Euler equations:

$$\begin{cases} \nabla_{\mathrm{u}}\varepsilon + (\varepsilon + p)\Theta = 0 \\ \nabla_{\perp}p + (\varepsilon + p)\mathrm{a} = 0 \,, \end{cases} \tag{13.38}$$

where $\nabla_{\mathrm{u}} = \mathrm{u}\cdot\nabla$, $\Theta = \nabla\cdot\mathrm{u}$, $\nabla_{\perp\mu} = \Delta_{\mu}{}^{\nu}\nabla_{\nu}$, and $\mathrm{a} = \mathrm{u}\cdot\nabla\mathrm{u}$ (more formulas on kinematics of relativistic fluids are collected in App. 1).

The higher-order corrections to the energy–momentum tensor involve the transport coefficients of the fluid. These are phenomenological parameters that encode the microscopic properties of the underlying system. Listing them order by order requires to classify all transverse tensors (possibly limited to traceless and Weyl-covariant if the microscopic theory is conformally invariant) and this depends on the space–time dimension.[13] In the context of field theories, the transport coefficients can be determined from studying correlation functions of the energy–momentum tensor at finite temperature in the low-frequency and low-momentum regime (see for example [54]).

Equilibrium and Perfect Equilibrium Studying fluids at *equilibrium* on *non-trivial* backgrounds can provide information on their transport properties. A fluid in global thermodynamic equilibrium[14] is described by a stationary solution[15] of the relativistic equations of motion (13.35), assuming that such solutions exist. Finding solutions to these equations is generally a hard task, in particular because most of the transport coefficients are unknown. As it will become clear in a short while, the concept of *perfect equilibrium* provides a natural way out, giving access to non-trivial information about transport properties.

The prototype example, where global thermodynamic description applies, is the one of an inertial fluid in Minkowski background with globally defined constant temperature, energy density and pressure. In this case, irrespective of whether the fluid itself is viscous, its energy–momentum tensor, evaluated at the solution, takes the zeroth-order (perfect) form (13.37) because all derivatives of the hydrodynamic variables vanish. On the one hand, this equilibrium situation is easy to handle because the relevant equations are the zeroth-order ones, (13.38); on the other hand, it does not allow to learn anything about transport properties because the effect of

[13] We recommend [50,51] for a recent account of that subject. Insightful information was also made available thanks to the developments on fluid/gravity correspondence [52,53].

[14] This should not be confused with a steady state, where we have stationarity due to a balance between external driving forces and internal dissipation. Such situations will not be discussed here.

[15] It is admitted that a non-relativistic fluid is stationary when its velocity field is time-independent. This is of course an observer-dependent statement. For relativistic fluids, one could make this more intrinsic saying that the velocity field commutes with a globally defined time-like Killing vector, assuming that the later exists. Note also that statements about global thermodynamic equilibrium in gravitational fields are subtle and the subject still attracts interest [55].

transport is washed out by the geometry itself. If we insist keeping Minkowski as a background, the only way, which would give access to the transport coefficients, is to perturb the fluid away from its global equilibrium configuration.

Although naive, the equilibrium paradigm in Minkowski has the virtue to suggest an alternative general method that may fit certain classes of fluids. It indeed raises a less naive question: are there other situations of fluids on gravitational backgrounds, where the hydrodynamic description is also perfect *i.e.* the energy–momentum tensor, in equilibrium, takes the perfect form (13.37) solving Eqs. (13.38)?

As anticipated, we call these special configurations *perfect-equilibrium states*. For these configurations to exist, *all* terms in (13.36), except for the first one, must vanish, either because the transport coefficients are zero, or because the corresponding tensors vanish kinematically—requiring in particular a special relationship between the fluid's velocity and the background geometry. It should be stressed that the fluid in perfect equilibrium is *not* perfect—the equilibrium is.

At this stage of the presentation, the question to answer is whether fluids exist, which can exhibit, on certain backgrounds, perfect-equilibrium configurations. Holography and the methods discussed in Sects. 13.2 and 13.3 for finding exact bulk solutions provide the tools for this analysis. The strategy to follow is straightforward:

- Choose a class of backgrounds possessing a time-like Killing vector ξ.
- Assume perfect equilibrium and show that indeed perfect Euler Eqs. (13.38) are solved for a conformal fluid *i.e.* for a fluid such that $\varepsilon = 2p$. A hint for solving them is to impose that the fluid velocity field u is aligned with ξ.
- Impose the "self-duality" condition (13.34) and restrict the family of backgrounds at hand. The three-dimensional geometries obtained in that way are called *perfect geometries* because their Cotton–York tensor is of the perfect-fluid form.
- Use the Fefferman–Graham expansion to reconstruct the four-dimensional bulk geometry, hoping indeed that Eq. (13.34) acts as an integrability condition, allowing for resummation of the series into an exact Einstein space. This is crucial for sustaining the claim that we are describing a holographic conformal fluid behaving *exactly* as a perfect fluid.

If this procedure goes through with genuinely non-trivial geometries, it enables us to probe transport properties of the holographic fluid despite its global equilibrium state: all transport coefficients coupled to Weyl-covariant, traceless and transverse tensors $\mathscr{T}_{\mu\nu}$ that are non-vanishing and whose divergence is also non-vanishing, when evaluated in the perfect-equilibrium solution, must be zero. We call such tensors *dangerous tensors*. Listing them requires the knowledge of the specific perfect geometry and of the kinematic configuration of the fluid.[16] Any fluid, which would have non-vanishing corresponding transport coefficient, would not be in equilibrium in the configuration at hand. This may occur for transport

[16]More data are available on the dangerous tensors in certain classes of geometries in [23].

coefficients of any order in the expansion of the energy–momentum tensor, as dangerous tensors appear at arbitrarily large derivative order. Therefore the insight gained in this manner on the transport properties of the holographic fluid, concerns usually infinite series of coefficients. This is a non-trivial piece of information about the conformal fluid at hand, and a statement about the underlying microscopic theory.

There are non-trivial backgrounds (Minkowski space being a trivial example) where no dangerous tensors are present. However, one can also find a large class of backgrounds with a unique time-like Killing vector field, which have infinitely many non-zero dangerous tensors; those allow to probe an infinite number of transport coefficients. It is not clear at present whether all these backgrounds exhaust the perfect geometries. Nevertheless, the question of whether our analysis regarding all possible transport coefficients is exhaustive or not requires more work. It is clear that further insight on this matter can only be gained by perturbing the perfect-equilibrium state.

Perfect Equilibrium in Papapetrou–Randers Backgrounds A stationary three-dimensional metric can be written in the generic form ($x = \left(x^1, x^2\right)$ and $i, j, \ldots = 1, 2$)

$$\mathrm{d}s^2 = B(x)^2 \left(-(\mathrm{d}t - b_i(x)\mathrm{d}x^i)^2 + a_{ij}(x)\mathrm{d}x^i \mathrm{d}x^j\right) , \tag{13.39}$$

where B, b_i, a_{ij} are space-dependent but time-independent functions. These metrics were introduced by Papapetrou in [56]. They will be called hereafter *Papapetrou–Randers* because they are part of an interesting network of relationships involving the Randers form [57]. These metrics admit a generically *unique* time-like Killing vector, $\xi \equiv \partial_t$, with norm $\|\xi\|^2 = -B(x)^2$.

At this stage of the analysis, we would like to restrict ourselves to the case where the Killing vector is normalized, *i.e.* where B is constant and can therefore be consistently set to 1. This is a severe limitation, because it excludes equilibrium situations where the temperature or the chemical potential are x-dependent.[17] However, it illustrates the onset of perfect equilibrium configurations, and allows to establish a wide class of perfect geometries, intimately connected with holography.

In the background (13.39) (with $B = 1$), the vector $\xi = \partial_t$, satisfies

$$\nabla_{(\mu}\xi_{\nu)} = 0 , \quad \xi_\mu \xi^\mu = -1 .$$

[17]Remember that inside a stationary gravitational field, under certain conditions, global thermodynamic equilibrium requires $T\sqrt{-g_{00}}$ be constant [58]. Here $\sqrt{-g_{00}} = B$. Holographically, if the rescaling of the boundary metric by $B(x)$ (as in (13.39)) is accompanied with an appropriate rescaling of the energy–momentum tensor, the bulk geometry is unaffected, and $B(x)$ is generated by a bulk diffeomorphism.

We leave as an exercise to show that congruences defined by ξ have vanishing acceleration, shear and expansion (see App. 1), but non-zero vorticity[18] $\omega = \frac{1}{2}\mathrm{d}\xi \Leftrightarrow \omega_{\mu\nu} = \nabla_\mu \xi_\nu$. Then, it is easy to show that a solution of the perfect Euler equations (13.38), for a conformal fluid is:

$$\mathrm{u} = \xi\,, \quad \varepsilon = 2p = \text{constant}\,, \quad T = \text{constant}\,, \quad s = \text{constant}\,. \tag{13.40}$$

Therefore a fluid in perfect equilibrium will align its velocity field[19] u with the vector $\xi = \partial_t$, while thermalize at everywhere-constant p and T. Fluid worldlines form a shearless and expansionless geodesic congruence.

The normalized three-velocity one-form of the fluid at perfect equilibrium is

$$\mathrm{u} = -\mathrm{d}t + \mathrm{b}\,, \tag{13.41}$$

where $\mathrm{b} = b_i \mathrm{d}x^i$. We will often write the metric (13.39) as

$$\mathrm{d}s^2 = -\mathrm{u}^2 + \mathrm{d}\ell^2\,, \quad \mathrm{d}\ell^2 = a_{ij}\,\mathrm{d}x^i \mathrm{d}x^j\,. \tag{13.42}$$

A conformal fluid in perfect equilibrium on Papapetrou–Randers backgrounds has the energy–momentum tensor

$$T^{(0)}_{\mu\nu}\,\mathrm{d}x^\mu \mathrm{d}x^\nu = p\left(2\mathrm{u}^2 + \mathrm{d}\ell^2\right) \tag{13.43}$$

with the velocity form being given by (13.41) and p constant. We will adopt the convention that hatted quantities will be referring to the two-dimensional positive-definite metric a_{ij}, therefore $\hat{\nabla}$ for the covariant derivative and $\hat{R}_{ij}\,\mathrm{d}x^i \mathrm{d}x^j = \frac{\hat{R}}{2}\mathrm{d}\ell^2$ for the Ricci tensor built out of a_{ij}. We collect in App. 2 some useful formulas regarding Papapetrou–Randers backgrounds and the kinematics of fluids at perfect equilibrium.

Let us close this chapter by insisting once more on the meaning of the perfect-equilibrium configuration (13.40) for a conformal fluid that *is not a priori* perfect. For this configuration to be effectively realized, *all* higher-derivative

---

[18]Vorticity is inherited from the fact that ∂_t is not hypersurface-orthogonal. For this very same reason, Papapetrou–Randers geometries may in general suffer from global hyperbolicity breakdown. This occurs whenever regions exist, where constant-t surfaces cease being space-like, and potentially exhibit closed time-like curves. All these issues were discussed in detail in [20–22].

[19]One important point to note is that in perfect equilibrium we have no frame ambiguity in defining the velocity field. Since the velocity field is geodesic and is aligned with a Killing vector field of unit norm, it describes a unique local frame where all forces (like those induced by a temperature gradient) vanish.

corrections in (13.36) must be absent. It is easy to check that this is indeed the case for the first corrections, which in the 2 + 1-dimensional case under consideration read:

$$T^{\mu\nu}_{(1)} = -2\eta\sigma^{\mu\nu} - \zeta_{\mathrm{H}}\eta^{\rho\lambda(\mu}u_\rho\sigma_\lambda{}^{\nu)} \,. \tag{13.44}$$

The first term in (13.44) involves the shear viscosity η, which is a dissipative transport coefficient. The second is present in systems that break parity and involves the non-dissipative rotational-Hall-viscosity coefficient ζ_{H}. Notice that the bulk-viscosity term $\zeta\Delta^{\mu\nu}\Theta$ or the anomalous term $\tilde{\zeta}\Delta^{\mu\nu}\eta^{\alpha\beta\gamma}u_\alpha\nabla_\beta u_\gamma$ cannot appear in a conformal fluid because they are tracefull, namely for conformal fluids $\zeta = \tilde{\zeta} = 0$. Since the fluid congruence is shearless, the first corrections (13.44) vanish. Demanding that higher-order corrections also vanish, on the one hand, sets constraints on the transport coefficients coupled to the dangerous tensors that can be constructed with the vorticity only; on the other hand, it leaves free many other coefficients, which couple to tensors vanishing because of the actual kinematic state of the fluid. If the transport coefficients coupled to the dangerous tensors are non-zero, the geodesic fluid congruence with constant temperature is not a solution of the full Euler equations (13.35). The resolution of the latter alters the above perfect equilibrium state, leading in general to $\mathrm{u} = \xi + \delta\mathrm{u}(\mathrm{x})$ and $T = T_0 + \delta T(\mathrm{x})$. Such an excursion will be stationary or not depending on whether the non-vanishing corrections to the perfect energy–momentum tensor are non-dissipative or dissipative.

13.4.2 Perfect-Cotton Geometries and Their Bulk Ascendants

The Strategy The analysis presented in Sect. 13.4.1 is useful if there exist conformal fluids, which are indeed in perfect equilibrium on a Papapetrou–Randers background. This is not guaranteed *a priori* since it requires infinite classes of transport coefficients to vanish. Holography provides the appropriate tools for addressing this problem. The strategy has already been described above, and the remaining two steps are the following:

1. Impose condition (13.34) with perfect energy–momentum tensor and hence restrict the Papapetrou–Randers geometries to those which have a Cotton–York tensor of the perfect-fluid form (13.43):

$$C_{\mu\nu} = \frac{c}{2}(3u_\mu u_\nu + g_{\mu\nu}) \,, \tag{13.45}$$

where c is a constant with the dimension of an energy density.[20] This form is known in the literature as Petrov class $\mathrm{D_t}$.[21] Notice that the existence of perfect geometries is an issue unrelated to holography.
2. Sum the Fefferman–Graham series expansion. It turns out that the bulk geometries obtained in this way are exact solutions of Einstein's equations: perfect-Cotton geometries are boundaries of 3 + 1-dimensional exact Einstein spaces, and the resulting boundary energy–momentum tensor is also of the perfect-fluid form. This shows that the assumption of perfect equilibrium is well motivated, and the "self-duality" condition (13.34) does indeed ensure integrability.

Classification of the Perfect Papapetrou–Randers Geometries Consider a metric of the form (13.39) with $B(x) = 1$. Requiring its Cotton–York tensor (13.69) to be of the form (13.45) is equivalent to impose the conditions:

$$\hat{\nabla}^2 q + q(\delta - q^2) = 2c \,, \tag{13.46}$$

$$a_{ij}\left(\hat{\nabla}^2 q + \frac{q}{2}(\delta - q^2) - c\right) = \hat{\nabla}_i \hat{\nabla}_j q \,, \tag{13.47}$$

$$\hat{R} + 3q^2 = \delta \tag{13.48}$$

with δ being a constant relating the curvature of the two-dimensional base space, $\hat{R}$, with the vorticity strength q (see App. 2 for definitions and formulas).

It is remarkable that perfect-Cotton geometries *always* possess an extra space-like Killing vector. To prove[22] this we rewrite (13.46) and (13.47) as

$$\left(\hat{\nabla}_i \hat{\nabla}_j - \frac{1}{2} a_{ij} \hat{\nabla}^2\right) q = 0 \,. \tag{13.49}$$

Any two-dimensional metric can be locally written as

$$\mathrm{d}\ell^2 = 2\mathrm{e}^{2\Omega(z,\bar{z})} \mathrm{d}z\, \mathrm{d}\bar{z} \,, \tag{13.50}$$

where z and $\bar{z}$ are complex-conjugate coordinates. Plugging (13.50) in (13.49) we find that the non-diagonal equations are always satisfied (tracelessness of the Cotton–York tensor), while the diagonal ones read:

$$\partial_z^2 q = 2\partial_z \Omega \partial_z q \,, \quad \partial_{\bar{z}}^2 q = 2\partial_{\bar{z}} \Omega \partial_{\bar{z}} q \,.$$

---

[20] We recall that ε has dimensions of energy density or equivalently $(\text{length})^{-3}$, therefore the energy–momentum tensor and the Cotton–York tensor have the same natural dimensions.

[21] The subscript t stands for *time-like* and refers to the nature of the vector u. For an exhaustive review on Petrov & Segre classification of three-dimensional geometries see [59] (useful references are also [60–62]).

[22] I thank Jakob Gath for clarifying this point.

The latter can be integrated to obtain

$$\partial_z q = \mathrm{e}^{2\Omega - 2\bar{C}(\bar{z})}\ , \quad \partial_{\bar{z}} q = \mathrm{e}^{2\Omega - 2C(z)} \tag{13.51}$$

with $C(z)$ an arbitrary holomorphic function and $\bar{C}(\bar{z})$ its complex conjugate. Trading these functions for

$$w(z) = \int \mathrm{e}^{2C(z)} \mathrm{d}z\ , \quad \bar{w}(\bar{z}) = \int \mathrm{e}^{2\bar{C}(\bar{z})} \mathrm{d}\bar{z}\ ,$$

and introducing new coordinates (X, Y) as

$$X = w(z) + \bar{w}(\bar{z})\ , \quad Y = i\left(\bar{w}(\bar{z}) - w(z)\right)\ ,$$

we find using (13.51) that the vorticity strength depends only on X: $q = q(X)$. Hence, (13.50) reads:

$$\mathrm{d}\ell^2 = \frac{1}{2} \partial_X q \left(\mathrm{d}X^2 + \mathrm{d}Y^2\right)\ .$$

This condition enforces the existence of an extra Killing vector. Finally we note that (13.48) can be obtained by differentiating (13.46) with respect to X.

The presence of the space-like isometry actually simplifies the perfect-Cotton conditions for Papapetrou–Randers metrics. Without loss of generality, we take the space-like Killing vector to be ∂_y and write the metric as

$$\mathrm{d}s^2 = -\left(\mathrm{d}t - b(x)\,\mathrm{d}y\right)^2 + \frac{\mathrm{d}x^2}{G(x)} + G(x)\mathrm{d}y^2\ . \tag{13.52}$$

Thus

$$q = -\partial_x b\ ,$$

and (13.46)–(13.48) can be solved in full generality. The solution is written in terms of 6 arbitrary parameters c_i:

$$b(x) = c_0 + c_1 x + c_2 x^2\ , \tag{13.53}$$

$$G(x) = c_5 + c_4 x + c_3 x^2 + c_2 x^3\left(2c_1 + c_2 x\right)\ . \tag{13.54}$$

It follows that the vorticity strength takes the linear form

$$q(x) = -c_1 - 2c_2 x\ , \tag{13.55}$$

and the constants c and δ are given by:

$$c = -c_1^3 + c_1 c_3 - c_2 c_4 \ , \tag{13.56}$$

$$\delta = 3c_1^2 - 2c_3 \ . \tag{13.57}$$

Finally, the Ricci scalar of the two-dimensional base space is given by

$$\hat{R} = -2\left(c_3 + 6c_2 x(c_1 + c_2 x)\right) \ ,$$

and using (13.67) one can easily find the form of the three-dimensional scalar curvature as well. Not all the six parameters c_i correspond to physical quantities: some of them can be just reabsorbed in a change of coordinates. In particular, we set here $c_0 = 0$ by performing the diffeomorphism $t \to t + p\,y$, with constant p, which does not change the form of the metric.

The Bulk Duals of the Perfect Geometries At this stage, the reader may wonder what the interpretation of the parameters c_i is. It is more convenient to answer that question after unravelling the Einstein metrics that fit the boundary data (13.43), and (13.52) (with (13.53) and (13.54)). As already advertised, with these boundary data, the Fefferman–Graham series is resummable because (13.43) and (13.45) satisfy the "self-duality" condition (13.34) with $w = {}^{-c}/_{\varepsilon}$. The resulting exact Einstein space reads, in Eddington–Finkelstein coordinates (where $g_{rr} = 0$ and $g_{r\mu} = -u_\mu$):

$$\begin{aligned} \mathrm{d}s^2 = &-2\mathrm{u}\left(\mathrm{d}r - \frac{1}{2k^2}G(x)\partial_x q\,\mathrm{d}y\right) + \rho^2 k^2 \mathrm{d}\ell^2 \\ &-\left(r^2k^2 + \frac{\delta}{2k^2} - \frac{q^2}{4k^2} - \frac{1}{\rho^2}\left(2Mr + \frac{qc}{2k^6}\right)\right)\mathrm{u}^2 \end{aligned} \tag{13.58}$$

with

$$\mathrm{u} = -\mathrm{d}t + b\,\mathrm{d}y \ , \tag{13.59}$$

$$\rho^2 = r^2 + \frac{q^2}{4k^4} \ . \tag{13.60}$$

The various quantities appearing in (13.58)–(13.60), $b(x)$, $G(x)$, $q(x)$, c and δ, are reported in Eqs. (13.53)–(13.57). Notice also that a coordinate transformation is needed in order to recast (13.58) in Boyer–Lindqvist coordinates, and a further one to move to the canonical Fefferman–Graham frame (13.23). Details can be found in [23], which we will not present here because they lie beyond the main scope of these lectures. Even though r is not the Fefferman–Graham radial coordinate, in the limit $r \to \infty$, they both coincide. It is easy then to see that the boundary geometry is indeed the stationary Papapetrou–Randers metric (13.42), (13.52), and that the boundary energy–momentum tensor is of the perfect-fluid form with

$$\varepsilon = \frac{Mk^2}{4\pi G_\mathrm{N}} .$$

Upon performing coordinate transformations and parameter redefinitions, one can show that for $c_4 \neq 0$, the bulk metrics at hand belong to the general class of Plebañski–Demiañski type D, analyzed in [63]. For vanishing c_4, depending on the other parameters, one finds the flat-horizon solution of [64], or the rotating topological black hole of [65], or a set of metrics, which were found (but still not fully studied) in [23]. All these solutions are AdS black holes, which have mass M, nut charge n and angular velocity a. The acceleration parameter, present in Plebañski–Demiañski [63] is missing here. Actually, this parameter is an obstruction to perfect-Cotton boundary (*i.e.* to $\mathrm{D_t}$ Petrov–Segre class), and this is why it does not appear in our classification (see also [66]).

For all these metrics, the horizon is spherical, flat or hyperbolic.[23] The isometry group contains at least the time-like Killing vector ∂_t and the space-like Killing vector ∂_y. In the absence of rotation, two extra Killing fields appear, which together with ∂_y generate $SU(2)$, Heisenberg or $SL(2,\mathbb{R})$. The bulk metric is then a foliation over Bianchi IX, II or VIII homogeneous geometries.

From the explicit form of the bulk space–time metric (13.58), we observe that it can have a curvature singularity when $\rho^2 = 0$. The locus of this singularity will then be at

$$r = 0, \quad q(x) = 0 .$$

It also has an ergosphere, where the Killing vector ∂_t becomes null,[24] at $r(x)$ solution of

$$r^2k^2 + \frac{\delta}{2k^2} - \frac{q^2}{4k^2} - \frac{1}{\rho^2}\left(2Mr + \frac{qc}{2k^6}\right) = 0 .$$

We will not pursue any further this discussion on the bulk geometries. A thorough analysis of horizons, singularities or closed time-like curves can be found in the already quoted literature. A last comment concerning these black holes should however be made in relation with their symmetries: they are stationary and possess

[23]This is a local property. In the flat or hyperbolic cases, a quotient by a discrete subgroup of the isometry group is possible and allows to reshape the global structure, making the horizon compact without conical singularities (a two-torus for example).

[24]The Killing vector ∂_t is time-like and normalized at the boundary, where it coincides with the velocity field of the fluid, but its norm gets altered along the holographic coordinate, towards the horizon.

at least an additional spatial isometry. This is a consequence of the perfect-Cotton structure of their boundary, and this is consistent with the rigidity theorem in $3+1$ dimensions, which requires all stationary black hole solutions in flat space–time to have an axial symmetry. However, as far as we are aware, it is not known if this theorem is valid for $3+1$-dimensional asymptotically AdS stationary black holes. The above analysis appears thus as an indirect and somehow unexpected hint in favor of the rigidity theorem beyond asymptotically flat space–times.

Let us end this paragraph with an example, which generalizes (in Lorentzian signature) the case (13.18) presented in Sect. 13.2.3: the AdS Kerr–Taub–NUT with spherical horizon. In Boyer–Lindqvist coordinates this reads:

$$ds^2 = \frac{\rho^2}{\Delta_r}dr^2 - \frac{\Delta_r}{\rho^2}(dt + \beta d\varphi)^2 + \frac{\rho^2}{\Delta_\vartheta}d\vartheta^2 + \frac{\sin^2\vartheta\Delta_\vartheta}{\rho^2}(a dt + \alpha d\varphi)^2 \tag{13.61}$$

with

$$\begin{aligned}
\rho^2 &= r^2 + (n - a\cos\vartheta)^2 , \\
\Delta_r &= k^2r^4 + r^2(1 + k^2a^2 + 6k^2n^2) - 2Mr + (a^2 - n^2)(1 + 3k^2n^2) , \\
\Delta_\vartheta &= 1 + k^2 a\cos\vartheta(4n - a\cos\vartheta)
\end{aligned}$$

and

$$\begin{aligned}
\beta &= -\frac{2(a - 2n + a\cos\vartheta)}{\Xi}\sin^2\vartheta/2 , \\
\alpha &= -\frac{r^2 + (n-a)^2}{\Xi} , \\
\Xi &= 1 - k^2a^2 .
\end{aligned}$$

Back to the Boundaries and Transport Properties The boundary physics depends on the subset of those parameters among the c_is, which are non-trivial. The boundary metric is in general a function of two parameters, n and a, whereas M appears in the boundary energy–momentum tensor. The bulk isometry group is conserved. Thus, in the absence of rotation parameter $a = 0$, the boundary is a homogeneous and stationary space–time: squashed S^3 (including e.g. Gödel space), squashed Heisenberg or squashed AdS_3. The fluid undergoes a homogeneous rotation (*i.e.* without center, *monopolar*) with constant vorticity strength q.

For non-vanishing a, the boundary space–time is stationary but has only spatial axial symmetry. The vorticity is a superposition of a *monopole* and a *dipole* , and the fluid has now a cyclonic rotation around the poles on top of the uncentered one.

We give for illustration the boundary metric of the Kerr–Taub–NUT space–time with spherical horizon (13.61):

$$ds^2_{\text{bry.}} = -(dt + \beta d\varphi)^2 + \frac{1}{k^2\Delta_\vartheta}\left(d\vartheta^2 + \frac{\Delta_\vartheta^2}{\Xi^2}\sin^2\vartheta\, d\varphi^2\right) . \tag{13.62}$$

For vanishing a, $\mathrm{d}\ell^2$ is an ordinary two-sphere and $\mathrm{b} = -\beta \mathrm{d}\varphi$ is a Dirac-monopole-like potential. Switching-on a deforms axially the base space $\mathrm{d}\ell^2$, while it adds a dipole contribution to b. From the perspective of transport in holographic fluids, the purpose is to list the dangerous tensors carried by this kind of boundaries. The more tensors we have, the more information we gain on vanishing transport coefficients: since the energy–momentum tensor that emerges holographically is perfect, any transport coefficient coupled to a dangerous tensor is necessarily zero.

For the boundary metric (13.62), the vorticity strength, the Cotton prefactor and the scalar curvature read:

$$
\begin{aligned}
q &= 2k^2(n - a\cos\vartheta)\,,\\
c &= 2k^4 n\left(1 + k^2\left(4n^2 - a^2\right)\right)\,,\\
R &= 2k^2\left(1 + k^2n^2 + 10k^2 na\cos\vartheta + k^2a^2\left(1 - 5\cos^2\vartheta\right)\right)\,.
\end{aligned}
$$

We observe that, on the one hand, the nut charge n is responsible for the $2+1$-dimensional boundary not being conformally flat. The ordinary rotation parameter a, on the other hand, introduces a ϑ-dependence in q and R. This betrays the breaking of homogeneity due to a: when a vanishes, the boundary is an squashed S^3 with $SU(2)\times\mathbb{R}$ isometry, which is a homogeneous space–time, and all of its scalars are constants.[25]

Coming back to the discussion on the dangerous tensors, we expect them to be more numerous when less symmetry is present. Indeed, for vanishing a, all scalars are constant and both the Riemann and the Cotton are combinations of $u_\mu u_\nu$ and $g_{\mu\nu}$ with constant coefficients. Any covariant derivative acting on those will be algebrised in a similar fashion. Thus

- all hydrodynamic scalars are constants,
- all hydrodynamic vectors are of the form Au_μ with constant A, and
- all hydrodynamic tensors are of the form $Bu_\mu u_\nu + Cg_{\mu\nu}$ with constant B and C.

Hence there exists no traceless transverse tensor that can correct the hydrodynamic energy–momentum tensor in perfect equilibrium. In other words, there is no dangerous tensor. Therefore, in the case of monopolar geometries, the symmetry is too rich and in such a highly symmetric kinematical configuration, the fluid dynamics cannot be sensitive to any dissipative or non-dissipative coefficient. As soon as a dipole component is added ($a \neq 0$), a space-dependence emerges in the various scalars and tensors, and infinitely many dangerous tensors appear, which provide valuable information on the vanishing transport coefficients of the holographic fluid.

---

[25]This family includes Gödel space–time (see [67, 68] for more information). The important issue of closed time-like curves emerges as a consequence of the lack of global hyperbolicity. This was discussed in [20–22], in relation with holographic fluids. When the bulk geometry has hyperbolic horizon, this caveat can be circumvented.

The above discussion provides a guide for the subsequent analysis. To have access to more transport coefficients, we must perturb the geometry in a way organized e.g. as a multipolar expansion: the higher the multipole in the geometry, the richer the spectrum of transport coefficients that can contribute, if non-vanishing, to the state of the fluid. No exact Einstein spaces are however available beyond dipole configuration (Kerr).[26] Thus, this programme lies outside of the present framework, as it requires to work with perturbed bulk Einstein spaces, and handle fluid perturbations potentially bringing the fluid away from perfect equilibrium.

13.5 Monopolar Boundaries and Topologically Massive Gravity

Monopolar geometries have been mentioned in Sect. 13.4.2 around the example (13.62), which appears as the boundary of Taub–NUT Schwarzschild AdS black hole with spherical horizon. This terminology is justified by the fact that the vorticity strength q is constant (like the strength of the magnetic field on a sphere surrounding a Dirac monopole). Within the perfect-Cotton Papapetrou–Randers geometries (13.52), there is a whole class of monopolar boundaries, obtained by setting $c_2 = 0$ in (13.55). With constant q, using the general equations (13.67) and (13.68) for Papapetrou–Randers, as well as (13.45)–(13.48) for perfect-Cotton geometries, we find:

$$
\begin{aligned}
R &= \delta - \frac{5q^2}{2} , \\
R_{\mu\nu}\, \mathrm{d}x^\mu \mathrm{d}x^\nu &= \frac{\delta - q^2}{2} \mathrm{u}^2 + \left(\frac{\delta}{2} - q^2\right) \mathrm{d}s^2 , \\
C_{\mu\nu}\, \mathrm{d}x^\mu \mathrm{d}x^\nu &= \frac{q}{4} \left(\delta - q^2\right) \left(3\mathrm{u}^2 + \mathrm{d}s^2\right) .
\end{aligned}
$$

These expressions can be combined into

$$
R_{\mu\nu} - \frac{R}{2} g_{\mu\nu} + \lambda g_{\mu\nu} = \frac{1}{\mu} C_{\mu\nu} \tag{13.63}
$$

with

$$
\lambda = \frac{\delta}{6} - \frac{5q^2}{12}, \quad \mu = \frac{3q}{2} .
$$

---

[26] In 1919, Weyl exhibited multipolar Ricci-flat solutions, which do not seem extendible to the Einstein case (see [69] for details).

Expression (13.63) shows that monopolar geometries solve the topologically massive gravity equations [24] for appropriate constants λ and μ. This is not surprising, as it is a known fact that, for example, squashed anti-de-Sitter or squashed three-spheres solve topologically massive gravity equations [59–62]. However, what is worth stressing here is that reversing the argument and requiring a generic Papapetrou–Randers background (13.39) to solve (13.63) leads *necessarily* to a monopolar geometry. We leave as an exercise to set that result.[27]

As already advertised, the topological mass term (resulting from the Chern–Simons action in (13.3)) appears explicitly, in the cases under consideration, as part of topologically massive gravity equations. The reader might be puzzled by this connection. The $2+1$-dimensional geometries analyzed here are not supposed to carry any gravity degree of freedom since they are ultimately designed to serve as holographic boundaries. Hence, the emergence of topologically massive gravity should not *a priori* be considered as a sign of dynamics. Nevertheless, as for the general "self-dual" case (Eqs. (13.34) obtained by varying (13.3)), we should leave open the option of introducing some topologically massive graviton dynamics on the boundary. This approach should not be confused with that of some recent works [70, 71], where topologically massive gravity and its homogeneous solutions play the role of *bulk* geometries. Investigating the interplay between these two viewpoints might be of some relevance.

13.6 Outlook

Modified versions of Einstein's gravity are of interest primarily in cosmology. The aim of the present lectures is to set a bridge with a somewhat less expected area of applications, namely holography. Prior to holography we actually find, in four-dimensional Euclidean framework, quaternionic spaces. These, from the Fefferman–Graham viewpoint, require a boundary condition, which is obtained holographically as the extremization of

$$S = S_{\text{holographic matter}} + S_{\text{Chern–Simons}}\,. \tag{13.64}$$

Assuming homogeneity for the boundary metric, further restricts (13.64) to the topologically massive gravity action, as shown in the last paragraph of these notes. Although, at this stage, only the extremum of this action is relevant, investigating

---

[27]Use the expression for the Ricci tensor for Papapetrou–Randers geometries (13.68), impose tracelessness and extract λ. Then use (13.69) and (13.54) and conclude that q must be constant and related to μ. Combine these results and reach the conclusion that all solutions are fibrations over a two-dimensional space with metric $\mathrm{d}\ell^2$ of constant curvature $\hat{R} = 6\lambda - {}^{2\mu^2}\!/_9$. They are thus homogeneous spaces of either positive (S^2), null ($\mathbb{R}^2$) or negative curvature (H_2).

boundary graviton dynamics in holographic set-ups might prove interesting in the future.

Translating the bulk Weyl self-duality condition into boundary data opens up the possibility to make it applicable for Lorentzian-signature bulk and boundary geometries. This sort of integrability requirement is not necessary, however, and many Einstein spaces exist, which do not satisfy (13.34).[28] Investigating further the relationships amongst the boundary energy–momentum tensor and the boundary Cotton tensor may be instructive in the case of exact Einstein spaces, which fall outside of the class studied here.[29] This could be useful both for understanding the underlying gravitational structure and for studying transport properties in conformal holographic fluids.

Besides potential generalizations of (13.34), appears also here the issue of the form of the boundary metric and of the energy–momentum tensor. Our analysis has been limited to (i) stationary Papapetrou–Randers boundary geometries (13.39) with $B = 1$, and (ii) perfect-fluid-like boundary energy–momentum tensors. These options make operational the determination of vanishing transport coefficients by imposing perfect equilibrium, which turns out to exist holographically. We may however scan more general situations as many more exact Einstein spaces exist that deserve to be analyzed. We have already quoted in Sect. 13.4.2 the Plebañski–Demiański Einstein stationary solutions [63], for which the acceleration parameter is a source of deviation from the perfect-Cotton boundary geometry. Non-stationary spaces provide equally interesting laboratories for further investigation (see footnote 29). Finally, on the Euclidean side, a great deal of techniques (isomonodromic deformations, twistors, ...) have been developed for finding the families of quaternionic spaces quoted in [9, 10, 13, 15, 16, 18, 29] (see also [73] for a review). Among these, the Calderbank–Pedersen two-Killing family [19] is particularly interesting, because it includes the Euclidean Weyl-self-dual version[30] of the Kerr–Taub–NUT (13.61). Since this family contains more self-dual metrics than our exhaustive analysis of Sect. 13.4.2 has revealed, these metrics must necessarily lead to a non-perfect boundary energy–momentum tensor, potentially combined with a Papapetrou–Randers boundary geometry with non-constant B. Although this discussion is valid in the Euclidean and not all Euclidean solutions admit a real-time continuation, it should help clarifying the landscape of self-duality holographic properties, and possibly be useful for Lorentzian extensions.

Last, but very intriguing, comes the limitation in the dimension. We have been analyzing four dimensional bulk geometries because our guideline was self-duality, which indeed exists in this (Euclidean) framework. It can however be generalized in eight-dimensional spaces. There, it is known that the octonionic symbols Ψ^{ABCD} allow to define a duality map: $\tilde{\mathscr{R}}^{AB} = \Psi^{ABCD}\mathscr{R}_{CD}$. Reducing the Riemann two-form

[28] As usual with instantons, self-duality selects ground states, but exact excited states can also exist.

[29] Recently this was discussed for a non-stationary solution of Einstein's equations [72].

[30] In this case, (13.21) is traded for $M = n\left(1 - k^2\left(4n^2 - a^2\right)\right)$ (see also [74]).

$\mathscr{R}_{AB}$, which belongs to the **28** of $SO(8)$, with respect to $\text{Spin}_7 \subset SO(8)$ leads to a self-dual component $\mathscr{S}_{\mathbf{21}}$ and an anti-self-dual one $\mathscr{A}_{\mathbf{7}}$. Equations (13.11) and (13.12) are now traded for

$$\mathscr{S}_{\mathbf{21}} = W^{\mathbf{168}}\phi_{\mathbf{21}} + s\phi_{\mathbf{21}} + W^{\mathbf{105}}\chi_{\mathbf{7}} ,$$
$$\mathscr{A}_{\mathbf{7}} = W^{\mathbf{27}}\chi_{\mathbf{7}} + s\chi_{\mathbf{7}} + S^{\mathbf{35}}\phi_{\mathbf{21}} ,$$

where the singlet s is the scalar curvature, $S^{\mathbf{35}}$ is the traceless Ricci, and the $W^{\mathbf{I}}$ are the three irreducible components of the Weyl tensor. Riemann self-dual gravitational instantons, obtained by setting $\mathscr{A}_{\mathbf{7}} = 0$, are known to exist [75–79]. Those are Ricci flat. The question is still open to find Weyl self-dual Einstein spaces, by demanding $S^{\mathbf{35}} = 0$ and $W^{\mathbf{27}} = 0$. From the boundary perspective, $W^{\mathbf{27}} = 0$ could be interpreted as the extremization requirement for (13.64) with respect to the seven-dimensional boundary metric, the Chern–Simons being now the seven-dimensional one [80].

Acknowledgements I wish to thank the organizers of the 7th Aegean summer school *Beyond Einstein's theory of gravity*, where these lectures were delivered. The material presented here is borrowed from recent or on-going works realized in collaboration with M. Caldarelli, C. Charmousis, J.–P. Derendinger, J. Gath, R. Leigh, A. Mukhopadhyay, A. Petkou, V. Pozzoli, K. Sfetsos, K. Siampos and P. Vanhove. I also benefited from interesting discussions with I. Bakas, D. Klemm, N. Obers and Ph. Spindel. The feedback from the Southampton University group was also valuable during a recent presentation of this work in their seminar. This research was supported by the LABEX P2IO, the ANR contract 05-BLAN-NT09-573739, the ERC Advanced Grant 226371.

Appendix 1: On Vector-Field Congruences

We consider a manifold endowed with a space–time metric of the generic form

$$\mathrm{d}s^2 = g_{\mu\nu}\mathrm{d}x^{\mu}\mathrm{d}x^{\nu} = \eta_{\mu\nu}E^{\mu}E^{\nu}$$

(to avoid inflation of indices we do not distinguish between flat and curved ones). Consider now an arbitrary time-like vector field u, normalised as $u^{\mu}u_{\mu} = -1$, later identified with the fluid velocity. Its integral curves define a congruence which is characterised by its acceleration, shear, expansion and vorticity (see e.g. [81, 82]):

$$\nabla_{\mu}u_{\nu} = -u_{\mu}a_{\nu} + \frac{1}{D-1}\Theta\Delta_{\mu\nu} + \sigma_{\mu\nu} + \omega_{\mu\nu}$$

with[31]

$$
\begin{aligned}
a_\mu &= u^\nu \nabla_\nu u_\mu, \quad \Theta = \nabla_\mu u^\mu , \\
\sigma_{\mu\nu} &= \frac{1}{2} \Delta_\mu{}^\rho \Delta_\nu{}^\sigma \left(\nabla_\rho u_\sigma + \nabla_\sigma u_\rho\right) - \frac{1}{D-1} \Delta_{\mu\nu} \Delta^{\rho\sigma} \nabla_\rho u_\sigma \\
&= \nabla_{(\mu} u_{\nu)} + a_{(\mu} u_{\nu)} - \frac{1}{D-1} \Delta_{\mu\nu} \nabla_\rho u^\rho , \\
\omega_{\mu\nu} &= \frac{1}{2} \Delta_\mu{}^\rho \Delta_\nu{}^\sigma \left(\nabla_\rho u_\sigma - \nabla_\sigma u_\rho\right) = \nabla_{[\mu} u_{\nu]} + u_{[\mu} a_{\nu]} .
\end{aligned}
$$

The latter allows to define the vorticity form as

$$
2\omega = \omega_{\mu\nu}\, \mathrm{dx}^\mu \wedge \mathrm{dx}^\nu = \mathrm{du} + \mathrm{u} \wedge \mathrm{a} . \tag{13.65}
$$

The time-like vector field u has been used to decompose any tensor field on the manifold in transverse and longitudinal components. The decomposition is performed by introducing the longitudinal and transverse projectors:

$$
U^\mu{}_\nu = -u^\mu u_\nu, \quad \Delta^\mu{}_\nu = u^\mu u_\nu + \delta^\mu_\nu , \tag{13.66}
$$

where $\Delta_{\mu\nu}$ is also the induced metric on the surface orthogonal to u. The projectors satisfy the usual identities:

$$
U^\mu{}_\rho U^\rho{}_\nu = U^\mu{}_\nu, \quad U^\mu{}_\rho \Delta^\rho{}_\nu = 0, \quad \Delta^\mu{}_\rho \Delta^\rho{}_\nu = \Delta^\mu{}_\nu, \quad U^\mu{}_\mu = 1, \quad \Delta^\mu{}_\mu = D-1 ,
$$

and similarly:

$$
u^\mu a_\mu = 0, \quad u^\mu \sigma_{\mu\nu} = 0, \quad u^\mu \omega_{\mu\nu} = 0, \quad u^\mu \nabla_\nu u_\mu = 0, \quad \Delta^\rho{}_\mu \nabla_\nu u_\rho = \nabla_\nu u_\mu .
$$

Appendix 2: Papapetrou–Randers Backgrounds and Aligned Fluids

In this appendix, we collect a number of useful expressions for stationary Papapetrou–Randers three-dimensional geometries (13.39) with $B = 1$, and for fluids in perfect equilibrium on these backgrounds. The latter follow geodesic congruences, aligned with the normalized Killing vector ∂_t, with velocity one-form given in (13.41).

[31] Our conventions are: $A_{(\mu\nu)} = 1/2\left(A_{\mu\nu} + A_{\nu\mu}\right)$ and $A_{[\mu\nu]} = 1/2\left(A_{\mu\nu} - A_{\nu\mu}\right)$.

We introduce the inverse two-dimensional metric a^{ij}, and b^i such that

$$a^{ij}a_{jk} = \delta^i_k \,, \quad b^i = a^{ij}b_j \,.$$

The three-dimensional metric components read:

$$g_{00} = -1 \,, \quad g_{0i} = b_i \,, \quad g_{ij} = a_{ij} - b_i b_j \,,$$

and those of the inverse metric:

$$g^{00} = a^{ij}b_i b_j - 1 \,, \quad g^{0i} = b^i \,, \quad g^{ij} = a^{ij} \,.$$

Finally,

$$\sqrt{|g|} = \sqrt{a} \,,$$

where a is the determinant of the symmetric matrix with entries a_{ij}.

Using (13.41) and (13.65) we find that the vorticity of the aligned fluid can be written as the following two-form (the acceleration term is absent here)

$$\omega = \frac{1}{2}\omega_{\mu\nu}\mathrm{d}x^\mu \wedge \mathrm{d}x^\nu = \frac{1}{2}\mathrm{db} \,.$$

The Hodge-dual of $\omega_{\mu\nu}$ is

$$\psi^\mu = \eta^{\mu\nu\rho}\omega_{\nu\rho} \Leftrightarrow \omega_{\nu\rho} = -\frac{1}{2}\eta_{\nu\rho\mu}\psi^\mu \,.$$

In $2+1$ dimensions it is aligned with the velocity field:

$$\psi^\mu = q u^\mu \,,$$

where, in our set-up,

$$q(x) = -\frac{\epsilon^{ij}\partial_i b_j}{\sqrt{a}} \,.$$

It is a static scalar field that we call the *vorticity strength*, carrying dimensions of inverse length. Together with $\hat{R}(x)$—the curvature of the two-dimensional metric $\mathrm{d}\ell^2$ introduced in (13.42), the above scalar carries all relevant information for the curvature of the Papapetrou–Randers geometry. We quote for latter use the three-dimensional curvature scalar:

$$R = \hat{R} + \frac{q^2}{2} \,, \tag{13.67}$$

the three-dimensional Ricci tensor

$$R_{\mu\nu}\,\mathrm{d}x^\mu\mathrm{d}x^\nu = \frac{q^2}{2}\mathrm{u}^2 + \frac{\hat{R}+q^2}{2}\mathrm{d}\ell^2 - \mathrm{u}\,\mathrm{d}x^\rho u^\sigma \eta_{\rho\sigma\mu}\nabla^\mu q\;, \tag{13.68}$$

as well as the three-dimensional Cotton–York tensor:

$$\begin{aligned} C_{\mu\nu}\,\mathrm{d}x^\mu\mathrm{d}x^\nu = & \frac{1}{2}\left(\hat{\nabla}^2 q + \frac{q}{2}(\hat{R}+2q^2)\right)\left(2\mathrm{u}^2+\mathrm{d}\ell^2\right) \\ & -\frac{1}{2}\left(\hat{\nabla}_i\hat{\nabla}_j q\,\mathrm{d}x^i\mathrm{d}x^j + \hat{\nabla}^2 q\,\mathrm{u}^2\right) \\ & -\frac{\mathrm{u}}{2}\mathrm{d}x^\rho u^\sigma \eta_{\rho\sigma\mu}\nabla^\mu(\hat{R}+3q^2)\,. \end{aligned} \tag{13.69}$$

References

1. E.T. Newman, L. Tamburino, T.J. Unti, Empty-space generalization of the Schwarzschild metric. J. Math. Phys. **4**, 915 (1963)
2. T. Eguchi, A.J. Hanson, Self-dual solutions to Euclidean gravity. Ann. Phys. **120**, 82 (1979)
3. T. Eguchi, A.J. Hanson, Gravitational instantons. Gen. Rel. Grav. **11**, 315 (1979)
4. M.F. Atiyah, N.J. Hitchin, Low-energy scattering of non-abelian monopoles. Phys. Lett. **107A**, 21 (1985)
5. R.S. Ward, Integrable and solvable systems, and relations among them. Phil. Trans. R. Soc. Lond. **A315**, 451 (1985)
6. G.W. Gibbons, C.N. Pope, $\mathbb{CP}_2$ as a gravitational instanton. Comm. Math. Phys. **61**, 239 (1978)
7. R.S. Ward, Self-dual space–times with cosmological constant. Comm. Math. Phys. **78**, 1 (1980)
8. C.R. Lebrun, $\mathscr{H}$-space with a cosmological constant. Proc. R. Soc. Lond. **A380**, 171 (1982)
9. H. Pedersen, Eguchi–Hanson metrics with cosmological constant. Class. Quant. Grav. **2**, 579 (1985)
10. H. Pedersen, Einstein metrics, spinning top motions and monopoles. Math. Ann. **274**, 35 (1986)
11. H. Pedersen, Y.S. Poon, Hyper-Kähler metrics and a generalization of the Bogomolny equations. Comm. Math. Phys. **117**, 569 (1988)
12. H. Pedersen, Y.S. Poon, Kähler surfaces with zero scalar curvature. Class. Quant. Grav. **7**, 1707 (1990)
13. M. Przanowski, Killing vector fields in self-dual, Euclidean Einstein spaces with $\Lambda \neq 0$. J. Math. Phys. **32**, 1004 (1991)
14. K.P. Tod, A comment on a paper of Pedersen and Poon. Class. Quant. Grav. **8**, 1049 (1991)
15. K.P. Tod, Self-dual Einstein metrics from the Painlevé VI equation. Phys. Lett. **A190**, 221 (1994)
16. N.J. Hitchin, Twistor spaces, Einstein metrics and isomonodromic deformations. J. Differ. Geom. **42**, 30 (1995)
17. R. Maszczyk, L.J. Mason, N.M.J. Woodhouse, Self-dual Bianchi metric and Painlevé transcendents. Class. Quant. Grav. **11**, 65 (1994)
18. D.M.J. Calderbank, H. Pedersen, Self-dual spaces with complex structures, Einstein–Weyl geometry and geodesics. Ann. Inst. Fourier **50**, 921 (2000)
19. D.M.J. Calderbank, H. Pedersen, Self-dual Einstein metrics with torus symmetry. J. Differ. Geom. **60**, 485 (2002)
20. R.G. Leigh, A.C. Petkou, P.M. Petropoulos, Holographic three-dimensional fluids with non-trivial vorticity. Phys. Rev. **D85**, 086010 (2012) [arXiv:1108.1393 [hep-th]]

21. R.G. Leigh, A.C. Petkou, P.M. Petropoulos, Holographic fluids with vorticity and analogue gravity systems. JHEP **1211**, 121 (2012) [arXiv:1205.6140 [hep-th]]
22. M.M. Caldarelli, R.G. Leigh, A.C. Petkou, P.M. Petropoulos, V. Pozzoli, K. Siampos, Vorticity in holographic fluids. Proc. Sci. **Corfu11**, 076 (2012) [arXiv:1206.4351 [hep-th]]
23. A. Mukhopadhyay, A.C. Petkou, P.M. Petropoulos, V. Pozzoli, K. Siampos, Holographic perfect fluidity, Cotton energy–momentum duality and transport properties. JHEP **04**, 136 (2014) [arXiv:1309.2310 [hep-th]]
24. S. Deser, R. Jackiw, S. Templeton, Topologically massive gauge theories. Ann. Phys. **140**, 372 (1982); Erratum-ibid. **185**, 406 (1988); Three-dimensional massive gauge theories. Phys. Rev. Lett. **48**, 975 (1982)
25. M. Cahen, R. Debever, L. Defrise, A complex vectorial formalism in general relativity. J. Math. Mech. **16**, 761 (1967)
26. M.F. Atiyah, N.J. Hitchin, I.M. Singer, Self-duality in four dimensional Riemannian geometry. Proc. Roy. Soc. Lond. **A362**, 425 (1978)
27. T. Eguchi, P.B. Gilkey, A.J. Hanson, Gravitation, gauge theories and differential geometry. Phys. Rept. **66**, 213 (1980)
28. G.W. Gibbons, S.W. Hawking, Classification of gravitational instanton symmetries. Commun. Math. Phys. **66**, 291 (1979)
29. C.R. Lebrun, Counter-examples to the generalized positive action conjecture. Commun. Math. Phys. **118**, 591 (1988)
30. K. Zoubos, Holography and quaternionic Taub–NUT. JHEP **0212**, 037 (2002) [arXiv:hep-th/0209235]
31. K. Zoubos, A Conformally invariant holographic two point function on the Berger sphere. JHEP **0501**, 031 (2005) [arXiv:hep-th/0403292]
32. C. Fefferman, C.R. Graham, Conformal invariants, in *Elie Cartan et les mathématiques d'aujourd'hui*, Astérisque, 1985, numéro hors série Soc. Math. France, Paris, 95
33. C. Fefferman, C.R. Graham, The ambient metric, arXiv:0710.0919 [math.DG]
34. K. Skenderis, S.N. Solodukhin, Quantum effective action from the AdS/CFT correspondence. Phys. Lett. **B472**, 316 (2000) [arXiv:hep-th/9910023]
35. S. de Haro, K. Skenderis, S.N. Solodukhin, Holographic reconstruction of spacetime and renormalization in the AdS/CFT correspondance. Commun. Math. Phys. **217**, 595 (2001) [arXiv:hep-th/0002230]
36. I. Papadimitriou, K. Skenderis, Thermodynamics of asymptotically locally AdS spacetimes. JHEP **0508**, 004 (2005) [arXiv:hep-th/0505190]
37. D.S. Mansi, A.C. Petkou, G. Tagliabue, Gravity in the $3+1$-split formalism I: holography as an initial value problem. Class. Quant. Grav. **26**, 045008 (2009) [arXiv:0808.1212 [hep-th]]
38. D.S. Mansi, A.C. Petkou, G. Tagliabue, Gravity in the $3+1$-split formalism II: self-duality and the emergence of the gravitational Chern–Simons in the boundary. Class. Quant. Grav. **26**, 045009 (2009) [arXiv:0808.1213 [hep-th]]
39. G.W. Gibbons, C.N. Pope, The positive action conjecture and asymptotically Euclidean metrics in quantum gravity. Commun. Math. Phys. **66**, 267 (1979)
40. F. Bourliot, J. Estes, P.M. Petropoulos, Ph. Spindel, Gravitational instantons, self-duality and geometric flows. Phys. Rev. **D81**, 104001 (2010) [arXiv:0906.4558 [hep-th]]
41. F. Bourliot, J. Estes, P.M. Petropoulos, Ph. Spindel, G3-homogeneous gravitational instantons. Class. Quant. Grav. **27**, 105007 (2010) [arXiv:0912.4848 [hep-th]]
42. P.M. Petropoulos, V. Pozzoli, K. Siampos, Self-dual gravitational instantons and geometric flows of all Bianchi types. Class. Quant. Grav. **28**, 245004 (2011) [arXiv:1108.0003 [hep-th]]
43. V.A. Belinsky, G.W. Gibbons, D.N. Page, C.N. Pope, Asymptotically Euclidean Bianchi IX metrics in quantum gravity. Phys. Lett. **76B**, 433 (1978)
44. R.G. Leigh, A.C. Petkou, Gravitational duality transformations on (A)dS$_4$. JHEP **0711**, 079 (2007) [arXiv:0704.0531 [hep-th]]
45. S. de Haro, A.C. Petkou, Holographic aspects of electric–magnetic dualities. J. Phys. Conf. Ser. **110**, 102003 (2008) [arXiv:0710.0965 [hep-th]]

46. S. de Haro, Dual gravitons in AdS_4 / CFT_3 and the holographic cotton tensor. JHEP **0901**, 042 (2009) [arXiv:0808.2054 [hep-th]]
47. O. Miskovic, R. Olea, Topological regularization and self-duality in four-dimensional anti-de Sitter gravity. Phys. Rev. **D79**, 124020 (2009) [arXiv:0902.2082 [hep-th]]
48. I. Bakas, Energy-momentum/Cotton tensor duality for AdS_4 black holes. JHEP **0901**, 003 (2009) [arXiv:0809.4852 [hep-th]]
49. I. Bakas, Duality in linearized gravity and holography. Class. Quant. Grav. **26**, 065013 (2009) [arXiv:0812.0152 [hep-th]]
50. P. Romatschke, New developments in relativistic viscous hydrodynamics. Int. J. Mod. Phys. **E19**, 1 (2010) [arXiv:0902.3663 [hep-ph]]
51. P. Kovtun, Lectures on hydrodynamic fluctuations in relativistic theories. J. Phys. **A45**, 473001 (2012) [arXiv:1205.5040 [hep-th]]
52. V.E. Hubeny, S. Minwalla, M. Rangamani, The fluid/gravity correspondence, arXiv:1107.5780 [hep-th]
53. M. Rangamani, Gravity and hydrodynamics: lectures on the fluid-gravity correspondence. Class. Quant. Grav. **26**, 224003 (2009) [arXiv:0905.4352 [hep-th]]
54. G.D. Moore, K.A. Sohrabi, Kubo formulæ for second-order hydrodynamic coefficients. Phys. Rev. Lett. **106**, 122302 (2011) [arXiv:1007.5333 [hep-ph]]
55. N. Banerjee, J. Bhattacharya, S. Bhattacharyya, S. Jain, S. Minwalla, T. Sharma, Constraints on fluid dynamics from equilibrium partition functions. JHEP **1209**, 046 (2012) [arXiv:1203.3544 [hep-th]]
56. A. Papapetrou, Champs gravitationnels stationnaires à symétrie axiale. Ann. Inst. H. Poincaré **A4**, 83 (1966)
57. G. Randers, On an asymmetrical metric in the four-space of general relativity. Phys. Rev. **59**, 195 (1941)
58. L.D. Landau, E.M. Lifchitz, *Physique Théorique*, vol. 5 *Physique Statistique* §27, MIR 1969
59. D.D.K. Chow, C.N. Pope, E. Sezgin, Classification of solutions in topologically massive gravity. Class. Quant. Grav. **27**, 105001 (2010) [arXiv:0906.3559 [hep-th]]
60. G. Guralnik, A. Iorio, R. Jackiw, S.Y. Pi, Dimensionally reduced gravitational Chern–Simons term and its kink. Ann. Phys. **308**, 222 (2003) [arXiv:hep-th/0305117]
61. D. Grumiller, W. Kummer, The classical solutions of the dimensionally reduced gravitational Chern–Simons theory. Ann. Phys. **308**, 211 (2003) [arXiv:hep-th/0306036]
62. G. Moutsopoulos, P. Ritter, An exact conformal symmetry ansatz on Kaluza–Klein reduced TMG. Gen. Rel. Grav. **43**, 3047 (2011) [arXiv:1103.0152 [hep-th]]
63. J.F. Plebañski, M. Demiañski, Rotating, charged, and uniformly accelerating mass in general relativity. Ann. Phys. (NY) **98**, 98 (1976)
64. N. Alonso-Alberca, P. Meessen, T. Ortin, Supersymmetry of topological Kerr–Newman–Taub–NUT–AdS space–times. Class. Quant. Grav. **17**, 2783 (2000) [arXiv:0003071 [hep-th]]
65. D. Klemm, V. Moretti, L. Vanzo, Rotating topological black holes. Phys. Rev. **D57**, 6127 (1998) [Erratum-ibid. **D60** (1999) 109902] [arXiv:gr-qc/9710123]
66. D. Klemm, A. Maiorana, Fluid dynamics on ultrastatic spacetimes and dual black holes (2014). arXiv:1404.0176 [hep-th]
67. A.K. Raychaudhuri, S.N. Guha Thakurta, Homogeneous space–times of the Gödel type. Phys. Rev. **D22**, 802 (1980)
68. M.J. Rebouças, J. Tiomno, Homogeneity of Riemannian space–times of Gödel type. Phys. Rev. **D28**, 1251 (1983)
69. H. Stephani, D. Kramer, M. MacCallum, C. Hoenselaers, E. Herlt, *Exact solutions to Einstein's Field Equations*. Cambridge Monographs on Mathematical Physics (CUP, Cambridge, 2003)
70. D. Anninos, W. Li, M. Padi, W. Song, A. Strominger, Warped AdS_3 black holes. JHEP **0903**, 130 (2009) [arXiv:0807.3040 [hep-th]]
71. D. Anninos, S. de Buyl, S. Detournay, Holography for a de Sitter–Esque geometry. JHEP **1105**, 003 (2011) [arXiv:1102.3178 [hep-th]]
72. G.B. de Freitas, H.S. Reall, Algebraically special solutions in AdS/CFT (2014). arXiv:1403.3537 [hep-th]

73. P.M. Petropoulos, P. Vanhove, Gravity, strings, modular and quasimodular forms. Ann. Math. Blaise Pascal **19**, 379 (2012) [arXiv:1206.0571 [math-ph]]
74. K. Behrndt, G. Dall'Agata, D. Lüst, S. Mahapatra, Intersecting six-branes from new seven manifolds with G_2 holonomy. JHEP **0208**, 027 (2002) [arXiv:hep-th/0207117]
75. B.S. Acharya, M. O'Loughlin, Self-duality in $D \leq$ eight-dimensional Euclidean gravity. Phys. Rev. **D55**, 4521 (1997) [arXiv:hep-th/9612182]
76. E.G. Floratos, A. Kehagias, Eight-dimensional self-dual spaces. Phys. Lett. **B427**, 283 (1998) [arXiv:hep-th/9802107]
77. I. Bakas, E.G. Floratos, A. Kehagias, Octonionic gravitational instantons. Phys. Lett. **B445**, 69 (1998) [arXiv:hep-th/9810042]
78. A. Bilal, J.-P. Derendinger, K. Sfetsos, (Weak) G_2 holonomy from self-duality, flux and supersymmetry. Nucl. Phys. **B628**, 112 (2002) [arXiv:hep-th/0111274]
79. R. Hernandez, K. Sfetsos, An eight-dimensional approach to G_2 manifolds. Phys. Lett. **B536**, 294 (2002) [arXiv:hep-th/0202135]
80. J. Zanelli, Introduction to Chern–Simons theories. Contribution to the 7th Aegean summer school, Paros (GR), September 2013
81. J. Ehlers, Contributions to the relativistic mechanics of continuous media. Gen. Rel. Grav. **25**, 1225 (1993)
82. H. van Elst, C. Uggla, General relativistic $1 + 3$-orthonormal frame approach revisited. Class. Quant. Grav. **14**, 2673 (1997) [arXiv:gr-qc/9603026]

Chapter 14
Beyond Supergravity in AdS-CFT: An Application to Jet Quenching

Diana Vaman

Abstract These notes are dedicated to the study of jet quenching in the strongly coupled limit using gauge/string duality. We are interested in corrections to the infinite coupling $\lambda=\infty$ result for the jet stopping, in powers of $\lambda^{-1/2}$. To estimate these corrections we need to go beyond supergravity in AdS-CFT, and include all higher-derivative corrections to the supergravity action which arise from the string α' expansion. For the particular type of "jets" that we study, the expansion in $\lambda^{-1/2}$ is well behaved for jets whose stopping distance ℓ_{stop} is in the range $\lambda^{-1/6}\ell_{\text{max}} \ll \ell_{\text{stop}} \lesssim \ell_{\text{max}}$, but the expansion breaks down for jets created in such a way that $\ell_{\text{stop}} \ll \lambda^{-1/6}\ell_{\text{max}}$. The reason for the breakdown of the $\lambda^{-1/2}$ expansion is caused by the excitation of massive string states. In particular, consider "jets" which are dual to high-momentum gravitons. In the black brane background the gravitons, which are closed string states, get stretched into relatively large classical strings by tidal forces. These stringy excitations of the graviton are not contained in the supergravity approximation, but the jet stopping problem can nonetheless still be solved by drawing on various string-theory methods (the eikonal approximation, the Penrose limit, string quantization in pp-wave backgrounds) to obtain a probability distribution for the late-time classical string loops.

14.1 A Brief Introduction and Overview

Inspired by the observation of (and rapidly growing body of experimental information on) jet quenching in relativistic heavy ion collisions, there has for many years been an interest in the theory of jet quenching and what can be learned about that theory by studying interesting limiting cases. One of the simplest-to-pose thought experiments is this: How far does a very-high momentum excitation (the potential precursor of a would-be jet) travel in a thermal QCD medium before it loses energy, stops, and thermalizes in the medium? And how does the answer to that question depend on the effective strength α_s of the strong coupling?

D. Vaman (✉)
Department of Physics, University of Virginia, Box 400714, Charlottesville, Virginia 22904, USA
e-mail: dv3h@virginia.edu

E. Papantonopoulos (ed.), *Modifications of Einstein's Theory of Gravity at Large Distances*, Lecture Notes in Physics 892, DOI 10.1007/978-3-319-10070-8_14

This question can be addressed from first principles in various theoretical limits. One such limit is that of weak coupling, which in principle applies to asymptotically large temperatures T and jet energies E, for which the relevant running values of α_s are small. In that limit, the stopping distance ℓ_{stop} for a high-energy parton ($E \gg T$) scales with energy as $E^{1/2}$, up to logarithms [1]. A contrasting limit of interest occurs when the running values of α_s relevant to jet stopping are all large.[1] This problem is not very tractable from first principles in QCD itself, but, through gauge-gravity duality, progress can be made for QCD-like plasmas with gravity duals, such as $\mathcal{N}=4$ super Yang Mills (SYM) theory. For some years, people have considered various ways to study analogs of jet stopping in such plasmas, namely the stopping distance for various types of localized, high-momentum excitations. The exact stopping distance depends on details of exactly how the "jet" is prepared, but universally these studies have found that the maximum possible stopping distance ℓ_{max} scales with energy as $E^{1/3}$, in contrast to the weak-coupling scaling of $E^{1/2}$. [3–8][2] have used gauge-gravity duality to study this problem in the strong coupling limit $\lambda \equiv N_c g_{YM}^2 \to \infty$ of large-N_c, $\mathcal{N}=4$ supersymmetric Yang-Mills (SYM) and related QCD-like plasmas. This is an interesting theoretical result because it teaches us that the scaling of jet stopping with energy depends on the strength of the coupling. It remains an open question (which we will not answer here) how $E^{1/3}$ starts to move toward $E^{1/2}$ as one lowers the coupling, and vice versa.

The stopping distance of high-momentum, localized excitations travelling through the plasma depends on more than just the energy of the excitation. Depending on exactly how one creates the excitation (the "jet"), one may get stopping distances ℓ_{stop} significantly smaller than the maximum ℓ_{max}. As an example from weak coupling, imagine that we spread out the total energy and momentum E of the jet among 10 partons, each having energy $E/10$, rather than putting it all into a single parton of energy E. Each of the 10 partons has lower energy than the single one and so will stop sooner; so the stopping distance for the high-momentum excitation depends on how many high-energy partons we use in the initial state. In the weak-coupling case, the maximum stopping distance ℓ_{max} corresponds to the particular initial state where all the energy is concentrated into a single initial parton.

[1]In a weak-coupling analysis, the two running couplings relevant to jet stopping are, roughly, $\alpha_s(T)$ and $\alpha_s(Q_\perp)$, where $Q_\perp \sim (\hat{q}E)^{1/4}$ grows slowly with energy and is the scale of the typical relative momentum between two daughter partons when a high energy parton splits through hard bremsstrahlung or pair production. ($\hat{q} \sim \alpha_s^2 T^3$ is a scale characteristic of the plasma that parametrizes transverse momentum diffusion of high-energy partons.) A third limiting case of interest, not addressed here, is where $\alpha_s(T)$ is large but $\alpha_s(Q_\perp)$ is small. See, for example, Liu, Rajagopal, and Wiedemann [2].

[2]See Sin and Zahed [9] for the earliest attempt we are aware of to discuss jet stopping in the context of gauge-gravity duality. See also [10]. In our work, we will only consider analogs of light-particle jets and will not study the heavy-particle case. For $\lambda=\infty$ analysis of the latter, see, for example, [11, 12] and references therein.

In the strong-coupling case, we cannot speak of individual partons, but the stopping distance again depends on how we prepare the initial high-momentum excitation. In our work [6, 7, 13], we create the initial excitation in a way that is analogous to what you would get if a high-momentum, slightly virtual photon (or graviton or other massless particle) decayed hadronically in the quark-gluon plasma, as depicted in Fig. 14.1. Alternatively, one could consider the decay of a high-momentum on-shell W boson (also depicted). For these methods of creating "jets," one finds that the maximum possible stopping distance scales as

$$\ell_{\text{max}} \sim \frac{E^{1/3}}{T^{4/3}}. \tag{14.1}$$

As we will review later, it turns out that the stopping distance may be made smaller than (14.1) by varying the virtuality $-q^2 \equiv -q_\mu q^\mu$ of the virtual photon (or equivalently the mass-squared M_{w}^2 of the on-shell W boson) [4, 7]. The important point is that there is a range of stopping distances $\ell_{\text{stop}} \lesssim \ell_{\text{max}}$ for our "jets," depending on the details of how those excitations are created.

Most top-down studies of jet stopping using gauge-gravity duality have studied the infinite color and infinite coupling limit, $N_c{=}\infty$ and $\lambda{=}\infty$, where $\lambda \equiv N_c g_{\text{YM}}^2$ is the 't Hooft coupling. To understand the true high-energy behavior, however, it is important to study the corrections to these limits. As an example, Fig. 14.2 shows two different scenarios one might imagine for the maximum stopping distance ℓ_{max} for strongly-coupled $\mathcal{N}{=}4$ SYM.

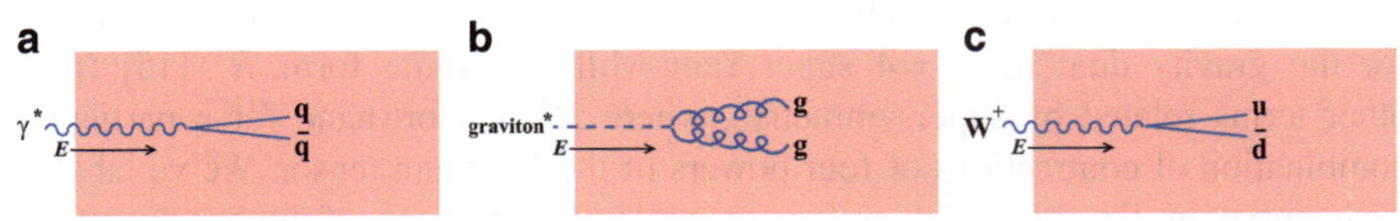

Fig. 14.1 Examples of the decay of a very high-energy (**a**) slightly virtual photon, (**b**) slightly virtual graviton, or (**c**) on-shell W^+ boson, inside a standard-model quark-gluon plasma, producing high-momentum partons moving to the right. In the context of $\mathcal{N}{=}4$ super Yang Mills, the q, u, and $\bar{d}$ above represent adjoint-color fermions or scalars carrying R charge. For strong coupling, of course, one should not picture perturbatively, as in this figure, the high-momentum excitation created in the plasma by the decay

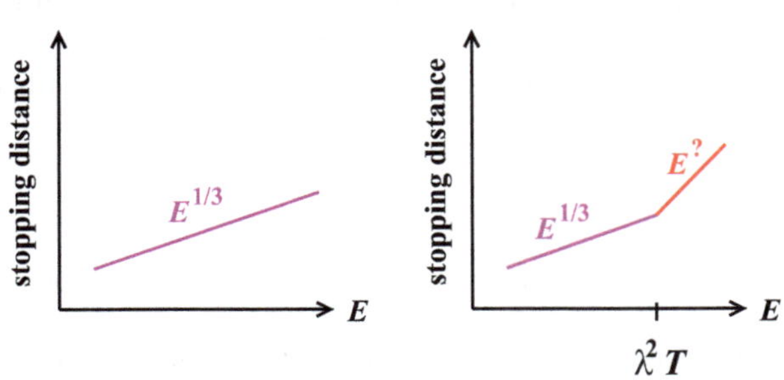

Fig. 14.2 Examples of two different scenarios for the high-energy ($E \gg T$) behavior of the maximum jet stopping distance $\ell_{\text{max}}(E)$ which are indistinguishable with $\lambda{=}\infty$ calculations

One is that $\ell_{\max}$ grows like $E^{1/3}$ at high energy, up to arbitrarily high energies. The other is that it starts growing like $E^{1/3}$ at high energy $E \gg T$, but then crosses over to some different power-law behavior once E exceeds some positive power of λ times T (e.g. $\lambda^2 T$ in the figure). In the latter case, $E^{1/3}$ would not be the true behavior for arbitrarily large E and large but finite λ. But there is no way to tell the difference between these two scenarios if one only has $\lambda{=}\infty$ calculations!

For this reason, we must analyze the parametric size of finite-λ corrections to jet stopping distances. Consequently, in the AdS/CFT we must go beyond the supergravity limit, and use the full fledged type IIB string theory dual.

Here we address the question of what happens when λ is large but not infinite, while keeping $N_c{=}\infty$.[3] In the holographic dual, this will require considering the effect of string corrections to the supergravity action. These corrections correspond to higher-derivative terms in the supergravity action, such as the 4th power of the Riemann curvature. Formally, the effects of higher and higher derivative corrections to the supergravity action are suppressed by more and more factors of $1/\sqrt{\lambda}$, but these suppressions might be compensated by large factors of E/T in the jet stopping problem. For this reason we will find that finite λ effects are sizable, as opposed to other results discussed in the literature where finite λ corrections are subleading and yield small corrections.

For the particular type of "jet" excitations that we study, Fig. 14.3 summarizes our findings. This figure depicts the parametric importance of corrections as a function of the $\lambda{=}\infty$ result ℓ_{stop} for the stopping length of the jet. The straight lines on this log-log plot represent simple power-law dependencies on the stopping distance. The first correction to the ten-dimensional low-energy supergravity action for the gravity dual to $\mathcal{N}{=}4$ super-Yang-Mills is of the form R^4 [15] (plus other terms related by supersymmetry), where R^4 is short-hand for a particular combination of contractions of four powers of the Riemann tensor. We've labeled each curve in Fig. 14.3 with a few examples of the type of higher-derivative correction that contributes to each. The precise meaning of the importance of an operator, denoted by the vertical axis, will be explained in Sect. 14.2.3. It is not quite the same thing as the relative change in stopping distance due to that operator, but, when the "importance" is small, the effect on the stopping distance will also be small. Finally, we stress that Fig. 14.3 assumes the high-energy limit $E \gg \lambda^{1/2} T$. If $E \ll \lambda^{1/2} T$ (which is equivalent to $\lambda^{-1/6} \ell_{\max} \ll 1/T$), then the corrections to $\lambda{=}\infty$ jet stopping results remain small from $\ell_{\text{stop}} \sim \ell_{\max}$ all the way down to $\ell_{\text{stop}} \sim 1/T$, which is the smallest jet stopping distance that we will consider.[4]

---

[3]For a discussion of one potential source of $1/N_c$ corrections to jet propagation, see Shuryak, Yee, and Zahed [14].

[4]At a technical level, we define where the jet stops [6, 7] following Chesler et al. [5, 16] as the location where the jet's energy and momentum and charge first begin to evolve hydrodynamically. Since hydrodynamics is an effective theory only on distance scales $\gg 1/T$ at strong coupling, it does not make sense to apply this definition to stopping distances small compared to $1/T$.

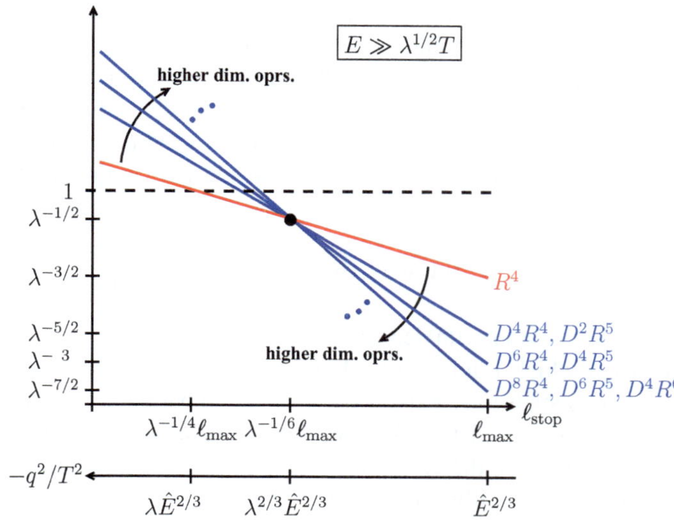

Fig. 14.3 A parametric picture of the relative importance of higher-derivative corrections to the low-energy supergravity action as a function of the stopping distance $\ell_{\rm stop}$ (using the $\lambda=\infty$ result for $\ell_{\rm stop}$). The axis are both logarithmic, and an importance of 1 indicates that the individual correction would, by itself, significantly modify the $\lambda=\infty$ analysis. Our measure of "importance" is explained in Sect. 14.2.3. Also shown, as an alternative horizontal axis, is the four-dimensional virtuality $-q^2$ of the source that created the jet, where $\hat{E} \equiv E/T$

We emphasize that finding the *parametric* dependence of corrections shown in Fig. 14.3 does not depend on knowing details of the precise form of higher-derivative corrections to the supergravity action, nor on details of their precise effects on the AdS$_5$-Schwarzschild background. Such details are not known for corrections involving high powers of curvature. And though we have taken care in Fig. 14.3 to only depict higher-derivative corrections that actually appear as string corrections to Type IIB supergravity,[5] which is the case relevant to $\mathcal{N}=4$ SYM, our qualitative results do not depend on these details either.

In fact, given the types of "jets" that we study, the maximum stopping distance scale $\ell_{\max}$ given by (14.1) will only be defined parametrically. It is the distance scale beyond which the amount of charge that a highly-penetrating jet deposits in the medium, on average, begins to fall exponentially. One may define a related scale $\ell_{\rm tail}$ by characterizing this exponential fall-off as

$$\text{deposition}(x^3) \sim \text{prefactor} \times e^{-x^3/\ell_{\rm tail}} \qquad \text{for} \qquad x^3 \gg \ell_{\max} \tag{14.2}$$

[5]For a nice summary of higher-dimensional gravitational corrections in Type II supergravity generated by tree-level string amplitudes (i.e. in the $N_c=\infty$ limit), see Table 1 of Stieberger [17]. Though not relevant to the $N_c=\infty$ case we are discussing, a nice discussion of corrections generated from one-loop string amplitudes may be found in Richards [18].

for a jet moving in the x^3 direction. Figure 14.3 indicates that the expansion in higher-derivative corrections should be well-behaved around ℓ_{max}, and correspondingly the corrections to ℓ_{tail} should be well-behaved. We explicitly computed the leading, R^4 correction to ℓ_{tail} in [13]. The precise result depends on details of the type of source used to initially create the jets. As an example, here is the result corresponding to creating a jet in the $\mathcal{N}=4$ SYM plasma via the decay of a high momentum, slightly off-shell graviton:

$$\ell_{\text{tail}} = \ell_{\text{tail}}^{\lambda=\infty}\left[1 + 47.162\,\lambda^{-3/2} + O(\lambda^{-5/2})\right]. \tag{14.3}$$

This result for ℓ_{tail} increases with decreasing λ.

The conclusion which follows from Fig. 14.3 is that the formal expansion in $1/\sqrt{\lambda}$ (which corresponds to an expansion in the string parameter α' on the gravity side) breaks down for some jets and is safe for others, depending on the stopping distance ℓ_{stop} of the jet (and therefore on the virtuality $-q^2$). The corrections to the $\lambda=\infty$ result are parametrically small for $\lambda^{-1/6}\ell_{\text{max}} \ll \ell_{\text{stop}} \lesssim \ell_{\text{max}}$. In particular, corrections to the maximum stopping distance $\ell_{\text{max}} \propto E^{1/3}$ are small. But the interesting case is when jets are created in such a way that

$$T^{-1} \ll \ell_{\text{stop}} \lesssim \lambda^{-1/6}\ell_{\text{max}}, \tag{14.4a}$$

which is

$$T^{-1} \ll \ell_{\text{stop}} \lesssim \frac{\left(E/\sqrt{\lambda}\right)^{1/3}}{T^{4/3}}. \tag{14.4b}$$

For $\ell_{\text{stop}} \sim \lambda^{-1/6}\ell_{\text{max}}$, all the corrections are the same size, and so the formal expansion in powers of $1/\sqrt{\lambda}$ has broken down. Yet the individual corrections are all small (of relative importance $\lambda^{-1/2}$) for that ℓ_{stop}. From Fig. 14.3, we cannot tell whether the *sum* of the corrections to $\lambda=\infty$ will remain small for $\ell_{\text{stop}} \lesssim \lambda^{-1/6}\ell_{\text{max}}$ or whether, instead, the $\lambda=\infty$ calculation becomes useless there.

Perhaps the relative size of the total correction to the $\lambda=\infty$ result flattens out, as depicted in Fig. 14.4a. Perhaps the corrections sum to give rapid (e.g. exponential) growth, as in Fig. 14.4b. Perhaps they sum to give rapid suppression, as in Fig. 14.4c. Figuring out what happens for $\ell_{\text{stop}} \ll \lambda^{-1/6}\ell_{\text{max}}$ involves a full string-theory analysis of the problem.

Before continuing it is useful to first explain one other qualitative feature of the $\lambda=\infty$ calculation. Excitations created in the field theory correspond to excitations created on the boundary of AdS_5-Schwarzschild, which then fall towards the black brane over time, such as depicted in Fig. 14.5. The 3-space distance that this excitation travels before falling into the horizon matches the stopping distance of

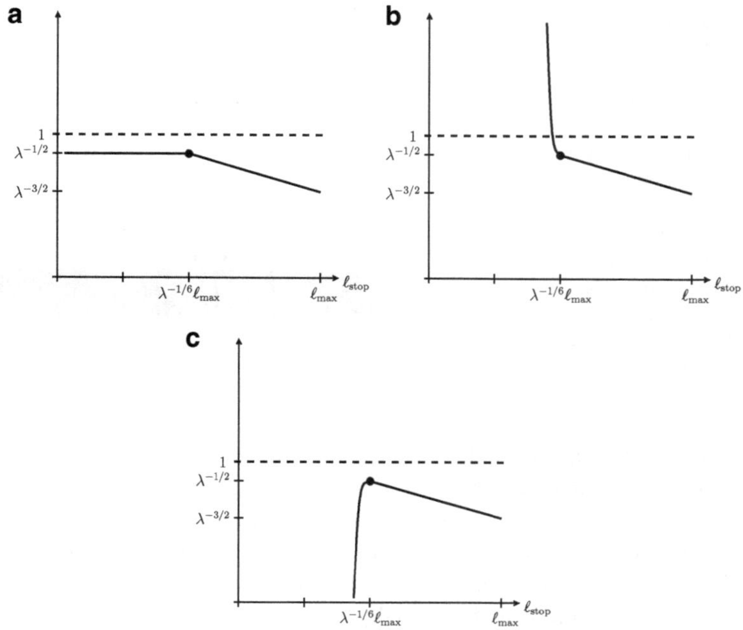

Fig. 14.4 Like Fig. 14.3 but showing some different behaviors that the total correction (summing all higher-derivative corrections) might conceivably have

the corresponding excitation in $\mathcal{N}=4$ SYM.[6] For $\ell_{\text{stop}} \ll \ell_{\max}$, which includes the region (14.4) of interest, there is a nice simplification. On the gravity side, the excitation falling in Fig. 14.5 turns out to be a spatially *small* wavepacket which can be treated in the geometric optics approximation. The wavepacket's motion is the same (up to parametrically small corrections) as that of a five-dimensional "particle" traveling in the AdS_5-Schwarzschild geometry, and so it follows a geodesic whose trajectory is easily calculated in terms of the 4-momentum q_μ of the excitation. (See Sect. 14.2.2.2 for more detail.)

In the strong-coupling limit $\lambda=\infty$ of the field theory, the AdS/CFT correspondence reduces to one between the field theory and the infrared limit of the string theory, which is a supergravity theory. The quanta of the supergravity fields correspond to string states that are massless in flat ten-dimensional spacetime, such as the graviton. For $\lambda=\infty$, the well-known gravitational dual of finite-temperature $\mathcal{N} = 4$ SYM is Type IIB supergravity in an $(\text{AdS}_5\text{-Schwarzschild})\times S^5$ background.

The classical wave packet falling in Fig. 14.5 is a localized, classical excitation of the supergravity fields. For the sake of specificity, consider the case where it is

[6]See [7] for a discussion in the context of the present paper, but this correspondence is implicit in the earlier work of [3–5].

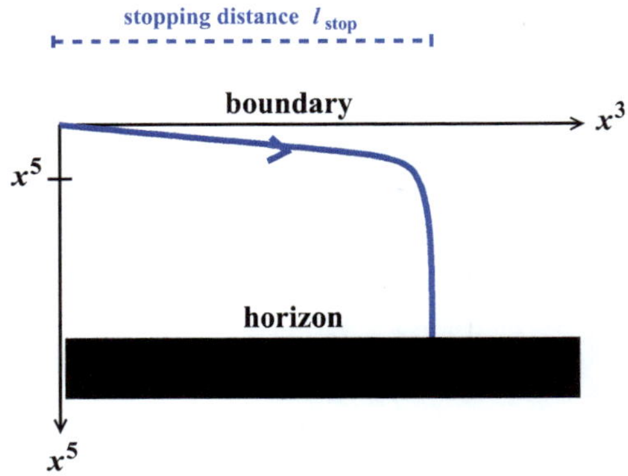

Fig. 14.5 Qualitative sketch of the motion though AdS_5-Schwarzschild of a wave packet with high 3-momentum in the x^3 direction. As measured by x^0, the particle takes infinitely long to reach the horizon. Of special importance is the parametric scale $x^5_\star$ in the fifth dimension, where the trajectory turns over and beyond which progress in x^3 rapidly slows to a stop

an excitation of the background gravitational field. A graviton is really a closed string whose internal degrees of freedom are in their ground state. Because of the gravitational field from the black brane, this closed string will feel tidal forces as it falls, which will try to stretch the string in some directions and squeeze it in others. As the graviton gets further from the boundary (and so closer to the black brane), the tidal forces will increase, and eventually they will become large enough to excite the internal string degrees of freedom of the graviton. It is the excitation of these string degrees of freedom that is responsible for the breakdown of the expansion in Fig. 14.3 in the problem region (14.4).

In the problematic case (14.4) where $\ell_{\text{stop}} \ll \lambda^{-1/6}\ell_{\text{max}}$, the tidal forces are strong enough to stretch that loop of string to become classically large before the stopping distance is reached. This is why stringy corrections cannot be ignored in that case, explaining the breakdown of the expansion in Fig. 14.3. (In contrast, the tidal forces are not strong enough to excite the graviton's internal degrees of freedom soon enough when $\ell_{\text{stop}} \gg \lambda^{-1/6}\ell_{\text{max}}$.) Though the resulting classical string loop will be large compared to the size of a graviton, we must ask how its size compares to the stopping distance ℓ_{stop}. We find that the ratio of (i) the stretched, classical string's size in the direction of motion x^3 to (ii) the stopping distance ℓ_{stop} is parametrically of order

$$\frac{(\delta x^3)_{\text{string}}}{\ell_{\text{stop}}} \sim \lambda^{-1/4} \ln^{1/2}\left(\frac{\lambda^{-1/6}\ell_{\text{max}}}{\ell_{\text{stop}}}\right). \tag{14.5}$$

Because of the $\lambda^{-1/4}$, this ratio is typically parametrically small for large but finite λ, and we argue that the stretching of the graviton into a string (and the accompanying breakdown of the formal expansion in $1/\sqrt{\lambda}$ in Fig. 14.3) then has sub-leading impact on $\lambda{=}\infty$ results for the stopping distance. But (14.5) also includes cases where the stretching of the string may play an important role: If one considers a situation where the argument of the logarithm in (14.5) is exponentially large, then the logarithm can be large enough to compensate for the factor of $\lambda^{-1/4}$.

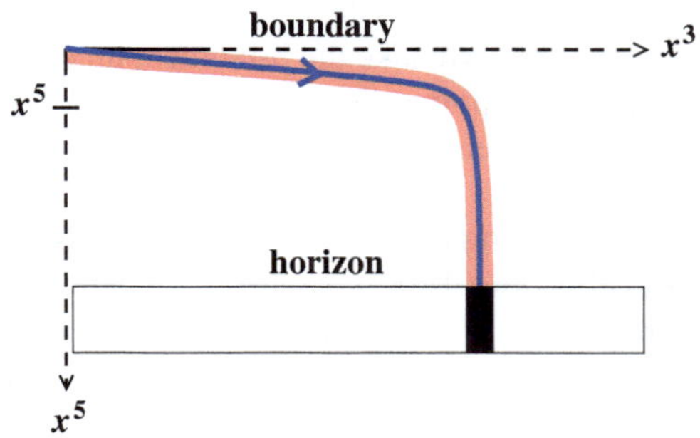

Fig. 14.6 The *(pink)* shaded area represents a narrow region of space-time around the null geodesic of Fig. 14.5. The AdS_5-Schwarzschild metric in this region may be approximated as a pp-wave background for the purpose of quantizing a small, falling loop of string that describes a graviton (or other particle) in the initial falling wavepacket

Since the tidal forces stretch a quantum string (the graviton) into a larger classical string, one may wonder whether or not it is possible to do a real, detailed calculation of the transition between the two. In general, it is not known how to quantize a string in an AdS_5-Schwarzschild background. But remember that our graviton is localized and so only probes a region of space-time near the geodesic depicted in Fig. 14.5. It is enough to consider only a narrow region of the space-time that lies near a null geodesic, as depicted in Fig. 14.6, and so we may treat the full background metric in an approximation (known as a Penrose limit) that treats displacements from the null geodesic as small. The resulting approximation to the background metric is an example of what is known as a pp-wave background, and it *is* known how to quantize a string in a pp-wave background. In particular, it is possible to calculate the probability distribution of the shape of the classical string loop. The methods we use are similar to previous works by other authors on the excitation of string modes in scattering processes and/or in pp-wave backgrounds [19–23].

14.2 Basic Framework and the Jets We Study

Our basic framework for studying jet stopping is to create high-momentum excitations of the strongly-coupled plasma by perturbing the plasma with high-momentum sources, as in [6, 7], and studying the response. As it will be reviewed below, for certain types of sources applied to the quantum field theory, the response in the gravity dual is the generation of a highly-localized and highly-oscillatory wave packet that moves through space while falling in the fifth dimension towards the black brane horizon. This wave packet has approximately well-defined five-dimensional position and momentum, and (for $\lambda=\infty$), its motion can be approximated (up to parametrically small corrections) by a geodesic—that is, by the trajectory that a particle would take through the AdS_5-Schwarzschild background. This particle (geometric optics) approximation makes the $\lambda{=}\infty$ calculation of stopping distances particularly simple and efficient. We take the 3-momentum of our excitations to be in the x^3 direction, and the stopping distance is simply given by how far in x^3 the corresponding geodesic travels before falling into the black brane horizon, as depicted in Fig. 14.5.[7]

An obvious effect of the higher-derivative corrections to the supergravity action on trajectories such as Fig. 14.5 stems from corrections made to the AdS$_5$-Schwarzschild background, but the dominant effects turn out to be the changes made to the equation of motion of the wave packets, which will no longer follow geodesics.

14.2.1 Notation

In these notes we will use Greek letters for four-dimensional space-time indices $(\mu, \nu = 0, 1, 2, 3)$. Lower-case roman letters (a, b) will be used for ten-dimensional indices. The first five of those 10 dimensions, corresponding to AdS$_5$-Schwarzschild when $\lambda{=}\infty$, will be represented by upper-case roman letters $(I, J = 0, 1, 2, 3, 5)$. The remaining five dimensions, corresponding to the compact 5-sphere S^5, will be indicated by dotted lower-case roman letters $(\dot{a}, \dot{b})$. When we use the adjective "five-dimensional" without further qualification, then we are referring to the noncompact dimensions—those of AdS$_5$-Schwarzschild.

We use the Poincare form of the AdS$_5$-Schwarzschild metric[8]

$$(ds)^2 = \frac{\mathbf{R}^2}{z^2}\left[-f(dt)^2 + (d\mathbf{x})^2 + f^{-1}(dz)^2\right], \tag{14.6}$$

where z is the coordinate x^5 of the fifth dimension, $\mathbf{R}$ is the radius of the 5-sphere (and will drop out of final results),

$$f \equiv 1 - \frac{z^4}{z_{\mathrm{h}}^4}, \tag{14.7}$$

the boundary is at $z{=}0$, and the horizon is at

$$z_{\mathrm{h}} = \frac{1}{\pi T}. \tag{14.8}$$

---

[7]For more discussion of why the distance the excitation travels before falling into the horizon should be identified with the stopping distance in the $3+1$ dimensional field theory problem, see the discussion in [7], as well as earlier discussions in the context of falling classical strings [3, 16].

[8]The coordinate used in [6, 7] is $u = z^2/z_{\mathrm{h}}^2$, which is $u = (z/2)^2$ when working in the units $2\pi T = 1$ used there.

The metric of four-dimensional flat space-time follows the mostly plus convention: $\eta_{\mu\nu} \equiv \mathrm{diag}(-1, +1, +1, +1)$.

14.2.2 Review of $\lambda = \infty$ Results

14.2.2.1 Set Up

The jet stopping problem is addressed using the set up given in [6, 7]. An external source is applied to the strongly-interacting gauge theory in order to create the initial high-energy high-momentum excitation. Specifically, we add a source term to the Lagrangian,

$$\mathscr{L} \to \mathscr{L} + \mathscr{N}\, O(x)\, e^{i\bar{k}\cdot x}\, \Lambda_L(x), \tag{14.9}$$

where $\mathscr{N}$ is an arbitrarily small source amplitude, $O(x)$ is a source operator,

$$\bar{k}^\mu \simeq (E, 0, 0, E) \tag{14.10}$$

is the large 4-momentum of the desired excitation, and $\Lambda_L(x)$ is a slowly varying envelope function that localizes the source to within a distance L of the origin in both x^3 and time. For example,

$$\Lambda_L(x) = e^{-\frac{1}{2}(x^0/L)^2} e^{-\frac{1}{2}(x^3/L)^2}. \tag{14.11}$$

L is chosen large compared to $1/E$ but small compared to the stopping distance we wish to measure. The small amplitude $\mathscr{N}$ is so that we can treat the external source as a small-perturbation to the strongly-interacting gauge theory, so that the source will never create more than one jet with energy E at a time.

The source operator $O(x)$ is a matter of choice. As an example, [6] found it convenient to focus on "jets" created by an external R-charge field (somewhat analogous to the excitation that would be created by the hadronic decay of a high-momentum W boson inside a standard-model quark-gluon plasma, but with isospin replaced by R charge). In that case the operator was $O(x) = j^\perp(x)$, where j^μ is a combination of R current operators. The details of the choice of source operator O are unimportant [7], however, as long as the operator has finite conformal dimension in the $\lambda = \infty$ limit.

Sometimes in previous work [6], the characteristic 4-momentum $\bar{k}$ of the source has been taken to be exactly light-like, $\bar{k} = (E, 0, 0, E)$. Because the source is confined to a space-time region of size L, the momentum components q^μ of the source are smeared out around $\bar{k}^\mu$ by an amount of order $1/L$, and so the typical magnitude of the virtuality $q^2 \equiv q^\mu q_\mu$ of the source is then of order $|q^2| \sim E/L \ll E^2$.

In later work, a calculational and conceptual simplification was found if one instead chooses the characteristic 4-momentum $\bar{k}$ to be just a little bit time-like,

$$\bar{k} = (E + \epsilon, 0, 0, E - \epsilon) \tag{14.12a}$$

with

$$\frac{1}{L} \ll \epsilon \ll E. \tag{14.12b}$$

The first inequality guarantees that the uncertainty in momentum does not overwhelm the size of ϵ. In this case, the source has an approximately well-defined virtuality in 4-momentum space q of

$$-q^2 \equiv -q^\mu q_\mu \simeq -\bar{k}^\mu \bar{k}_\mu \simeq 4\epsilon E. \tag{14.13}$$

This is the case where the response created in the dual gravity theory can be shown [7] to be a highly localized, highly oscillatory wave packet that falls in the fifth dimension toward the black brane horizon. The trajectory of the wave packet is the geodesic that would be followed by a massless five-dimensional *particle* traveling in the AdS_5-Schwarzschild background as in Fig. 14.5. Calculations using this particle picture [7] are much simpler and more efficient than calculations directly in terms of the five-dimensional field excitations [6].

14.2.2.2 The Geodesic

The five-dimensional mass m associated with a supergravity field, and therefore with the five-dimensional particle trajectory, is determined by the conformal dimension Δ of the field theory operator dual to that field.[9] We take Δ to be of order one. It was shown in [7] that, in the high-energy limit, this mass does not affect the stopping distance for sources described by (14.12) when $\ell_{\text{stop}} \ll \ell_{\text{max}}$, which will be our focus here. So we may ignore the five-dimensional mass and focus on the trajectories $dx^I dx_I = 0$ corresponding to null geodesics in AdS_5-Schwarzschild. The solution for such geodesics (for a metric that depends only on x^5 and has four-dimensional parity) is

$$x^\mu(x^5) = \int \sqrt{g_{55}}\, dx^5 \, \frac{g^{\mu\nu} q_\nu}{(-q_\alpha g^{\alpha\beta} q_\beta)^{1/2}}, \tag{14.14}$$

[9]It is important to note that the masses of five-dimensional fields in the gravity dual have nothing to do with the masses of four-dimensional excitations in the $\mathcal{N}=4$ SYM field theory. The five-dimensional mass m is not the "mass of a jet."

where the 4-momentum q_α with lower index is conserved in five-dimensional motion and is given by the 4-momentum (14.12) of our source, $q_\alpha = \eta_{\alpha\beta}\bar{k}^\beta$.

Taking the integral in (14.14) all the way to the horizon for μ=3 (the direction of the jet), and using the metric (14.6), gives the geometric optics approximation to the stopping distance

$$\ell_{\rm stop} \simeq \int_0^{z_{\rm h}} dz \, \frac{|\boldsymbol{q}|}{\sqrt{-q^2 + \frac{z^4}{z_{\rm h}^4}|\boldsymbol{q}|^2}}, \tag{14.15}$$

where we have used rotation invariance to rewrite q_3 as $|\boldsymbol{q}|$. Here and throughout this review we use the symbol q^2 for the 4-virtuality of the source,

$$q^2 \equiv q_\mu \eta^{\mu\nu} q_\nu < 0. \tag{14.16}$$

We restrict attention to the case $-q^2 \ll E^2$ as in (14.13), as this is the case which generates stopping distances large compared to $1/T$. In this limit, the integral in (14.15) is dominated by small values of z, of order

$$z_\star \sim z_{\rm h} \left(\frac{-q^2}{|\boldsymbol{q}|^2}\right)^{1/4} \sim z_{\rm h} \left(\frac{-q^2}{E^2}\right)^{1/4} \ll z_{\rm h}, \tag{14.17}$$

corresponding to the parametric scale labelled $x_\star^5$ in Fig. 14.5. Neglecting parametrically small corrections, we may replace the upper limit of integration by infinity in (14.15) to get[10]

$$\ell_{\rm stop} \simeq \frac{\Gamma^2(\frac{1}{4})}{(4\pi)^{1/2}} \left(\frac{E^2}{-q^2}\right)^{1/4} \frac{1}{2\pi T}. \tag{14.18}$$

The validity of (14.18) is restricted to the range of validity of the geometric optics approximation. For a detailed discussion, see [7]. Here, for the sake of simplicity of this review, we will just give a quick, crude way to see the limit of applicability from the result (14.18) itself. By the uncertainty principle, the components of the source's 4-momentum will be smeared out by $1/L$, where L is the source size. Consequently, the virtuality $-q^2$ given by (14.13) will only be (approximately) well-defined when $\epsilon \gg 1/L$ and so when

$$-q^2 \gg \frac{E}{L}. \tag{14.19}$$

[10] See [7] for this explicit result, but the parametric behavior $\ell_{\rm stop} \sim (E^2/-q^2)^{1/4}$, within its range of validity, was found earlier by Hatta, Iancu and Mueller [4].

But the result (14.18) from the geometric optics approximation is only meaningful if $-q^2$ is approximately well defined. Combining (14.18) and (14.19) requires

$$\frac{\ell_{\text{stop}}^4 T^4}{E} \ll L\,. \tag{14.20}$$

On the other hand, it wouldn't be sensible to try to measure a stopping distance unless we choose a source size that is smaller than the distance we wish to measure. So L needs to satisfy

$$\frac{\ell_{\text{stop}}^4 T^4}{E} \ll L \ll \ell_{\text{stop}}\,. \tag{14.21}$$

Choosing such an L is possible exactly when $\ell_{\text{stop}} \ll \ell_{\text{max}}$ with ℓ_{max} given by (14.1).

The chance of propagating excitations created by sources like (14.9) to distances $\gg \ell_{\text{max}}$ is negligible. But showing this convincingly requires abandoning the geometric optics analysis and doing a wave analysis, as in [4, 6]. For most of this discussion, we stick to the region $\ell_{\text{stop}} \ll \ell_{\text{max}}$ where the geometric optics approximation is valid, and then make only parametric extrapolations to the boundary $\ell_{\text{stop}} \sim \ell_{\text{max}}$ of the range of validity.

Even for $\ell_{\text{stop}} \ll \ell_{\text{max}}$, the geometric approximation eventually breaks down at sufficiently large $z \gg z_\star$. At that point, however, the wave packet is falling essentially straight down towards the horizon, and the fact that it can no longer be treated as a particle no longer matters to how far it travels in x^3. The stopping distance is determined by the behavior of the trajectory for $z \sim z_\star$. (That is, $z \ll z_\star$ and $z \gg z_\star$ give parametrically small contributions to the stopping length.)

14.2.2.3 Other Authors' Methods for Describing "jets"

There is a long history of considering jet-like states that are dual to classical strings falling towards the horizon in the gravity theory (as well as a history of using geodesics to help understand the strings' motion) [3–5]. One difference is that the maximum stopping distance for these states is parametrically smaller than for the states we consider—ℓ_{max} for the states related to classical strings scales as $\lambda^{-1/6}E^{1/3}T^{-4/3}$ rather than the $E^{1/3}T^{-4/3}$ of (14.1). It's amusing to note that, perhaps coincidentally, $\lambda^{-1/6}E^{1/3}T^{4/3}$ is the same stopping distance scale where corrections become problematical in our Fig. 14.3. In any case, we will not attempt here to study $1/\lambda$ corrections to previous results based on classical strings.

Yet another, recent method for creating a gluon-like jet is to generate it as a beam of synchrotron radiation from a heavy quark that is forced into circular motion [8]. These gluon-like jets (under certain conditions) penetrate a distance of order ℓ_{max} given by (14.1). We will not attempt to study the $1/\lambda$ corrections in this synchrotron problem, but we would not be surprised if they work out similar to the $\ell_{\text{stop}} \sim \ell_{\text{max}}$ case in our analysis.

Finally, since coupling does not run with scale in $\mathcal{N}=4$ SYM, we are treating the coupling as large at *all* scales relevant to energy loss. This is in contrast to programs, such as [2], that try to isolate the soft effects of a strongly-coupled medium on weakly-coupled hard bremsstrahlung or pair-production vertices.[11] For work on $1/\lambda$ corrections in that context, see [25].

14.2.3 Determining the Importance of Corrections

In this section we describe the measure used to decide whether or not higher-derivative corrections to the supergravity action can invalidate the $\lambda=\infty$ result for finite but large λ and large energy. We have seen above that the $\lambda=\infty$ stopping distance is generated by the behavior of the particle trajectory for $z \sim z_\star$ given by (14.17). So the simple way to address our question is to check whether or not higher-derivative corrections make significant changes to the trajectory for $z \sim z_\star$. The "importance" represented by the vertical axis of Fig. 14.3 is just the relative effect on the trajectory at $z \sim z_\star$.

The relative effects of higher-derivative corrections increase with increasing z (see Appendix 1), and so, at high energies, there is always a point $z \gg z_\star$ where, in the geometric optics approximation, the expansion in effects of higher-derivative corrections goes bad. In some cases, this will occur for z's large enough that the geometric optics approximation has already broken down there anyway. But in all cases, we work under the following, physically reasonable assumption:

Assumption 1. Once the five-dimensional wave packet has stopped moving significantly in x^3, so that it is falling essentially straight toward the horizon, then it will thereafter continue falling essentially straight toward the horizon and will not move significantly in x^3 again.

Assumption 2. $1/\lambda$ corrections do not significantly modify the (approximate) equality between (i) the late-time x^3 position of the five-dimensional wave packet as it approaches the horizon and (ii) the position where the jet stops and thermalizes in the four-dimensional field theory as measured, for example, by the center of the late-time diffusing distribution of R charge.

Under these simplifying assumptions we study the jet stopping distance by analyzing corrections to how far the five-dimensional wave packet travels.

14.2.4 The Choice of Source Operator

The source operator $O(x)$ used to generate the jet via (14.9) determines the type of field that is excited in the dual theory—that is, the type of wave packet whose trajectory is depicted by Fig. 14.5. To keep our analysis simple, we choose a class of

[11] For a very brief summary of the relevant scales for the coupling, see, for example, [24].

source operators that are dual to fields which are five-dimensional scalars after the S^5 reduction. (This restriction can be relaxed as in [13] where a concrete example of an explicit calculation of a leading correction was given.) Moreover, in ten-dimensional language, it is convenient to mainly focus on the purely gravitational terms in the supergravity action (about which the most is known). For convenience we consider a ten-dimensional field of the form

$$h_{ab}(x,y) = \begin{cases} \phi(x)\, Y_{ab}(y), & a,b \in \{6,7,8,9,10\}; \\ 0, & \text{otherwise,} \end{cases} \tag{14.22}$$

where x is the AdS_5-Schwarzschild coordinate, y is the S^5 coordinate, ϕ is a five-dimensional scalar field, and $Y_{\dot{a}\dot{b}}(y)$ is any *traceless* tensor spherical harmonic on S^5. We crudely summarize this choice by writing the field as $h_{\dot{a}\dot{b}}$ when thinking of it as a ten-dimensional field and as ϕ when thinking of it as a five-dimensional field.

Source operators dual to the fields (14.22) are those operators of the form[12]

$$O \sim \mathrm{tr}(\lambda\lambda\bar{\lambda}\bar{\lambda}X^k), \qquad k = 0, 1, 2, \cdots \tag{14.23}$$

obtainable as supersymmetry descendants $Q^2\bar{Q}^2$ of $\mathrm{tr}(X^{k+4})$, where the X are the adjoint-color scalar fields of $\mathcal{N}=4$ SYM and λ are the gluinos. The source operators (14.23) have conformal dimension $\Delta = k + 6$ and carry non-trivial R charge (and so the jets they produce will carry R charge). Different choices k in (14.23) produce different representations $(2, k, 2)$ of the SU(4) R symmetry, which correspond to different types of tensor harmonics $Y_{\dot{a}\dot{b}}(y)$ in (14.22). None of the details of R charge representations, which element we choose, or tensor harmonics matter for what follows.

For simplicity we treat the conformal dimension Δ of the source operator as parametrically of order one, $\Delta \sim 1$. See [7] for a discussion in the context of $\lambda{=}\infty$ stopping distances of what happens when $\Delta \gg 1$ but with Δ still parametrically small compared to powers of E and λ.

14.3 The R^4 Correction

Though our final qualitative conclusions do not depend on the exact pattern of higher-derivative supergravity corrections that arise from Type IIB string theory, it is convenient to start the discussion with the first one that does, which is R^4.

---

[12]See, for example, Table 7 of the review by D'Hoker and Freedman [26]. Here X^k is shorthand for any symmetric product $X^{(i_1}X^{i_2}\cdots X^{i_k)}$ of k factors of the three complex scalar fields X^1, X^2, X^3 of $\mathcal{N}=4$ SYM.

14.3.1 R^4 Term in the Ten-Dimensional Supergravity Action

The first corrections to low-energy supergravity arise from the low-energy limit of the string-string scattering amplitude, for which $N_c = \infty$ corresponds to the tree-level amplitude[13]

$$R \to R + \tfrac{1}{8}\zeta(3)\,\alpha'^3 \left[C^{hmnk}C_{pmnq}C_h{}^{rsp}C^q{}_{rsk} + \tfrac{1}{2}C^{hkmn}C_{pqmn}C_h{}^{rsp}C^q{}_{rsk}\right] \tag{14.24}$$

or equivalently (via Bianchi identities)

$$R \to R + \tfrac{1}{8}\zeta(3)\,\alpha'^3 \left[C^{hmnk}C_{pmnq}C_h{}^{rsp}C_{krs}{}^q + \tfrac{1}{4}C^{hkmn}C_{pqmn}C_h{}^{prs}C^q{}_{krs}\right], \tag{14.25}$$

where R is the Ricci scalar in this equation, and α' is the string tension. The usual duality relation between the string tension and 't Hooft coupling is [29]

$$\frac{\alpha'}{\mathbf{R}^2} = \lambda^{-1/2} \equiv (g_{\rm YM}^2 N_c)^{-1/2}, \tag{14.26}$$

where $\mathbf{R}$ is the S^5 radius.

14.3.2 The ϕ Equation of Motion

We want to extract the corresponding linearized equation of motion for our five-dimensional scalar field $\phi \sim h_{\dot a \dot b}$ in the AdS$_5$-Schwarzschild background. By "linearized," we mean linearized in ϕ, not in the background metric. That means that we want terms in the action (14.25) that are quadratic in ϕ.

The C^4 correction term is suppressed by $\alpha'^3 \propto \lambda^{-3/2} \ll 1$, and so the only way we can get an unsuppressed correction is if there are compensating factors of the large energy E associated with the ϕ wave packet created by the source. The dominant correction is the one with the most powers of E. Powers of E will arise from derivatives (with indices in AdS$_5$-Schwarzschild) hitting ϕ. *The crucial observation is that one only needs to focus on the pieces of the C^4 term in the action (14.25) that are quadratic in ϕ and have as many five-dimensional derivatives acting on ϕ as possible.*

One way to get a factor of $\phi \sim h_{\dot a \dot b}$ is to consider the piece of the Weyl tensor C_{ijkl} that involves two five-dimensional derivatives of the S^5 metric fluctuation $h_{\dot a \dot b}$, e.g.

$$C_{I\dot a J\dot b} \simeq -\tfrac{1}{2}\nabla_I \nabla_J h_{\dot a \dot b}. \tag{14.27}$$

[13] Equation (14.24) is nicely summarized in Eqs. (3.1–3) of [27] and originates from [15, 28].

Other terms, with fewer derivatives acting on $h_{\dot{a}\dot{b}}$, will be suppressed because they do not generate as many factors of E. Similarly, a term involving S^5 derivatives like $\nabla_{\dot{m}}\nabla_{\dot{n}}h_{\dot{a}\dot{b}}$ arising from $C_{\dot{m}\dot{a}\dot{n}\dot{b}}$ will be suppressed compared to (14.27) because S^5 derivatives of ϕ do not yield factors of E. The dominant terms in C^4 that contribute to the linearized ϕ equation of motion will therefore be those terms that have two factors of the form (14.27) and two factors of the background Weyl tensor $\bar{C}_{ijkl}$ of (AdS$_5$-Schwarzschild)$\times S^5$. Now use the fact that the background Weyl tensor $\bar{C}_{ijkl}$ in this case vanishes unless all of its indices live in AdS$_5$-Schwarzschild.[14] The dominant C^4 terms in the action then have the form

$$\#\alpha'^3(\nabla\nabla\phi)(\nabla\nabla\phi)\bar{C}\bar{C}, \tag{14.28}$$

where # indicates some coefficient and the suppressed indices on ∇ and $\bar{C}$ are all five-dimensional indices and contracted.

Such terms only arise from the first term in brackets in (14.25) and not the second. For $\frac{1}{4}C^{hkmn}C_{pqmn}C_h{}^{prs}C^q{}_{krs}$, getting two factors of the form (14.27) would require evaluating background Weyl tensors $\bar{C}$ with at least one S^5 index, which gives zero.

The five-dimensional equation of motion for ϕ corresponding to (14.28) has the schematic form

$$[\nabla\nabla + \#(\alpha')^3\nabla\nabla\bar{C}\bar{C}\nabla\nabla]\phi = 0, \tag{14.29}$$

where we have dropped the five-dimensional mass term (determined by the conformal dimension of the source operator and arising in part from S^5 derivatives on ϕ in the leading-order supergravity action). As mentioned earlier, the mass is ignorable when computing the stopping distance for $\ell_{\text{stop}} \ll \ell_{\text{max}}$.

14.3.3 The WKB Approximation and the Point-Particle Approximation

Next consider solutions to the five-dimensional equation of motion (14.29) that have large, definite 4-momentum q_μ:

$$\phi = \Phi(x^5)\, e^{iq_\alpha x^\alpha}. \tag{14.30}$$

At high energy, we can make a WKB-like approximation and re-write (14.30) as

$$\phi = e^{iS(x^5)}\, e^{iq_\alpha x^\alpha} \tag{14.31}$$

[14]This is a statement about the uncorrected, i.e. $\lambda{=}\infty$, (AdS$_5$-Schwarzschild)$\times S^5$ background, and does not account for corrections to that background due to α'^3C^4. But this is good enough for figuring out the leading correction to the ϕ equation of motion.

where $S(x^5)$ is large. As discussed in [6, 7], this approximation only works sufficiently far from the boundary: $z \gg z_{\rm WKB}$ where $z_{\rm WKB} \sim 1/\sqrt{-q^2}$. But for $\ell_{\rm stop} \ll \ell_{\rm max}$ (i.e. $-q^2 \gg E^{2/3}T^{4/3}$) that is good enough to analyze the stopping distance, which is dominated by $z \sim z_\star$ given by (14.17).

Now use (14.31) in the equation of motion (14.29). 4-space derivatives ∇_μ will give the largest contribution when they hit the phase factor in (14.31) and bring down a factor of the large q_α rather than hitting something involving the background metric. x^5 derivatives ∇_5 will give the largest contribution when they hit the $e^{iS(x^5)}$ in (14.31) and bring down a large factor of $i\partial_5 S$ rather than hitting something involving the background metric. So, in the large-energy limit, the dominant terms correspond to replacing

$$\nabla_I \to iQ_I \equiv i(q_\mu, q_5) \tag{14.32}$$

in the equation of motion for ϕ, where

$$q_5 \equiv \frac{\partial S}{\partial x^5}\,. \tag{14.33}$$

We capitalize Q_I just as a way of notationally emphasizing that it is a 5-vector momentum. The result of (14.32) is to replace the equation of motion (14.29) by

$$-Q^I Q_I + \tfrac{1}{4}\zeta(3)\,\alpha'^3 Q^H Q^K Q_P Q_Q C_H{}^{RSP} C^Q{}_{RSK} \simeq 0. \tag{14.34}$$

Before we discuss the parametric size of the α'^3 correction to the $\lambda{=}\infty$ dispersion relation $Q^I Q_I \simeq 0$, well briefly review how to turn a dispersion relation like (14.34) into a particle trajectory using the geometric optics approximation.

The point-particle or geometric optics approximation consists of approximating the wave packets as simultaneously having (i) well defined 5-momentum Q_I, satisfying (14.34) above, and (ii) well defined 5-position (x^μ, x^5). One way to get the particle equation of motion is to start from the WKB approximation to $\phi(x)$ and to make a wave packet[15] $\Lambda_L(x)$:

$$\phi(x) \sim \int d^4q\; e^{iq_\alpha x^\alpha + i\int q_5(q,x^5)\,dx^5}\,\tilde{\Lambda}_L(-q). \tag{14.35}$$

This integral can be done by saddle point methods, and the saddle point condition is

$$0 = \frac{\partial}{\partial q_\mu}\Big[iq_\alpha x^\alpha + i\int q_5(q,x^5)\,dx^5\Big], \tag{14.36}$$

[15] See [7] for a $\lambda{=}\infty$ discussion of when the wave packet is small enough to treat as a particle. The summary is that L can be chosen appropriately so that everything is fine at $z \sim z_\star$ when by convolving with an appropriate localized boundary source function $\ell_{\rm stop} \ll \ell_{\rm max}$.

which gives

$$x^{\mu} = -\int dx^5 \,\frac{\partial q_5}{\partial q_\mu}. \tag{14.37}$$

Formally, we may then use this expression to find the generalization

$$\ell_{\rm stop} \simeq -\int_0^{z_{\rm h}} dz\, \frac{\partial q_5}{\partial |\boldsymbol{q}|} \tag{14.38}$$

of the stopping distance integral (14.15). This integral will require care in interpretation in the region where z is large enough that the expansion in higher-derivative corrections breaks down. Our focus will be on the relative importance of higher-derivative corrections in the integrand at $z \sim z_\star$.

14.3.4 The Relative Importance of the C^4 Correction

In cases where C^4 effects are a small correction, the dispersion relation (14.34) can be solved iteratively. That is, first solve the $\lambda{=}\infty$ equation $Q^I Q_I = 0$ for q_5, and then plug that solution into the correction term and solve

$$Q^I Q_I = \tfrac{1}{4}\,\zeta(3)\,\alpha'^3 Q^H Q^K Q_P Q_Q C_H{}^{RSP} C^Q{}_{RSK}\Big|_{\text{null } Q_I} \tag{14.39a}$$

for q_5. Explicitly evaluating the AdS$_5$-Schwarzschild Weyl tensor,[16] one finds

$$Q^H Q^K Q_P Q_Q C_H{}^{RSP} C^Q{}_{RSK}\Big|_{\text{null } Q_I} = 24\,\frac{z^{12}|\boldsymbol{q}|^4}{(z_{\rm h}\mathbf{R})^8}\,. \tag{14.39b}$$

Eq. (14.39), and the arguments leading up to it, give the leading high-energy terms of the C^4-corrected dispersion relation.

Here is a schematic and general argument which parametrically reproduces ((14.39b). The left-hand side of (14.39b) has the form

$$g^{\bullet\bullet} g^{\bullet\bullet} g^{\bullet\bullet} g^{\bullet\bullet} g^{\bullet\bullet} g^{\bullet\bullet} Q_\bullet Q_\bullet Q_\bullet Q_\bullet C_{\bullet\bullet\bullet\bullet} C_{\bullet\bullet\bullet\bullet}, \tag{14.40}$$

where we will use bullets ($\bullet$) to denote five-dimensional indices without focusing on the details of how they are contracted. (i) The four powers of $Q_\bullet$ produce four powers of E (as long as they are not contracted with each other). (ii) The six powers

[16] $(C_{0101}, C_{1212}, C_{0505}, C_{1515}) = (f, 1, -3, -f^{-1}) {\times} \mathbf{R}^2/z_{\rm h}^4$, with all other components determined by symmetry.

of the inverse metric give six powers of $z^2/\mathbf{R}^2$ for $z \ll z_h$, for a total of $z^{12}/\mathbf{R}^{12}$. (iii) Finally, consider $C_{\bullet\bullet\bullet\bullet}$ for small z. In AdS$_5$-Schwarzschild the Weyl tensor has size

$$C_{IJKL} \sim \frac{\mathbf{R}^2}{z^4} \times O\left(\frac{z^4}{z_h^4}\right) \sim \frac{\mathbf{R}^2}{z_h^4}. \tag{14.41}$$

Considerations of (i) through (iii) above yield

$$g^{\bullet\bullet}g^{\bullet\bullet}g^{\bullet\bullet}g^{\bullet\bullet}g^{\bullet\bullet}g^{\bullet\bullet}Q_\bullet Q_\bullet Q_\bullet Q_\bullet C_{\bullet\bullet\bullet\bullet}C_{\bullet\bullet\bullet\bullet} \sim \left(\frac{z^2}{\mathbf{R}^2}\right)^6 \times E^4 \times \left(\frac{\mathbf{R}^2}{z_h^4}\right)^2 \sim \frac{z^{12}E^4}{(z_h\mathbf{R})^8} \tag{14.42}$$

for small z, consistent with the exact result (14.39b).

Then solving the five-dimensional dispersion relation (14.39) for $q_5(q_\mu, x^5)$ gives

$$q_5 \simeq \sqrt{g_{55}\left(-q_\mu g^{\mu\nu} q_\nu + \frac{\varepsilon z^{12}|\boldsymbol{q}|^4}{z_h^8\mathbf{R}^2}\right)}, \tag{14.43}$$

where

$$\varepsilon \equiv \frac{24}{\mathbf{R}^6} \times \tfrac{1}{4}\,\zeta(3)\,\alpha'^3 = \frac{6\,\zeta(3)}{\lambda^{3/2}} \tag{14.44}$$

is small.

At this point, we could measure the parametric importance of the C^4 correction simply by comparing the relative sizes of the $\varepsilon z^{12}|\boldsymbol{q}|^4/z_h^8\mathbf{R}^2$ and $-q_\mu g^{\mu\nu} q_\nu$ terms in (14.43) at $z \sim z_\star$. But, for the sake of being slightly more explicit, let's first use (14.43) to get the stopping distance integral (14.38):

$$\ell_{\rm stop} \simeq \int_0^{z_h} dz\, \frac{|\boldsymbol{q}|\left[1 - \frac{2\varepsilon z^{10}}{z_h^8}|\boldsymbol{q}|^2\right]}{\sqrt{-q^2 + \frac{z^4}{z_h^4}|\boldsymbol{q}|^2 + \frac{\varepsilon z^{10}}{z_h^8}|\boldsymbol{q}|^4 f}}, \tag{14.45}$$

where $q^2 \equiv q_\mu \eta^{\mu\nu} q_\nu$ denotes the 4-momentum virtuality. The $\lambda{=}\infty$ result (14.15) corresponds to $\varepsilon = 0$. There are various features of the integrand in (14.45) that need to be discussed, but first let's look at the relative size of the C^4 correction at $z_\star$ (14.17). Under the square root in the denominator, the $-q^2$ and $z^4|\boldsymbol{q}|^2/z_h^4$ terms are the same size at $z \sim z_\star$—that's how $z_\star$ was determined in the first place. Since $z_\star \ll z_h$, we have $f \simeq 1$, and the relative size of the correction term is

$$\text{Importance}(C^4) \sim \frac{\frac{\varepsilon z_\star^{10}}{z_h^8}E^4}{-q^2} \sim \frac{(-q^2)^{3/2}}{\lambda^{3/2} E T^2}. \tag{14.46a}$$

Using (14.1) and (14.18), this may be rewritten as

$$\text{Importance}(C^4) \sim \left(\frac{\lambda^{-1/4}\ell_{\text{max}}}{\ell_{\text{stop}}}\right)^6 \sim \lambda^{-1/2}\left(\frac{\lambda^{-1/6}\ell_{\text{max}}}{\ell_{\text{stop}}}\right)^6, \qquad (14.46\text{b})$$

which gives the R^4 (red) line in Fig. 14.3.

The numerator correction in (14.45) is less important at $z \sim z_\star$. The relative size of its correction to the $\lambda=\infty$ integrand is

$$\frac{\frac{\varepsilon z^{10}}{z_h^8}E^2}{1} \qquad (14.47)$$

which is smaller than (14.46a) at $z \sim z_\star$ by a factor of $-q^2/E^2 \ll 1$. One seemingly disturbing feature of the numerator correction is its sign for large enough z. For

$$z \gg \left(\frac{\lambda^{3/4}T}{E}\right)^{1/5} z_h \qquad (14.48)$$

(which is much larger than $z_\star$) the integrand is large and negative. And so the integral (14.45) if blindly integrated up to z_h as written, yields a pathological result. But it can be shown that the expansion in higher-derivative corrections breaks down well before one reaches z's as large as (14.48) [13]. Following our assumptions from Sect. 14.2.3, we therefore stick to (14.46) as the measure of the importance of C^4 corrections.

As far as C^4 corrections are concerned in regard with Fig. 14.3 we only require the parametric information (14.46) on the importance of the C^4 correction at $z \sim z_\star$. One might be tempted to attempt to extract an exact size for the leading correction from the explicit integral (14.45). We show in Appendix 2 why this fails.

14.4 The $D^{2n}R^4$ Corrections

Next we consider the first sequence of higher and higher derivative corrections to the ten-dimensional supergravity dual, by looking at R^4 terms with higher and higher powers of covariant derivatives.

14.4.1 Review: 4-Point String Amplitude

Just like the R^4 interaction in supergravity arises from the low-energy limit of graviton-graviton scattering in string theory, the $D^{2n}R^4$ operators arise by looking more generally at the energy/momentum dependence of that scattering. At tree level (appropriate for $N_c{=}\infty$), the energy dependence of the amplitude is captured by an overall factor

$$T(s,t,u) = -\frac{\Gamma(-\alpha' s/4)\,\Gamma(-\alpha' t/4)\,\Gamma(-\alpha' u/4)}{\Gamma(1+\alpha' s/4)\,\Gamma(1+\alpha' t/4)\,\Gamma(1+\alpha' u/4)}, \tag{14.49}$$

where s, t, and u are the Mandelstam variables (in 10 dimensions). This result is an "on-shell" result, which means it is derived for string scattering in a flat-space background with the external momenta on-shell. That is, the result assumes $q_a q^a = 0$ for each of the four ten-dimensional external momenta, which means $s{+}t{+}u = 0$. Expanding $T(s,t,u)$ in powers of momenta gives [30]

$$\begin{aligned} T &= \frac{64}{\alpha'^3 stu} \exp\left[\sum_{n=1}^{\infty} \frac{2\,\zeta(2n+1)}{2n+1} \left(\frac{\alpha'}{4}\right)^{2n+1} (s^{2n+1} + t^{2n+1} + u^{2n+1})\right] \\ &= \frac{3}{\sigma_3} + 2\,\zeta(3) + \zeta(5)\,\sigma_2 + \tfrac{2}{3}\,\zeta^2(3)\,\sigma_3 + \tfrac{1}{2}\,\zeta(7)\,(\sigma_2)^2 + \tfrac{2}{3}\,\zeta(3)\,\zeta(5)\,\sigma_2\sigma_3 + \cdots, \end{aligned} \tag{14.50}$$

where

$$\sigma_k \equiv \left(\frac{\alpha'}{4}\right)^k (s^k + t^k + u^k). \tag{14.51}$$

The first term in the expansion (14.50) corresponds to scattering that occurs through the interchange of an intermediate graviton, and this process is already accounted for by the usual Einstein-Hilbert piece R of the low-energy supergravity action. The second term in (14.50), when generalized to curved space, gives the R^4 interaction previously discussed in Sect. 14.3. In order, the remaining terms give interactions that are schematically of the form $D^4 R^4$, $D^6 R^4$, etc.

The parametric size of the effects of these interactions on the wave packet trajectory gives a measure of their importance as depicted in Fig. 14.3. One may wonder why think about a derivative expansion $D^{2n} R^4$, and worry about where that expansion breaks down, when it is known that the expansion sums up to (14.49). There are two reasons. First, in those cases where the expansion is breaking down, the "on-shell" assumption $q^a q_a = 0$ for the graviton momenta also breaks down. So it is safest to not make any explicit assumptions about the detailed form of the $D^{2n} R^4$ interactions. Secondly, this provides a useful warm-up to more generally analyzing higher-derivative corrections $D^{2n} R^m$, which, as shown in Fig. 14.3, are equally important when the derivative expansion breaks down for $D^{2n} R^4$. As advertised, everything goes wrong at the same time, and so the explicit formula (14.49) for the 4-point amplitude is not particularly useful then.

14.4.2 Factors of $\alpha' Q Q$

Earlier we found a five-dimensional dispersion relation for the linearized scalar field ϕ with schematic form

$$Q^I Q_I = \alpha'^3 g^{\bullet\bullet} g^{\bullet\bullet} g^{\bullet\bullet} g^{\bullet\bullet} g^{\bullet\bullet} g^{\bullet\bullet} Q_{\bullet} Q_{\bullet} Q_{\bullet} Q_{\bullet} C_{\bullet\bullet\bullet\bullet} C_{\bullet\bullet\bullet\bullet}\Big|_{\text{null } Q_I} . \tag{14.52}$$

This arose from terms in the Lagrangian quadratic in ϕ, with five-dimensional form

$$\alpha'^3 (\nabla\nabla\phi)(\nabla\nabla\phi) C C \tag{14.53}$$

- coming from the ten-dimensional $\alpha'^3 C^4$.

For now, consider what would happen if we went from $\alpha'^3 C^4$ to something of the form $\alpha'^4 D^2 C^4$. Naively, we might think that the largest contribution arises from the case where both of the new derivatives have five-dimensional indices and hit ϕ's, modifying the right-hand-side of (14.52) to include an additional factor of

$$\alpha' g^{\bullet\bullet} Q_{\bullet} Q_{\bullet}. \tag{14.54}$$

Here the new indices might contract with the other indices in (14.52) or with each other. If we then note that Q_μ grows like E, we might at first guess that the parametric size of the additional factor (14.54) could be as large as

$$\alpha' \times \frac{z^2}{\mathbf{R}^2} \times E \times E, \tag{14.55}$$

but this is an overestimate. If all six Q's in

$$\alpha'^4 g^{\bullet\bullet} g^{\bullet\bullet} g^{\bullet\bullet} g^{\bullet\bullet} g^{\bullet\bullet} g^{\bullet\bullet} g^{\bullet\bullet} Q_{\bullet} Q_{\bullet} Q_{\bullet} Q_{\bullet} Q_{\bullet} Q_{\bullet} C_{\bullet\bullet\bullet\bullet} C_{\bullet\bullet\bullet\bullet} \tag{14.56}$$

are contracted with indices of the two Weyl tensor factors, the result must vanish because C_{IJKL} is anti-symmetric in IJ and KL. As a result, two of the Q's must contract with each other, and so the cost of the factor (14.54) is

$$\alpha' Q^I Q_I \tag{14.57}$$

instead of (14.55). In the $\lambda{=}\infty$ calculation, $Q^I Q_I = 0$. In our calculation here, the effects discussed earlier arising from the C^4 correction change this to (14.39),

$$Q^I Q_I \sim \alpha'^3 \frac{z^{12} E^4}{(z_h \mathbf{R})^8} . \tag{14.58}$$

So the size of the factor (14.54) at $z \sim z_\star$ is

$$\alpha' Q^I Q_I \big|_{z\sim z_\star} \sim \alpha'^4 \frac{z_\star^{12} E^4}{(z_h \mathbf{R})^8} \sim \frac{(-q^2)^3}{\lambda^2 E^2 T^4} \sim \left(\frac{\lambda^{-1/6} \ell_{\max}}{\ell_{\text{stop}}} \right)^{12} . \tag{14.59}$$

The term in (14.54) is not, in fact, the dominant contribution for large ℓ_{stop} simply because of the suppression from having to contract the Q's. But let's focus on this type of contribution for a moment longer. First, (14.59) tells us that this particular contribution from $\alpha'^4 D^2 C^4$ becomes just as important as $\alpha'^3 C^4$ when $\ell_{\text{stop}} \sim \lambda^{-1/6} \ell_{\max}$, and so this is our first example of the breakdown of the expansion in higher-derivative corrections depicted in Fig. 14.3. If we add yet another factor of $\alpha' D^2$ to go to $\alpha'^5 D^4 C^4$, and consider just the contributions of the form (14.54) for that factor as well, then we will get another factor of (14.59), which will also not be suppressed at $\ell_{\text{stop}} \lesssim \lambda^{-1/6} \ell_{\max}$. Finally, note that all of the effects discussed so far are arising from $\alpha' Q^I Q_I$ factors, which is just the dominant piece of ten-dimensional $\alpha' q^a q_a$ factors. These are precisely the sort of factors that are left out of standard string theory "on-shell" results for higher-derivative corrections $D^{2n} R^4$, but they become important for $\ell_{\text{stop}} \lesssim \lambda^{-1/6} \ell_{\max}$.

14.4.3 The Dominant Factors

If we add a factor of $\alpha' D^2$ and neither derivative hits a ϕ, then there will be no powers of E to compensate the suppression from α'. The dominant terms come from the case where one derivative hits a ϕ and the other hits the background field:

$$\alpha'^4 g^{\bullet\bullet} g^{\bullet\bullet} g^{\bullet\bullet} g^{\bullet\bullet} g^{\bullet\bullet} g^{\bullet\bullet} g^{\bullet\bullet} Q_\bullet Q_\bullet Q_\bullet Q_\bullet Q_\bullet D_\bullet C_{\bullet\bullet\bullet\bullet} C_{\bullet\bullet\bullet\bullet} . \tag{14.60}$$

Since the background field depends only on x^5, it is natural to consider the case where the derivative $D_\bullet$ hitting the background Weyl tensor is a D_5. The contribution of terms involving other components D_μ of $D_\bullet$ hitting the background is slightly more subtle and is included in Appendix 3.

Having the $D_\bullet$ which hits the background Weyl tensor be D_5 means (by four-dimensional parity invariance) that one of the $Q_\bullet$'s must be q_5. Parametrically, a D_5 on the background Weyl tensor has size z^{-1} for $z \ll z_h$ (such as $z \sim z_\star$). So the factor of $\alpha' D^2$ has cost

$$\alpha' g^{55} q_5 \, D_5(\text{on bkgd}) \sim \alpha' \times \frac{z^2}{\mathbf{R}^2} \times q_5 \times z^{-1} . \tag{14.61}$$

The factors in (14.53) can be thought off as a 4-point amplitude with the two ϕ factors being legs 1 and 3 and the two C factors being legs 2 and 4. Then the cost shown above corresponds to (the curved background generalization of) an $\alpha' s$ or $\alpha' u$ factor in the string amplitude expansion (14.50).

For the size of q_5, we can just take the $\lambda{=}\infty$ result from $Q^I Q_I \simeq 0$,

$$q_5 \simeq \sqrt{g_{55}(-q_\mu g^{\mu\nu} q_\nu)} \simeq \sqrt{-q^2 + \frac{z^4}{z_h^4}|\boldsymbol{q}|^2}. \tag{14.62}$$

This is just (14.43) with ε ignored. At $z \sim z_\star$, the two terms under the square root in (14.62) have comparable size, and so

$$q_5\big|_{z\sim z_\star} \sim \sqrt{-q^2}. \tag{14.63}$$

The cost (14.61) of $\alpha' D^2$ is then

$$\alpha' D^2|_{z\sim z_\star} \sim \frac{\alpha' z_\star \sqrt{-q^2}}{\mathbf{R}^2} \sim \frac{(-q^2)^{3/4}}{\lambda^{1/2} E^{1/2} T} \sim \left(\frac{\lambda^{-1/6}\ell_{\max}}{\ell_{\text{stop}}}\right)^3. \tag{14.64}$$

This indeed dominates over (14.59) in the regime $\ell_{\text{stop}} \gg \lambda^{-1/6}\ell_{\max}$ where the expansion in higher-derivative corrections has not already broken down. Multiplying the importance (14.46) of C^4 by any number of factors (14.64) gives

$$\text{Importance}(D^{2n}C^4) \sim \lambda^{-1/2}\left(\frac{\lambda^{-1/6}\ell_{\max}}{\ell_{\text{stop}}}\right)^{3n+6}, \tag{14.65}$$

which is shown by the $D^{2n}R^4$ curves in Fig. 14.3. For more details distinguishing R^m and C^m see [13].

14.5 Higher Powers of Curvature

Similar considerations yield

$$\text{Importance}(D^{2k+2n}C^{4+k}) \sim \lambda^{-1/2}\left(\frac{\lambda^{-1/6}\ell_{\max}}{\ell_{\text{stop}}}\right)^{6k+3n+6}, \tag{14.66}$$

as shown in Fig. 14.3. When considering higher-derivative terms A in the supergravity Lagrangian, the subset with the dominant effect for a given engineering dimension $\dim A$ has importance

$$\lambda^{-1/2}\left(\frac{\lambda^{-1/6}\ell_{\max}}{\ell_{\text{stop}}}\right)^{\frac{3}{2}\dim A-6}. \tag{14.67}$$

We've now seen the basic structure of corrections that give rise to Fig. 14.3, but there are still a few details to clear up. So far, we have considered only powers of the Weyl curvature tensor. In Appendix 3, we show that it will not matter, qualitatively, if supergravity interactions instead involved the full Riemann tensor at some order in derivatives.

14.6 Discussion of $1/\sqrt{\lambda}$ Expansion and Reasons for Its Failure

In this section we explain the reasons for the failure of the $1/\sqrt{\lambda}$ expansion in the problematic region (14.4). First, recall the AdS/CFT identifications [29]

$$g_{\text{string}} = \frac{\lambda}{4\pi N_c} = \frac{g_{\text{YM}}^2}{4\pi}, \tag{14.68}$$

$$\frac{1}{2\pi \mathbb{T} \mathbf{R}^2} = \frac{\alpha'}{\mathbf{R}^2} = \lambda^{-1/2} \equiv (g_{\text{YM}}^2 N_c)^{-1/2}, \tag{14.69}$$

where g_{string} is the string loop expansion parameter. The string tension $\mathbb{T}$ sets the mass scale for massive string excitations, and so $\alpha' \to 0$ corresponds to taking the scale for massive string excitations to infinity. For $\lambda = \infty$, the strongly-coupled four-dimensional quantum field theory is dual to the infrared limit of the ten-dimensional string theory, namely supergravity, in the appropriate background. For large but finite λ, massive string modes are not completely ignorable, and the effective supergravity theory of the massless modes gets corrections, in the form of higher-dimensional terms in its action, from integrating out the effects of the massive modes. Schematically, the effective supergravity Lagrangian becomes

$$\begin{aligned} \mathscr{L} \sim R &+ \left[\alpha'^3 R^4 + \alpha'^5 D^4 R^4 + \alpha'^6 D^6 R^4 + \cdots\right] \\ &+ \left[\alpha'^5 D^2 R^5 + \alpha'^6 D^4 R^5 + \alpha'^7 D^6 R^5 + \cdots\right] + \cdots, \end{aligned} \tag{14.70}$$

where we have focused just on the gravitational fields for simplicity. R represents factors of the Riemann tensor, and we have not shown numerical coefficients or how the indices contract. For $N_c = \infty$, there are no loop effects ($g_{\text{string}} = 0$), and accounting for the massive string modes in the effective theory is analogous to replacing the effects of the W boson by the Fermi 4-point interaction in electroweak theory. So, for example, the R^4 terms in (14.70) are calculated from string amplitudes for $2 \to 2$ graviton scattering and, crudely speaking, they correspond to processes which involve intermediate massive string states, as depicted schematically in Fig. 14.7. The R^5 terms similarly account for corrections to the 5-point graviton interaction, and so forth.

In our application, we are interested in the evolution of a high-energy excitation propagating through the soft AdS_5-Schwarzschild background. For simplicity of presentation, we focused earlier on the case where the excitation is in the

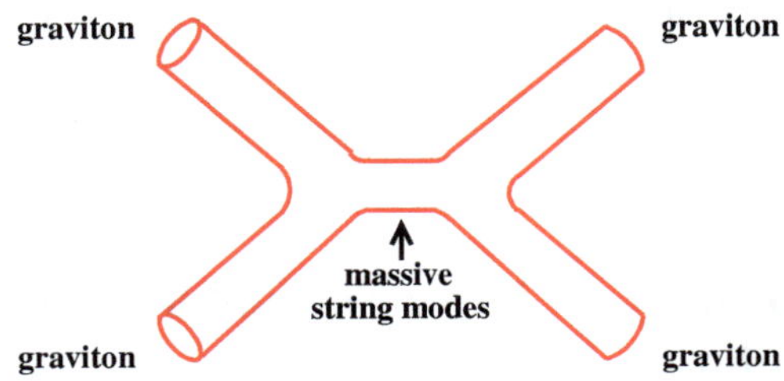

Fig. 14.7 A picture of massive string mode corrections to graviton-graviton scattering that are accounted for by the $D^m R^4$ corrections to the effective supergravity action

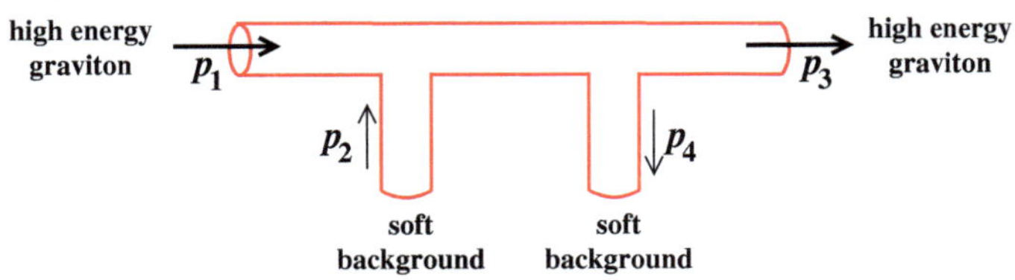

Fig. 14.8 A high-energy graviton, depicted as a string loop, interacting twice with the AdS$_5$-Schwarzschild background gravitational field

five-dimensional gravitational fields, though our conclusions will not be sensitive to this assumption. The relevant string scattering amplitudes are those where two of the external lines are the incoming and outgoing high-energy gravitons and the others are the soft background field. So, for a 4-point scattering amplitude such as Fig. 14.7, the relevant kinematic limit is that depicted in Fig. 14.8. With the notation used in that figure, the high-energy limit corresponds to potentially large $s = -(p_1 + p_2)_I(p_1 + p_2)^I$ but small $t = -(p_1 - p_3)_I(p_1 - p_3)^I$. [Here and throughout we may think of the p's as five-dimensional momenta in AdS$_5$-Schwarzschild rather than ten-dimensional momenta in (AdS$_5$-Schwarzschild)$\times S^5$ because in our problem there is no interesting dynamics associated with the 5-sphere S^5.] The $D^m R^4$ terms in (14.70) all become equally important in the jet stopping problem when this five-dimensional $\sqrt{s}$ becomes large enough at $z \sim z_\star$ to excite massive string modes in Fig. 14.8. The string mass scale is of order $1/\sqrt{\alpha'}$, and this condition

$$\sqrt{s_{\text{(5-dim)}}} \gtrsim \frac{1}{\sqrt{\alpha'}} \tag{14.71}$$

which is the same, as we have reviewed earlier, as the condition

$$\ell_{\text{stop}} \lesssim \lambda^{-1/6} \ell_{\text{max}}, \tag{14.72}$$

which is the problematic case (14.4) highlighted in the introduction. In this region, massive string states in the intermediate state in Fig. 14.8 are kinematically accessible and cannot be ignored.

As the high-energy excitation falls from the boundary to the horizon, as in Fig. 14.5, it does not just interact with the background field once or twice but does so over and over again, as depicted in Fig. 14.9. If the massive string states are kinematically accessible as in (14.71), then they cannot be neglected in any of the internal lines, which means in the effective theory language of (14.70) that all $D^m R^n$ terms will also become important. This is just what happens at the $\ell_{\text{stop}} \sim \lambda^{-1/6}\ell_{\max}$ point in Fig. 14.3, where all the corrections become the same size, corresponding to the threshold $\sqrt{s_{(5\text{-dim})}} \sim 1/\sqrt{\alpha'}$.

So it is the gravitational effect due to the presence of the black brane that contributes to massive string mode excitation. As a result, the effects of excited string modes are negligible at the boundary and become stronger as one moves away from it (and so closer to the black brane). At some distance from the boundary which we will review later, the gravitational effects of the black brane become strong enough that (14.71) is satisfied, which is when string modes may first be excited.

From the point of view of an effective theory (14.70) of gravitons, having all the correction terms become the same size (or worse), seems like an hopeless disaster for the purpose of computations. However, the picture of Fig. 14.9 suggests a different tack. What is happening is that the ten-dimensional gravitons which make up the classical excitation are really tiny (quantum) loops of string which are getting their internal string degrees of freedom excited as they fall in the background gravitational field. Specifically, internal degrees of freedom of a small object are affected by gravitational *tidal* forces, which try to compress the object in some directions and stretch it in others. In any case, consider the fate of a single graviton as depicted by Fig. 14.9: a high-momentum object moves through a soft background field. Various authors have previously studied applications of the eikonal approximation to string scattering [19,31]. The upshot is that Fig. 14.9 may be replaced by the evolution of a single string quantized in the classical background field (Figs. 14.10 and 14.11).

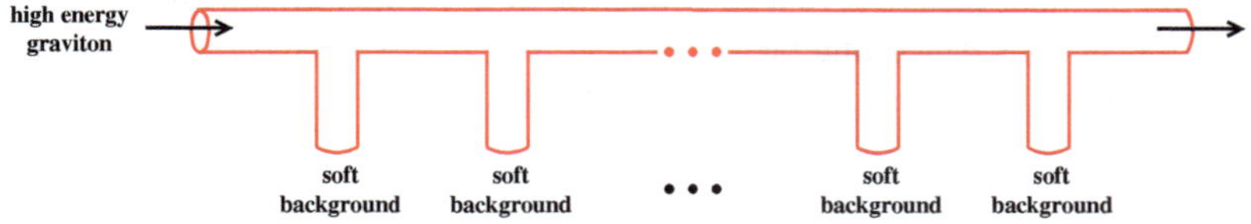

Fig. 14.9 Like Fig. 14.8 but with many interactions with the soft background field

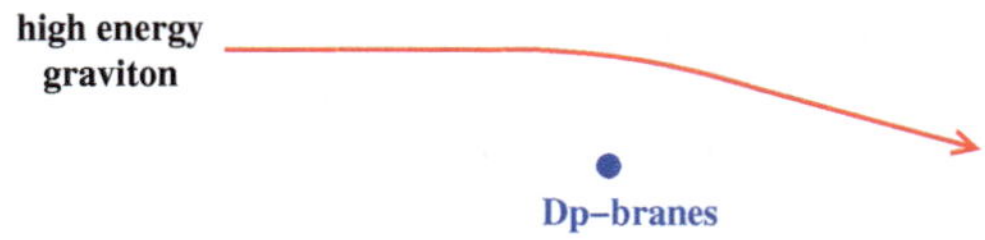

Fig. 14.10 A high energy massless string mode, such as a graviton, deflected by the gravitational field sourced by a stack of Dp-branes. The plane of the figure is a plane orthogonal to the Dp-branes. (So, for instance, a D1-brane could be visualized as a line extending out of the page)

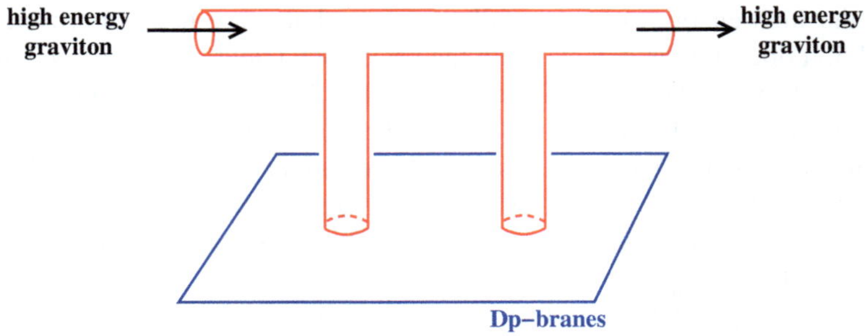

Fig. 14.11 The 2nd-order scattering amplitude for a graviton to elastically scatter from a stack of Dp-branes, calculated as a string scattering amplitude with two connections to the Dp-branes. This is only a topological picture: As described in Fig. 14.10, the motion of the graviton is orthogonal to the Dp-branes in the problem studied by D'Appollonio et al. [19]

14.7 The Penrose Limit

We begin by taking the Penrose limit to describe a narrow region around the null geodesic reviewed earlier in (14.2.2.2) and depicted in Fig. 14.6.[17] For an overview of taking Penrose limits, see [32–34].

The null geodesic can be written as

$$dx^{\mu} = \frac{g^{\mu\nu} q_{\nu}}{\omega}\, du, \tag{14.73a}$$

where u is an affine parameter for the geodesic determined by[18]

$$du = \omega \frac{\sqrt{g_{55}}\, dx^5}{(-q_{\alpha} g^{\alpha\beta} q_{\beta})^{1/2}}, \tag{14.73b}$$

which can be integrated to give u as a function of x^5. The normalization of u is just convention, and the cancelling factors of $\omega = -q_0$ in (14.73) have just been chosen to give u dimensions of length. In the metric (14.6), u is given by

[17]Penrose limits have previously been studied in AdS$_5$-Schwarzschild by Pando Zayas and Sonnenschein [35], but the null geodesic studied was different. Their geodesic fell straight toward the horizon in AdS$_5$-Schwarzschild, corresponding to $\boldsymbol{q}=0$ in our problem rather than $|\boldsymbol{q}| \simeq E$. Their geodesics also have non-trivial motion on the 5-sphere S^5. In our application, no dynamical evolution of the S^5 degrees of freedom takes place, and the S^5 string degrees of freedom simply remain in a quantum state given by the S^5 harmonic of the supergravity field of interest.

[18]This u should not be confused with the coordinate $u \equiv (z/z_{\rm h})^2$ used in earlier work by some of the authors [6, 7].

$$u = \mathbf{R}^2 \int \frac{dz}{z^2(1 - f|\boldsymbol{q}|^2/\omega^2)^{1/2}}, \tag{14.74}$$

which is linearly divergent as $z \to 0$. We take u as running from $u(z{=}0) = -\infty$ at the boundary to $u(z{=}\infty) = 0$ at the black brane singularity, and so

$$u(z) = -\mathbf{R}^2 \int_z^{\infty} \frac{dz}{z^2(1 - f|\boldsymbol{q}|^2/\omega^2)^{1/2}}. \tag{14.75}$$

The important thing to remember in what follows is that *late* times correspond to *small* negative values of u (which we will later also call τ).

The reference geodesic $\bar{x}^\mu(x^5)$ starts at the origin $\bar{x}^\mu = 0$ on the boundary. The 4-positions in AdS$_5$-Schwarzschild are measured relative to this geodesic by defining

$$\Delta x^\mu \equiv x^\mu - \bar{x}^\mu(x^5). \tag{14.76}$$

We choose $\boldsymbol{q}$ to be in the x^3 direction. Changing coordinates from x^5 and Δx^0 to $u{=}u(x^5)$ and

$$v \equiv \frac{q_\mu\, \Delta x^\mu}{\omega} = -\Delta x^0 + \frac{|\boldsymbol{q}|}{\omega}\, \Delta x^3 \tag{14.77}$$

puts the AdS$_5$-Schwarzschild metric (14.6) into the form

$$(ds)^2 = 2\, du\, dv + \frac{\mathbf{R}^2}{z^2}\Bigg[(dx^1)^2 + (dx^2)^2 + \frac{(\omega^2 - f|\boldsymbol{q}|^2)}{\omega^2}\, (d\Delta x^3)^2 + 2f\, \frac{|\boldsymbol{q}|}{\omega}\, dv\, d\Delta x^3 - f\,(dv)^2 \Bigg], \tag{14.78}$$

where f is now implicitly a function of u. The Penrose limit consists of keeping only the terms in the metric that would dominate after a scaling of coordinates

$$u \to u, \quad v \to \gamma^{-2} v \quad x^i \to \gamma^{-1} x^i \tag{14.79}$$

for very large γ. This is analogous to making what would be a very large boost ($u \to \gamma u$, $v \to \gamma^{-1} v$) in flat space, and so looking at physics close to the light cone, followed by rescaling all coordinates by a factor of γ^{-1}. For (14.78), the resulting limit is

$$(ds)^2_{\text{pp}} = 2\, du\, dv + \frac{\mathbf{R}^2}{z^2}\Bigg[(dx^1)^2 + (dx^2)^2 + \frac{(\omega^2 - f|\boldsymbol{q}|^2)}{\omega^2}\, (d\Delta x^3)^2 \Bigg], \tag{14.80}$$

which is a particular example of a pp-wave metricin Rosen coordinates.

The metric (14.80) has the schematic form

$$(ds)^2 = 2\,du\,dv + \sum_i \kappa_i(u)\,(dy_i)^2. \tag{14.81}$$

It is useful to normalize the last term by switching to Brinkmann coordinates

$$\hat{y}_i \equiv y_i \sqrt{\kappa_i(u)} \tag{14.82}$$

and

$$\hat{v} \equiv v - \tfrac{1}{2} \sum_i \partial_u (\ln \sqrt{\kappa_i})\, \hat{y}_i^2 \tag{14.83}$$

to give

$$(ds)^2 = 2\,du\,d\hat{v} + \sum_i (d\hat{y}_i)^2 + \Big(\sum_i \frac{\partial_u^2 \sqrt{\kappa_i}}{\sqrt{\kappa_i}}\, \hat{y}_i^2\Big)(du)^2\,. \tag{14.84}$$

In our case, using (14.74) to rewrite

$$\partial_u = \frac{z^2}{\mathbf{R}^2}\Big(1 - f\frac{|\boldsymbol{q}|^2}{\omega^2}\Big)^{1/2} \partial_z, \tag{14.85}$$

the metric in Brinkmann coordinates is

$$ds^2_{\rm pp} = 2\,du\,d\hat{v} + (d\hat{x}^1)^2 + (d\hat{x}^2)^2 + (d\Delta\hat{x}^3)^2 + \mathscr{G}(u, \hat{x}^1, \hat{x}^2, \Delta\hat{x}^3)\,(du)^2 \tag{14.86}$$

with

$$\mathscr{G}(u, \hat{x}^1, \hat{x}^2, \Delta\hat{x}^3) = \mathscr{G}_1(u)\left[(\hat{x}^1)^2 + (\hat{x}^2)^2\right] + \mathscr{G}_3(u)\,(\Delta\hat{x}^3)^2, \tag{14.87a}$$

$$\mathscr{G}_1(u) = \mathscr{G}_2(u) = \frac{\partial_u^2(z^{-1})}{z^{-1}} = \frac{z^3 f' |\boldsymbol{q}|^2}{2\mathbf{R}^4\omega^2} = -2\frac{z^6 |\boldsymbol{q}|^2}{z_{\rm h}^4 \mathbf{R}^4 \omega^2} \simeq -2\frac{z^6}{z_{\rm h}^4 \mathbf{R}^4}\,, \tag{14.87b}$$

$$\mathscr{G}_3(u) = \frac{\partial_u^2\big[z^{-1}(\omega^2 - f|\boldsymbol{q}|^2)^{1/2}\big]}{z^{-1}(\omega^2 - f|\boldsymbol{q}|^2)^{1/2}} = \frac{z^3(f' - zf'')|\boldsymbol{q}|^2}{2\mathbf{R}^4\omega^2} = 4\frac{z^6|\boldsymbol{q}|^2}{z_{\rm h}^4\mathbf{R}^4\omega^2} \simeq 4\frac{z^6}{z_{\rm h}^4\mathbf{R}^4}\,. \tag{14.87c}$$

Here, primes denote derivatives with respect to z, and $z{=}z(u)$ is implicitly a function of u, determined by inverting (14.75). The metric (14.86) would be flat if not for the $\mathscr{G}_i$, which arise from tidal forces. Note that these tidal terms vanish for null geodesics in pure AdS ($f{=}1$, or equivalently $z_{\rm h}{\to}\infty$), in agreement with general arguments that the Penrose limit of AdS is flat Minkowski space [32]. They also vanish in AdS_5-Schwarzschild if $\boldsymbol{q}=0$ (i.e. if the excitation fell straight down in the x^5 direction as a function of time x^0).

14.8 Quantizing the Falling Closed String

14.8.1 Overview

The string quantization in a pp-wave background proceeds straightforwardly [19–23]. The bosonic sector is described by a σ-model in the pp-wave background metric:

$$S = -\frac{1}{4\pi\alpha'}\int d\tau\, d\sigma \sqrt{-\gamma}\,\gamma^{\alpha\beta}(\partial_\alpha X^I)(\partial_\beta X^J)\, g_{IJ}(X), \tag{14.88}$$

where γ is the world-sheet metric and X^I are the world-sheet fields corresponding to the coordinates. For the pp-wave space-time metric (14.86), this takes the form

$$S = S_0 - \frac{1}{4\pi\alpha'}\int d\tau\, d\sigma \sqrt{-\gamma}\,\gamma^{\alpha\beta}(\partial_\alpha U)(\partial_\beta U)\,\mathcal{G}(U, \Delta\hat{X}), \tag{14.89}$$

where S_0 is the Minkowski string action. Identifying world-sheet time τ with the affine parameter u for the pp-wave space-time metric (14.86) then gives a constraint equation for $\partial_\alpha \hat{V}$ and gives the light-cone gauge Lagrangian

$$\begin{aligned} L &= \frac{p^u}{2}\int_0^{2\pi}\frac{d\sigma}{2\pi}\sum_i \Big(\partial_\tau\Delta\hat{X}^i\,\partial_\tau\Delta\hat{X}^i - (\alpha' p^u)^{-2}\partial_\sigma\Delta\hat{X}^i\,\partial_\sigma\Delta\hat{X}^i + \mathcal{G}_i(\tau)\,\Delta\hat{X}^i\,\Delta\hat{X}^i\Big) \\ &= \frac{p^u}{2}\sum_i\sum_{n=-\infty}^{\infty}\Big(\partial_\tau\Delta\hat{X}^{i\,*}_n\,\partial_\tau\Delta\hat{X}^i_n - \omega^2_{i,n}(\tau)\,\Delta\hat{X}^{i\,*}_n\,\Delta\hat{X}^i_n\Big) \end{aligned} \tag{14.90}$$

for the $\Delta\hat{X}^i$, where $p^u = p_{\hat{v}} \simeq E$ and $\Delta\hat{X}^i = \sum_n \Delta\hat{X}^i_n(\tau)\,e^{in\sigma}$ and

$$\omega^2_{i,n}(\tau) \equiv \frac{n^2}{(\alpha' p^u)^2} - \mathcal{G}_i(\tau). \tag{14.91}$$

(We have suppressed the fields corresponding to S^5 coordinates because they will play no role in our discussion.) We have chosen a convention where τ has units of time and σ is dimensionless.

Each mode $\Delta\hat{X}^i_n$ of the string is a time-dependent harmonic oscillator problem with classical frequency $\omega_{i,n}(\tau)$. The $\Delta\hat{X}^1_n$ and $\Delta\hat{X}^2_n$ modes are tidally compressed as the string moves away from the boundary since $\mathcal{G}_1 = \mathcal{G}_2$ is negative in (14.87b), so that the curvature $\omega_{1,n}(\tau)$ of the harmonic oscillator potential increases with time

$\tau = u$. In contrast, the $\Delta\hat{X}_n^3$ oscillators are tidally stretched, since $\mathcal{G}_3$ is positive in (14.87c):

$$\omega_{3,n}^2(\tau) = \frac{n^2}{(\alpha' p^u)^2} - \mathcal{G}_3(\tau). \tag{14.92}$$

When the string gets far enough from the boundary (i.e. at late enough times τ), the $\mathcal{G}_3$ term in (14.92) becomes dominant and $\Delta\hat{X}_n^3$ oscillators become unstable. Physically, this is when tidal forces come to dominate over string tension. Using (14.87c), this instability occurs when $z \geq z_n$, where

$$z_n \simeq \left(\frac{n z_{\rm h}^2 \mathbf{R}^2}{2\alpha' E}\right)^{1/3} = \lambda^{1/6}\left(\frac{n z_{\rm h}^2}{2E}\right)^{1/3} = n^{1/3} z_1. \tag{14.93}$$

The instability for the center-of-mass mode $n=0$ is not particularly interesting: it has nothing to do with exciting internal degrees of freedom of the string and just reflects the slight spread of the falling wavepacket in Fig. 14.5 due to curvature effects. Disregarding the $n \neq 0$ modes, the first tidal instability kicks in at $z \simeq z_1$.

Recall that $z_\star$, defined earlier in (14.17), characterizes the scale where the x^3 motion of the bulk excitation in Fig. 14.5 is coming to a stop (no significant motion for $z \gg z_\star$). Using (14.18) and (14.1), the ratio of the instability scale z_n to $z_\star$ is

$$\frac{z_n}{z_\star} \sim \frac{n^{1/3}\lambda^{1/6}(z_{\rm h}^2/E)^{1/3}}{z_{\rm h}(-q^2/E^2)^{1/4}} \sim \frac{n^{1/3}\ell_{\rm stop}}{\lambda^{-1/6}\ell_{\rm max}}. \tag{14.94}$$

So the tidal instability kicks in before the stopping distance is reached ($z_1 \lesssim z_\star$) precisely when we are in the interesting regime $\ell_{\rm stop} \lesssim \lambda^{-1/6}\ell_{\rm max}$ (14.4) identified in earlier work as the case where stringy corrections become important. Correspondingly, the modes which become tidally unstable for $z \lesssim z_\star$ are $n \lesssim n_\star$ with

$$n_\star \sim \left(\frac{\ell_{\rm stop}}{\lambda^{-1/6}\ell_{\rm max}}\right)^{-3}. \tag{14.95}$$

Although the instability develops at $z = z_n$, the modes $n \lesssim n_\star$ do not have time to stretch significantly until $z \sim z_\star$. This fact is suggested by the geodesic picture in Fig. 14.15: For $z \ll z_\star$, the impact of the black hole on the evolution in Fig. 14.15c is negligible, and so the evolution at those times is well approximated by the pure AdS case of Fig. 14.15a and b, for which the proper size of the string remains constant.

Once a given mode becomes unstable, the quantum mechanics of that mode will be somewhat analogous to a standard quantum mechanics thought experiment: What is the longest time that an idealized pencil can be balanced on its tip before it falls? Because of the uncertainty principle, the pencil cannot be started simultaneously at rest and perfectly vertical, and so it must fall. The pencil might be started in a

Gaussian wavepacket chosen to maximize the average fall time. But it is clear that, once the top of the pencil has fallen a macroscopic distance, *classical* mechanics will suffice to describe its subsequent motion: at that time, its position and momentum may, to excellent approximation, be considered as simultaneously well defined. For late enough times t_1, the pencil's motion for $t > t_1$ is approximately classical, with the only effect of its initial wave packet at $t{=}0$ being to determine a classical probability distribution for the pencil end's position at t_1. The corresponding momentum at t_1 is given (to excellent approximation) by the momentum the pencil would have picked up falling classically to that position from vertical.

For the string modes $n \lesssim n_\star$ of interest, the analogous transformation from a quantum description to a probability distribution for classical configurations occurs when $z \gg z_\star$.

14.8.2 Solution of the Time-Dependent Harmonic Oscillators

14.8.2.1 Basics

The distinction between ΔX^i (the difference between X^i and the reference geodesic) and X^i does not affect the $n \neq 0$ modes that are our focus. Similarly for the normalized coordinates $\Delta\hat{X}^i$. Therefore we make our notation a little less cumbersome and henceforth write $\Delta\hat{X}^i_n$ as simply $\hat{X}^i_n$ (for $n \neq 0$).

Each of the real degrees of freedom $\sqrt{2}\,\mathrm{Re}\,\hat{X}^i_n$ and $\sqrt{2}\,\mathrm{Im}\,\hat{X}^i_n$ in (14.90) have a harmonic oscillator Lagrangian of the form

$$L = \tfrac{1}{2} m\big(\dot{x}^2 - \omega^2(\tau)\, x^2\big), \tag{14.96}$$

with the translation $m \to p^u \simeq E$ and $\omega^2(\tau) \to \omega^2_{i,n}(\tau)$. The squared frequency $\omega^2(\tau)$ starts at a non-zero value $\omega^2(-\infty)$ and then changes with time τ. The quantum mechanical solution to such time-dependent harmonic oscillator problems has a long history. Useful explicit formulas for wave functions may be found in [36], with applications to strings in pp-wave backgrounds in [20, 22]. In our case, the harmonic oscillators all start in their ground state (the string state describing a graviton) at early times ($\tau \to -\infty$) and so start with Gaussian wave functions. For a time-dependent harmonic oscillator that starts as a Gaussian $\psi(x) \propto \exp[-x^2/4\sigma^2]$ at some time τ_0, one may check that the Schrödinger equation

$$i\dot{\psi} = \Big[-\frac{1}{2m}\,\partial_x^2 + \frac{1}{2}\, m\,\omega^2(\tau)\, x^2\Big]\psi \tag{14.97}$$

is solved by

$$\psi(x,\tau) \propto \frac{1}{\sqrt{\chi(t)}}\, \exp\Big[\frac{i}{2}\,\frac{\dot{\chi}(t)}{\chi(t)}\, m x^2\Big], \tag{14.98}$$

where the complex-valued function $\chi(\tau)$ satisfies the classical equation of motion

$$\ddot{\chi} = -\omega^2(\tau)\,\chi \tag{14.99}$$

with initial conditions

$$\chi(\tau_0) = 1, \qquad \dot{\chi}(\tau_0) = \frac{i}{2m\sigma^2}\,. \tag{14.100}$$

In our case, where we start in the early-time ground state, that's

$$\chi(\tau_0) = 1, \qquad \dot{\chi}(\tau_0) = i\,\omega(-\infty) \tag{14.101}$$

with $\tau_0 \to -\infty$.

The corresponding probability distribution $|\psi(x,\tau)|^2$ for x is just a Gaussian distribution

$$\text{Prob}(x,\tau) = \frac{e^{-x^2/2x_{\text{rms}}^2(\tau)}}{(2\pi)^{1/2}\,x_{\text{rms}}(\tau)} \tag{14.102}$$

with width

$$x_{\text{rms}}(\tau) = \frac{|\chi(\tau)|}{\sqrt{2m\,\omega(-\infty)}} = |\chi(\tau)|\,x_{\text{rms}}(-\infty). \tag{14.103}$$

Using (14.92), $x_{\text{rms}}(-\infty)$ corresponds to $\sqrt{\alpha'/2n}$.

The remaining task is to solve the classical equation of motion (14.99) for χ. For the case of interest $\hat{X}_n^3$, using (14.17), (14.85), (14.87c), (14.92), and $\omega \simeq |\boldsymbol{q}| \simeq E$, the χ equation may be put into the form

$$\frac{d^2\chi}{d\bar{\tau}^2} = -4(\xi^6 - \bar{z}^6)\chi, \tag{14.104a}$$

$$\frac{d\bar{z}}{d\bar{\tau}} = \bar{z}^2(1+\bar{z}^4)^{1/2}, \tag{14.104b}$$

where

$$\bar{z} \equiv \frac{z}{z_\star}, \tag{14.105}$$

$$\bar{\tau} \equiv \frac{z_\star^3}{z_{\text{h}}^2\mathbf{R}^2}\,\tau, \tag{14.106}$$

$$\xi = \xi_n \equiv n^{1/3}\xi_1 \equiv \left(\frac{n}{2}\right)^{1/3}\frac{z_{\text{h}}^2/z_\star}{\lambda^{-1/6}z_{\text{h}}^{4/3}E^{1/3}}. \tag{14.107}$$

In these variables, the initial conditions (14.101) on χ are

$$\chi(\bar{\tau}_0) = 1, \qquad \frac{d\chi}{d\bar{\tau}}(\bar{\tau}_0) = 2i\xi^3 . \tag{14.108}$$

Note from (14.1) and (14.18) that, parametrically,

$$\xi_1 \sim \frac{\ell_{\text{stop}}}{\lambda^{-1/6}\ell_{\text{max}}}, \tag{14.109}$$

and so the smallness of ξ_1 is a specific measure of how far we are into the interesting regime (14.4) of $\ell_{\text{stop}} \lesssim \lambda^{-1/6}\ell_{\text{max}}$. The modes $n \lesssim n_\star$ of interest to us correspond to $\xi_n \lesssim 1$.

Using the above equations, one may check that the string does not stretch significantly in proper size at early times $z \ll z_\star$ ($\bar{z} \ll 1$).But it is what the string does at late times which is more relevant.

14.8.2.2 Late-Time Behavior

For $z \gg z_\star$ (which is $\bar{z} \gg 1$), the $\bar{z}$ equation (14.104b) gives

$$\bar{z} = (-3\bar{\tau})^{-1/3}, \tag{14.110}$$

remembering that our convention is that τ is negative and that $\tau(z{=}\infty) = 0$. Note from (14.110) that $-\bar{\tau}$ is very small at the horizon $z = z_{\text{h}}$, where $-\bar{\tau} \sim (z_{\text{h}}/z_\star)^{-3} \sim (-q^2/E^2)^{3/4} \sim (\ell_{\text{stop}}T)^{-3}$.

Plugging (14.110) into the χ equation (14.104a) yields late-time ($\bar{z} \gg 1$) solutions $\chi \propto (-\bar{\tau})^{-1/3}$ and $\chi \propto (-\bar{\tau})^{4/3}$. The dominant solution will be

$$\chi \propto (-\bar{\tau})^{-1/3}. \tag{14.111}$$

Though χ is a complex-valued function whose purpose is to track the evolution of the wavepacket, exactly the same arguments as above give that a classical trajectory would have late-time behavior $x \propto (-\bar{\tau})^{-1/3}$. That means that $\dot{x} \propto (-\bar{\tau})^{-4/3}$, and so x and $\dot{x}$ both become large at late times, justifying a classical description at late times.[19] The classical relation between the two is determined by $x \propto (-\bar{\tau})^{-1/3}$ to be

[19]For example, at late times the exponential in the wavepacket (14.98) becomes $\exp[iS]$, where $S \propto x^2/(-\tau)$. The WKB condition $|\partial_x^2 S| \ll (\partial_x S)^2$ is satisfied as $\tau \to 0$ for $x \propto (-\tau)^{-1/3}$. We will see shortly that the proportionality constant in (14.111) is of order 1 for the modes $n \lesssim n_\star$ of interest. If one keeps track parametrically of all the proportionality constants in the exponential $\exp[iS]$, one finds more specifically that the WKB condition is satisfied when $-\bar{\tau} \ll 1$ (i.e. $\bar{z} \gg 1$).

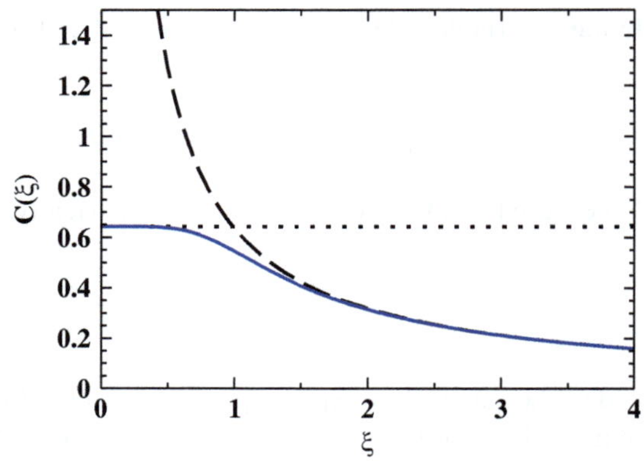

Fig. 14.12 The proportionality constant $C(\xi)$ in (14.113), which determines the late-time width of the probability distribution for the amplitude of a string mode. The *sloping dashed curve* shows the large-ξ approximation (14.114a)

$$\frac{\dot{x}}{x} \simeq \frac{1}{-3\bar{\tau}} \simeq \bar{z}^3. \tag{14.112}$$

One may extract the proportionality constant in the late-time behavior (14.111) by solving (14.104) numerically with initial conditions (14.108) for χ, matching the late time behavior of the numerical solution to (14.111), and repeating the calculation for earlier and earlier values of τ_0 in order to take the $\tau_0 \to -\infty$ limit. The late-time behavior is

$$|\chi(\bar{\tau})| \to \frac{C(\xi)}{(-\bar{\tau})^{1/3}} \tag{14.113}$$

with $C(\xi)$ given by Fig. 14.12.[20] We show in Appendix 4 that the limiting behavior for large ξ is

$$C(\xi) \simeq \frac{\Gamma(\frac{5}{6})}{\pi^{1/2}\xi} \qquad \text{for } \xi \gg 1, \tag{14.114a}$$

shown as a dashed curve in the figure. In the opposite limit of $\xi \ll 1$, our numerical results approach a constant

$$C(\xi) \simeq 0.6428 \qquad \text{for } \xi \ll 1. \tag{14.114b}$$

From (14.92), (14.103) and (14.113), and remembering that the analogs of x are $\sqrt{2}\,\mathrm{Re}\,X^i_n$ and $\sqrt{2}\,\mathrm{Im}\,X^i_n$, the late time probability distribution of mode amplitudes $\hat{X}^3_n$ is given by a Gaussian with width

[20] For numerical work, it is mildly convenient to eliminate $\bar{\tau}$ and express all of the relevant equations solely in terms of $\bar{z}$, giving $\bar{z}^4(1+\bar{z}^4)\chi'' + 2\bar{z}^3(1+2\bar{z}^4)\chi' = -4(\xi^6 - \bar{z}^6)\chi$ and $\chi'(\bar{z}_0) = 2i\xi^3/\bar{z}_0^2$ (with $\bar{z}_0 \to 0$) and $|\chi(\bar{z})| \to 3^{1/3}\,C(\xi)\,\bar{z}$ (as $\bar{z} \to \infty$).

$$|\hat{X}_n^3|_{\rm rms} \simeq \frac{C(\xi_n)}{(-\bar{\tau})^{1/3}} \left(\frac{\alpha'}{2n}\right)^{1/2} \qquad \text{for } -\bar{\tau} \ll 1. \tag{14.115}$$

Using (14.110), that may be rewritten as

$$|\hat{X}_n^3|_{\rm rms} \simeq 3^{1/3}\, C(\xi_n)\, \frac{z}{z_\star} \left(\frac{\alpha'}{2n}\right)^{1/2} \qquad \text{for } z \gg z_*, \xi_n. \tag{14.116}$$

As in (14.112), the corresponding momenta in this classical regime are related to the mode amplitudes $\hat{X}_n^3$ by

$$\partial_{\bar{\tau}} \hat{X}_n^3 \simeq \frac{\hat{X}_n^3}{-3\bar{\tau}} \simeq \frac{z^3}{z_\star^3}\, \hat{X}_n^3. \tag{14.117}$$

Using (14.80) and (14.82), the conversion between the normalized coordinate $\Delta\hat{x}^3$ and the displacement Δx^3 from the reference geodesic in Poincare coordinates is

$$\Delta x^3 = \frac{z}{\mathbf{R}} \left(1 - f \frac{|\boldsymbol{q}|^2}{\omega^2}\right)^{-1/2} \simeq \frac{z}{\mathbf{R}} \left(\frac{z_\star^4 + z^4}{z_{\rm h}^4}\right)^{-1/2} \Delta\hat{x}^3, \tag{14.118}$$

which is

$$\Delta x^3 \simeq \frac{z_{\rm h}^2}{z\mathbf{R}}\, \Delta\hat{x}^3 \qquad \text{for } z \gg z_\star. \tag{14.119}$$

So, from (14.116), the amplitudes of the stretched modes in the Poincare coordinate system are

$$|X_n^3|_{\rm rms} \simeq 3^{1/3}\, C(\xi_n)\, \frac{z_{\rm h}^2}{z_\star \mathbf{R}} \left(\frac{\alpha'}{2n}\right)^{1/2} = \frac{3^{1/3}\, C(\xi_n)\, z_{\rm h}^2}{(2n)^{1/2} \lambda^{1/4} z_\star} \tag{14.120}$$

for fixed τ (and so fixed z) in the classical regime. Using (14.18) and (14.17), this may be written as

$$|X_n^3|_{\rm rms} \simeq \frac{3^{1/3} (8\pi)^{1/2}\, C(\xi_n)}{n^{1/2}\, \Gamma^2(\frac{1}{4})}\, \lambda^{-1/4} \ell_{\rm stop} \tag{14.121}$$

for $z \gg z_\star, \xi_n$.

Note that fixed-τ (i.e. fixed-z) slices of the string worldsheet look different than fixed-x^0 slices of the string worldsheet, which is why our depiction of the string at various times x^0 in Fig. 14.15c were not horizontal.

14.8.3 The Size of the String at Late Times

Parametrically, the average size (14.121) of each stretched mode X_n^3 in Poincare coordinates is just the $\lambda^{-1/4}\ell_{\rm stop}$ which can be easily found with a back-of-the-envelope argument (see [37]). The total average size of the string is however given by summing all the modes. A convenient measure of the size scale of the string is the average rms deviation from the center of the string,

$$(\delta X^3)_{\rm rms} \equiv \left\langle \overline{\left(X^3 - \overline{X^3}\right)^2} \right\rangle^{1/2}, \tag{14.122}$$

where overlines indicate averaging over the string worldsheet position σ and the angle brackets indicate averaging over the late-time classical probability distribution for each mode amplitude. This is given by

$$(\delta X^3)_{\rm rms} = \left(2\sum_{n=1}^{\infty}\langle|X_n^3|^2\rangle\right)^{1/2}. \tag{14.123}$$

$\langle|X_n^3|^2\rangle$ is just the square of what we called $|X_n^3|_{\rm rms}$ in (14.121). Combining the limiting forms (14.114) with (14.121), and recalling from (14.107) that $\xi_n = n^{1/3}\xi_1$,

$$|X_n^3|_{\rm rms}^2 \simeq \left(\frac{3^{1/3}(8\pi)^{1/2}}{\Gamma^2(\frac{1}{4})}\lambda^{-1/4}\ell_{\rm stop}\right)^2 \times \begin{cases} \frac{[C(0)]^2}{n}, & n \ll n_\star; \\ \frac{\Gamma^2(\frac{5}{6})}{\pi\xi_1^2 n^{5/3}}, & n \gg n_\star, \end{cases} \tag{14.124}$$

where $C(0)$ is given by (14.114b). The sum in (14.123) is therefore convergent at large n and is dominated by a logarithm coming from $n = 1$ up to $n \sim n_\star$. At leading order in inverse powers of that logarithm,

$$(\delta X^3)_{\rm rms} \simeq |X_1^3|_{\rm rms}\sqrt{2\ln n_\star} \simeq \frac{3^{1/3}4\sqrt{\pi}\,C(0)}{\Gamma^2(\frac{1}{4})}\lambda^{-1/4}\ell_{\rm stop}\sqrt{\ln n_\star}. \tag{14.125}$$

Using (14.95), this may be rewritten as

$$(\delta X^3)_{\rm rms} \simeq 0.8660\,\lambda^{-1/4}\ell_{\rm stop}\,\ln^{1/2}\!\left(\frac{\lambda^{-1/6}\ell_{\rm max}}{\ell_{\rm stop}}\right), \tag{14.126}$$

where a parametric expression for the argument of the logarithm is adequate if we are only keeping track of the logarithmic term.

Ignoring the numerical constant in front, (14.126) is the parametric result (14.5) that we presented in the introduction.

The logarithm in (14.125) arises because of (i) the logarithmic UV divergence associated with the bosonic modes in the ground state, combined with (ii) the fact that the bosonic mode amplitudes with $n \ll n_\star$ all grow by an equal large factor from tidal stretching (and no longer cancel against fermionic modes in their physical consequences), while those with $n \gg n_\star$ do not grow significantly in comparison.

At this stage one must check that the string is not so big, or so far away from the reference geodesic, that the Penrose limit taken in Sect. 14.7 breaks down. This amounts to verifying that the $dv\, d\Delta x^3$ and $(dv)^2$ terms in the AdS$_5$-Schwarzschild metric (14.78), which were dropped in the Penrose limit, are parametrically small compared to the $du\, dv$ term,

$$\frac{f\mathbf{R}^2}{z^2}\frac{|\boldsymbol{q}|}{\omega}\,|dv\, d\Delta x^3| \quad \text{and} \quad \frac{f\mathbf{R}^2}{z^2}\,(dv)^2 \quad \ll \quad |du\, dv|, \tag{14.127}$$

for the string motions that we have found. Dividing both sides by $|du\, dv|$ and using $\omega \simeq |\boldsymbol{q}|$, we rewrite these conditions as

$$\frac{f\mathbf{R}^2}{z^2}\left|\frac{d\Delta x^3}{du}\right| \quad \text{and} \quad \frac{f\mathbf{R}^2}{z^2}\left|\frac{dv}{du}\right| \quad \ll \quad 1. \tag{14.128}$$

These conditions on the string motion are further analyzed in Appendix 6, where we find that the condition on $|d\Delta x^3/du|$ is the strongest and requires

$$\lambda^{-1/4}\sqrt{\ln n_\star} \ll 1 \tag{14.129}$$

in order for our earlier analysis to be valid. Using our result (14.125), this condition may be written as

$$(\delta X^3)_{\text{rms}} \ll \ell_{\text{stop}}. \tag{14.130}$$

That is, the Penrose limit only breaks down if one considers the extreme case (to be discussed in a moment) where the string becomes as large as the stopping distance itself.

14.9 Discussion

In our scheme for creating "jets," we have seen different behaviors in the dual theory depending on the virtuality (and so the stopping distance) of the jet. For $\ell_{\text{stop}} \gg \lambda^{-1/6}\ell_{\text{max}}$, the gravitons (or other massless string modes) composing the excitation in the gravity description remain gravitons until after the excitation has stopped moving in the x^3 direction, and there is no difficulty in using the supergravity approximation for the calculation. For $\ell_{\text{stop}} \ll \lambda^{-1/6}\ell_{\text{max}}$, each graviton is instead stretched into a classical string loop. However, provided that

$$\lambda^{-1/4}\ln^{1/2}\!\left(\frac{\lambda^{-1/6}\ell_{\text{max}}}{\ell_{\text{stop}}}\right) \ll 1, \tag{14.131}$$

the size of that string remains small compared to the stopping distance ℓ_{stop}. The string remains close to its reference geodesic, and so corrections to the $\lambda=\infty$ description of the jet stopping are parametrically small (if one only attempts to resolve details on size scales large compared to the size of the string). However, if instead

$$\lambda^{-1/4}\,\ln^{1/2}\left(\frac{\lambda^{-1/6}\ell_{\max}}{\ell_{\text{stop}}}\right) \gtrsim 1, \tag{14.132}$$

then the string loop will stretch out to be parametrically as large as the stopping distance itself. Our quantum analysis of the string breaks down in this case (because of the failure of the Penrose limit), but we can see what happens qualitatively by tracking what happens to the $\lambda^{-1/4}\log^{1/2} \ll 1$ results as we increase the logarithm towards $\lambda^{-1/4}\log^{1/2} \sim 1$.

In particular, a nice way to visualize what happens is to follow the classical evolution of a closed string that initially starts with a proper size Σ of order $\sqrt{\alpha' \ln n_\star}$, which is roughly the initial rms size from the modes $n \lesssim n_\star$ which become classically excited. Increasing the logarithm towards $\lambda^{-1/4}\sqrt{\ln n\star} \sim 1$ is equivalent to increasing Σ towards $\sim \mathbf{R}$. Figure 14.13 compares examples of such evolution for the cases (a) $\lambda^{-1/4}\log^{1/2} \ll 1$ and (b) $\lambda^{-1/4}\log^{1/2} \sim 1$. The interesting feature of Fig. 14.13b is that, at late times, the string looks like the original picture advocated by Gubser et al. [3] of gluon jets as dual to the evolution of a trailing, folded classical string falling in AdS_5-Schwarzschild. Our string is a folded closed string, as depicted in the cartoon of Fig. 14.14a, whereas the one studied by Gubser et al. was a folded infinite open string, as depicted by Fig. 14.14b.[21] However, the left end of the string in these figures, which is very close to the horizon, does not play a significant role in the effect on the boundary theory, and so the physics of these two situations is approximately the same.

Historically, the original motivation of our own method for posing "jet" stopping problems [6], outlined in the introduction to this paper and motivated by Fig. 14.1, was to give a precise field theory problem in $\mathcal{N}=4$ SYM that could be solved, beginning to end, using gauge-gravity duality. It has not previously been know how to precisely set up a problem in $\mathcal{N}=4$ SYM that corresponds to earlier studies of jets [3–5] that made use of classical strings in the gravity description. It is interesting to now make contact between our approach and Gubser et al.'s classical string approach, in the particular limit (14.132), which can be roughly rewritten as

$$T^{-1} \ll \ell_{\text{stop}} \lesssim T^{-4/3}\left(\frac{E}{\sqrt{\lambda}}\right)^{1/3} e^{-O(\lambda^{1/2})}, \tag{14.133}$$

[21] More precisely, Gubser et al. first considered a folded open string that stretched out from beyond the horizon, as in Fig. 1 of [3]. But in actual calculations, they focused on the trailing infinite folded string, as in Fig. 2 of that reference.

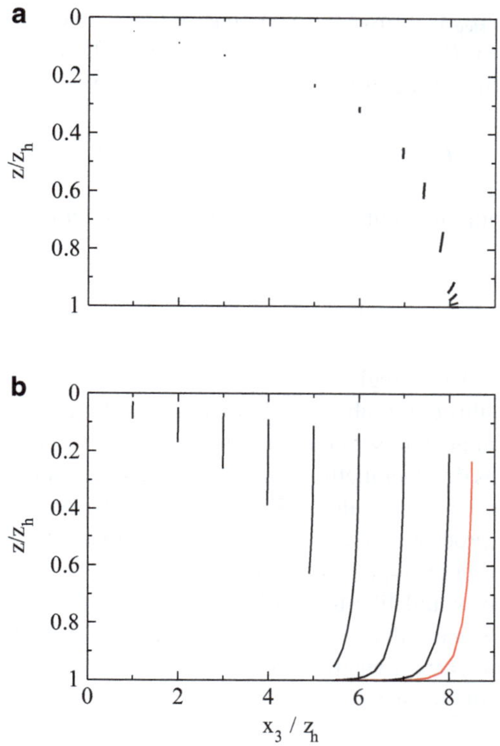

Fig. 14.13 Examples of numerical solutions of the evolution of a falling classical string loop that starts near the boundary with proper size (**a**) $\Sigma \ll \mathbf{R}$ and (**b**) $\Sigma \sim \mathbf{R}$. These are snapshots of the string at fixed x^0. See Appendix C of [37] for details of the initial condition

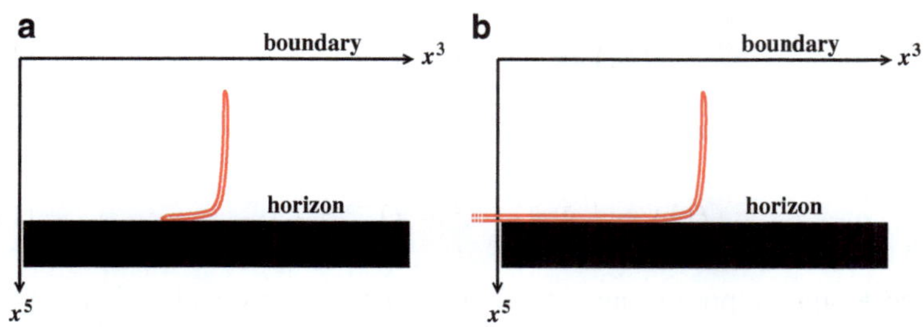

Fig. 14.14 Schematic pictures of classical folded strings. (**a**) A closed folded string produced by extreme tidal stretching of a graviton in our method of generating "jets" in the case of $\lambda^{-1/4}\log^{1/2} \gtrsim 1$ (14.132), and (**b**) the infinite, folded open string considered by Gubser et al. [3]. In the latter case, the trailing string continues to get closer and closer to the horizon as $x^3 \to -\infty$

where the $T^{-1} \ll \ell_{\text{stop}}$ is thrown in to emphasize that we've always been assuming $-q^2 \ll E^2$ and so $T^{-1} \ll \ell_{\text{stop}}$ throughout, and where O means "of order." Alternatively, in terms of the virtuality $-q^2$ of the source of our "jet," (14.133) is

$$E^2 \gg -q^2 \gtrsim T^{4/3} E^{2/3} \lambda^{2/3} e^{+O(\lambda^{1/2})}. \tag{14.134}$$

This window of stopping lengths only appears once the jet energy is large enough that

$$E \gg T\sqrt{\lambda}\, e^{+O(\lambda^{1/2})}. \tag{14.135}$$

Even though there is a region of overlap (14.133) of our results with strings that look similar to those of Gubser et al., there are still important differences once we get out of this range. Gubser et al. found a *maximum* stopping distance of order $T^{-4/3}(E/\sqrt{\lambda})^{1/3}$, as do other methods that also model excitations with semi-infinite classical strings in the gravity dual [5]. In contrast, the types of excitations that we create, through processes like Fig. 14.1, have a parametrically larger maximum stopping distance of order $\ell_{\max} \sim T^{-4/3} E^{1/3}$.

To conclude, we would like to highlight a remaining mystery concerning the λ-dependence of stopping distances: How does the $E^{1/3}$ scaling of the maximum stopping distance at $\lambda{=}\infty$ transition to $E^{1/2}$ at small λ? One might naively guess the scaling to be of the form[22]

$$\ell_{\max} \propto E^{f(\lambda)} \tag{14.136}$$

for some function $f(\lambda)$ with $f(0) = \frac{1}{2}$ and $f(\infty) = \frac{1}{3}$. One might further hope that $f(\lambda)$ has relatively simple expansions around $\lambda{=}0$ and $\lambda{=}\infty$. For example, perhaps

$$f(\lambda) = \tfrac{1}{2} + \#\lambda + \#\lambda^2 + \cdots \tag{14.137}$$

and

$$f(\lambda) = \tfrac{1}{3} + \#\lambda^{-3/2} + \#\lambda^{-5/2} + \cdots, \tag{14.138}$$

where the # signs represent numerical coefficients. The details of the expansion don't matter—one can imagine there could be different powers of λ than shown above, or factors of $\ln\lambda$ in the expansion, and so forth. But take (14.138) as an example. Then, expanding (14.136) around $\lambda{=}\infty$,

[22]We ignore here logarithmic energy dependence in the prefactor of the exponential. The small-λ scaling is really $(E/\ln E)^{1/2}$, which is equivalent to including a log-of-log energy dependence in the exponent f: $(E/\ln E)^{1/2} = \exp[\frac{1}{2} - \frac{1}{2}\ln\ln E]$.

$$\ell_{\max} \propto E^{1/3}\big[1 + \#\lambda^{-3/2} \ln E + O(\lambda^{-5/2})\big]. \tag{14.139}$$

We might therefore expect if we compute something that is related to the maximum stopping distance, like $\ell_{\rm tail}$, the corrections in powers of $1/\lambda$ should also come with powers of $\ln E$, as above. But there is no sign of a $\ln E$ factor in our concrete formulae for $\ell_{\rm tail}$ (14.3). Perhaps the exponent $f(\infty) = \frac{1}{3}$ does not receive corrections until a yet-higher power in λ, but we are unsure how a $\ln E$ could arise in yet-higher-order calculations of the shift of the quasi-normal mode pole. Or perhaps the exponent does not have an expansion in powers of $1/\lambda$ but instead behaves like $\frac{1}{3} + \#e^{-\#\lambda}$ for large λ. Or perhaps the tail scale $\ell_{\rm tail}$ is a misleading stand-in for $\ell_{\max}$, as is known to happen in the case of $\Delta \gg 1$ [7]. Whatever the resolution, given the absence of a $\ln E$ in our result for $\ell_{\rm tail}$, the question of how $E^{1/3}$ begins to make its way towards $E^{1/2}$ (and vice versa) remains an open question.

Appendix 1: What Happens for $z \gg z_\star$?

By making the reasonable assumptions that we outlined in Sect. 14.2.3, we have managed to analyze the question of when corrections become important by focusing on particle trajectories at $z \sim z_\star$. As z increases beyond this scale, the forward progress of the trajectory slows to a stop.We previously asserted that at some scale $z_{\rm bad} \gg z_\star$ the expansion in higher-derivative corrections would eventually break downeven if the expansion was well behaved at $z_\star$.

Start by considering the cost of an $\alpha' D^2$ factor, which we analyzed in Sect. 14.4.3 for $z \sim z_\star$. At larger z, with $z_\star \lesssim z \ll z_{\rm h}$, (14.62) gives

$$q_5\big|_{z \gtrsim z_\star} \sim \frac{z^2 E}{z_{\rm h}^2}. \tag{14.140}$$

Then the cost (14.61) of each $\alpha' D^2$ factor is

$$\alpha' D^2|_{z \gtrsim z_\star} \sim \frac{\alpha' z q_5}{\mathbf{R}^2} \sim \frac{z^3 E T^2}{\lambda^{1/2}}. \tag{14.141}$$

This cost is unsuppressed for $z \gtrsim z_{\rm bad}$, where

$$z_{\rm bad} \sim \frac{\lambda^{1/6}}{E^{1/3} T^{2/3}}. \tag{14.142}$$

The same constraint arises from the other important corrections that we analyzed. For instance, the cost of adding an $\alpha'^2 D^2 C$ factor was $z^6 E^2/\lambda z_{\rm h}^4$, which also becomes unsuppressed at the same $z \gtrsim z_{\rm bad}$.

Note that the requirement $\ell_{\rm stop} \gg \lambda^{-1/6} \ell_{\max}$ for the expansion in corrections to be well-behaved in Fig. 14.3 is the same condition as requiring $z_{\rm bad} \gg z_\star$.

Appendix 2: Why (14.46) Cannot Precisely Determine $\Delta\ell_{\text{stop}}$

Consider the safe region $\ell_{\text{stop}} \gg \lambda^{-1/6}\ell_{\max}$ of Fig. 14.3, where the effects of higher-dimensional supergravity interactions should be suppressed. The R^4 corrections then dominate the corrections at $z \sim z_\star$. We might then be tempted to use the explicit form (14.25) of the R^4 correction, combined with the particle-based formula (14.38) for the stopping distance, to explicitly calculate the first correction to the $\lambda{=}\infty$ result (14.18) for the stopping distance. In this appendix, we discuss why that does not work.

We start with (14.45),

$$\ell_{\text{stop}} \simeq \int_0^{z_h} dz\, \frac{|\boldsymbol{q}|\left[1 - \frac{2\varepsilon z^{10}}{z_h^8}|\boldsymbol{q}|^2\right]}{\sqrt{-q^2 + \frac{z^4}{z_h^4}|\boldsymbol{q}|^2 + \frac{\varepsilon z^{10}}{z_h^8}|\boldsymbol{q}|^4 f}}. \tag{14.143}$$

First we explain why the potentially sign-changing behavior of the numerator correction in the R^4-corrected formula (14.143) for the stopping distance could be ignored. The disturbing features of this correction arise in the z range given by (14.48),

$$z \gg z_{\text{disturbing}} \sim \left(\frac{\lambda^{3/4}T}{E}\right)^{1/5} z_h. \tag{14.144}$$

This difficulty only arises at all if $z_{\text{disturbing}} < z_h$, which requires

$$E \gg \lambda^{3/4}T. \tag{14.145}$$

Now compare (14.144) and (14.142):

$$z_{\text{disturbing}} \sim \left(\frac{E}{\lambda^{1/8}T}\right)^{2/15} z_{\text{bad}}. \tag{14.146}$$

The inequality (14.145) then gives

$$z_{\text{disturbing}} \gg z_{\text{bad}}, \tag{14.147}$$

and so the numerator correction cannot be believed in the range of z for which it becomes disturbing.

Dropping the numerator correction from (14.143) leaves

$$\ell_{\text{stop}} \simeq \int_0^{z_h} dz\, \frac{|\boldsymbol{q}|}{\sqrt{-q^2 + \frac{z^4}{z_h^4}|\boldsymbol{q}|^2 + \frac{\varepsilon z^{10}}{z_h^8}|\boldsymbol{q}|^4 f}}, \tag{14.148}$$

For simplicity, in what follows we will just analyze the case $E \gg \lambda^{3/4}T$.

In the integrand, look at the expression under the square root in the denominator. The relative importance of the εz^{10} term grows with increasing z. For $z \gg z_\star$, the z^4 term under the square root dominates over the $-q^2$ term, so we should compare the εz^{10} term to the z^4 term. These are the same size at a scale $z_{\star\star} \gg z_\star$ given by

$$z_{\star\star} \sim \frac{z_{\rm h}^{2/3}}{\varepsilon^{1/6}|\boldsymbol{q}|^{1/3}} \sim \frac{\lambda^{1/4}}{E^{1/3}T^{2/3}} \sim \frac{\lambda^{1/4}}{l_{\rm max}T^2}, \tag{14.149}$$

assuming that $z_{\star\star} \ll z_{\rm h}$ so that $f \simeq 1$. But $z_{\star\star} \ll z_{\rm h}$ follows from (14.149) and our consideration of $E \gg \lambda^{3/4}T$.

Now calculate the correction $\Delta\ell$ to the stopping distance by subtracting the $\lambda{=}\infty$ result (14.15) from (14.148),

$$\Delta\ell \equiv \ell_{\rm stop} - \ell_{\rm stop}^{\lambda=\infty} \simeq \int_0^{z_{\rm h}} dz \left[\frac{|\boldsymbol{q}|}{\sqrt{-q^2 + \frac{z^4}{z_{\rm h}^4}|\boldsymbol{q}|^2 + \frac{\varepsilon z^{10}}{z_{\rm h}^8}|\boldsymbol{q}|^4 f}} - \frac{|\boldsymbol{q}|}{\sqrt{-q^2 + \frac{z^4}{z_{\rm h}^4}|\boldsymbol{q}|^2}}\right]. \tag{14.150}$$

This integral is dominated by $z \sim z_{\star\star}$. So, to explicitly calculate $\Delta\ell$ will require trusting the integrand at $z \sim z_{\star\star}$ given by (14.149). Compare this to the z scale (14.142) where the expansion in supergravity corrections breaks down:

$$z_{\star\star} \sim \lambda^{1/12} z_{\rm bad} \gg z_{\rm bad}. \tag{14.151}$$

So we cannot trust (14.150) in the range of z where we want to use it to get an explicit result for $\Delta\ell$.

Appendix 3: Other Higher-Derivative Terms

In Sect. 14.4.3, we addressed the $\alpha' Q^5 D_5$ piece of $\alpha' D^2$ when studying the importance of $D^{2n}C^4$. We also recycled our conclusion from that analysis when later considering applying extra powers of derivatives to $D^{2k}C^{4+k}$. The dominant terms involved $\alpha' Q^I D_I$ where the D_I hits a background Weyl tensor. We motivated focusing on $Q^5 D_5$ by noting that the background Weyl tensor depends only of the x^5 coordinate. If the D's were ordinary derivatives instead of covariant derivatives, that would be the end of the story. However, the other components D_μ of the covariant derivative do not vanish when applied to the background Weyl tensor. In fact, they are parametrically of order $1/z$, just like D_5. As a result, for example,

$$\alpha' Q^3 D_3 = \alpha' Q_3 g^{33} D_3 \sim \alpha' \times E \times \frac{z^2}{\mathbf{R}^2} \times \frac{1}{z} \sim \frac{Ez}{\lambda^{1/2}} \tag{14.152}$$

is actually parametrically *larger* than the derivative

$$\alpha' Q^5 D_5 \sim \frac{q_5 z}{\lambda^{1/2}} \tag{14.153}$$

considered in the main text (14.61).

So why doesn't this lead to much larger results for the importance of $D^{2n} R^4$ and other operators than shown in Fig. 14.3? Our answer requires thinking about how the indices of the background Weyl tensor C_{JKLM} hit by $\alpha' Q^I D_I$ contract with everything else.

Because C_{IJKL} depends only on x^5, a non-zero value for $Q^\mu D_\mu C_{IJKL}$ arises only from the terms of D involving the Christoffel symbols:

$$Q^\mu D_\mu C_{IJKL} = -Q^\mu \Gamma^{\bar{I}}{}_{I\mu} C_{\bar{I}JKL} - Q^\mu \Gamma^{\bar{J}}{}_{J\mu} C_{I\bar{J}KL} - \cdots . \tag{14.154}$$

Now write

$$\Gamma = \Gamma^{(\mathrm{AdS})} + \Delta\Gamma, \tag{14.155}$$

where $\Gamma^{(\mathrm{AdS})}$ is the zero-temperature, purely AdS expression for the connection Γ. The difference between AdS and AdS$_5$-Schwarzschild is the difference between taking $f = 1$ and $f = 1-(z/z_\mathrm{h})^4$ in the metric (14.6). As a result, the $\Delta\Gamma$ piece of (14.155) is suppressed compared to the $\Gamma^{(\mathrm{AdS})}$ piece by order $(z/z_\mathrm{h})^4$. For studying the dominant corrections at $z \sim z_\star \ll z_\mathrm{h}$, we should therefore focus on $\Gamma^{(\mathrm{AdS})}$. In particular,

$$\alpha' Q^\mu \Delta\Gamma^{\bar{I}}{}_{I\mu} \sim \alpha' E \times \frac{z^2}{\mathbf{R}^2} \times \frac{1}{z}\left(\frac{z}{z_\mathrm{h}}\right)^4 \sim \frac{E z^5}{\lambda^{1/2} z_\mathrm{h}^4} \tag{14.156}$$

is always less important at $z \sim z_\star$ than the $\alpha' Q^5 D_5$ term (14.153) that we considered in the main text.

So now focus on Γ^{AdS}:

$$Q^\mu D_\mu C_{IJKL} \simeq -Q^\mu (\Gamma^{\bar{I}}{}_{I\mu})^{\mathrm{AdS}} C_{\bar{I}JKL} - Q^\mu (\Gamma^{\bar{J}}{}_{J\mu})^{\mathrm{AdS}} C_{I\bar{J}KL} - \cdots . \tag{14.157}$$

Because AdS space has four-dimensional Lorentz invariance, the μ index on Q^μ above must pass through to contract with something else. For example,

$$Q^\mu (\Gamma^{\bar{I}}{}_{I\mu})^{\mathrm{AdS}} C_{\bar{I}JKL} \times (\text{other stuff})^{IJKL}$$
$$\simeq \frac{1}{z} Q^\mu C_{\mu JKL} \times (\text{other stuff})^{5JKL} - \frac{1}{z} C^5{}_{JKL} Q_\mu \times (\text{other stuff})^{\mu JKL} \tag{14.158}$$

and

$$Q^{\mu}(\Gamma^{\bar{J}}{}_{J\mu})^{\mathrm{AdS}}C_{I\bar{J}KL}\times(\text{other stuff})^{IJKL}$$
$$\simeq \frac{1}{z}\,Q^{\mu}C_{I\mu KL}\times(\text{other stuff})^{I5KL}-\frac{1}{z}\,C_{I}{}^{5}{}_{KL}Q_{\mu}\times(\text{other stuff})^{I\mu KL} \tag{14.159}$$

But now recall that our dominant terms already had every C contracted with two Q's. So the "other stuff" above had the form

$$(\text{other stuff})^{IJKL}\sim Q^{I}Q^{K}(\text{something})^{JL}, \tag{14.160}$$

and these terms were dominant because both Q_I and Q_K were parametrically of order E when contracted with the Weyl tensor C_{IJKL}. We are currently worried about the possibility that the Q_μ factor above is also of order E. Now look at the first term in (14.158). The $Q_\mu Q_I Q_K \sim E^3$ is contracted in such a way that it instead gives $Q_\mu Q_5 Q_K \sim q_5 E^2 \ll E^3$, which is not problematical. The second term in (14.158) contracts two Q's together to give a factor of $Q_\mu \eta^{\mu\nu} Q_\nu \sim q^2$ instead of an E^2, and so it also is suppressed. Next look at the first term in (14.159). There we have $Q^\mu Q^I Q^K C_{I\mu KL}$. Up to terms which are suppressed by $q_5 \ll E$, this is the same as $Q^J Q^I Q^K C_{IJKL}$, which vanishes by the symmetry of the Weyl tensor. Finally, look at the second term of (14.159), which involves

$$Q_{\mu}(\text{something})^{\mu L}. \tag{14.161}$$

For the dominant terms analyzed in the main text of this paper, the "something" is made up of factors of Q and QQC. If $(\text{something})^{\mu L}$ gives a factor of Q^μ, then two of our Q's that were supposed to be giving factors of E will instead give a factor of $-q^2 \ll E$. If $(\text{something})^{\mu L}$ gives a factor of $Q_N Q_P C^{N\mu P\bullet}$, then we'll get a suppression as before because of the symmetry of C.

Appendix 4: Large ξ Behavior of $C(\xi)$

For large ξ, the $\bar{z}^6$ term in the differential equation (14.104a) for ξ can be ignored until $\bar{z} \gg 1$. At that point, however, we may use the simple large-$\bar{z}$ result (14.110) for $\bar{z}$. Substituting this into (14.104a) gives

$$\frac{d^2\chi}{d\bar{\tau}^2}=-4\left[\xi^6-\frac{1}{(-3\bar{\tau})^2}\right]\chi, \tag{14.162}$$

whose solution is

$$\chi=(-\pi\xi^3\bar{\tau})^{1/2}H^{(2)}_{5/6}(-2\xi^3\bar{\tau}). \tag{14.163}$$

The late-time behavior $\tau \to 0$ is

$$|\chi| \simeq \frac{\Gamma(\frac{5}{6})}{\pi^{1/2}\xi(-\bar{\tau})^{1/3}}. \tag{14.164}$$

Appendix 5: A Back-of-the-Envelope Estimate

In this appendix we give a parametric estimate of the amount of tidal stretching of the string compared to the size of the stopping distance ℓ_{stop}. Here the only thing we need to know is that the stopping distance given by following a null geodesic as in Fig. 14.5 is proportional to a power of the slope dx^3/dx^5 of that geodesic where it starts, at the boundary. The more downward-directed one starts the trajectory in Fig. 14.5, the less distance it will travel in x^3 before reaching the horizon.

Now interpret the trajectory of Fig. 14.5 as a trajectory for the center of mass of a tiny, falling loop of string. Once the string gets far enough from the boundary that the tidal forces dominate over the string tension, then the string tension becomes ignorable, and different pieces of the string will fall independently along their own geodesics, the string stretching accordingly. Imagine plotting two such geodesics, for the two bits of the string loop that are most separated. The separation of those geodesics is a measure of the extent of the tidally-stretched loop of string as it falls towards the horizon. The *proper* size of the string should start out of order the quantum mechanical size Σ of the graviton, which is roughly set by dimensional analysis in terms of the string tension $\mathbb{T}$ as

$$\Sigma_{\text{graviton}} \sim \mathbb{T}^{-1/2} \sim \sqrt{\alpha'}, \tag{14.165}$$

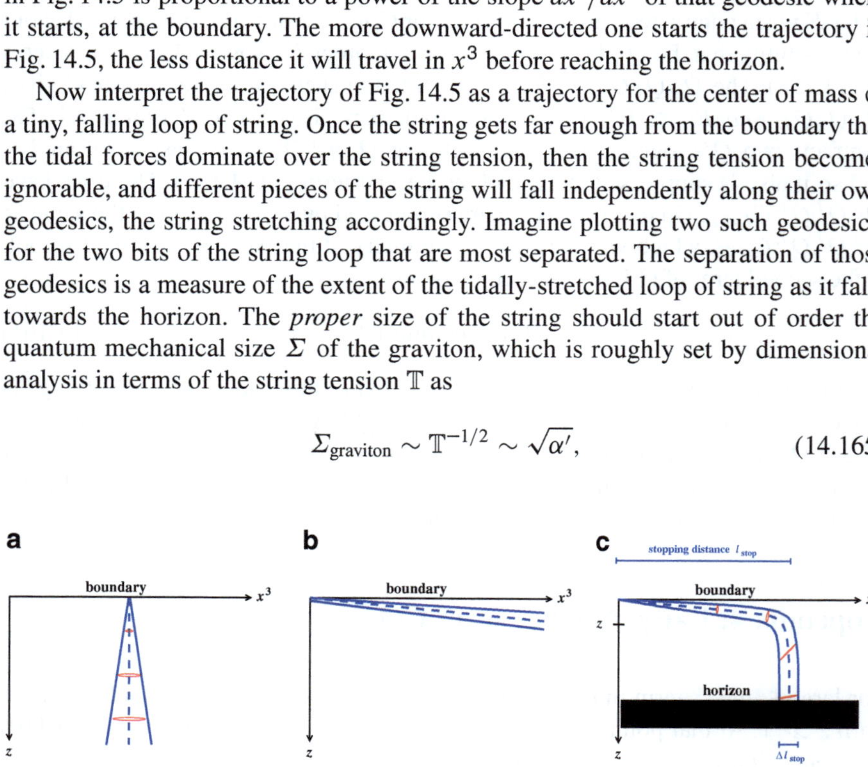

Fig. 14.15 (**a**) Parallel geodesics in AdS_5 in the Lorentz frame where the excitation is at rest in 3-space. These geodesics maintain a constant proper separation as they fall into the bulk, and this separation should be thought of as of order the characteristic size ($\sim \sqrt{\alpha'}$) of the closed quantum string loop describing the graviton (or other massless string mode). The narrow red loops are meant to be suggestive of the closed string loop. (**b**) The same picture boosted to the original Lorentz frame. (**c**) A picture of how those geodesics evolve in AdS_5-Schwarzschild rather than AdS_5. The early-time behavior is the same as (**b**). [For classical oscillating string solutions, the strings depicted in (**a**) may be thought of as snapshots at moments when the string's proper extent in x^3 is at, say, maximum (or half-maximum or whatever). Such solutions would similarly oscillate in the $z \ll z_\star$ part of (**b**) but not in the $z \gg z_\star$ part, where tidal forces dominate over tension]

where $\alpha' = 1/2\pi\mathbb{T}$ is the string slope parameter.

Very close to the boundary, the tidal forces due to the black hole are negligible, and the closed loop of string is in its ground state. We can set up our two geodesics above so that, correspondingly, they maintain constant proper separation Σ_{graviton} near the boundary, where the AdS$_5$-Schwarzschild metric approaches a purely AdS$_5$ metric. To see how this works, imagine making a four-dimensional boost from (i) the plasma rest frame, in which we create an excitation with large 4-momentum $q^\mu = (\omega, 0, 0, q_3) \simeq (E, 0, 0, E)$ and relatively small 4-virtuality $-q^2 \ll E^2$, to (ii) the excitation's initial rest frame, where the 4-momentum is instead $(\sqrt{-q^2}, 0, 0, 0)$. The Lorentz boost factor for this transformation is

$$\gamma = \sqrt{\frac{\omega^2}{-q^2}} \simeq \sqrt{\frac{E^2}{-q^2}} \gg 1. \tag{14.166}$$

In AdS$_5$, the trajectory in the new frame will drop straight down away from the boundary, as depicted by the dashed line in Fig. 14.15a.

Now consider the graviton as an extended object with proper size Σ. The two straight solid null lines in Fig. 14.15a depict the extent of the graviton in AdS$_5$ in the excitation's rest frame at early times. In pure AdS$_5$ null geodesics are straight lines. We parametrize the two solid lines of Fig. 14.15a as

$$x^I = \left(\gamma_+, 0, 0, \pm\beta_+\gamma_+, 1\right) z \tag{14.167}$$

with $\beta_+ \ll 1$ and $\gamma_+ \equiv (1 - \beta_+^2)^{-1/2} \simeq 1$. Because of the warp factor in the metric, these two lines are parallel and maintain constant proper separation $\sqrt{\Delta x^3\, g_{33}\, \Delta x^3} = 2\beta_+\gamma_+\mathbf{R} \simeq 2\beta_+\mathbf{R}$ as a function of the rest-frame time. Setting this proper separation to be of order the graviton size Σ given by (14.165) then gives

$$\beta_+ \sim \frac{\Sigma_{\text{graviton}}}{\mathbf{R}} \sim \frac{\sqrt{\alpha'}}{\mathbf{R}} \sim \lambda^{-1/4}, \tag{14.168}$$

and (14.167) gives

$$x^I \sim \left(1, 0, 0, \pm\lambda^{-1/4}, 1\right) z. \tag{14.169}$$

Now boost back to the original plasma frame using (14.166) to get the early-time trajectories depicted by solid lines in Fig. 14.15b: $x^I \sim \left(\gamma(1 \pm \lambda^{-1/4}), 0, 0, \gamma(1 \pm \lambda^{-1/4}), 1\right) z$, where we have used $\gamma \gg 1$ (14.166). Then

$$\left.\frac{\Delta(dx^3/dz)}{dx^3/dz}\right|_{\text{initial}} \sim \lambda^{-1/4}. \tag{14.170}$$

As discussed before, the stopping distance (which requires a calculation in the full AdS_5-Schwarzschild metric) covered by a null geodesic is power-law related to this initial slope, and so the difference $\Delta\ell_{\text{stop}}$ in how far the two bits of string travel also has the same small size (14.170) relative to ℓ_{stop}:

$$\frac{\Delta\ell_{\text{stop}}}{\ell_{\text{stop}}} \sim \lambda^{-1/4}. \tag{14.171}$$

Appendix 6: Checking the Penrose Limit: Details

In order to check the validity of the Penrose limit, here we characterize the string by following null geodesics that roughly trace different bits of string and which deviate slightly from our reference geodesic. This approximation amounts to ignoring the tension in the string as in Appendix 5 (for an alternative check of the Penrose limit outside of this approximation see [37]).

From the null geodesic formula and the metric (14.6), the x^3 coordinate for such geodesics is given by

$$\frac{dx^3}{dz} = \frac{\hat{q}_3}{\sqrt{1 - f\hat{\boldsymbol{q}}^2}}, \tag{14.172}$$

where

$$\hat{q}_\mu \equiv \frac{q_\mu}{\omega} = (-1, \hat{\boldsymbol{q}}). \tag{14.173}$$

Remembering that $\Delta x^\mu \equiv x^\mu - \bar{x}^\mu(z)$ is the deviation relative to the reference geodesic, we have

$$\frac{d\Delta x^3}{dz} = \frac{\hat{q}_3}{\sqrt{1 - f\hat{\boldsymbol{q}}^2}} - \frac{\bar{\hat{q}}_3}{\sqrt{1 - f\bar{\hat{\boldsymbol{q}}}^2}}. \tag{14.174}$$

Expand to first order in $\Delta\hat{q}_3 \equiv \hat{q}_3 - \bar{\hat{q}}_3$:

$$\frac{d\Delta x^3}{dz} \simeq \frac{\Delta\hat{q}_3}{(1 - f\bar{\hat{\boldsymbol{q}}}^2)^{3/2}}. \tag{14.175}$$

Then using (14.85) (and defining u with respect to the reference geodesic $\bar{x}$),

$$\frac{f\mathbf{R}^2}{z^2}\frac{d\Delta x^3}{du} \simeq \frac{f\,\Delta\hat{q}_3}{1 - f\hat{\boldsymbol{q}}^2}. \tag{14.176}$$

Since $1 - f\hat{\boldsymbol{q}}^2 \simeq (z_\star^4 + z^4)/z_{\rm h}^4$, the combination (14.176) is largest for $z \lesssim z_\star$, and the $d\Delta x^3/du$ condition in (14.128) requires

$$\Delta\hat{q}_3 \ll \frac{z_\star^4}{z_{\rm h}^4} \tag{14.177}$$

for the Penrose limit. Use (14.18) to relate this to the stopping distance:

$$\ell_{\rm stop} \sim \frac{z_{\rm h}^2}{z_\star} \sim z_{\rm h}\left(\frac{E^2}{-q^2}\right)^{1/4} \sim \frac{z_{\rm h}}{(1-\hat{q}_3)^{1/4}}, \tag{14.178}$$

so that

$$\Delta\ell_{\rm stop} \sim \frac{z_{\rm h}\,\Delta\hat{q}_3}{(1-\hat{q}_3)^{5/4}} \sim \frac{\Delta\hat{q}_3\,\ell_{\rm stop}}{1-\hat{q}_3} \sim \Delta\hat{q}_3\,\ell_{\rm stop}\,\frac{z_{\rm h}^4}{z_\star^4}. \tag{14.179}$$

Combining (14.177) and (14.179) gives the condition

$$\Delta\ell_{\rm stop} \ll \ell_{\rm stop} \tag{14.180}$$

quoted in (14.130).

Now turn to the condition on dv/du in (14.128). The definition (14.77) of v gives

$$dv = \bar{\hat{q}}_\mu\, d(\Delta x^\mu) = -d(\Delta x^0) + \bar{\hat{q}}_3\, d(\Delta x^3), \tag{14.181}$$

and so we need a formula for $d(\Delta x^0)$. The analog of (14.172) is

$$\frac{dx^0}{dz} = \frac{-f^{-1}\hat{q}_0}{\sqrt{1-f\hat{\boldsymbol{q}}^2}}, \tag{14.182}$$

with expansion

$$\frac{d\Delta x^0}{dz} \simeq \frac{\hat{q}_3\,\Delta\hat{q}_3}{(1-f\bar{\hat{q}}^2)^{3/2}}. \tag{14.183}$$

Combining (14.175), (14.181), and (14.183), gives $dv/du \simeq 0$. We therefore have to go back and make our expansions to second-order in $\Delta\hat{\boldsymbol{q}}$. The result is

$$\frac{dv}{dz} \simeq -\frac{(\Delta\hat{\boldsymbol{q}}_\perp)^2}{2(1-f\bar{\hat{q}}_3^2)^{1/2}} - \frac{(\Delta\hat{q}_3)^2}{2(1-f\bar{\hat{q}}_3^2)^{3/2}}, \tag{14.184}$$

and so

$$\frac{f\mathbf{R}^2}{z^2}\left|\frac{dv}{du}\right| \simeq \frac{f(\Delta\hat{\boldsymbol{q}}_\perp)^2}{2} + \frac{f(\Delta\hat{q}_3)^2}{2(1-f\bar{\hat{q}}_3^2)} . \tag{14.185}$$

The corresponding condition on dv/du in (14.128) is then

$$f(\Delta\hat{\boldsymbol{q}}_\perp)^2 \quad \text{and} \quad \frac{f(\Delta\hat{q}_3)^2}{(1-f\bar{\hat{q}}_3^2)} \quad \ll \quad 1. \tag{14.186}$$

The first condition is strongest for $z \ll z_\mathrm{h}$ and the second for $z \lesssim z_\star$, giving

$$|\Delta\hat{\boldsymbol{q}}_\perp| \quad \text{and} \quad |\Delta\hat{q}_3|\,\frac{z_\mathrm{h}^2}{z_\star^2} \quad \ll \quad 1. \tag{14.187}$$

Using (14.179), the condition involving $\Delta\hat{q}_3$ becomes

$$\Delta\ell_\mathrm{stop} \ll \ell_\mathrm{stop}\frac{z_\mathrm{h}^2}{z_\star^2} . \tag{14.188}$$

Since $z_\star \ll z_\mathrm{h}$, this is weaker than the previous condition (14.180).

Lastly, consider the other condition, $|\Delta\boldsymbol{q}_\perp| \ll 1$ in (14.187). To estimate $|\Delta\boldsymbol{q}_\perp|$, return to the arguments of Appendix 5, but now, in the rest frame, include an initial proper displacement of the two geodesics in $\boldsymbol{x}^\perp$ of the same parametric size as the initial proper displacement in x^3. Following through the argument, one finds $x^I \simeq \Big(\gamma(1+\beta\beta_+), \boldsymbol{\beta}_\perp, \gamma(\beta+\beta_+), 1\Big)\,z$with $\beta_\perp \sim \beta_+$. Then

$$\Delta\hat{q}_3 = \Delta\frac{q_3}{q_0} = \Delta\frac{dx^3/dz}{dx^0/dz} \simeq \Delta\frac{\gamma(\beta+\beta_+)}{\gamma(1+\beta\beta_+)} \simeq \frac{\beta_+}{\gamma^2} \tag{14.189}$$

and

$$\Delta\hat{\boldsymbol{q}}_\perp = \Delta\frac{\boldsymbol{q}_\perp}{q_0} = \Delta\frac{d\boldsymbol{x}^\perp/dz}{dx^0/dz} \simeq \Delta\frac{\boldsymbol{\beta}_\perp}{\gamma(1+\beta\beta_+)} \simeq \frac{\boldsymbol{\beta}_\perp}{\gamma}, \tag{14.190}$$

so that

$$\frac{|\Delta\hat{\boldsymbol{q}}_\perp|}{|\Delta\hat{q}_3|} \sim \gamma \sim \sqrt{\frac{E^2}{-q^2}} \sim \frac{z_\mathrm{h}^2}{z_\star^2} . \tag{14.191}$$

So, using (14.179),

$$|\Delta\hat{\boldsymbol{q}}_\perp| \sim |\Delta\hat{q}_3|\,\frac{z^2}{z_\star^2} \sim \frac{\Delta\ell_\mathrm{stop}}{\ell_\mathrm{stop}}\,\frac{z_\star^2}{z_\mathrm{h}^2} . \tag{14.192}$$

The condition $|\Delta \hat{\boldsymbol{q}}_\perp| \ll 1$ is therefore the same as the previous condition (14.188) and so is also weaker than (14.180).

References

1. P.B. Arnold, S. Cantrell, W. Xiao, Stopping distance for high energy jets in weakly-coupled quark-gluon plasmas. Phys. Rev. D **81**, 045017 (2010) [arXiv:0912.3862 [hep-ph]]
2. H. Liu, K. Rajagopal, U.A. Wiedemann, Calculating the jet quenching parameter from AdS/CFT. Phys. Rev. Lett. **97**, 182301 (2006) [hep-ph/0605178]
3. S.S. Gubser, D.R. Gulotta, S.S. Pufu, F.D. Rocha, Gluon energy loss in the gauge-string duality. JHEP **0810**, 052 (2008) [arXiv:0803.1470 [hep-th]]
4. Y. Hatta, E. Iancu, A.H. Mueller, Jet evolution in the $\mathcal{N}=4$ SYM plasma at strong coupling. JHEP **0805**, 037 (2008) [arXiv:0803.2481 [hep-th]]
5. P.M. Chesler, K. Jensen, A. Karch, L.G. Yaffe, Light quark energy loss in strongly-coupled $\mathcal{N}=4$ supersymmetric Yang-Mills plasma. Phys. Rev. D **79**, 125015 (2009) [arXiv:0810.1985 [hep-th]]
6. P. Arnold, D. Vaman, Jet quenching in hot strongly coupled gauge theories revisited: 3-point correlators with gauge-gravity duality. JHEP **1010**, 099 (2010) [arXiv:1008.4023 [hep-th]]
7. P. Arnold, D. Vaman, Jet quenching in hot strongly coupled gauge theories simplified. JHEP **1104**, 027 (2011) [arXiv:1101.2689 [hep-th]]
8. P.M. Chesler, Y.-Y. Ho, K. Rajagopal, Shining a gluon beam through quark-gluon plasma. Phys. Rev. D **85**, 126006 (2012) [arXiv:1111.1691 [hep-th]]
9. S.-J. Sin, I. Zahed, Holography of radiation and jet quenching. Phys. Lett. B **608**, 265 (2005) [hep-th/0407215]
10. M. Spillane, A. Stoffers, I. Zahed, Jet quenching in shock waves. JHEP **1202**, 023 (2012) [arXiv:1110.5069 [hep-th]]; A. Stoffers, I. Zahed, arXiv:1110.2943 [hep-th]
11. C.P. Herzog, A. Karch, P. Kovtun, C. Kozcaz, L.G. Yaffe, Energy loss of a heavy quark moving through N = 4 supersymmetric Yang-Mills plasma. JHEP **0607**, 013 (2006) [hep-th/0605158]
12. J. Casalderrey-Solana, D. Teaney, Heavy quark diffusion in strongly coupled $\mathcal{N}=4$ Yang-Mills. Phys. Rev. D **74**, 085012 (2006) [hep-ph/0605199]
13. P. Arnold, P. Szepietowski, D. Vaman, Coupling dependence of jet quenching in hot strongly-coupled gauge theories. JHEP **1207**, 024 (2012) [arXiv:1203.6658 [hep-th]]
14. E. Shuryak, H.-U. Yee, I. Zahed, Self-force and synchrotron radiation in odd space-time dimensions. Phys. Rev. D **85**, 104007 (2012) [arXiv:1111.3894 [hep-th]]
15. D.J. Gross, E. Witten, Superstring modifications of Einstein's equations. Nucl. Phys. B **277**, 1 (1986)
16. P.M. Chesler, K. Jensen, A. Karch, Jets in strongly-coupled $\mathcal{N}=4$ super Yang-Mills theory. Phys. Rev. D **79**, 025021 (2009) [arXiv:0804.3110 [hep-th]]
17. S. Stieberger, Constraints on tree-level higher order gravitational couplings in superstring theory. Phys. Rev. Lett. **106**, 111601 (2011) [arXiv:0910.0180 [hep-th]]
18. D.M. Richards, The one-loop five-graviton amplitude and the effective action. JHEP **0810**, 042 (2008) [arXiv:0807.2421 [hep-th]]
19. G. D'Appollonio, P. Di Vecchia, R. Russo, G. Veneziano, High-energy string-brane scattering: Leading eikonal and beyond. JHEP **1011**, 100 (2010) [arXiv:1008.4773 [hep-th]]
20. G. Papadopoulos, J.G. Russo, A.A. Tseytlin, Solvable model of strings in a time dependent plane wave background. Class. Quant. Grav. **20**, 969 (2003) [hep-th/0211289]
21. G.T. Horowitz, A.R. Steif, Space-time singularities in string theory. Phys. Rev. Lett. **64**, 260 (1990); Strings in strong gravitational fields. Phys. Rev. D **42**, 1950 (1990)
22. E.G. Gimon, L.A. Pando Zayas, J. Sonnenschein, Penrose limits and RG flows. JHEP **0209**, 044 (2002) [hep-th/0206033]

23. H.J. de Vega, N.G. Sanchez, Strings falling into space-time singularities. Phys. Rev. D **45**, 2783 (1992); H.J. de Vega, M. Ramon Medrano, N.G. Sanchez, Classical and quantum strings near space-time singularities: Gravitational plane waves with arbitrary polarization. Class. Quant. Grav. **10**, 2007 (1993)
24. P. Arnold, D. Vaman, Some new results for 'jet' stopping in AdS/CFT. arXiv:1106.1680, An abridged version appears in J. Phys. G **38**, 124175 (2011)
25. N. Armesto, J.D. Edelstein, J. Mas, Jet quenching at finite 't Hooft coupling and chemical potential from AdS/CFT. JHEP **0609**, 039 (2006) [hep-ph/0606245]
26. E. D'Hoker, D.Z. Freedman, (author) Supersymmetric gauge theories and the AdS / CFT correspondence, in *Strings, Branes and Extra Dimensions: TASI 2001: Proceedings*, ed. by Steven S. Gubser, Joseph D. Lykken (River Edge, NJ, World Scientific, 2004) [hep-th/0201253]
27. A. Buchel, J.T. Liu, A.O. Starinets, Coupling constant dependence of the shear viscosity in $\mathcal{N}=4$ supersymmetric Yang-Mills theory. Nucl. Phys. B **707**, 56 (2005) [hep-th/0406264]
28. M.T. Grisaru, D. Zanon, Sigma model superstring corrections to the Einstein-Hilbert action. Phys. Lett. B **177**, 347 (1986); M.D. Freeman, C.N. Pope, M.F. Sohnius, K.S. Stelle, Higher order sigma model counterterms and the effective action for superstrings. Phys. Lett. B **178**, 199 (1986); Q-H. Park, D. Zanon, More on σ-model β functions and low-energy effective actions. Phys. Rev. D **35**, 4038 (1987)
29. O. Aharony, S.S. Gubser, J.M. Maldacena, H. Ooguri, Y. Oz, Large N field theories, string theory and gravity. Phys. Rept. **323**, 183 (2000) [hep-th/9905111]
30. M.B. Green, P. Vanhove, Low-energy expansion of the one loop type II superstring amplitude. Phys. Rev. D **61**, 104011 (2000) [hep-th/9910056]
31. D. Amati, M. Ciafaloni, G. Veneziano, Superstring collisions at planckian energies. Phys. Lett. B **197**, 81 (1987); Higher order gravitational deflection and soft bremsstrahlung in planckian energy superstring collisions. Nucl. Phys. B **347**, 550 (1990). Effective action and all order gravitational eikonal at Planckian energies. Nucl. Phys. B **403**, 707 (1993); D.N. Kabat, M. Ortiz, Eikonal quantum gravity and planckian scattering. Nucl. Phys. B **388**, 570 (1992) [hep-th/9203082]; L. Cornalba, M.S. Costa, J. Penedones, R. Schiappa, Eikonal approximation in AdS/CFT: from shock waves to four-point functions. JHEP **0708**, 019 (2007) [hep-th/0611122]; L. Cornalba, M.S. Costa, J. Penedones, Eikonal approximation in AdS/CFT: resumming the gravitational loop expansion. JHEP **0709**, 037 (2007) [arXiv:0707.0120 [hep-th]]; S.B. Giddings, R.A. Porto, The gravitational S-matrix. Phys. Rev. D **81**, 025002 (2010) [arXiv:0908.0004 [hep-th]]
32. M. Blau, J.M. Figueroa-O'Farrill, G. Papadopoulos, Penrose limits, supergravity and brane dynamics. Class. Quant. Grav. **19**, 4753 (2002) [hep-th/0202111]
33. M. Blau, Penrose waves and penrose limits. Lecture given at 2004 Saalburg/Wolfesdorf Summer School, http://www.blau.itp.unibe.ch/Lecturenotes.html
34. M. Blau, D. Frank, S. Weiss, Fermi coordinates and Penrose limits. Class. Quant. Grav. **23**, 3993 (2006) [hep-th/0603109]
35. L.A. Pando Zayas, J. Sonnenschein, On Penrose limits and gauge theories. JHEP **0205**, 010 (2002) [hep-th/0202186]
36. S.P. Kim, C.H. Lee, Nonequilibrium quantum dynamics of second order phase transitions. Phys. Rev. D **62**, 125020 (2000) [hep-ph/0005224]
37. P. Arnold, P. Szepietowski, D. Vaman, G. Wong, Tidal stretching of gravitons into classical strings: application to jet quenching with AdS/CFT. JHEP **1302**, 130 (2013) [arXiv:1212.3321 [hep-th]]

Index

E. Papantonopoulos (ed.), *Modifications of Einstein's Theory of Gravity at Large Distances*, Lecture Notes in Physics 892, DOI 10.1007/978-3-319-10070-8

MIX
Papier aus verantwortungsvollen Quellen
Paper from responsible sources
FSC® C105338

If you have any concerns about our products,
you can contact us on
ProductSafety@springernature.com

In case Publisher is established outside the EU,
the EU authorized representative is:
Springer Nature Customer Service Center GmbH
Europaplatz 3, 69115 Heidelberg, Germany

Printed by Libri Plureos GmbH
in Hamburg, Germany